MANAGING, CONTROLLING, AND IMPROVING QUALITY

Douglas C. Montgomery
Arizona State University

Cheryl L. Jennings
Bank of America

Michele E. Pfund
Arizona State University

JOHN WILEY & SONS WILEY, INC.

VP & Executive Publisher:	Don Fowley
Acquisitions Editor:	Jennifer Welter
Editorial Assistant:	Alexandra R. Spicehandler
Senior Designer:	Kevin Murphy
Photo Editor:	Sheena Goldstein
Marketing Manager:	Christopher Ruel
Production Manager:	Janis Soo
Senior Production Editor:	Joyce Poh

Cover credit © Diamond Sky Images/Getty Images

This book was set in 10/12 Palatino Roman by Thomson Digital and printed and bound by R.R. Donnelley.
The cover was printed by R.R. Donnelley.

This book is printed on acid free paper.

Library of Congress Cataloging-in-Publication Data
Montgomery, Douglas C.
Managing, controlling, and improving quality / Douglas C. Montgomery, Cheryl L. Jennings, Michele E. Pfund.
p. cm.
Includes index.
ISBN 978-0-471-69791-6 (cloth : alk. paper) 1. School management and organization. 2. Educational change.
3. Educational leadership. I. Jennings, Cheryl L. II. Pfund, Michele E. III. Title.
LB2900.5.M66 2010
371.2'07—dc22

2010000766

Printed in the United States of America
10 9 8 7 6 5 4 3 2 1

Preface

Motivation

Managing, controlling and improving quality is a critical activity in modern business organizations. Quality is directly linked to productivity, competitiveness, customer satisfaction, business growth, elimination of waste, and other non-value added activities, and overall business success. Cycle time and throughput is just as important in a hospital emergency room as it is in a semiconductor factory. Defects and errors don't occur just in factories, they occur in transactional and service businesses such as banks, insurance companies, and hospitals. Even your local (or state) government has a keen interest in improving service quality in operations such as issuing drivers licenses and motor vehicle registration. The U.S. Navy has had an intensive quality improvement program for many years.

Approach

This book presents an organized approach to quality management, control, and improvement. Because quality problems usually are the outcome of uncontrolled or excessive variability in product or service characteristics that are critical to the customer, statistical tools and other analytical methods play an important role in solving these problems. However, these techniques need to be implemented within a management structure that will ensure success. We focus on both the management structure and the statistical and analytical tools. Our approach to organizing and presenting this material is based on many years of teaching, research, and professional practice across a wide range of business and industrial settings.

Intended Audience

This book has been written at an introductory level and does not have extensive prerequisites in terms of mathematical or statistical background. Anyone with basic (noncalculus) statistics knowledge should be able to read the book. It is suitable as a textbook for a first course on quality management that intends to expose students to the range of techniques required to successfully control and improve quality. Students in technology programs, community college programs, or universities and undergraduate students in business or management should find this text accessible and focused on the topics that they need to gain an introductory understanding of this very important field.

Hallmark Features

- **An organized approach to quality management, control, and improvement;** focusing on both the statistical and analytical tools, and the management structure needed to implement these techniques effectively.
- **Accessible presentation:** Algebra-based, no complicated equations, and each topic is explained effectively. Example problems and thorough solutions help students new to the topic. Accessible to students in business and technology curricula, but rigorous enough for engineering students.

- **Mini-Cases** at the beginning of each chapter help motivate student interest and understanding of the importance and value of quality management tools and techniques.
- **Software implementation:** Both Minitab® and SPC XL are used within the book to demonstrate modern software implementation of statistical methods in quality control and improvement. Samples of output, and tips for using the software are included.
- **Examples and exercises** demonstrate a broad range of applications, and are presented at a diverse level to allow students to start at a level accessible to them, and build upon their skills with further problems.
- **Conversational tone** appeals to students.
- **Appropriate depth and breadth of coverage:** The subject of quality management and control is covered in depth.
- **Six sigma and DMAIC:** Presents in detail six sigma and the define, measure, analyze, improve, and control (DMAIC) process for quality improvement and implementation.
- **Designed experiments:** Demonstrates the value of designed experiments as a quality improvement tool.
- **Workplace implementation** is addressed.

Chapter Topics and Organizaton

Chapters 1 and 2 present the basic philosophy and tools of quality management. We discuss the various dimensions of quality; show the important links between variability and quality, productivity and quality, and quality and cost; discuss the legal aspect of quality and liability; present the philosophy of Deming, Juran, and Feigenbaum; discuss the merits of quality systems and standards and the Malcolm Baldrige National Quality Award; and present in some detail six-sigma and the define, measure, analyze, improve, and control (DMAIC) process for quality improvement. We believe that DMAIC is an excellent way to implement and manage quality improvement in an organization, regardless of the organization's involvement with six-sigma.

Chapter 3 is an introduction to the tools and techniques for quality control and improvement. The important tools covered include Pareto charts, cause-and-effect analysis, scatter diagrams, defect concentration diagrams, methods for data collection, flow charting, operations process charting and analysis, value stream mapping, the basic concepts of control charts, and we provide an illustration of a designed experiment. These tools help us identify and eliminate sources of variability. We also discuss quality improvement in transactional and service businesses.

Chapter 4 is a review of the statistical background and a convenient summary of the statistical methods useful in quality control and improvement. It contains methods for data both numerically and graphically, key probability distributions, and basic statistical inference. Many instructors and students will want to make use of computers in this and subsequent chapters. We use both Minitab and SPC XL to demonstrate how modern software implements these methods.

Minitab is a widely used general statistics software package. It may be the most widely used general package for quality control and improvement. To request a set including this book with the Student Version of Minitab at an additional cost, contact your local Wiley representative at www.wiley.com/college/rep. *Note*: The student version of Minitab has limited functionality and does not include DOE capability. If your students will need DOE capability, they may download the fully functional 30-day trial for the Minitab software at www.minitab.com, or they may purchase a fully functional time-limited version of Minitab from e-academy.com.

SPC XL and DOE PRO are Microsoft® Excel® Add-Ins developed by SigmaZone.com and co-owned with Air Academy Associates. To request a set including with this book access to time-limited versions of SPC XL and DOE PRO at an additional cost, contact your local Wiley representative at www.wiley.com/college/rep.

Chapter 5 gives a detailed presentation of control charts for measurements data (data that are expressed on a numerical scale). The $\bar{x}$ and R charts are discussed in detail, and examples are provided to show how to set up and operate these charts. Control charts are extremely important not only as a monitoring technique, but also in providing diagnostic information about how to reduce variability in a process. Consequently, we discuss several variations of the basic control chart, including the $\overline{X}$ and S charts individuals charts, and the CUSUM and EWMA control charts. We also discuss process capability.

Chapter 6 continues the presentation of control charts, concentrating on control charts for attributes (quality characteristics expressed as a go/no go decision). Charts for nonconforming (defective) product, defects, and defects per unit are discussed and illustrated. The use of computer software for construction and operation of these charts is included.

Chapter 7 presents techniques for lot-by-lot acceptance sampling. Methods for designing sampling plans to obtain specified statistical performance are given. We also discuss and provide illustrations of using standard sampling plans, including MIL STD 105E, the Dodge–Romig plans, and MIL STD 414. We also discuss some very useful specialized techniques; continuous sampling plans, chain sampling, and skip-lot sampling plans.

Designed experiments are the topic of Chapter 8. This may be the most important chapter in the book, from a methodology point of view, because designed experiments are generally recognized as the most powerful of the quality improvement tools. Designed experiments give the practitioner an opportunity to introduce purposeful changes in the variables of a system to learn which variables produce the most important effects and what levels of these variables should be employed to achieve optimum performance. Designed experiments are equally important in developing new products. This chapter discusses and illustrates the analysis of variance, factorial designs, the 2^k factorial system, fractional factorials, and response surface methods and designs. We show how designed experiments can be used to make product and processes robust to external sources of variability such as environmental factors (in general, noise factors) and variability transmitted from components.

Chapter 9 introduces the basic concepts of reliability, and discusses how these concepts are used in the design of new products and the improvement of existing ones. Reliability, one of the key dimensions of quality, is essentially quality over time. This chapter discusses basic reliability definitions including failure distributions and the concept of failure rate. The exponential, normal, and

Weibull distributions are introduced as models of the lifetime of a system. The notions of availability and maintainability are also discussed.

Website

The website for this book is at www.wiley.com/college/montgomery, and contains the following resources:

For both Students and Instructors:
Data Sets: Data Sets for all example problems and exercises in the text.

For Instructors only:
Instructor Solutions Manual: Complete solutions to all problems in the book.
Image Gallery: Figures from the text in electronic format, for easy creation of your own lecture slides or class presentations.
Lecture Slides: Lecture slides developed by the authors, which instructors may customize to fit their own course.

The contents of the instructor section of the book website are for instructor use only and are password-protected. Visit the Instructor Companion Site portion of the website located at www.wiley.com/college/montgomery to register for a password.

Computer Software

SPC XL and DOE PRO (Microsoft® Excel® Add-Ins) and Minitab are used in the text to demonstrate modern software implementation of summarizing data. Versions of these commercial software intended for educational use are available to students from Wiley*, packaged at an additional cost with the textbook:

SPC XL and DOE PRO
Excel Add-Ins for statistical analysis and design of experiments functionality developed by SigmaZone.com and co-owned with Air Academy Associates. Access is provided to a time-limited version of these software for educational purposes.

Student Version of Minitab
Note: The student version of Minitab has limited functionality and does not include DOE capability. If your students will need DOE capability, they may download the fully functional 30-day trial for the Minitab software at www.minitab.com, or they may purchase a fully functional time-limited version of Minitab from e-academy.com.

To request a set including with this book at an additional cost either a time-limited version of SPC XL and DOE PRO, or the Student Version of Minitab, contact your local Wiley representative at www.wiley.com.

---

* Sets including this software are available to adoptions in the continental United States and Canada only.

Acknowledgments

We thank the many colleagues and instructors who have helped in the development of this text, and have offered valuable feedback and suggestions to help us improve this text including:

Mohamed Aboul-Seoud, Rensselaer Polytechnic Institute

Stanley F. Bullington, Mississippi State University

Geoff Foster, North Carolina A & T State University

Scott Metlen, University of Idaho

Quinton J. Nottingham, Pamplin College of Business

Rama Shankar, Illinois Institute of Technology and Daley College

Carrie Steinlicht, South Dakota State University

Mathew P. Stephens, Purdue University

Douglas C. Montgomery
Cheryl L. Jennings
Michele E. Pfund

Table of Contents

Chapter Seven Lot-by-Lot Acceptance Sampling Procedures 281

Chapter Eight Process Design and Improvement with Designed Experiments 336

chapter One Introduction to Quality

Chapter Outline

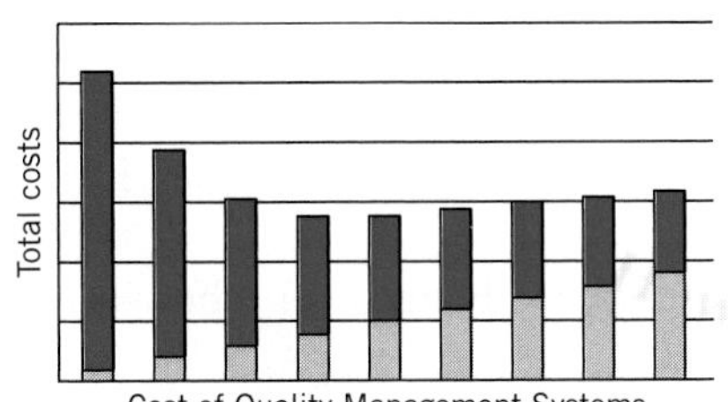

Cost of Quality Management Systems
■ Cost of production failure (Internal & external)
□ Cost of Quality program

Is an ounce of prevention really worth a pound of cure?

Philip B. Crosby authored a landmark book in 1980 entitled *Quality is Free*. This book challenged the assertion that it would cost more to produce higher-quality goods and provided an argument that it would cost more to produce a low-quality product.

The adjacent diagram illustrates the tradeoff. Firms that do not invest in quality management programs often have poor product quality. As a result, they incur a high cost of product failure. These costs may be internal costs (such as the cost of 100% inspection of rework) or external costs (warranty repairs, liability claims, or recalls).

Indeed, firms that *do* invest in quality management programs pay a price to maintain these programs (quality system program costs, inspections, design reviews, training and education). However, because they invest in quality up front, these firms also reap the significant benefit of experiencing much lower costs related to service defects. As demonstrated in the diagram, service defect costs tend to decline rapidly for firms investing in robust quality management systems.

Thus it is clear that for firms experiencing significant service defect costs, an ounce of prevention really is worth a pound of cure.

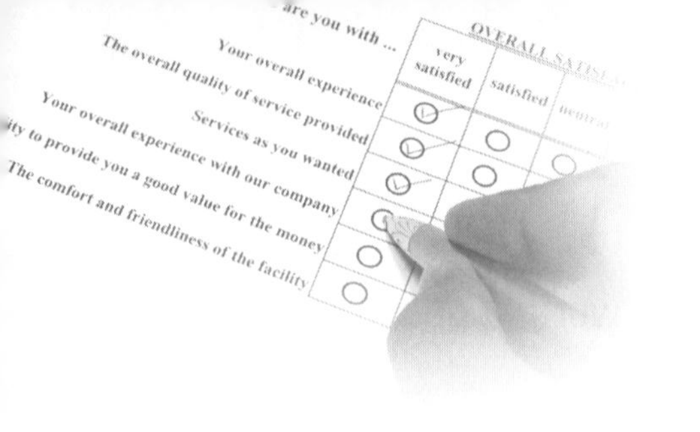

Chapter Overview and Learning Objectives

This book is about the use of statistical methods and other problem-solving techniques to improve the **quality** of the products used by our society. These products consist of **manufactured goods** such as automobiles, computers, and clothing, as well as **services** such as the generation and distribution of electrical energy, public transportation, banking, retailing, and health care. Quality improvement methods can be applied to any area within a company or organization, including manufacturing, process development, engineering design, finance and accounting, marketing, distribution and logistics, customer service, and field service of products. This text presents the technical tools that are needed to achieve quality improvement in these organizations.

In this chapter we give the basic definitions of quality, quality improvement, and other quality engineering terminology. We also discuss the historical development of quality improvement methodology and overview the essential linkage between quality and productivity, quality cost, and the legal and liability aspects of quality.

After careful study of this chapter you should be able to do the following:

1. Define and discuss quality and quality improvement
2. Discuss the different dimensions of quality
3. Discuss the evolution of modern quality improvement methods
4. Discuss the role that variability and statistical methods play in controlling and improving quality
5. Explain the links between quality and productivity and between quality and cost
6. Discuss product liability

1.1 The Meaning of Quality and Quality Improvement

Quality

We may define **quality** in many ways. Most people have a conceptual understanding of quality as relating to one or more desirable characteristics that a product or service should possess. The American Society for Quality describes "quality" as a subjective term for which each person or sector has its own definition. Although this type of conceptual understanding is certainly a useful starting point, we will give a more precise and useful definition.

Quality has become one of the most important consumer decision factors in the selection among competing products and services. The phenomenon is widespread, regardless of whether the consumer is an individual, an industrial organization, a retail store, a hospital, a bank or financial institution, or a military defense program. Consequently, understanding and improving quality are key factors leading to business success, growth, and competitiveness. There is a substantial return on investment from improved quality and from successfully employing quality as an integral part of overall business strategy. In this section we provide operational definitions of quality and quality improvement. We begin with a brief discussion of the different dimensions of quality and some basic terminology.

1.1.1 DIMENSIONS OF QUALITY

The quality of a product can be described and evaluated in several ways. It is often very important to differentiate these different **dimensions of quality.** Garvin (1987) provides an excellent discussion of eight components or dimensions of quality. We summarize his key points concerning these dimensions of quality as follows:

1. **Performance** (Will the product do the intended job?)	Potential customers usually evaluate a product to determine if it will perform certain specific functions and determine how well it performs them. For example, you could evaluate spreadsheet software packages for a PC to determine which data manipulation operations they perform. You may discover that one outperforms another with respect to execution speed.
2. **Reliability** (How often does the product fail?)	Complex products, such as many appliances, automobiles, or airplanes, will usually require some repair over their service life. For example, you should expect that an automobile will require occasional repair, but if the car requires frequent repair, we say that it is unreliable. There are many industries in which the customer's view of quality is greatly impacted by the reliability dimension of quality.
3. **Durability** (How long does the product last?)	This is the effective service life of the product. Customers obviously want products that perform satisfactorily over a long period of time. The automobile and major appliance industries are examples of businesses where this dimension of quality is very important to most customers.
4. **Serviceability** (How easy is it to repair the product?)	There are many industries in which the customer's view of quality is directly influenced by how quickly and economically a repair or routine maintenance activity can be accomplished. Examples include the appliance and automobile industries and many types of service industries. (How long did it take a credit card company to correct an error in your bill?)
5. **Aesthetics** (What does the product look like?)	This is the visual appeal of the product, often taking into account factors such as style, color, shape, packaging alternatives, tactile characteristics, and other sensory features. For example, soft-drink beverage manufacturers have relied on the visual appeal of their packaging to differentiate their product from other competitors. In the service sector, this is the physical appearance of the facility.
6. **Features** (What does the product do?)	Usually, customers associate high quality with products that have added features; that is, those that have features beyond the basic performance of the competition. For example, you might consider a spreadsheet software package to be of superior quality if it had built-in statistical analysis features while its competitors did not.
7. **Perceived Quality** (What is the reputation of the company or its product?)	In many cases, customers rely on the past reputation of the company concerning quality of its products. This reputation is directly influenced by failures of the product that are highly visible to the public or that require product recalls, and by how the customer is treated when a quality-related problem with the product is reported. Perceived quality, customer loyalty, and repeated business are *(Continued)*

	closely interconnected. For example, if you make regular business trips using a particular airline, and the flight almost always arrives on time and the airline company does not lose or damage your luggage, you will probably prefer to fly on that carrier instead of its competitors.
8. **Conformance to Standards** (Is the product made exactly as the designer intended?)	We usually think of a high-quality product as one that exactly meets the requirements placed on it. For example, how well does the hood fit on a new car? Is it perfectly flush with the fender height, and is the gap exactly the same on all sides? Manufactured parts that do not exactly meet the designer's requirements can cause significant quality problems when they are used as the components of a more complex assembly. An automobile consists of several thousand parts. If each one is just slightly too big or too small, many of the components will not fit together properly, and the vehicle (or its major subsystems) may not perform as the designer intended.
9. **Responsiveness**	How long they did it take the service provider to reply to your request for service? How willing were they to be helpful? How promptly was your request handled?

These dimensions are adequate to describe most business and industrial situations, although in the service sector we could add some additional ones.

10. **Professionalism**	This is the knowledge and skills of the service provider, and relate to the competency of the organization to provide the required services.
11. **Attentiveness**	Customers generally want caring, personalized attention from their service providers. Customers want to feel that their needs and concerns are important and are being addressed.

Quality means fitness for use.

We see from the foregoing discussion that quality is indeed a multifaceted entity. Consequently, a simple answer to questions such as "What is quality?" or "What is quality improvement?" is not easy. The **traditional** definition of quality is based on the viewpoint that products and services must meet the requirements of those who use them.

There are two general aspects of fitness for use: **quality of design** and **quality of conformance.** All goods and services are produced in various grades or levels of quality. These variations in grades or levels of quality are intentional, and, consequently, the appropriate technical term is quality of design. For example, all automobiles have as their basic objective providing safe transportation for the consumer. However, automobiles differ with respect to size, appointments, appearance, and performance. These differences are the result of intentional design differences among the types of automobiles. These design differences include the types of materials used in construction, specifications on the components, reliability obtained through engineering development of engines and drive trains, and other accessories or equipment.

The quality of conformance is how well the product conforms to the specifications required by the design. Quality of conformance is influenced by a number of factors, including the choice of manufacturing processes, the training and supervision of the workforce, the types of process controls, tests, and inspection activities that are employed, the extent to which these procedures are followed, and the motivation of the workforce to achieve quality.

Unfortunately, this definition has become associated more with the conformance aspect of quality than with design. This is in part due to the lack of formal education most designers and engineers receive in quality engineering methodology. This also leads to much less focus on the customer and more of a "conformance-to-specifications" approach to quality, regardless of whether the product, even when produced to standards, is actually "fit for use" by the customer. Also, there is still a widespread belief that quality is a problem that can be dealt with solely in manufacturing, or that the only way quality can be improved is by "gold-plating" the product.

We prefer the **modern** definition of quality in the adjacent display:

Quality is inversely proportional to variability.

Note that this definition implies that if variability[1] in the important characteristics of a product decreases, the quality of the product increases.

As an example of the operational effectiveness of this definition, a few years ago, one of the automobile companies in the United States performed a comparative study of a transmission that was manufactured in a domestic plant and by a Japanese supplier. An analysis of warranty claims and repair costs indicated that there was a striking difference between the two sources of production, with the Japanese-produced transmission having much lower costs, as shown in Figure 1.1. As part of the study to discover the cause of this difference in cost and performance, the company selected random samples of transmissions from each plant, disassembled them, and measured several critical quality characteristics.

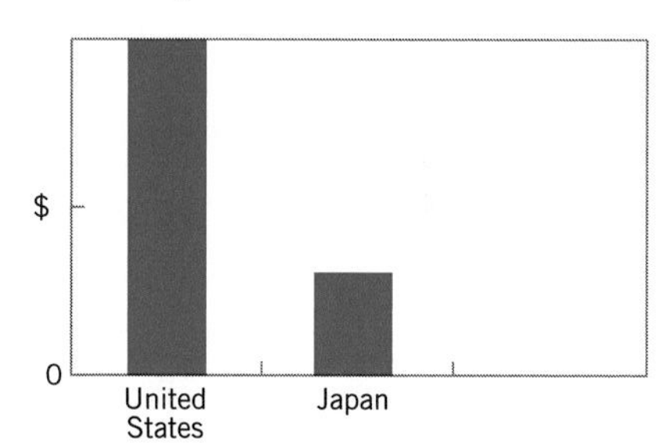

Figure 1.1 Warranty Costs for Transmissions

Figure 1.2 is generally representative of the results of this study. Note that both distributions of critical dimensions are centered at the desired or target value. However, the distribution of the critical characteristics for the transmissions manufactured in the United States takes up about 75% of the width of the specifications, implying that very few nonconforming units would be produced. In fact, the plant was producing at a quality level that was quite good, based on the generally accepted view of quality within the company. In contrast, the Japanese plant produced transmissions for which the same critical characteristics take up only about 25% of the specification band. As a result, there is considerably less variability in the critical quality characteristics of the Japanese-built transmissions in comparison to those built in the United States.

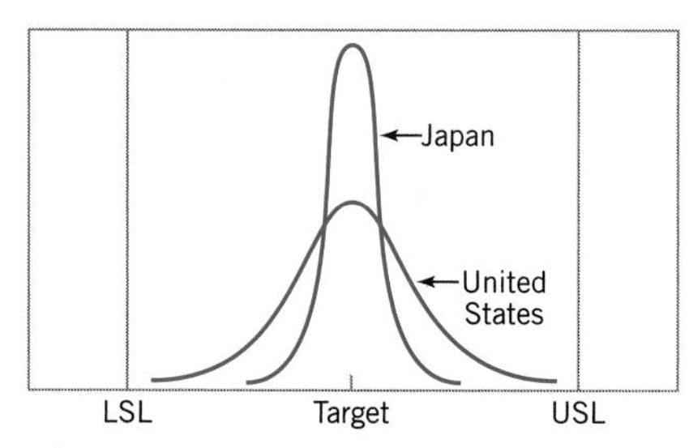

Figure 1.2 Distributions of Critical Dimensions for Transmissions

This is a very important finding. Jack Welch, the retired chief executive officer of General Electric, has observed that your customer doesn't see the mean of your process (the target in Figure 1.2), he only sees the variability around that target that you have not removed. In almost all cases, this variability has significant customer impact.

There are two obvious questions here: Why did the Japanese do this? How did they do this? The answer to the "why" question is obvious from examination of Figure 1.1. Reduced variability has directly translated into lower costs (the Japanese fully understood the point

---

[1] We are referring to unwanted or harmful variability. There are situations in which variability is actually good. As Professor Bob Hogg has pointed out, "I really like Chinese food, but I don't want to eat it every night."

made by Welch). Furthermore, the Japanese-built transmissions shifted gears more smoothly, ran more quietly, and were generally perceived by the customer as superior to those built domestically. Fewer repairs and warranty claims means less **rework** and the reduction of wasted time, effort, and money. Thus, quality truly is inversely proportional to variability. Furthermore, it can be communicated very precisely in a language that everyone (particularly managers and executives) understands—namely, money.

Quality improvement is the reduction of variability in processes and products.

How did the Japanese do this? The answer lies in the systematic and effective use of the methods described in this book. It also leads to the adjacent definition of **quality improvement.**

Excessive variability in process performance often results in **waste.** For example, consider the wasted money, time, and effort associated with the repairs represented in Figure 1.1 or the amount of time wasted in correction an erroneous posting to a credit card account. Therefore, an alternate and frequently very useful definition is that quality improvement is the **reduction of waste.** This definition is particularly effective in **service industries,** where there may not be as many things that can be directly measured (like the transmission critical dimensions in Figure 1.2). In service industries, a quality problem may be an error or a mistake, the correction of which requires effort and expense. Wasted effort in service processes often results in longer waiting times for the customer. By improving the service process, this wasted effort expense can be avoided.

We now present some quality terminology that is used throughout the book.

1.1.2 QUALITY TERMINOLOGY

Quality Characteristics

Every product possesses a number of elements that jointly describe what the user or consumer thinks of as quality. These parameters are often called **quality characteristics.** Sometimes these are called **critical-to-quality (CTQ)** characteristics. Quality characteristics may be of several types:

1. **Physical:** length, weight, voltage, viscosity
2. **Sensory:** taste, appearance, color
3. **Time Orientation:** reliability, durability, serviceability

In some service operations the CTQ, may be the issue of satisfaction that the customer has experienced as a result of using one or more services. Note that the different types of quality characteristics can relate directly or indirectly to the dimensions of quality discussed in the previous section.

Quality Engineering

Quality engineering is the set of operational, managerial, and technical activities that a company uses to ensure that the quality characteristics of a product are at the nominal or required levels and that the variability around these desired levels is minimum. The techniques discussed in the book form much of the basic methodology used by quality professionals to achieve these goals.

Most organizations find it difficult (and expensive) to provide the customer with products that have quality characteristics that are always identical from unit to unit, or are at levels that match customer expectations. A major reason for this is **variability.** There is a certain amount of variability in every product; consequently, no two products are ever identical. For example, the thickness of the blades on a jet turbine engine impeller is not identical even on the same impeller. Blade thickness will also differ between impellers. If this variation in blade thickness is small, then it may have no impact on the customer. However, if the variation is large, then the customer may perceive the unit to be undesirable and unacceptable. Sources of this variability include differences in materials, differences in the performance and operation of the manufacturing equipment, and differences in the way the operators perform their tasks. This line of thinking led to the previous definition of quality improvement.

Role of Statistics

Since variability can only be described in statistical terms, **statistical methods** play a central role in quality improvement efforts. In the application of statistical methods to quality engineering, it is fairly typical to classify data on quality characteristics as either **attributes** or **variables** data. Variables data are usually continuous measurements, such as length, voltage, or viscosity. Attributes data, on the other hand, are usually discrete data, often taking the form of counts, such as the number of loan applications that could not be properly processed because of missing required information, or the number of emergency room arrivals that have to wait more than 30 minutes to receive medical attention. We will describe statistical-based quality engineering tools for dealing with both types of data.

Specifications

Quality characteristics are often evaluated relative to **specifications.** For a manufactured product, the specifications are the desired measurements for the quality characteristics of the components and subassemblies that make up the product, as well as the desired values for the quality characteristics in the final product. For example, the diameter of a shaft used in an automobile transmission cannot be too large or it will not fit into the mating bearing—nor can it be too small, resulting in a loose fit, causing vibration, wear, and early failure of the assembly. In the service industries, specifications are typically in terms of the maximum amount of time to process an order or to provide a particular service. This is sometimes called **cycle time**.

Cycle Time is an important measurement in service industries.

A value of a measurement that corresponds to the desired value for that quality characteristic is called the **nominal** or **target value** for that characteristic. These target values are usually bounded by a range of values that, most typically, we believe will be sufficiently close to the target so as to not impact the function or performance of the product if the quality characteristic is in that range. The largest allowable value for a quality characteristic is called the **upper specification limit (USL),** and the smallest allowable value for a quality characteristic is called the **lower specification limit (LSL).** Some quality characteristics have specification limits on only one side of the target. For example, the compressive strength of a component used in an automobile bumper likely has a target value and a lower specification limit, but not an upper specification limit.

Many aspects of Product Quality are determined by the design.

Specifications are usually the result of the design process for the product. Traditionally, product designers have arrived at a product design configuration through the use of engineering science principles, which often results in the designer specifying the target values for the critical design parameters. Then prototype construction and testing follow. This testing is often done in a very unstructured manner, without the use of statistically based experimental design procedures, and without much interaction with or knowledge of the manufacturing processes that must produce the component parts and final product. However, through this general procedure, the specification limits are usually determined by the design engineer. Then the final product is released to manufacturing. We refer to this as the **over-the-wall** approach to design.

Problems in product quality usually are greater when the over-the-wall approach to design is used. In this approach, specifications are often set without regard to the inherent variability that exists in materials, processes, and other parts of the system, which results in components or products that are nonconforming; that is, **nonconforming products** are those that fail to meet one or more of its specifications. A specific type of failure is called a **nonconformity.** A nonconforming product is not necessarily unfit for use; for example, a detergent may have a concentration of active ingredients that is below the lower specification limit, but it may still perform acceptably if the customer uses a greater amount of the product. A nonconforming product is considered **defective** if it has one or more **defects,** which are nonconformities that are serious enough to significantly affect the safe or effective use of the product. Obviously, failure on the part of a company to improve its manufacturing processes can also cause nonconformities and defects.

The over-the-wall design process has been the subject of much attention in the past 25 years. CAD/CAM systems have done much to automate the design process and to more effectively translate specifications into manufacturing activities and processes. Design for manufacturability and assembly has emerged as an important part of overcoming the inherent problems with the over-the-wall approach to design, and most engineers receive some background on those areas today as part of their formal education. The recent emphasis on **concurrent engineering** has stressed a team approach to design, with specialists in manufacturing, quality engineering, and other disciplines working together with the product designer at the earliest stages of the product design process. Furthermore, the effective use of the quality improvement methodology in this book, at all levels of the process used in technology commercialization and product realization, including product design, development, manufacturing, distribution, and customer support, plays a crucial role in quality improvement.

1.2 A Brief History of Quality Control and Improvement

Quality always has been an integral part of virtually all products and services. However, our awareness of its importance and the introduction of formal methods for quality control and improvement have been an evolutionary development. Table 1.1 presents a timeline of some of the important milestones in this evolutionary process. We will briefly discuss some of the events on this timeline.

Table 1.1

A Timeline of Quality Methods

1700–1900	Quality is largely determined by the efforts of an individual craftsman. Eli Whitney introduces standardized, interchangeable parts to simplify assembly.
1875	Frederick W. Taylor introduces "Scientific Management" principles to divide work into smaller, more easily accomplished units—the first approach to dealing with more complex products and processes. The focus was on productivity. Later contributors were Frank and Lillian Gilbreth and Henry Gantt.
1900–1930	Henry Ford—the assembly line—further refinement of work methods to improve productivity and quality; Ford developed mistake-proof assembly concepts, self-checking, and in-process inspection.
1901	First standards laboratories established in Great Britain.
1907–1908	AT&T begins systematic inspection and testing of products and materials.
1908	W. S. Gosset (writing as "Student") introduces the *t*-distribution—results from his work on quality control at Guinness Brewery.
1915–1919	WWI—British government begins a supplier certification program.
1919	Technical Inspection Association is formed in England; this later becomes the Institute of Quality Assurance.
1920s	AT&T Bell Laboratories forms a quality department—emphasizing quality, inspection and test, and product reliability. B. P. Dudding at General Electric in England uses statistical methods to control the quality of electric lamps.
1922–1923	R. A. Fisher publishes series of fundamental papers on designed experiments and their application to the agricultural sciences.
1924	W. A. Shewhart introduces the control chart concept in a Bell Laboratories technical memorandum.
1928	Acceptance-sampling methodology is developed and refined by H. F. Dodge and H. G. Romig at Bell Labs.
1931	W. A. Shewhart publishes *Economic Control of Quality of Manufactured Product*—outlining statistical methods for use in production and control chart methods.
1932	W. A. Shewhart gives lectures on statistical methods in production and control charts at the University of London.
1932–1933	British textile and woolen industry and German chemical industry begin use of designed experiments for product/process development.
1933	The Royal Statistical Society forms the Industrial and Agricultural Research Section.
1938	W. E. Deming invites Shewhart to present seminars on control charts at the U.S. Department of Agriculture Graduate School.
1940	The U.S. War Department publishes a guide for using control charts to analyze process data.
1940–1943	Bell Labs develop the forerunners of the military standard sampling plans for the U.S. Army.
1942	In Great Britain, the Ministry of Supply Advising Service on Statistical Methods and Quality Control is formed.

(Continued)

Table 1.1

Continued

1942–1946	Training courses on statistical quality control are given to industry; more than 15 quality societies are formed in North America.
1944	*Industrial Quality Control* begins publication.
1946	The American Society for Quality Control (ASQC) is formed as the merger of various quality societies. The International Standards Organization (ISO) is founded. W. E. Deming is invited to Japan by the Economic and Scientific Services Section of the U.S. War Department to help occupation forces in rebuilding Japanese industry. The Japanese Union of Scientists and Engineers (JUSE) is formed.
1946–1949	W. E. Deming is invited to give statistical quality control seminars to Japanese industry.
1948	G. Taguchi begins study and application of experimental design.
1950	Deming begins education of Japanese industrial managers; statistical quality control methods begin to be widely taught in Japan. K. Ishikawa introduces the cause-and-effect diagram.
1950s	Classic texts on statistical quality control by Eugene Grant and A. J. Duncan appear.
1951	A. V. Feigenbaum publishes the first edition of his book, *Total Quality Control.* JUSE establishes the Deming Prize for significant achievement in quality control and quality methodology.
1951+	G. E. P. Box and K. B. Wilson publish fundamental work on using designed experiments and response surface methodology for process optimization; focus is on chemical industry. Applications of designed experiments in the chemical industry grow steadily after this.
1954	Joseph M. Juran is invited by the Japanese to lecture on quality management and improvement. British statistician E. S. Page introduces the cumulative sum (CUSUM) control chart.
1957	J. M. Juran and F. M. Gryna's *Quality Control Handbook* is first published.
1959	*Technometrics* (a journal of statistics for the physical, chemical, and engineering sciences) is established; J. Stuart Hunter is the founding editor. S. Roberts introduces the exponentially weighted moving average (EWMA) control chart. The U.S. manned spaceflight program makes industry aware of the need for reliable products; the field of reliability engineering grows from this starting point.
1960	G. E. P. Box and J. S. Hunter write fundamental papers on 2^{k-p} factorial designs. The quality control circle concept is introduced in Japan by K. Ishikawa.
1961	National Council for Quality and Productivity is formed in Great Britain as part of the British Productivity Council.
1960s	Courses in statistical quality control become widespread in industrial engineering academic programs. Zero defects (ZD) programs are introduced in certain U.S. industries.
1969	*Industrial Quality Control* ceases publication, replaced by *Quality Progress* and the *Journal of Quality Technology* (Lloyd S. Nelson is the founding editor of *JQT*).
1970s	In Great Britain, the NCQP and the Institute of Quality Assurance merge to form the British Quality Association.

(Continued)

Table 1.1

Continued

1975–1978	Books on designed experiments oriented toward engineers and scientists begin to appear. Interest in quality circles begins in North America—this grows into the total quality management (TQM) movement.
1980s	Experimental design methods are introduced to and adopted by a wider group of organizations, including electronics, aerospace, semiconductor, and the automotive industries. The works of Taguchi on designed experiments first appear in the United States.
1984	The American Statistical Association (ASA) establishes the Ad Hoc Committee on Quality and Productivity; this later becomes a full section of the ASA. The journal *Quality and Reliability Engineering International* appears.
1986	G. E. P. Box and others visit Japan, noting the extensive use of designed experiments and other statistical methods.
1987	ISO publishes the first quality systems standard. Motorola's six-sigma initiative begins.
1988	The Malcolm Baldrige National Quality Award is established by the U.S. Congress. The European Foundation for Quality Management is founded; this organization administers the European Quality Award.
1989	The journal *Quality Engineering* appears.
1990s	ISO 9000 certification activities increase in U.S. industry; applicants for the Baldrige award grow steadily; many states sponsor quality awards based on the Baldrige criteria.
1995	Many undergraduate engineering programs require formal courses in statistical techniques, focusing on basic methods for process characterization and improvement.
1997	Motorola's Six-Sigma approach spreads to other industries.
1998	The American Society for Quality Control becomes the American Society for Quality (see www.asq.org), attempting to indicate the broader aspects of the quality improvement field.
2000s	ISO 9000:2000 standard is issued. Supply-chain management and supplier quality become even more critical factors in business success. Quality improvement activities expand beyond the traditional industrial setting into many other areas including financial services, health care, insurance, and utilities.

The development of standardized and interchangeable parts was a key idea in making units of manufactured product more similar to each other, reducing variability. Frederick W. Taylor introduced some principles of scientific management as mass production industries began to develop prior to 1900. Taylor pioneered dividing work into tasks so that the product could be manufactured and assembled more easily. His work led to substantial improvements in productivity. Also, because of standardized production and assembly methods, the quality of manufactured goods was positively impacted as well. However, along with the standardization of work methods came the concept of work standards—a standard time to accomplish the work, or a specified number of units that must be produced per period. Frank and

Lillian Gilbreth and others extended this concept to the study of motion and work design. Much of this had a positive impact on productivity, but it often did not sufficiently emphasize the quality aspect of work. Furthermore, if carried to extremes, work standards have the risk of halting innovation and continuous improvement, which we recognize today as being a vital aspect of all work activities.

Statistical methods and their application in quality improvement have had a long history. In 1924, Walter A. Shewhart of the Bell Telephone Laboratories developed the statistical control chart concept. Many people consider this the formal beginning of the quality control and improvement field. Toward the end of the 1920s, Harold F. Dodge and Harry G. Romig, both of Bell Telephone Laboratories, developed statistically based acceptance-sampling as an alternative to 100% inspection. By the middle of the 1930s, statistical quality-control methods were in wide use at Western Electric, the manufacturing arm of the Bell System. However, the value of statistical quality control was not widely recognized by industry.

World War II saw a greatly expanded use and acceptance of statistical quality-control concepts in manufacturing industries. Wartime experience made it apparent that statistical techniques were necessary to control and improve product quality. The American Society for Quality Control (now the American Society for Quality) was formed in 1946. This organization promotes the use of quality improvement techniques for all types of products and services. It offers a number of conferences, technical publications, and training programs in quality assurance. The 1950s and 1960s saw the emergence of reliability engineering, the introduction of several important textbooks on statistical quality control, and the viewpoint that quality is a way of managing the organization.

In the 1950s, designed experiments for product and process improvement were first introduced in the United States. The initial applications were in the chemical industry. These methods were widely exploited in the chemical industry, and they are often cited as one of the primary reasons that the U.S. chemical industry is one of the most competitive in the world and has lost little business to foreign companies. The spread of these methods outside the chemical industry was relatively slow until the late 1970s or early 1980s, when many Western companies discovered that their Japanese competitors had been systematically using designed experiments since the 1960s for process improvement, new process development, evaluation of new product designs, improvement of reliability and field performance of products, and many other aspects of product design, including selection of component and system tolerances. This discovery sparked further interest in statistically designed experiments and resulted in extensive efforts to introduce the methodology in engineering and development organizations in industry, as well as in academic engineering curricula.

Importance of Statistical Methods

Since 1980, there has been a profound growth in the use of statistical methods for quality and overall business improvement in the United States. This has been motivated, in part, by the widespread loss of business and markets suffered by many domestic companies that began during the 1970s. For example, the U.S. automobile industry was nearly destroyed by foreign competition during this period. One domestic automobile company estimated its operating losses at nearly $1 million *per hour* in 1980. The adoption and use of statistical methods have played a central role in the re-emergence of U.S. industry. Various management systems have also emerged as frameworks in which to implement quality improvement. This book provides an introduction to the technical tools and managerial concept that are the basis of modern quality control and improvement.

1.3 Statistical Methods for Quality Control and Improvement

Statistical process control, design of experiments, acceptance-sampling.

This textbook provides guidance on statistical technology useful in quality improvement. Specifically, we focus on three major areas: **statistical process control, design of experiments,** and (to a lesser extent) **acceptance-sampling.** In addition to these techniques, a number of other statistical and analytical tools are useful in analyzing quality problems and improving the performance of processes. The role of some of these tools is illustrated in Figure 1.3, which presents a **process** as a system with a set of inputs and an output. In the case of a manufacturing process, the controllable input factors $x_1, x_2, \ldots, x_p$ are process variable such as temperatures, pressures, feed rates, and other process variables. The inputs $z_1, z_2, \ldots, z_q$ are uncontrollable (or difficult to control) inputs, such as environmental factors or properties of raw materials provided by an external supplier. The production process transforms the input raw materials, component parts, and subassemblies into a finished product that has several quality characteristics. The output variable y is a quality characteristic—that is, a measure of process and product quality. This model can also be used to represent **nonmanufacturing** or **service processes.** For example, consider a process in a financial institution that processes automobile loan applications. The inputs are the loan applications, which contain information about the customer and his/her credit history, the type of car to be purchased, its price, and the loan amount. The controllable factors are the type of training that the loan officer receives, the specific rules and

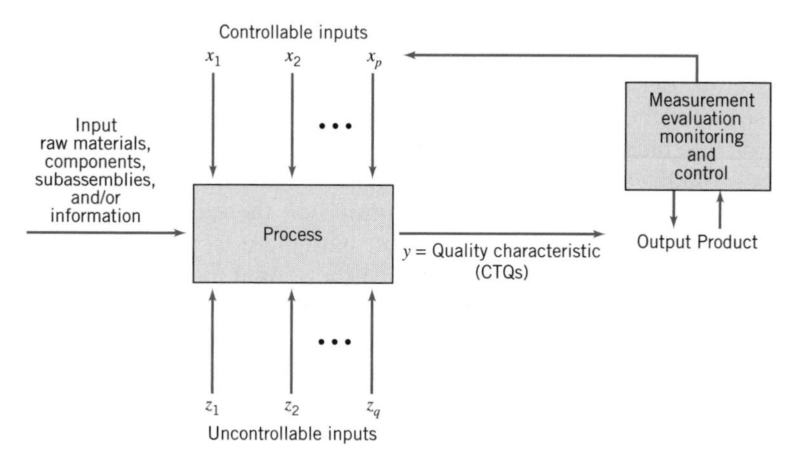

Figure 1.3 Production Process Inputs and Outputs

policies that the bank imposed on these loans, and the number of people working as loan officers at each time period. The uncontrollable factors include prevailing interest rates, the amount of capital available for these types of loans in each time period, and the number of loan applications that require processing each period. The output quality characteristics include whether or not the loan is funded, the number of funded loans that are actually accepted by the applicant, and the cycle time—that is, the length of time that the customer waits until a decision on his/her loan application is made. In service systems, cycle time is often a very important CTQ.

Control Charts

A **control chart** is one of the primary techniques of **statistical process control (SPC).** A typical control chart is shown in Figure 1.4. This chart plots the averages of measurements of a quality characteristic in samples taken from the process versus time (or the sample number). The chart has a center line (CL) and upper and lower control limits (UCL and LCL in Figure 1.4). The center line represents where this process characteristic should fall if there are no unusual sources of variability present. The control limits are determined from some simple statistical considerations that we will discuss in Chapters 4, 5, and 6. Classically, control charts are applied to the output variable(s) in a system such as in Figure 1.4. However, in some cases they can be usefully applied to the inputs as well.

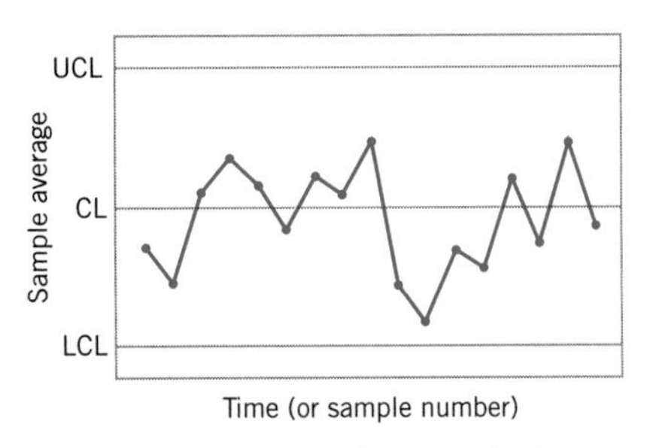

Figure 1.4 A Typical Control Chart

The control chart is a very useful **process monitoring technique**; when unusual sources of variability are present, sample averages will plot outside the control limits. This is a signal that some investigation of the process should be made and corrective action taken to remove these unusual sources of variability. Systematic use of a control chart is an excellent way to reduce variability.

A **designed experiment** is extremely helpful in discovering the key variables influencing the quality characteristics of interest in the process. A designed experiment is an approach to systematically varying the controllable input factors in the process and determining the effect these factors have on the output product parameters. Statistically designed experiments are invaluable in reducing the variability in the quality characteristics and in determining the levels of the controllable variables that optimize process performance. Often significant breakthroughs in process performance and product quality also result from using designed experiments.

Factorial Design

One major type of designed experiment is the **factorial design**, in which factors are varied together in such a way that all possible combinations of factor levels are tested. Figure 1.5 shows two possible factorial designs for the process in Figure 1.3, for the cases of $p = 2$ and $p = 3$ controllable factors. In Figure 1.5*a*, the factors have two levels, low and high, and the four possible test combinations in this factorial experiment form the corners of a square. In Figure 1.5*b*, there are three factors each at two levels, giving an experiment with eight test combinations arranged at the corners of a cube. The distributions at the corners of the cube represent the process performance at each combination of the controllable factors x_1, x_2, and x_3. It is clear that some combinations of factor levels produce better results than others. For example, increasing x_1 from low to high increases the average level of the process output and could shift it off the target value (T). Furthermore, process variability

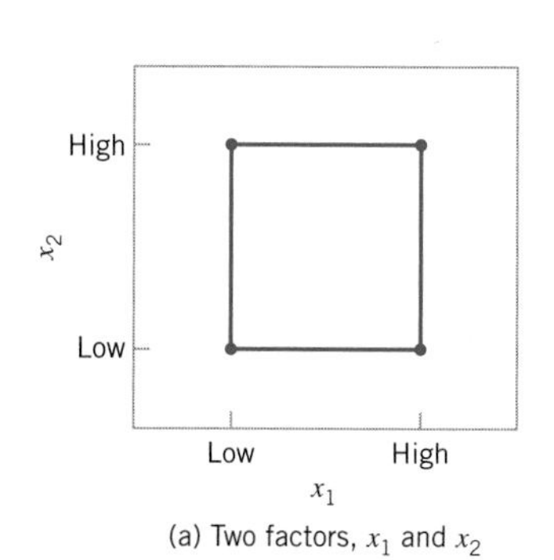

(a) Two factors, x_1 and x_2

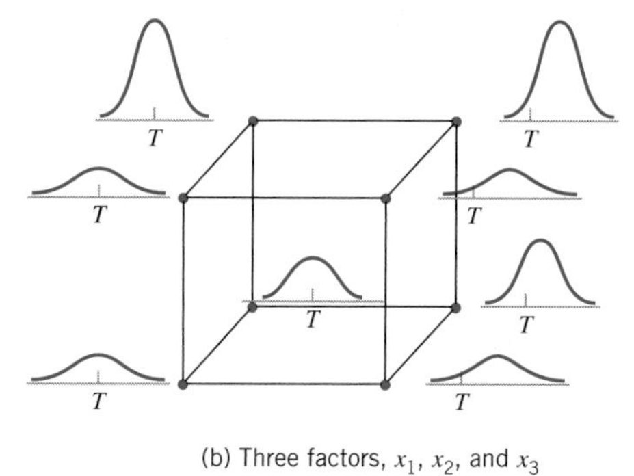

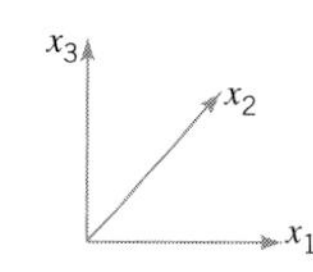

(b) Three factors, x_1, x_2, and x_3

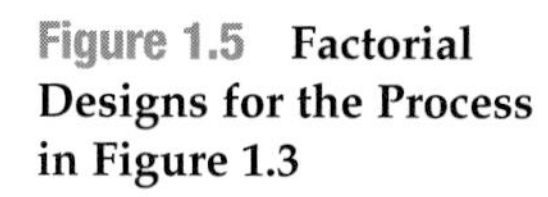
Figure 1.5 Factorial Designs for the Process in Figure 1.3

seems to be substantially reduced when we operate the process along the back edge of the cube, where x_2 and x_3 are at their high levels.

Designed experiments are an off-line improvement technique.

Designed experiments are a major **off-line** quality-control tool, because they are often used during development activities and the early stages of manufacturing, rather than as a routine **on-line** or **in-process** procedure. They play a crucial role in reducing variability.

Designed experiments lead to a model of the process.

Once we have identified a list of important variables that affect the process output, it is usually necessary to model the relationship between the influential input variables and the output quality characteristics. Statistical techniques useful in constructing such models include regression analysis and time series analysis. Detailed discussions of designed experiments, regression analysis, and time series modeling are in Montgomery (2005), Montgomery, Peck, and Vining (2006), and Box, Jenkins, and Reinsel (1994).

When the important variables have been identified and the nature of the relationship between the important variables and the process output has been quantified, then an on-line statistical process-control technique for monitoring and surveillance of the process can be employed with considerable effectiveness. Techniques such as control charts can be used to monitor the process output and detect when changes in the inputs are required to bring the process back to an in-control state. The models that relate the influential inputs to process outputs help determine the nature and magnitude of the adjustments required. In many processes, once the dynamic nature of the relationships between the inputs and the outputs are understood, it may be possible to routinely adjust the process so that future values of the product characteristics will be approximately on target. This routine adjustment is often called **engineering control, automatic control,** or **feedback control.**

The third area of quality control and improvement that we discuss is **acceptance-sampling.** This is closely connected with inspection and testing of product, which is one of the earliest aspects of quality control, dating back to long before statistical methodology was developed for quality improvement. Inspection can occur at many points in a process. Acceptance-sampling, defined as the inspection and classification of a sample of units selected at random from a larger batch or lot and the ultimate decision about disposition of the lot, usually occurs at two points: incoming raw materials or components and final production.

Figure 1.6 Variations of Acceptance-Sampling

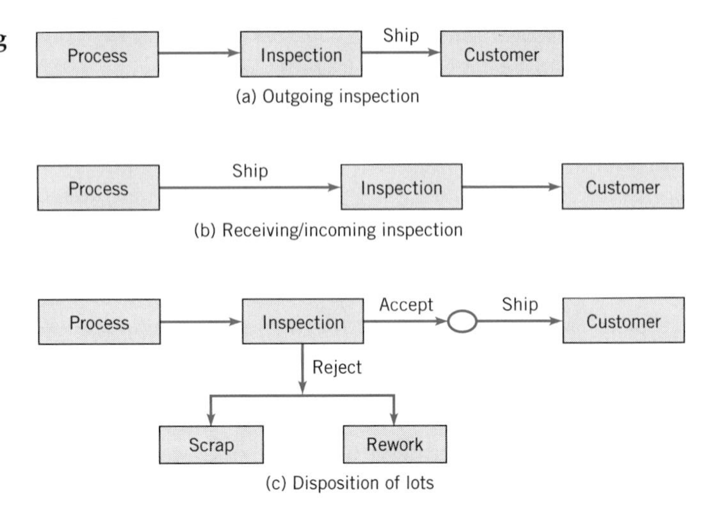

Several different variations of acceptance-sampling are shown in Figure 1.6. In Figure 1.6*a*, the inspection operation is performed immediately following production, before the product is shipped to the customer. This is usually called **outgoing inspection.** Figure 1.6*b* illustrates **incoming inspection;** that is, a situation in which lots of batches of product are sampled as they are received from the supplier. Various lot-dispositioning decisions are illustrated in Figure 1.6*c*. Sampled lots may either be accepted or rejected. Items in a rejected lot are typically either scrapped or recycled, or they may be reworked or replaced with good units. This latter case is often called **rectifying inspection.**

Modern quality assurance systems usually place less emphasis on acceptance-sampling and attempt to make statistical process control and designed experiments the focus of their efforts. Acceptance-sampling tends to reinforce the conformance-to-specification view of quality and does not have any feedback into either the production process or engineering design or development that would necessarily lead to quality improvement.

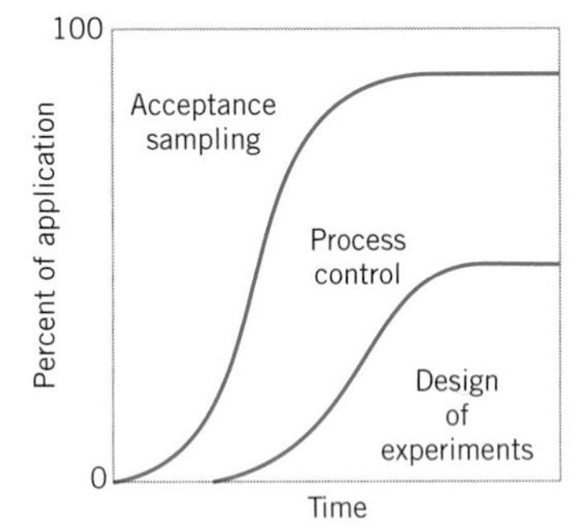

Figure 1.7 Phase Diagram of the Use of Quality-Engineering Methods

Figure 1.7 shows the typical evolution in the use of these techniques in most organizations. At the lowest level of maturity, management may be completely unaware of quality issues, and there is likely to be no effective organized quality improvement effort. Frequently there will be some modest applications of acceptance-sampling and inspection methods, usually for incoming parts and materials. The first activity as maturity increases is to intensify the use of sampling inspection. The use of sampling will increase until it is realized that quality cannot be inspected or tested into the product.

At that point, the organization usually begins to focus on process improvement. Statistical process control and experimental design potentially have major impacts on manufacturing, product design activities, and process development. The systematic introduction of these methods usually marks the start of substantial quality, cost, and productivity improvements in the organization. At the highest levels of maturity, companies use designed experiments and statistical process control methods intensively and make relatively modest use of acceptance-sampling.

The primary **objective** of quality engineering efforts is the **systematic reduction of variability** in the key quality characteristics of the product. Figure 1.8 shows how this happens over time. In the early stages, when acceptance-sampling is the major technique in use, process "fallout," or units that do not conform to the specifications, constitute a high percentage of the process output. The introduction of statistical process control will stabilize the process and reduce the variability. However, it is not satisfactory just to meet requirements—further reduction of variability usually leads to better product performance and enhanced competitive position, as was vividly demonstrated in the automobile transmission example discussed earlier. Statistically designed experiments can be employed in conjunction with statistical process monitoring and control to minimize process variability in nearly all industrial settings.

Reducing variability improves quality

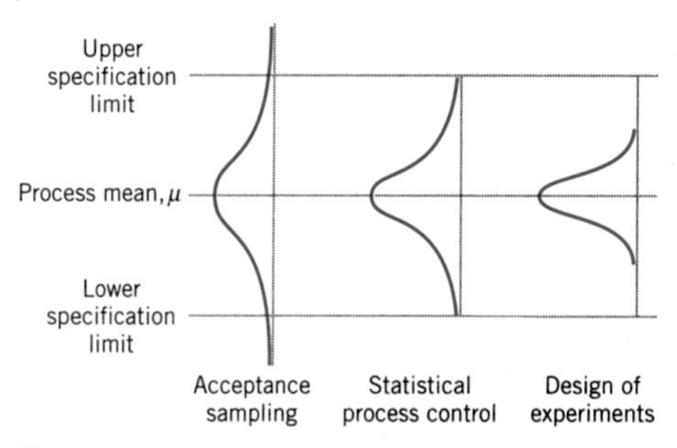

Figure 1.8 Application of Quality-Engineering Techniques and the Systematic Reduction of Process Variability

1.4 Quality and Productivity

Producing high-quality products in the modern industrial environment is not easy. A significant aspect of the problem is the rapid evolution of technology. The past 20 years have seen an explosion of technology in such diverse fields as electronics, metallurgy, ceramics, composite materials, biotechnology, and the chemical and pharmaceutical sciences that has resulted in many new products and services. For example, in the electronics field the development of the integrated circuit has revolutionized the design and manufacture of computers and many electronic office products. Basic integrated circuit technology has been supplanted by large-scale integration (LSI) and very large-scale integration (VLSI) technology, with corresponding developments in semiconductor design and manufacturing. When technological advances occur rapidly and new technologies are used quickly to exploit competitive advantages, the problems of designing and manufacturing products of superior quality are greatly complicated.

Often, too little attention is paid to achieving all dimensions of an optimal process: economy, efficiency, productivity, and quality. Effective quality improvement can be instrumental in increasing productivity and reducing cost. To illustrate, consider the manufacture of a mechanical component used in a copier machine. The parts are manufactured in a machining process at a rate of approximately 100 parts per day. For various reasons, the process is operating at a first-pass yield of about 75%. (That is, about 75% of the process output conforms to specifications, and about 25% of the output is nonconforming.) About 60% of the fallout (the 25% nonconforming) can be reworked into an acceptable product, and the rest must be scrapped. The direct manufacturing cost through this stage of production per part is approximately \$20. Parts that can be reworked incur an additional processing charge of \$4. Therefore, the manufacturing cost per good part produced is

$$\text{Cost/good part} = \frac{\$20(100) + \$4(15)}{90} = \$22.89$$

Note that the total yield from this process, after reworking, is 90 good parts per day.

An engineering study of this process reveals that excessive process variability is responsible for the extremely high fallout. A new statistical process-control procedure is implemented that reduces variability, and consequently the process fallout decreases from 25% to 5%. Of the 5% fallout produced, about 60% can be reworked, and 40% are scrapped. After the process-control program is implemented, the manufacturing cost per good part produced is

$$\text{Cost/good part} = \frac{\$20(100) + \$4(3)}{98} = \$20.53$$

Note that the installation of statistical process control and the reduction of variability that follows result in a 10.3% reduction in manufacturing costs. Furthermore, productivity is up by almost 10%; 98 good parts are produced each day as opposed to 90 good parts previously. This amounts to an increase in production capacity of almost 10%, without any additional investment in equipment, workforce, or overhead. Efforts to improve this process by other methods (such as Just-in-Time, lean manufacturing, etc.) are likely to be completely ineffective until the basic problem of excessive variability is solved.

1.5 Quality Costs

Quality costs are an important financial control tool.

Financial controls are an important part of business management. These financial controls involve a comparison of actual and budgeted costs, along with analysis and action on the differences between actual and budget. It is customary to apply these financial controls on a department or functional level. For many years, there was no direct effort to measure or account for the costs of the quality function. However, many organizations now formally evaluate the cost associated with quality. There are several reasons why the cost of quality should be explicitly considered in an organization. These include the following:

1. The increase in the cost of quality because of the increase in the complexity of manufactured products associated with advances in technology
2. Increasing awareness of life-cycle costs, including maintenance, spare parts, and the cost of field failures
3. Quality engineers' and managers' ability most effectively communicate quality issues in a way that management understands

As a result, quality costs have emerged as a financial control tool for management and as an aid in identifying opportunities for reducing quality costs.

Generally speaking, quality costs are those categories of costs that are associated with producing, identifying, avoiding, or repairing products that do not meet requirements. Many manufacturing and service organizations use four categories of quality costs: prevention

Table 1.2
Quality Costs

Prevention Costs	Internal Failure Costs
Quality planning and engineering	Scrap
New products review	Rework
Product/process design	Retest
Process control	Failure analysis
Burn-in	Downtime
Training	Yield losses
Quality data acquisition and analysis	Downgrading (off-specing)
Appraisal Costs	**External Failure Costs**
Inspection and test of incoming material	Complaint adjustment
Product inspection and test	Returned product/material
Materials and services consumed	Warranty charges
Maintaining accuracy of test equipment	Liability costs
	Indirect costs

costs, appraisal costs, internal failure costs, and external failure costs. These cost categories are shown in Table 1.2. We now discuss these categories in more detail.

Prevention Costs. Prevention costs are those costs associated with efforts in design and manufacturing that are directed toward the prevention of nonconformance. Broadly speaking, prevention costs are all costs incurred in an effort to "make it right the first time." The important subcategories of prevention costs:

Quality planning and engineering	Costs associated with the creation of the overall quality plan, the inspection plan, the reliability plan, the data system, and all specialized plans and activities of the quality-assurance function; the preparation of manuals and procedures used to communicate the quality plan; and the costs of auditing the system.
New products review	Costs of the preparation of bid proposals, the evaluation of new designs from a quality viewpoint, the preparation of tests and experimental programs to evaluate the performance of new products, and other quality activities during the development and preproduction stages of new products or designs.
Product/process design	Costs incurred during the design of the product or the selection of the production processes that are intended to improve the overall quality of the product. For example, an organization may decide to make a particular circuit component redundant because this will increase the reliability of the product by increasing the mean time between failures. Alternatively, it may decide to manufacture a component using process A rather than process B, because process A is capable of producing the product at tighter tolerances, which will result in fewer assembly and manufacturing problems. This may include a vendor's process, so the cost of dealing with other than the lowest bidder may also be a prevention cost. *(Continued)*

Process control	The cost of process-control techniques, such as control charts, that monitor the manufacturing process in an effort to reduce variation and build quality into the product.
Burn-in	The cost of preshipment operation of the product to prevent early-life failures in the field.
Training	The cost of developing, preparing, implementing, operating, and maintaining formal training programs for quality.
Quality data acquisition and analysis	The cost of running the quality data system to acquire data on product and process performance; also the cost of analyzing these data to identify problems. It includes the work of summarizing and publishing quality information for management.

Appraisal Costs. Appraisal costs are those costs associated with measuring, evaluating, or auditing products, components, and purchased materials to ensure conformance to the standards that have been imposed. These costs are incurred to determine the condition of the product from a quality viewpoint and ensure that it conforms to specifications. The major subcategories are shown as follow:

Inspection and test of incoming material	Costs associated with the inspection and testing of all material. This subcategory includes receiving inspection and test; inspection, test, and evaluation at the vendor's facility; and a periodic audit of the quality-assurance system. This could also include intraplant vendors.
Product inspection and test	The cost of checking the conformance of the product throughout its various stages of manufacturing, including final acceptance testing, packing and shipping checks, and any test done at the customer's facilities prior to turning the product over to the customer. This also includes life testing, environmental testing, and reliability testing.
Materials and services consumed	The cost of material and products consumed in a destructive test or devalued by reliability tests.
Maintaining accuracy of test equipment	The cost of operating a system that keeps the measuring instruments and equipment in calibration.

Internal Failure Costs. Internal failure costs are incurred when products, components, materials, and services fail to meet quality requirements and this failure is discovered prior to delivery of the product to the customer. These costs would disappear if there were no defects in the product. The major subcategories of internal failure costs follow.

Scrap	The net loss of labor, material, and overhead resulting from defective product that cannot economically be repaired or used.
Rework	The cost of correcting nonconforming units so that they meet specifications. In some manufacturing operations rework costs include additional operations or steps in the manufacturing process that are created to solve either chronic defects or sporadic defects. *(Continued)*

Retest	The cost of reinspection and retesting of products that have undergone rework or other modifications.
Failure analysis	The cost incurred to determine the causes of product failures.
Downtime	The cost of idle production facilities that results from nonconformance to requirements. The production line may be down because of nonconforming raw materials supplied by a supplier that went undiscovered in receiving inspection.
Yield losses	The cost of process yields that are lower than might be attainable by improved controls (for example, soft-drink containers that are overfilled because of excessive variability in the filling equipment).
Downgrading/off-specing	The price differential between the normal selling price and any selling price that might be obtained for a product that does not meet the customer's requirements. Downgrading is a common practice in the textile, apparel goods, and electronics industries. The problem with downgrading is that products sold do not recover the full contribution margin to profit and overhead as do products that conform to the usual specifications.

External Failure Costs. External failure costs occur when the product does not perform satisfactorily after it is delivered to the customer. These costs would also disappear if every unit of product conformed to requirements. Subcategories of external failure costs are shown in the following display:

Complaint adjustment	All costs of investigation and adjustment of justified complaints attributable to the nonconforming product.
Returned product/material	All costs associated with receipt, handling, and replacement of the nonconforming product or material that is returned from the field.
Warranty charges	All costs involved in service to customers under warranty contracts.
Liability costs	Costs or awards incurred from product liability litigation.
Indirect costs	In addition to direct operating costs of external failures, there are a significant number of indirect costs. These are incurred because of customer dissatisfaction with the level of quality of the delivered product. Indirect costs may reflect the customer's attitude toward the company. They include the costs of loss of business reputation, loss of future business, and loss of market share that inevitably results from delivering products and services that do not conform to the customer's expectations regarding fitness for use.

The Analysis and Use of Quality Costs. How large are quality costs? The answer, of course, depends on the type of organization and the success of their quality improvement effort. In some organizations quality costs are 4% or 5% of sales, whereas in others they can be as high as 35% or 40% of sales. Obviously, the cost of quality will be very different for a high-technology computer manufacturer than for a typical service industry, such as a department store or hotel chain. In most organizations, however, quality costs are higher than necessary, and

Table 1.3

Monthly Quality-Costs Information for Assembly of Printed Circuit Boards

Type of Defect	Percent of Total Defects	Scrap and Rework Costs
Insufficient solder	42	$37,500.00 (52%)
Misaligned components	21	12,000.00
Defective components	15	8,000.00
Missing components	10	5,100.00
Cold solder joints	7	5,000.00
All other causes	5	4,600.00
Totals	100%	$72,200.00

management should make continuing efforts to appraise, analyze, and reduce these costs.

The usefulness of quality costs stems from the **leverage effect;** that is, dollars invested in prevention and appraisal have a payoff in reducing dollars incurred in internal and external failures that exceeds the original investment. For example, a dollar invested in prevention may return $10 or $100 (or more) in savings from reduced internal and external failures.

Pareto analysis of quality costs

Quality-cost analyses have as their principal objective cost reduction through identification of improvement opportunities. This is often done with a **Pareto analysis.** The Pareto analysis consists of identifying quality costs by category, or by product, or by type of defect or nonconformity. For example, inspection of the quality-cost information in Table 1.3 concerning defects or nonconformities in the assembly of electronic components onto printed circuit boards reveals that insufficient solder is the highest quality cost incurred in this operation. Insufficient solder accounts for 42% of the total defects in this particular type of board and for almost 52% of the total scrap and rework costs. If the wave solder process can be improved, then there will be dramatic reductions in the cost of quality.

Reducing the cost of quality

How much reduction in quality costs is possible? Although the cost of quality in many organizations can be significantly reduced, it is unrealistic to expect it can be reduced to zero. Before that level of performance is reached, the incremental costs of prevention and appraisal will rise more rapidly than the resulting cost reductions. However, paying attention to quality costs in conjunction with a focused effort on variability reduction has the capability of reducing quality costs by 50% or 60% provided that no organized effort has previously existed. This cost reduction also follows the Pareto principle; that is, most of the cost reductions will come from attacking the few problems that are responsible for the majority of quality costs.

In analyzing quality costs and in formulating plans for reducing the cost of quality, it is important to note the role of prevention and appraisal. Many organizations devote far too much effort to appraisal and not enough to prevention. This is an easy mistake for an organization to make, because appraisal costs are often budget line items in manufacturing. On the other hand, prevention costs may not be routinely budgeted items. It is not unusual to find in the early stages of a quality-cost program that appraisal costs are eight or ten times the magnitude of prevention costs. This is probably an unreasonable ratio, as dollars spent in prevention have a much greater payback than do dollars spent in appraisal.

Generating quality cost information

Generating the quality-cost figures is not always easy, because most quality-cost categories are not a direct component in the accounting records of the organization. Consequently, it may be difficult to obtain extremely accurate information on the costs incurred with respect to the various categories. The organization's accounting system can provide information on those quality-cost categories that coincide with the usual business accounts—such as, for example, product testing and evaluation. In addition, many companies will have detailed information on various categories of failure cost. The information for cost categories for which exact accounting information is not available should be generated by using estimates, or, in some cases, by creating special monitoring and surveillance procedures to accumulate those costs over the study period.

The reporting of quality costs is usually done on a basis that permits straightforward evaluation by management. Managers want quality costs expressed in an index that compares quality cost with the opportunity for quality cost. Consequently, the usual method of reporting quality costs is in the form of a ratio in which the numerator is quality-cost dollars and the denominator is some measure of activity, such as hours of direct production labor, dollars of direct production labor, dollars of processing costs, dollars of manufacturing cost, dollars of sales, or (6) units of product.

Upper management may want a standard against which to compare the current quality-cost figures. It is difficult to obtain absolute standards and almost as difficult to obtain quality-cost levels of other companies in the same industry. Therefore, the usual approach is to compare current performance with past performance so that, in effect, quality-cost programs report variances from past performance. These trend analyses are primarily a device for detecting departures from standard and for bringing them to the attention of the appropriate managers. They are not necessarily in and of themselves a device for ensuring quality improvements.

Failure of quality cost initiatives

This brings us to an interesting observation: Some quality-cost collection and analysis efforts fail. That is, a number of companies have started quality-cost analysis activities, used them for some time, and then abandoned the programs as ineffective. There are several reasons why this occurs. Chief among these is failure to use quality-cost information as a mechanism for generating improvement opportunities. If we use quality-cost information as a scorekeeping tool only, and do not make conscious efforts to identify problem areas and develop improved operating procedures and processes, then the programs will not be totally successful.

Another reason why quality-cost collection and analysis does not lead to useful results is that managers become preoccupied with perfection in the cost figures. Overemphasis in treating quality costs as part of the accounting systems rather than as a management control tool is a serious mistake. This approach greatly increases the amount of time required to develop the cost data, analyze them, and identify opportunities for quality improvements. As the time required to generate and analyze the data increases, management becomes more impatient and less convinced of the effectiveness of the activity. Any program that appears to management to be going nowhere is likely to be abandoned.

A final reason for the failure of a quality-cost program is that management often underestimates the depth and extent of the commitment to prevention that must be made. The author has had numerous opportunities to examine quality cost data in many companies. In companies without effective quality improvement programs, the dollars allocated to prevention rarely exceed 1% to 2% of revenue. This must be increased to a threshold of about 5% to 6% of revenue, and these additional prevention dollars must be spent largely on the technical methods of quality improvement, and not on establishing programs such as TQM, Zero Defects, or other similar activities. If management is persistent in this effort, then the cost of quality will decrease substantially. These cost savings will typically begin to occur in one to two years, although it could be longer in some companies.

1.6 Legal Aspects of Quality

Consumerism and product liability are important reasons why quality assurance is an important business strategy. Consumerism is in part due to the seemingly large number of failures in the field of consumer products and the perception that service quality is declining. Highly visible field failures often prompt the questions of whether today's products are as good as their predecessors and whether manufacturers are really interested in quality. The answer to both of these questions is *yes*. Manufacturers are always vitally concerned about field failures because of heavy external failure costs and the related threat to their competitive position. Consequently, most producers have made product improvements directed toward reducing field failures. For example, solid-state and integrated-circuit technology has greatly reduced the failure of electronic equipment that once depended on the electron tube. Virtually every product line of today is superior to that of yesterday.

Consumer dissatisfaction and the general feeling that today's products are inferior to their predecessors arise from other phenomena. One of these is the explosion in the number of products. For example, a 1% field-failure rate for a consumer appliance with a production volume of 50,000 units per year means 500 field failures. However, if the production rate is 500,000 units per year and the field-failure rate remains the same, then 5,000 units will fail in the field. This is equivalent, in the total number of dissatisfied customers, to a 10% failure rate at the lower production level. Increasing production volume increases the **liability exposure** of the manufacturer. Even in situations in which the failure rate declines, if the production volume increases more rapidly than the decrease in failure rate, the total number of customers who experience failures will still increase.

A second aspect of the problem is that consumer tolerance for minor defects and aesthetic problems has decreased considerably, so that blemishes, surface-finish defects, noises, and appearance problems that were once tolerated now attract attention and result in adverse consumer reaction. Finally, the competitiveness of the marketplace forces many manufacturers to introduce new designs before they are fully evaluated and tested in order to remain competitive. These "early releases" of unproved designs are a major reason for new product quality failures. Eventually, these design problems are corrected, but the high failure rate connected with new products often supports the belief that today's quality is inferior to that of yesterday.

Product liability is a major social, market, and economic force. The legal obligation of manufacturers and sellers to compensate for injury or damage caused by defective products is not a recent phenomenon. The concept of product liability has been in existence for many years, but its emphasis has changed recently. The first major product liability case occurred in 1916 and was tried before the New York Court of Appeals. The court held that an automobile manufacturer had a product liability obligation to a car buyer, even though the sales contract was between the buyer and a third party—namely, a car dealer. The direction of the law has always been that manufacturers or sellers are likely to incur a liability when they have been unreasonably careless or negligent in what they have designed, or produced, or how they have produced it. In recent years, the courts have placed a more stringent rule in effect called **strict liability.** Two principles are characteristic of strict liability. The first is a strong responsibility for both manufacturer and merchandiser, requiring immediate responsiveness to unsatisfactory quality through product service, repair, or replacement of defective product. This extends into the period of actual use by the consumer. By producing a product, the manufacturer and seller must accept responsibility for the ultimate use of that product—not only for its performance, but also for its environmental effects, the safety aspects of its use, and so forth.

The second principle involves advertising and promotion of the product. Under strict product liability all advertising statements must be supportable by valid company quality or certification data, comparable to that now maintained for product identification under regulations for such products as automobiles.

These two strict product liability principles result in strong pressure on manufacturers, distributors, and merchants to develop and maintain a high degree of factually based evidence concerning the performance and safety of their products. This evidence must cover not only the quality of the product as it is delivered to the consumer, but also its durability or reliability, its protection from possible side effects or environmental hazards, and its safety aspects in actual use. A strong quality-assurance program can help management in ensuring that this information will be available, if needed.

1.7 Implementing Quality Improvement

Table 1.4

The Eight Dimensions of Quality from Section 1.1.1

1. Performance
2. Reliability
3. Durability
4. Serviceability
5. Aesthetics
6. Features
7. Perceived quality
8. Conformance to standards

In the past few sections we have discussed the philosophy of quality improvement, the link between quality and productivity, and both economic and legal implications of quality. These are important aspects of the management of quality within an organization. There are certain other aspects of the overall management of quality that warrant some attention.

Management must recognize that quality is a multifaceted entity, incorporating at least the eight dimensions we discussed in Section 1.1.1. For convenient reference, Table 1.4 summarizes these quality dimensions.

A critical part of the **strategic management of quality** within any business is the recognition of these dimensions by management and the selection of dimensions along which the business will compete. It will be very difficult to compete against companies that can successfully accomplish this part of the strategy.

A good example is the Japanese dominance of the videocassette recorder (VCR) market. The Japanese did not invent the VCR; the first units for home use were designed and produced in Europe and North America. However, the early VCRs produced by these companies were very unreliable and frequently had high levels of manufacturing defects. When the Japanese entered the market, they elected to compete along the dimensions of reliability and conformance to standards (no defects). This strategy allowed them to quickly dominate the market. In subsequent years, they expanded the dimensions of quality to include added features, improved performance, easier serviceability, improved aesthetics, and so forth. They used total quality as a competitive weapon to raise the entry barrier to this market so high that it was virtually impossible for a new competitor to enter.

Management must do this type of strategic thinking about quality. It is not necessary for the product to be superior in all dimensions of quality, but management must **select and develop** the "niches" of quality along which the company can successfully compete. Typically, these dimensions will be those that the competition has forgotten or ignored. The American automobile industry has been severely impacted by foreign competitors who expertly practiced this strategy.

Suppliers play a critical role in quality

The critical role of **suppliers** in quality management must not be forgotten. In fact, supplier selection and **supply chain management** may be the most critical aspects of successful quality management in industries such as automotive, aerospace, and electronics, where a very high percentage of the parts in the end item are manufactured by outside suppliers. Many companies have instituted formal supplier quality-improvement programs as part of their own **internal** quality-improvement efforts. Selection of suppliers based on **quality, schedule,** and **cost,** rather than on cost alone, is also a vital strategic management decision that can have a long-term significant impact on overall competitiveness.

It is also critical that management recognize that quality improvement must be a total, company-wide activity, and that every organizational unit *must* actively participate. Obtaining this participation is the responsibility of (and a significant challenge to) senior management. What is the role of the quality-assurance organization in this effect? The responsibility of quality assurance is to assist management in providing quality assurance for the companies' products. Specifically, the quality-assurance function is a technology warehouse that contains the skills and resources necessary to generate products of acceptable quality in the marketplace. Quality management also has the responsibility for evaluating and using quality-cost information for identifying improvement opportunities in the system, and for making these opportunities known to higher management. It is important to note, however, that the **quality function is not responsible for quality.** After all, the quality organization does not design, manufacture, distribute, or service the product. Thus, the responsibility for quality is distributed throughout the entire organization.

The responsibility for quality spans the entire organization. However, there is danger that if we adopt the philosophy that "quality is everybody's job," then quality will become nobody's job. This is why quality planning and analysis are important. Because quality improvement activities are so broad, successful efforts require, as an initial step, top management commitment. This commitment involves emphasis on the importance of quality, identification of the respective quality responsibilities of the various organizational units, and explicit accountability for quality improvement of all managers and employees in the company.

Quality management must involve quality planning, quality assurance, quality control and improvement.

Finally, strategic management of quality in an organization must involve all three components discussed earlier: **quality planning, quality assurance,** and **quality control and improvement.** Furthermore, *all* of the individuals in the organization must have an understanding of the basic tools of quality improvement. Central among these tools are the elementary statistical concepts that form the basis of process control and that are used for the analysis of process data. It is increasingly important that everyone in an organization, from top management to operating personnel, have an awareness of basic statistical methods and of how these methods are useful in manufacturing, engineering design, and development and in the general business environment. Certain individuals must have higher levels of skills; for example, those engineers and managers in the quality-assurance function would generally be experts in one or more areas of process control, reliability engineering, design of experiments, or engineering data analysis. However, the key point is the philosophy that statistical methodology is a language of communication about problems that enables management to mobilize resources rapidly and to efficiently develop solutions to such problems. Because six-sigma incorporates most of the elements for success that we have identified, it has proven to be a very effective framework for implementing quality improvement.

Important Terms and Concepts

Acceptance-sampling
Appraisal costs
Critical-to-quality (CTQ)
Dimensions of quality
Fitness for use
Internal and external failure costs
Nonconforming product or service
Prevention costs
Product liability
Quality assurance
Quality characteristics
Quality control and improvement
Quality engineering
Quality of conformance
Quality of design
Quality planning
Specifications
Variability

chapter One Exercises

1.1 Why is it difficult to define quality?

1.2 Briefly discuss the eight dimensions of quality. Does this improve our understanding of quality?

1.3 Select a specific product or service and discuss how the eight dimensions of quality impact its overall acceptance by consumers.

1.4 Is there a difference between quality for a manufactured product and quality for a service? Give some specific examples.

1.5 Can an understanding of the multidimensional nature of quality lead to improved product design or better service?

1.6 What are the internal customers of a business? Why are they important from a quality perspective?

1.7 What are the three primary technical tools used for quality control and improvement?

1.8 What is meant by the cost of quality?

1.9 Are internal failure costs more or less important than external failure costs?

1.10 Discuss the statement "Quality is the responsibility of the quality department."

1.11 Most of the quality management literature states that without top management leadership, quality improvement will not occur. Do you agree or disagree with this statement? Discuss why.

1.12 Explain why it is necessary to consider variability around the mean or nominal dimension as a measure of quality.

1.13 Suppose you had the opportunity to improve quality in a hospital. Which areas of the hospital would you look to as opportunities for quality improvement? What metrics would you use as measures of quality?

1.14 Suppose you had to improve service quality in a bank credit card application and approval process. What critical to-quality characteristics would you identify? How could you go about improving this system?

1.15 How would quality be defined in the following organization?

a. Health-care facility
b. Department store
c. Grocery store
d. University academic department

chapter Two Management Aspects of Quality

Chapter Outline

Six-Sigma saves millions at National Semiconductor

It is not surprising that building a new wafer fabrication plant is very expensive. Requiring more than 200 steps on 75 pieces of equipment, new fab construction can cost well over 1 billion dollars. It is important to keep in mind, however, that beyond just the initial factory construction, the entire semiconductor manufacturing process is very expensive also. National Semiconductor is a company that knows this all too well.

National Semiconductor has had excellent product quality with fewer than 10 defective parts per million. Given the high costs of manufacturing, however, the company decided to make any enhancements it could to lower its DPMs even more. The goals of its Six-Sigma effort were to reduce overall manufacturing costs while improving overall manufacturing processes.

The company created a Six-Sigma team of six members, who, over a nine-month period, reviewed product and workflow specifications, raw material sourcing, process simplifications, and other design changes. Following a define, measure, analyze, improve, and control (DMAIC) roadmap, the team was very successful. A 50% reduction in die patch waste and a 40% reduction in average per-unit cost for critical components (leadframes) were among the key results.

Motivated by such success, National Semiconductor went on to launch more than 30 other Six-Sigma projects, leading to tens of millions of dollars in cost savings. The company is also clear in noting that Six-Sigma has enhanced its problem solving capabilities and provided a logical path to follow in making key strategy decisions.

Chapter Overview and Learning Objectives

Technical tools drive quality control and improvement. However, these technical tools are most effective when they are implemented in a management framework that is focused on and that actively supports the improvement of products, processes, and systems. Several management philosophies and strategies have been developed to accomplish this. This chapter discusses these management techniques and provides an integrated view of how they fit together. The project-by-project approach to quality control and improvement is also discussed, with emphasis on the DMAIC (Define, Measure, Analyze, Improve, and Control) approach usually associated with six-sigma. DMAIC does not have to be linked to six-sigma, and we have found it to be an excellent way to manage projects and drive quality and business improvement. Several examples of improvement projects are given.

After careful study of this chapter, you should be able to do the following:

1. Describe the quality management philosophies of W. Edwards Deming, Joseph M. Juran, and Armand V. Feigenbaum
2. Discuss total quality management, six-sigma, the Malcolm Baldrige National Quality Award, and quality systems and standards
3. Understand the importance of selecting good projects for improvement activities
4. Explain the five steps of DMAIC
5. Know when and when not to use DMAIC

2.1 Introduction

There are many technical tools that are used in quality control and improvement activities; statistical process control, designed experiments, and reliability modeling and analysis, to name a few that will be discussed in this book. However, to be used most effectively, these techniques must be implemented within, and be part of, a management system that is focused on quality improvement. The management system of an organization must be organized to properly direct the overall quality improvement philosophy and ensure its deployment in all aspects of the business. The effective management of quality involves successful execution of three activities: **quality planning, quality assurance,** and **quality control and improvement.**

Quality Management

- *Quality Planning*
- *Quality Assurance*
- *Quality Control and Improvement*

Quality planning is a strategic activity, and it is just as vital to an organization's long-term business success as the product development plan, the financial plan, the marketing plan, and plans for the utilization of human resources. Without a strategic quality plan, an enormous amount of time, money, and effort will be wasted by the organization dealing with faulty designs, manufacturing defects, field failures, and customer complaints. Quality planning involves identifying customers, both external and those that operate internal to the business, and identifying their needs (this is sometimes called listening to the

voice of the customer [VOC]). Then products or services that meet or exceed customer expectations must be designed and developed. The eight dimensions of quality discussed in Section 1.1.1 are an important part of this effort. The organization must then determine how these products and services will be realized. Planning for quality improvement on a specific, systematic basis is also a vital part of this process.

Quality assurance is the set of activities that ensures that quality levels of products and services are properly maintained and that supplier and customer quality issues are properly resolved. Documentation of the quality system is an important component. Quality system documentation involves four components: policy, procedures, work instructions and specifications, and records. Policy generally deals with what is to be done and why, while procedures focus on the methods and personnel that will implement policy. Work instructions and specifications are usually product-, department-, tool-, or machine-oriented. Records are a way of documenting the policies, procedures, and work instructions that have been followed. Records are also used to track specific units or batches of product so that it can be determined exactly how they were produced. Records are often vital in providing data for dealing with customer complaints, corrective actions, and, if necessary, product recalls. Development, maintenance, and control of documentation are important quality assurance functions. One example of document control is ensuring that specifications and work instructions developed for operating personnel reflect the latest design and engineering changes. In other words, the purpose of a quality assurance activity is to "say what you are going to do, and do what you say."

Quality control and improvement involve the set of activities used to ensure that the products and services meet requirements and are improved on a continuous basis. Since variability is often a major source of poor quality, statistical techniques, including SPC and designed experiments, are the major tools of quality control and improvement. Quality improvement is often done on a project-by-project basis and involves teams led by personnel with specialized knowledge of statistical methods and experience in applying them. Projects should be selected so that they have significant business impact and are linked with the overall business goals for quality identified during the planning process. The techniques in this book are integral to successful quality control and improvement.

The next section provides a brief overview of some of the key elements of quality management. We discuss some of the important quality philosophies; quality systems and standards; the link between quality and productivity and quality and cost; and some aspects of implementation. The three aspects of quality planning, quality assurance, and quality control and improvement will be woven into the discussion.

Dimensions of Quality

- *Performance*
- *Reliability*
- *Durability*
- *Serviceability*
- *Aesthetics*
- *Features*
- *Perceived Quality*
- *Conformance to Standards*

2.2 Historical Development of Quality Philosophy

Many people have contributed to the statistical methodology of quality improvement. However, in terms of implementation and management philosophy, three individuals emerge as the leaders: W. E. Deming,

J. M. Juran, and A. V. Feigenbaum. We now briefly discuss the approaches and philosophy of those leaders in quality management.

W. EDWARDS DEMING

W. Edwards Deming (1900–1993) (Courtesy of National Semiconductor Corporation, © Catherine Karnow / © Corbis)

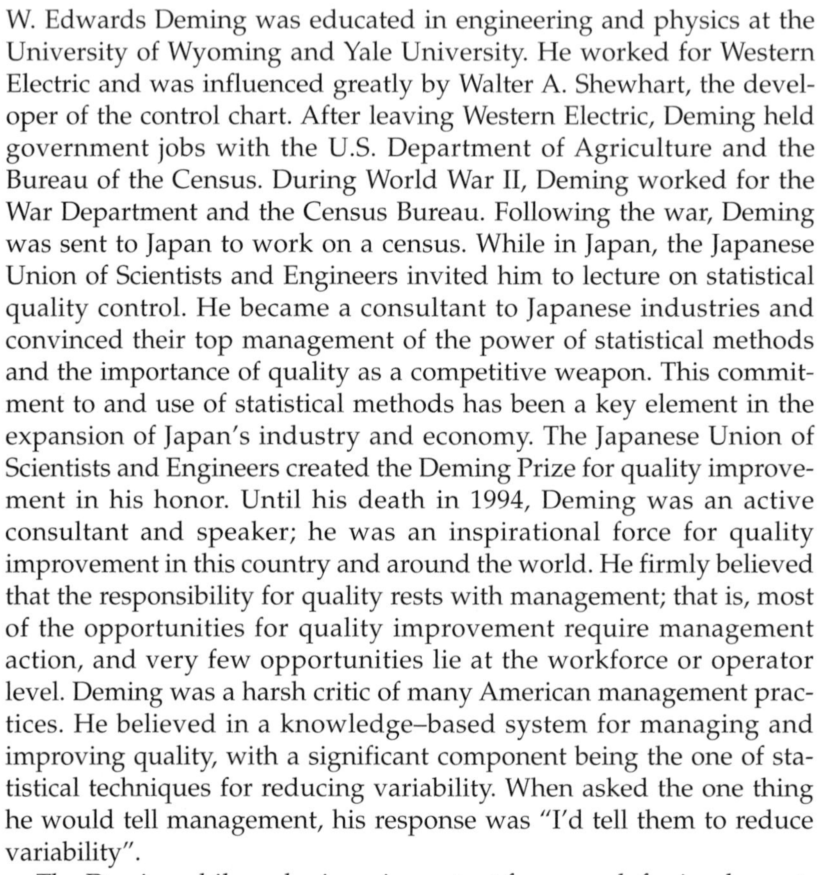

W. Edwards Deming was educated in engineering and physics at the University of Wyoming and Yale University. He worked for Western Electric and was influenced greatly by Walter A. Shewhart, the developer of the control chart. After leaving Western Electric, Deming held government jobs with the U.S. Department of Agriculture and the Bureau of the Census. During World War II, Deming worked for the War Department and the Census Bureau. Following the war, Deming was sent to Japan to work on a census. While in Japan, the Japanese Union of Scientists and Engineers invited him to lecture on statistical quality control. He became a consultant to Japanese industries and convinced their top management of the power of statistical methods and the importance of quality as a competitive weapon. This commitment to and use of statistical methods has been a key element in the expansion of Japan's industry and economy. The Japanese Union of Scientists and Engineers created the Deming Prize for quality improvement in his honor. Until his death in 1994, Deming was an active consultant and speaker; he was an inspirational force for quality improvement in this country and around the world. He firmly believed that the responsibility for quality rests with management; that is, most of the opportunities for quality improvement require management action, and very few opportunities lie at the workforce or operator level. Deming was a harsh critic of many American management practices. He believed in a knowledge–based system for managing and improving quality, with a significant component being the one of statistical techniques for reducing variability. When asked the one thing he would tell management, his response was "I'd tell them to reduce variability".

The Deming philosophy is an important framework for implementing quality and productivity improvement. This philosophy is summarized in his 14 points for management. The following display gives a brief statement and discussion of **Deming's 14 points:**

1. **Create a constancy of purpose focused on the improvement of products and services**	Deming was very critical of the short-term thinking of American management, which tends to be driven by quarterly business results and doesn't always focus on strategies that benefit the organization in the long run. Management should constantly try to improve product design and performance. This must include investment in research, development, and innovation which will have long-term payback to the organization.
2. **Adopt a new philosophy that recognizes we are in a different economic era**	Reject poor workmanship, defective products, or bad service. It costs as much to produce a defective unit as it does to produce a good one (and sometimes more). The cost of dealing with scrap, rework, and other losses created by defectives is an enormous drain on company resources. *(Continued)*

3. **Do not rely on mass inspection to "control" quality**	All inspection can do is sort out defectives, and at that point it is too late—the organization already has paid to produce those defectives. Inspection typically occurs too late in the process, is expensive, and is often ineffective. Quality results from prevention of defectives through process improvement, not inspection.
4. **Do not award business to suppliers on the basis of price alone, but also consider quality**	Price is a meaningful measure of a supplier's product only if it is considered in relation to a measure of quality. In other words, the total cost of the item must be considered, not just the purchase price. When quality is considered, the lowest bidder frequently is not the low-cost supplier. Preference should be given to suppliers who use modern methods of quality improvement in their business and who can demonstrate process control and capability. An adversarial relationship with suppliers is harmful. It is important to build effective, long-term relationships.
5. **Focus on continuous improvement**	Constantly try to improve the production and service system. Involve the workforce in these activities and make use of statistical methods (particularly the statistically based problem-solving tools discussed in this book).
6. **Practice modern training methods and invest in on-the-job training for all employees**	Everyone should be trained in the technical aspects of his or her job, and in modern quality- and productivity-improvement methods as well. The training should encourage all employees to practice these methods every day. Too often, employees are not encouraged to use the results of training, and management often believes employees do not need training or already should be able to practice the methods. Many organizations devote little or no effort to training.
7. **Improve leadership, and practice modern supervision methods**	Supervision should not consist merely of passive surveillance of workers but should be focused on helping the employees improve the system in which they work. The number one goal of supervision should be to improve the work system and the product.
8. **Drive out fear**	Many workers are afraid to ask questions, to report problems, or to point out conditions that are barriers to quality and effective production. In many organizations the economic loss associated with fear is large; only management can eliminate fear.
9. **Break down the barriers between functional areas of the business**	Teamwork among different organizational units is essential for effective quality and productivity improvement to take place.
10. **Eliminate targets, slogans, and numerical goals for the workforce**	A target such as "zero defects" is useless without a plan for the achievement of this objective. In fact, these slogans and "programs" are usually counterproductive. Work to improve the system and provide information on that.
11. **Eliminate numerical quotas and work standards**	These standards have historically been set without regard to quality. Work standards are often symptoms of management's inability to understand the work process and to provide an effective management system focused on improving this process.
12. **Remove the barriers that discourage employees from doing their jobs**	Management must listen to employee suggestions, comments, and complaints. The person who is doing the job knows the most about it and usually has valuable ideas about how to make the process work more effectively. The workforce is an important participant in the business, and not just an opponent in collective bargaining. *(Continued)*

13. **Institute an ongoing program of education for all employees**	Part 6 referred to job-related training. Here Deming was referring to more generalized and broad-based education. In general, education is valuable because it can enhance creativity and help drive innovation. Education is a way of making everyone partners in the business and the overall quality improvement process.
14. **Create a structure in top management that will vigorously advocate the first 13 points**	This structure must be driven from the very top of the organization. It must also include concurrent education/training activities and expedite application of the training to achieve improved business results. Everyone in the organization must know that continuous improvement is a common goal.

As we read Deming's 14 points we notice that there is a strong emphasis on **organizational change.** Also, the role of management in guiding this change process is of dominating importance. However, what should be changed, and how should this change process be started? For example, if we want to improve the yield of a semiconductor manufacturing process, what should we do? If we want to reduce the number of errors in processing purchase order requests, how do we proceed? It is in this area that statistical methods come into play most frequently. To improve the semiconductor process, we must determine which controllable factors in the process influence the number of defective units produced. To answer this question, we must collect data on the process and see how the system reacts to change in the process variables. Then actions to improve the process can be designed and implemented. Statistical methods, such as designed experiments and control charts, can contribute to these activities.

Deming frequently wrote and spoke about the **seven deadly diseases of management,** listed in Table 2.1. He believed that each disease was a barrier to the effective implementation of his philosophy.

Table 2.1

Deming's Seven Deadly Diseases of Management

1. **Lack of constancy of purpose**	Lack of constancy of purpose relates to the first of Deming's 14 points. Continuous improvement of products, processes, and services gives assurance to all stakeholders in the enterprise (employees, executives, investors, suppliers) that dividends and increases in the value of the business will continue to grow.
2. **Emphasis on short-term profits**	The second disease, too much emphasis on short-term profits, might make the "numbers" look good, but if this is achieved by reducing research and development investment, by eliminating employees' training, and by not deploying quality improvement activities, then irreparable long-term damage to the business is the ultimate result.
3. **Evaluation of performance, merit rating, and annual reviews of performance**	Deming believed that performance evaluation encouraged short-term performance, rivalries and fear, and discouraged effective teamwork. Performance reviews can leave employees bitter and discouraged, and they may feel unfairly treated, especially if they are working in an organization where their performance is impacted by system forces that are flawed and out of their control.

Table 2.1

Continued

4. **Mobility of top management**	Management mobility, refers to the widespread practice of job-hopping; that is, a manger spending very little time in the business function for which he or she is responsible. This often results in key decisions being made by someone who really doesn't understand the business. Managers often spend more time thinking about their next career move than about their current job and how to do it better. Frequent reorganizing and shifting management responsibilities is a barrier to constancy of purpose and often is a waste of resources that should be devoted to improving products and services. Bringing in a new chief executive officer to improve quarterly profits often leads to a business strategy that leaves a path of destruction throughout the business.
5. **Running a company on visible figures alone**	Management by visible figures alone (such as the number of defects, customer complaints, and quarterly profits) suggests that the really important factors that determine long-term organizational success are unknown and unknowable. As some evidence of this, of the 100 largest companies in 1900, only 16 still exist today, and of the 25 largest companies in 1900, only two are still among the top 25. Obviously, some visible figures are important; for example, suppliers and employees must be paid on time and the bank accounts must be managed. However, if visible figures alone were key determinates of success, it's likely that many more of the companies of 1900 still would be in business.
6. **Excessive medical costs**	Deming's cautions about excessive medical expenses are certainly prophetic: Health care costs may be the most important issue facing many sectors of business in the United States toady. For example, the medical costs for current and retired employees of United States automobile manufacturers General Motors, Ford, and Chrysler currently is estimated to be between \$1200 and \$1600 per vehicle, contrasted with \$250 to \$350 per vehicle for Toyota and Honda, two Japanese automobile manufacturers with extensive North American manufacturing and assembly operations.
7. **Excessive legal damage awards**	Liability and excessive damage awards, is also a major issue facing many organizations. Deming was fond of observing that the United States had more lawyers per capita than any other nation. He believed that government intervention likely would be necessary to provide effective long-term solutions to the medical cost and excessive liability awards problems.

Deming recommended the **Shewhart cycle,** shown in Figure 2.1, as a model to guide improvement. The four steps, **Plan-Do-Check-Act,** are often called the **PDCA cycle.** Sometimes the **Check** step is called **Study,** and the cycle becomes the **PDSA cycle.** In **Plan,** we propose a change in the system that is aimed at improvement. In **Do,** we carry out the change, usually on a small or pilot scale to ensure that we learn the results that will be obtained. **Check** consists of analyzing the results of the change to determine what has been learned about the changes that were carried out. In **Act,** we either adopt the change or, if it was unsuccessful, abandon it. The process is almost always iterative, and may require several cycles for solving complex problems.

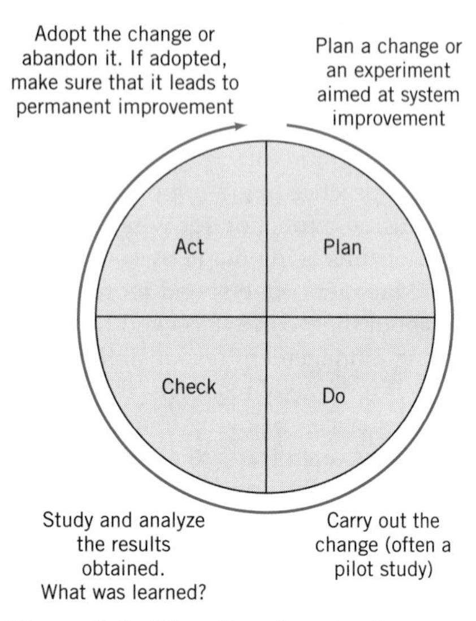

Figure 2.1 The Shewhart Cycle

In addition to Deming's 14 points and his seven deadly diseases of management, Deming wrote and lectured about an extensive collection of **obstacles to success.** Some of these include:

1. The belief that automation, computers, and new machinery will solve problems.
2. Searching for examples—trying to copy existing solutions.
3. The "our problems are different" excuse—not realizing that the principles that will solve problems are universal.
4. Obsolete schools, where graduates have not been taught how to successfully run businesses.
5. Poor teaching of statistical methods in industry: teaching tools without a framework for using them is going to be unsuccessful.
6. Reliance on inspection to produce quality.
7. Reliance on the "quality control department" to take care of all quality problems.
8. Blaming the workforce for problems.
9. False starts, such as broad teaching of statistical methods without a plan as to how to use them, quality circles, employee suggestion systems, and other forms of "instant pudding."
10. The fallacy of zero defects: Companies fail even though they produce products and services without defects. Meeting the specifications isn't the complete story in any business.
11. Inadequate testing of prototypes: A prototype may be a one-off article with artificially good dimensions, but without knowledge of variability, testing a prototype tells very little. This is a symptom of inadequate understanding of product design, development, and the overall activity of technology commercialization.
12. "Anyone that comes to help us must understand all about our business." This is bizarre thinking: there already are competent people in the organization who know everything about the business—except how to improve it. New knowledge and ideas (often from the outside) must be fused with existing business expertise to bring about change and improvement.

JOSEPH M. JURAN (1904–2008)

Juran was one of the founding fathers of the quality-control and improvement field. He worked for Walter A. Shewhart at AT&T Bell Laboratories and was at the leading edge of quality improvement throughout his career. Juran became the chief industrial engineer at Western Electric (part of the Bell System). He was an assistant administrator for the Lend-Lease Administration during World War II and played an important role in simplifying the administrative and paper work processes of that agency. After the war, he became the head of the Department of Administrative Engineering at New York University. He was invited to speak to Japanese industry leaders as they began their industrial transformation in the early 1950s. He also created an active consulting practice (the Juran Institute) and lectured widely through the American Management Association. He was the co-author (with Frank M. Gryna) of the *Quality Control Handbook,* a standard reference for quality methods and improvement since its initial publication in 1957.

Juran took a more strategic approach to quality management and improvement than Deming. He believed that most quality problems result from ineffective planning for quality. The Juran quality management philosophy focuses on three components: **planning**, **control**, and **improvement**. These are known as the **Juran Trilogy.** As we have noted previously, planning involves identifying external customers and determining their needs. Then products or services that meet these customer needs are designed and/or developed, and the processes for producing these products or services are then developed. The planning process should also involve planning for quality improvement on a regular (typically annual) basis. Control is employed by the operating forces of the business to ensure that the product or service meets the requirements. SPC is one of the primary tools of control. Improvement aims to achieve performance and quality levels that are higher than current levels. Juran emphasizes that improvement must be on a project-by-project basis. These projects are typically identified at the planning stage of the trilogy. Improvement can either be continuous (or incremental) or by breakthrough. Typically, a breakthrough improvement is the result of studying the process and identifying a set of changes that result in a large, relatively rapid improvement in performance. Designed experiments are an important tool that can be used to achieve breakthrough.

Juran Triology:

- Planning
- Control
- Improvement

ARMAND V. FEIGENBAUM (1922–)

Feigenbaum first introduced the concept of companywide quality control in his historic book *Total Quality Control* (first published in 1951). This book influenced much of the early philosophy of quality management in Japan in the early 1950s. In fact, many Japanese companies used the term "total quality control" to describe their efforts. He proposed a three-step approach to improving quality: quality leadership, quality technology, and organizational commitment. By **quality technology,** Feigenbaum means statistical methods and other technical and engineering methods, such as the ones discussed in this book.

Feigenbaum is concerned with organizational structure and a systems approach to improving quality. He proposed a 19-step improvement process, of which use of statistical methods was step 17. He initially suggested that much of the technical capability be concentrated in a specialized department. This is in contrast to the more modern view that knowledge and use of statistical tools need to be widespread. However, the organizational aspects of Feigenbaum's work are important, as quality improvement does not usually spring forth as a "grass-roots" activity; it requires a lot of management commitment to make it

Feigenbaum Approach:

- *Quality Leadership*
- *Quality Technology*
- *Organizational Commitment*

work. Feigenbaum was awarded the National medal of Technology and Innovation in 2007 for his work.

The brief descriptions of the philosophies of Deming, Juran, and Feigenbaum have highlighted both the common aspects and differences of their viewpoints. In one view, there are more similarities than differences among them, and the similarities are what are important. Together, they have laid the foundation for a modern **theory of quality**. All three of these pioneers stress the importance of quality as an essential competitive weapon, the important role that management must play in implementing quality improvement, and the importance of statistical methods and techniques in the "quality transformation" of an organization.

OTHER CONTRIBUTORS

Ichiro Ishikawa was a famous Japanese business leader who founded the influential Japanese Union of Scientists and Engineers (JUSE). His son, **Kaoru Ishikawa**, led JUSE during its early growth years and ultimately became one of the most important and influential leaders of the Japanese quality improvement movement. Kaoru Ishikawa developed many basic quality improvement tools, the most famous of which is the cause-and-effect diagram (also called the Ishikawa diagram). He believed strongly in training, and training became one of the primary missions of JUSE. Ishikawa was a pioneer in getting basic tools of process and quality improvement widely used throughout the workforce.

Philip Crosby was an executive with the International Telephone and Telegraph (ITT) Company who wrote two widely read books on quality; *Quality Is Free* (published in 1979), and *Quality without Tears* (published in 1984). Crosby believed that quality was a source of profit and opportunity for a company, and that by focusing on the cost of quality (see Section 1.5), areas where corrective actions are required could be identified and improvements obtained.

Crosby's books were timely, as the late 1970s and early 1980s was a period where many U.S. companies were searching for an effective strategy to improve quality. The books were easy to read and promised that improvement was possible by relying on the behavorial and motivational aspects of quality improvement. A central feature of this was the **zero defects** concept; that is, if people commit themselves and work hard to avoid errors, then defects can be eliminated. Crosby didn't advocate the widespread use of technical tools to solve quality problems. Because he made it look easy, and managers and executives are often looking for a "quick and easy fix," Crosby established a large consulting practice.

Crosby certainly deserves credit for making managers and executives aware that quality decisions are business decisions and that they carry economic consequences. However, his overall approach, relying on employee motivation and behavior modification, was unsuccessful, and his consulting ventures eventually failed. In most organizations, employees are already working hard and trying to avoid mistakes and errors. In most situations, only modest improvements are going to occur unless the appropriate technical tools are brought to bear on the right problems.

Genichi Taguchi is a Japanese engineer who believed that unwanted variability was a leading cause of poor quality. He argued that any deviation from the ideal or target value for a quality characteristic resulted in a loss, not just to the customer and the business, but to society. Taguchi used a **quadratic loss function** as shown in Figure 2.2 as a model for this situation. Notice that the losses begin to occur as soon as the quality characteristic deviates from the target. A more traditional view would be that losses only occur when the quality characteristic exceeds either the lower or upper specification (LSL and USL in Figure 2.2). In the Taguchi approach, some of this excess variability in products or processes occurs as a result of factors that are difficult or impossible to control, called **noise variables.** Taguchi suggested using designed experiments to determine the levels of the factors that could be well controlled to both optimize the target levels of quality characteristics and offset the variability transmitted from noise variables. He refers to this as **robust parameter design.** He proposed several novel techniques and approaches to solve the robust design problem. Taguchi's methods began to be introduced to U.S. industry in the early 1980s and for several years enjoyed considerable use, although the appropriateness of his methods was challenged by many experts in designed experiments and statistics.

Taguchi clearly identified an important problem. Making products and processes robust to uncontrollable sources of variability is a critical part of making any product successful. However, the technical methods he advocated to achieve robustness were shown to be inefficient and often ineffective. During the late 1980s and early 1990s considerable research was done on robust design problems and new methods introduced that solve these problems using better methods. These methods are discussed in Chapter 8.

Figure 2.2 **A Quadratic Loss Function**

2.3 Total Quality Management

Total quality management (TQM) is a strategy for implementing and managing quality improvement activities on an organizationwide basis. TQM began in the early 1980s with the philosophies of Deming and Juran as the focal point. It evolved into a broader spectrum of concepts and ideas, involving participative organizations and work culture, customer focus, supplier quality improvement, integration of the quality system with business goals, and many other activities to focus all elements of the organization around the quality improvement goal. Typically, organizations that have implemented a TQM approach to quality improvement have quality councils or high-level teams that deal with strategic quality initiatives, workforce-level teams that focus on routine production or business activities, and cross-functional teams that address specific quality improvement issues.

Total quality management (TQM) is a set of management practices aimed at instilling awareness of quality principles throughout the organization and ensuring that the customer requirements are consistently met or exceeded.

TQM has only had moderate success for a variety of reasons, but frequently because there is insufficient effort devoted to widespread utilization of the technical tools of variability reduction. Many organizations saw the mission of TQM as one of training. Consequently, many TQM efforts engaged in widespread training of the workforce in the philosophy of quality improvement and a few basic methods. This training was usually placed in the hands of human resources

Quality is free
During the heyday of TQM in the 1980s, another popular program was the **Quality is Free** initiative, in which management worked on identifying the cost of quality (or the cost of nonquality, as the Quality is Free devotees so cleverly put it). Indeed, identification of quality costs can be very useful (we discussed quality costs in Chapter 1), but the Quality is Free practitioners often had no idea about what to do to actually improve many types of complex industrial processes. In fact, the leaders of this initiative had no knowledge about statistical methodology and completely failed to understand its role in quality improvement. When TQM is wrapped around an ineffective program such as this, disaster is often the result.

departments, and much of it was ineffective. The trainers often had no real idea about *what* methods should be taught, and success was usually measured by the percentage of the workforce that had been "trained," not by whether any measurable impact on business results had been achieved. Some general reasons for the lack of conspicuous success of TQM include (1) lack of top-down, high-level **management commitment and involvement;** (2) inadequate use of statistical methods and insufficient recognition of variability reduction as a prime objective; (3) general as opposed to specific business-results-oriented objectives; and (4) too much emphasis on widespread **training** as opposed to focused technical **education.**

Another reason for the erratic success of TQM is that many managers and executives have regarded it as just another "program" to improve quality. During the 1950s and 1960s, programs such as **Zero Defects** and **Value Engineering** abounded, but they had little real impact on quality and productivity improvement.

2.4 Quality Systems and Standards

The International Standards Organization (founded in 1946 in Geneva, Switzerland), known as ISO, has developed a series of standards for quality systems. The first standards were issued in 1987. The current version of the standard is known as the ISO 9000 series. It is a generic standard, broadly applicable to any type of organization, and it is often used to demonstrate a supplier's ability to control its processes. The three standards of ISO 9000 are:

ISO 9000:2000 Quality Management System—Fundamentals and Vocabulary

ISO 9001:2000 Quality Management System—Requirements

ISO 9004:2000 Quality Management System—Guidelines for Performance Improvement

ISO 9000 is also an American National Standards Institute and ASQ standard.

The ISO 9001:2000 standard has eight clauses: (1) Scope, (2) Normative References, (3) Definitions, (4) Quality Management Systems, (5) Management Responsibility, (6) Resource Management, (7) Product (or Service) Realization, and (8) Measurement, Analysis, and Improvement. Clauses 4 through 8 are the most important, and their key components and requirements are shown in Table 2.2. To become certified under the ISO standard, a company must select a **registrar** and prepare for a **certification audit** by this registrar. There is no single independent authority that licenses, regulates, monitors, or qualifies registrars. As we will discuss later, this is a serious problem with the ISO system. Preparing for the certification audit involves many activities, including (usually) an initial or phase I audit that checks the present quality management system against the standard. This is usually followed by establishing teams to ensure that all components of the key clause are developed and implemented, training of

Table 2.2

ISO 9001:2000 Requirements

4.0	**Quality Management System**
4.1	General Requirements The organization shall establish, document, implement, and maintain a quality management system and continually improve its effectiveness in accordance with the requirements of the international standard.
4.2	Documentation Requirements Quality management system documentation will include a quality policy and quality objectives; a quality manual; documented procedures; documents to ensure effective planning, operation, and control of processes; and records required by the international standard.
5.0	**Management System**
5.1	Management Commitment a. Communication of meeting customer, statutory, and regulatory requirements b. Establishing a quality policy c. Establishing quality objectives d. Conducting management reviews e. Ensuring that resources are available
5.2	Top management shall ensure that customer requirements are determined and are met with the aim of enhancing customer satisfaction.
5.3	Management shall establish a quality policy.
5.4	Management shall ensure that quality objectives shall be established. Management shall ensure that planning occurs for the quality management system.
5.5	Management shall ensure that responsibilities and authorities are defined and communicated.
5.6	Management shall review the quality management system at regular intervals.
6.0	**Resource Management**
6.1	The organization shall determine and provide needed resources.
6.2	Workers will be provided necessary education, training, skills, and experience.
6.3	The organization shall determine, provide, and maintain the infrastructure needed to achieve conformity to product requirements.
6.4	The organization shall determine and manage the work environment needed to achieve conformity to product requirements.
7.0	**Product or Service Realization**
7.1	The organization shall plan and develop processes needed for product or service realization.
7.2	The organization shall determine requirements as specified by customers.
7.3	The organization shall plan and control the design and development for its products or services.
7.4	The organization shall ensure that purchased material or product conforms to specified purchase requirements.
7.5	The organization shall plan and carry out production and service under controlled conditions.
7.6	The organization shall determine the monitoring and measurements to be undertaken and the monitoring and measuring devices needed to provide evidence of conformity of products or services to determined requirements.
8.0	**Measurement, Analysis, and Improvement**
8.1	The organization shall plan and implement the monitoring, measurement, analysis, and improvement process for continual improvement and conformity to requirements.
8.2	The organization shall monitor information relating to customer perceptions.
8.3	The organization shall ensure that product that does not conform to requirements is identified and controlled to prevent its unintended use or delivery.
8.4	The organization shall determine, collect, and analyze data to demonstrate the suitability and effectiveness of the quality management system, including a. Customer satisfaction b. Conformance data c. Trend data d. Supplier data
8.5	The organization shall continually improve the effectiveness of the quality management system.

Adapted from the ISO 9001:2000 Standard. International Standards Organization, Geneva. Switzerland, 2003.

personnel, development of applicable documentation, and developing and installing all new components of the quality system that may be required. Then the certification audit takes place. If the company is certified, then periodic **surveillance audits** by the registrar continue, usually on an annual (or perhaps six-month) schedule.

Many organizations have required their suppliers to become certified under ISO 9000, or one of the standards that are more industry-specific. Examples of these industry-specific quality system standards are AS 9100 for the aerospace industry; ISO/TS 16949 and QS 9000 for the automotive industry; and TL 9000 for the telecommunications industry. Many components of these standards are very similar to those of ISO 9000.

The focus of ISO certification is quality assurance

Much of the focus of ISO 9000 (and of the industry-specific standards) is on formal documentation of the quality system; that is, on **quality assurance** activities. Organizations usually must make extensive efforts to bring their documentation in to line with the requirements of the standards; this is the Achilles' heel of ISO 9000 and other related or derivative standards. There is far too much effort devoted to documentation, paperwork, and bookkeeping and not nearly enough to actually reducing variability and improving processes and products. Furthermore, many of the third-party registrars, auditors, and consultants that work in this area are not sufficiently educated or experienced enough in the **technical** tools required for **quality improvement** or in how these tools should be deployed. They are all too often unaware of what constitutes modern engineering and statistical practice, and usually are familiar with only the most elementary techniques. Therefore, they concentrate largely on the documentation, record keeping, and paperwork aspects of certification. A functional quality system must have three components; quality planning, quality assurance, and quality control and improvement.

ISO certification does not ensure that high quality products are delivered to customers

There is also evidence that ISO certification or certification under one of the other industry-specific standards does little to prevent poor-quality products from being designed, manufactured, and delivered to the customer. For example, in 1999–2000, there were numerous incidents of rollover accidents involving Ford Explorer vehicles equipped with Bridgestone/Firestone tires. There were nearly 300 deaths in the United States alone attributed to these accidents, which led to a recall by Bridgestone/Firestone of approximately 6.5 million tires. Apparently, many of the tires involved in these incidents were manufactured at the Bridgestone/Firestone plant in Decatur, Illinois. In an article on this story in *Time* magazine (September 18, 2000), there was a photograph (p. 38) of the sign at the entrance of the Decatur plant which stated that the plant was "QS 9000 Certified" and "ISO 14001 Certified" (ISO 14001 is an environmental standard). Although the assignable causes underlying these incidents have not been fully discovered, there are clear indicators that despite quality systems certification, Bridgestone/Firestone experienced significant quality problems. ISO certification alone is no guarantee that good quality products are being designed, manufactured, and delivered to the customer. Relying on ISO certification is a strategic management mistake.

It has been estimated that ISO certification activities are approximately a *$40 billion annual business*, worldwide. Much of this money flows to the registrars, auditors, and consultants. This amount does not include all of the internal costs incurred by organizations to achieve registration, such as the thousands of hours of engineering and management effort, travel, internal training, and internal auditing. It is not clear whether any significant fraction of this expenditure has made its way to the bottom line of the registered organizations. Furthermore, there is no assurance that certification has any real impact on quality (as in the Bridgestone/Firestone tire incidents). Many quality engineering authorities believe that ISO certification is largely a waste of effort. Often, organizations would be far better off to "just say no to ISO" and spend a small fraction of that $40 billion on their quality systems and another larger fraction on meaningful variability reduction efforts, develop their own internal (or perhaps industry-based) quality standards, rigorously enforce them, and pocket the difference.

THE MALCOLM BALDRIGE NATIONAL QUALITY AWARD

The Malcolm Baldrige National Quality Award (MBNQA) was created by the U.S. Congress in 1987. It is given annually to recognize U.S. organizations for performance excellence. Awards are given to organizations in five categories: manufacturing, service, small business, health care, and education. Three awards may be given each year in each category. Many organizations compete for the awards, and many companies use the performance excellence criteria for self-assessment. The award is administered by NIST (the National Institute of Standards and Technology). Today there are many state quality awards, most of which are patterned on the MBNQA.

The performance excellence criteria and their interrelationships are shown in Figure 2.3. The point values for these criteria in the MBNQA are shown in Table 2.3. The criteria are directed towards results, where results are a composite of customer satisfaction and retention, market share and new market development, product/service quality, productivity and operational effectiveness, human resources development,

> MALCOLM BALDRIGE (1922–1987) was the U.S. Secretary of Commerce from 1980 to 1987. He had a business career that was characterized by great success. He believed that business must embrace a strong commitment to quality improvement, customer service, and gaining competitive advantage. He died in an accident while in office.

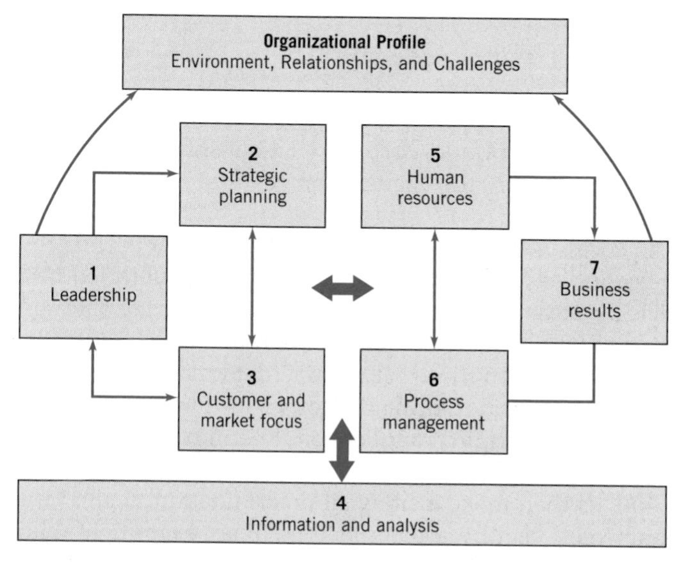

Figure 2.3 The Structure of the MBNQA, Performance Excellence Criteria

(Source: Foundation for the Malcolm Baldrige National Quality Award, 2002 Criteria for Performance Excellence)

Table 2.3

Performance Excellence Categories and Point Values

1	**Leadership**		120
	1.1 Leadership System	80	
	1.2 Company Responsibility and Citizenship	40	
2	**Strategic Planning**		85
	2.1 Strategy Development Process	40	
	2.2 Company Strategy	45	
3	**Customer and Market Focus**		85
	3.1 Customer and Market Knowledge	40	
	3.2 Customer Satisfaction and Relationship Enhancement	45	
4	**Information and Analysis**		90
	4.1 Measurement and Analysis of Performance	50	
	4.2 Information Management	40	
5	**Human Resource Focus**		85
	5.1 Work Systems	35	
	5.2 Employee Education, Training, and Development	25	
	5.3 Employee Well-Being and Satisfaction	25	
6	**Process Management**		85
	6.1 Management of Product and Service Processes	45	
	6.2 Management of Business Processes	25	
	6.3 Management of Support Processes	15	
7	**Business Results**		450
	7.1 Customer Results	125	
	7.2 Financial and Market Results	125	
	7.3 Human Resource Results	80	
	7.4 Organizational Results	120	
	Total Points		**1,000**

supplier performance, and public/corporate citizenship. The criteria are nonprescriptive. That is, the focus is on results, not the use of specific procedures or tools.

The MBNQA process is shown in Figure 2.4. An applicant sends the completed application to NIST. This application is then subjected to a first-round review by a team of Baldrige examiners. The board of Baldrige examiners consists of highly qualified volunteers from a variety of fields. Judges evaluate the scoring on the application to determine if the applicant will continue to **consensus.** During the consensus phase, a group of examiners who scored the original application determines a consensus score for each of the items. Once consensus is reached and a consensus report written, judges then make a site-visit determination. A site visit typically is a one-week visit by a team of four to six examiners who

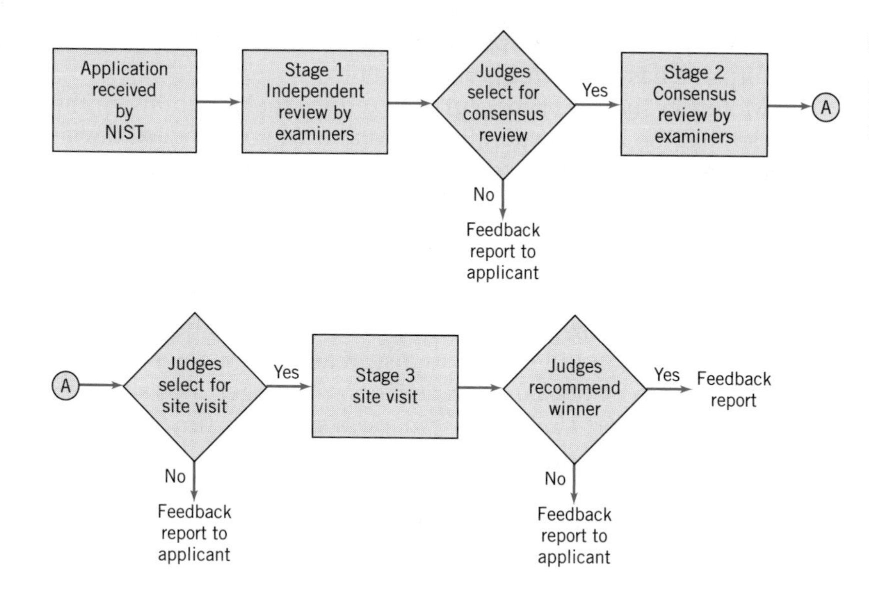

Figure 2.4 MBNQA Process

(Source: Foundation for the Malcolm Baldrige National Quality Award, 2002 Criteria for Performance Excellence.)

produce a site-visit report. The site-visit reports are used by the judges as the basis of determining the final MBNQA winners.

As shown in Figure 2.4, feedback reports are provided to the applicant at up to three stages of the MBNQA process. Many organizations have found these reports very helpful and use them as the basis of planning for overall improvement of the organization and for driving improvement in business results.

2.5 Six-Sigma

Products with many components typically have many opportunities for failure or defects to occur. Motorola developed the **Six-Sigma program** in the late 1980s as a response to the demand for their products. The focus of six-sigma is reducing variability in key product quality characteristics to the level at which failure or defects are extremely unlikely.

Six-Sigma is a business strategy that seeks to improve business performance by identifying and removing the causes of defects and errors and reducing variability in key quality characteristics.

Figure 2.5*a* shows a normal probability distribution as a model for a quality characteristic with the specification limits at three standard deviations on either side of the mean. Now it turns out that in this situation the probability of producing a product within these specifications is 0.9973, which corresponds to 2700 parts per million (ppm) defective. This is referred to as **three-sigma quality performance,** and it actually sounds pretty good. However, suppose we have a product that consists of an assembly of 100 independent components or parts and all 100 of these parts must be nondefective for the product to function satisfactorily. The probability that any specific unit of product is nondefective is

$$0.9973 \times 0.9973 \times \ldots \times 0.9973 = (0.9973)^{100} = 0.7631$$

That is, about 23.7% of the products produced under three-sigma quality will be defective. This is not an acceptable situation, because many products used by today's society are made up of many components. Even a relatively simple service activity, such as a visit by a family of four to a fast-food restaurant, can involve the assembly of several dozen components. A typical automobile has about 100,000 components and an airplane has between 1 and 2 million!

The Motorola six-sigma concept is to reduce the variability in the process so that the specification limits are at least six standard deviations from the mean. Then, as shown in Figure 2.5*a*, there will only be about 2 parts per *billion* defective. Under **six-sigma quality,** the probability that any specific unit of the hypothetical product above is nondefective is 0.9999998, or 0.2 ppm, a much better situation.

When the six-sigma concept was initially developed, an assumption was made that when the process reached the six-sigma quality level, the process mean was still subject to disturbances that could cause it to shift by as much as 1.5 standard deviations off target. This situation is shown in Figure 2.5*b*. Under this scenario, a six-sigma process would produce about 3.4 ppm defective.

There is an apparent inconsistency in this. Generally, we can only make predictions about process performance when the process is **stable;** that is, when the mean (and standard deviation, too) is **constant.** If the mean is drifting around, and ends up as much as 1.5 standard deviations off target, a prediction of 3.4 ppm defective may not be very reliable, because the mean might shift by *more* than the "allowed" 1.5 standard

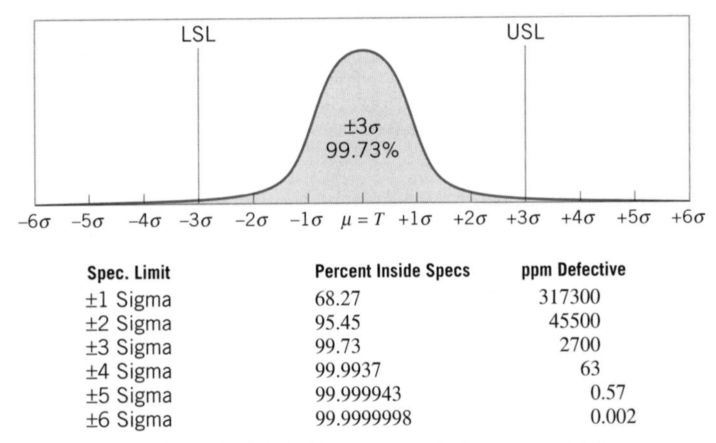

Spec. Limit	Percent Inside Specs	ppm Defective
±1 Sigma	68.27	317300
±2 Sigma	95.45	45500
±3 Sigma	99.73	2700
±4 Sigma	99.9937	63
±5 Sigma	99.999943	0.57
±6 Sigma	99.9999998	0.002

(a) Normal distribution centered at the target (T)

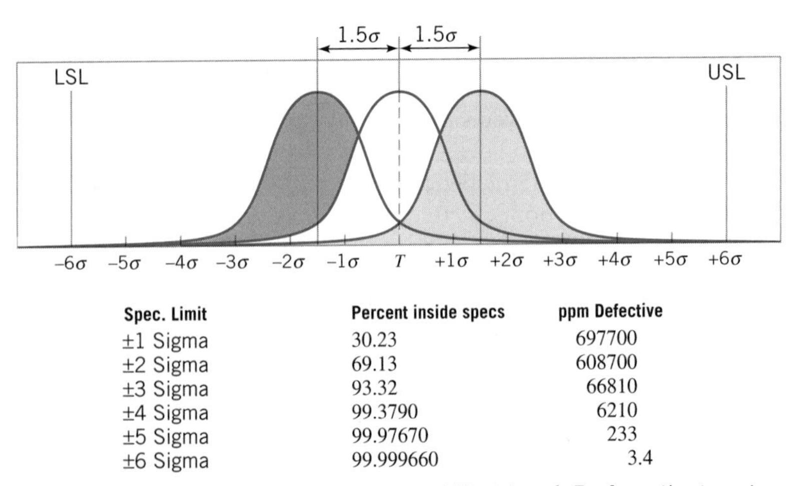

Spec. Limit	Percent inside specs	ppm Defective
±1 Sigma	30.23	697700
±2 Sigma	69.13	608700
±3 Sigma	93.32	66810
±4 Sigma	99.3790	6210
±5 Sigma	99.97670	233
±6 Sigma	99.999660	3.4

(b) Normal distribution with the mean shifted by ±1.5σ from the target

Figure 2.5 The Motorola Six-Sigma Concept

deviations. Process **performance** is not predictable unless the process **behavior** is stable.

However, no process or system is ever truly stable, and even in the best of situations, disturbances occur. These disturbances can result in the process mean shifting off-target, an increase in the process standard deviation, or both. The concept of a six-sigma process is one way to **model** this behavior. Like all models, it's probably not exactly right, but it has proven to be a useful way to think about process performance.

Motorola established six-sigma as both an objective for the corporation and as a focal point for process and product quality improvement efforts. In recent years, six-sigma has spread beyond Motorola and has come to encompass much more. It has become a program for improving corporate **business performance** by both improving quality and paying attention to reducing costs. Typical six-sigma projects are four to six months in duration and are selected for their potential impact in the business. The paper by Hoerl (2001) describes the components of a typical BB education program. Six-sigma uses a specific five-step problem-solving approach: Define, Measure, Analyze, Improve, and Control (DMAIC). The DMAIC framework utilizes control charts, designed experiments, process capability analysis, measurement systems capability studies, and many other basic statistical tools. The DMAIC approach is an extremely effective framework for improving processes. In section 2.3 we will give a complete presentation of DMAIC along with some examples.

The goals of six-sigma, a 3.4 ppm defect level, may seem artificially or arbitrarily high, but it is easy to demonstrate that even the delivery of relatively simple products or services at high levels of quality can lead to the need for six-sigma thinking. For example, consider the visit to a fast-food restaurant mentioned above. The customer orders a typical meal: a hamburger bun, meat, special sauce, cheese, pickle, onion, lettuce, and tomato, fries, and a soft drink. This product has ten components. Is 99% good quality satisfactory? If we assume that all ten components are independent, the probability of a good meal is

$$P(\text{Single meal good}) = (0.99)^{10} = 0.9044$$

which looks pretty good. There is better than a 90% chance that the customer experience will be satisfactory. Now suppose that the customer is a family of four. Again, assuming independence, the probability that all four meals are good is

$$P\{\text{All meal good}\} = (0.9044)^4 = 0.6690$$

This isn't so nice. The chances are only about two out of three that all of the family meals are good. Now suppose that this hypothetical family of four visits this restaurant once a month (this is about all their cardiovascular systems can stand!). The probability that all visits result in good meals for everybody is

$$P\{\text{All visits during the year good}\} = (0.6690)^{12} = 0.0080$$

What's Your "Belt"?

Companies involved in a six-sigma effort utilize specially trained individuals, called Green Belts (GBs), Black Belts (BBs), and Master Black Belts (MBBs), who lead teams focused on projects that have both quality and business (economic) impacts for the organization. The "belts" have specialized training and education on statistical methods and the quality and process improvement tools in this textbook that equips them to function as team leaders, facilitators, and problem solvers.

Consider the following statement from Jim Owens, chairman of heavy equipment manufacturer Caterpillar, Inc., who wrote in the 2005 annual company report:

". . . I believe that our people and world-class six-sigma deployment distinguish Caterpillar from the crowd. What an incredible success story six-sigma has been for Caterpillar! It is the way we do business—how we manage quality, eliminate waste, reduce costs, create new products and services, develop future leaders, and help the company grow profitably. We continue to find new ways to apply the methodology to tackle business challenges. Our leadership team is committed to encoding six-sigma into Caterpillar's "DNA" and extending its deployment to our dealers and suppliers—more than 500 of whom have already embraced the six-sigma way of doing business."

At the annual meeting of Bank of America in 2004, chief executive officer Kenneth D. Lewis told the attendees that the company had record earnings in 2003, had significantly improved the customer experience, and had raised its community development funding target to $750 billion over ten years. "Simply put, Bank of America has been making it happen," Lewis said. "And we've been doing it by following a disciplined, customer-focused and organic growth strategy." Citing the companywide use of six-sigma

(*Continued*)

This is obviously unacceptable. So, even in a very simple service system involving a relatively simple product, very high levels of quality and service are required to produce the desired high-quality experience for the customer.

Business organizations have been very quick to understand the potential benefits of six-sigma and to adopt the principles and methods. Between 1987 and 1993, Motorola reduced defectivity on its products by approximately 1300%. This success led to many organizations adopting the approach. Since its origins, there have been three generations of six-sigma implementations. **Generation I** six-sigma focused on defect elimination and basic variability reduction. Motorola is often held up as an exemplar of Generation I six-sigma. In **Generation II** six-sigma, the emphasis on variability and defect reduction remained, but now there was a strong effort to tie these efforts to projects and activities that improved business performance through cost reduction. General Electric is often cited as the leader of the Generation II phase of six-sigma.

In **Generation III,** six-sigma has the additional focus of creating value throughout the organization and for its stakeholders (owners, employees, customers, suppliers, and society at large). Creating value can take many forms: increasing stock prices and dividends, job retention or expansion, expanding markets for company products/services, developing new products/services that reach new and broader markets, and increasing the levels of customer satisfaction throughout the range of products and services offered.

As you can see from the adjacent display, many different kinds of businesses have embraced six-sigma and made it part of the culture of doing business. Caterpillar and Bank of America are good examples of Generation III six-sigma companies, because their implementations are focused on value creation for all stakeholders in the broad sense. Note Lewis's emphasis on reducing cycle times and reducing processing errors (items that will greatly improve customer satisfaction), and Owens's remarks on extending six-sigma to suppliers and dealers—the entire supply chain. Six-sigma has spread well beyond its manufacturing origins into areas including health care, many types of service business, and government/public service (the U.S. Navy has a strong and very successful six-sigma program). The reason for the success of six-sigma in organizations outside the traditional manufacturing sphere is that variability is everywhere, and where there is variability, there is an opportunity to improve business results. Some examples of situations where a six-sigma program can be applied to reduce variability, eliminate defects, and improve business performance include:

- Meeting delivery schedule and delivery accuracy targets
- Eliminating rework in preparing budgets and other financial documents
- Proportion of repeat visitors to an e-commerce Web site, or proportion of visitors that make a purchase
- Minimizing cycle time or reducing customer waiting time in any service system

- Reducing average and variability in days outstanding of accounts receivable
- Optimizing payment of outstanding accounts
- Minimizing stock-out or lost sales in supply chain management
- Minimizing costs of public accountants, legal services, and other consultants
- Improving inventory management (both finished goods and work-in-process)
- Improving forecasting accuracy and timing
- Improving audit processes
- Closing financial books, improving accuracy of journal entry and posting (a 3 to 4% error rate is fairly typical)
- Reducing variability in cash flow
- Improving payroll accuracy
- Improving purchase order accuracy and reducing rework of purchase orders

(*Continued*)

techniques for process improvement, he noted that in fewer than three years, Bank of America had "saved millions of dollars in expenses, cut cycle times in numerous areas of the company by half or more, and reduced the number of processing errors."

These statements are strong endorsements of six-sigma from two highly recognized business leaders that lead two very different types of organizations, manufacturing and financial services.

The structure of a six-sigma organization is shown in Figure 2.6. The lines in this figure identify the key links among the functional units. The **leadership team** is the executive responsible for that business unit and appropriate members of his/her staff and direct reports. This person has overall responsibility for approving the improvement projects undertaken by the six-sigma teams. Each project has a **champion,** a business leader whose job is to facilitate project identification and selection, identify black belts and other team members who are necessary for successful project completion, remove barriers to project completion, make sure that the resources required for project completion are available, and conduct

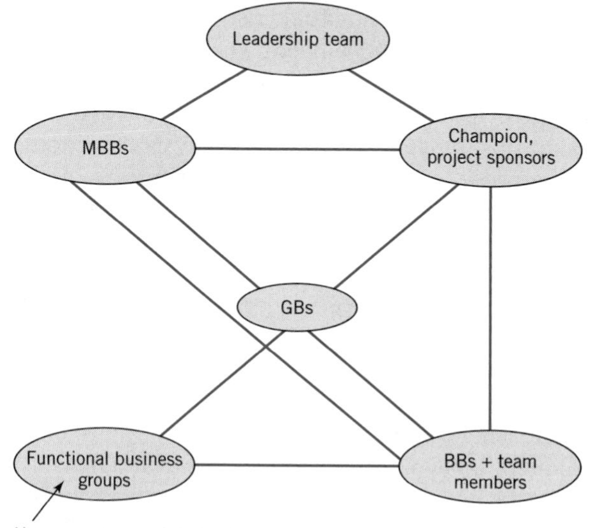

Figure 2.6 The Structure of a Six-Sigma Organization

(Adapted from R. D. Snee and R. W. Hoerl, *Six-Sigma Beyond the Factory Floor,* Upper Saddle River, NJ: Pearson Prentice Hall, 2005).

regular meetings with the team or the Black Belt to ensure that progress is being made and the project is on schedule. The champion role is not full-time, and champions often have several projects under their supervision. Black Belts are team leaders that are involved in the actual project completion activities. Team members often spend 25% of their time on the project, and may be drawn from different areas of the business, depending on project requirements. Green Belts typically have less training and experience in six-sigma tools and approaches than the Black Belts and may lead projects of their own under the direction of a champion or Black Belt or be part of a Black Belt–led team. A Master Black Belt is a technical leader and may work with the champion and the leadership team in project identification and selection, project reviews, consulting with Black Belts on technical issues, and training of Green Belts and Black Belts. Typically, the Black Belt and Master Black Belt roles are full-time.

2.6 Beyond Six-Sigma—DFSS and Lean

Design for Six-Sigma seeks to take customer requirements and process capabilities into consideration to design products and services that increase product and service effectiveness as perceived by the customer.

In recent years, two other tool sets have become identified with six-sigma, **lean systems,** and **design for six-sigma (DFSS)**. Many organizations regularly use one or both of these approaches as an integral part of their six-sigma implementation.

Design for six-sigma is an approach for taking the variability reduction and process improvement philosophy of six-sigma upstream from manufacturing or production into the design process, where new products (or services or service processes) are designed and developed. Broadly speaking, DFSS is a structured and disciplined methodology for the efficient commercialization of technology that results in new products, services, or processes. By a product, we mean anything that is sold to a consumer for use; by a service, we mean an activity that provides value or benefit to the consumer. DFSS spans the entire development process from the identification of customer needs to the final launch of the new product or service. Customer input is obtained through **voice of the customer (VOC)** activities designed to determine what the customer really wants, to set priorities based on actual customer wants, and to determine if the business can meet those needs at a competitive price that will enable it to make a profit. VOC data is usually obtained by customer interviews, by a direct interaction with and observation of the customer, through focus groups, by surveys, and by analysis of customer satisfaction data. The purpose is to develop a set of critical to quality requirements for the product or service. Traditionally, six-sigma is used to achieve **operational** excellence, while DFSS is focused on improving business results by increasing the sales revenue generated from new products and services and finding new applications or opportunities for existing ones. In many cases, an important gain from DFSS is the reduction of development lead time—that is, the cycle time to commercialize new technology and get the resulting new products to market. DFSS is directly focused on increasing value in the organization. Many of the tools that are used in operational six-sigma are also used in DFSS. The DMAIC process is also applicable, although some organizations and practitioners have slightly different approaches (DMADV, or Define, Measure, Analyze, Design, and Verify, is a popular variation).

Some organizations use **Quality Function Deployment** or **QFD** to focus the voice of the customer directly on the design of a product, service, or process. QFD was developed in Japan in the 1970s. It is basically a planning process and has proved useful in the design of new products and services, the improvement of existing ones, and the design of new processes. An essential component of QFD is the **house of quality.** This is essentially a matrix with rows corresponding to customer requirements and columns representing the technical response to these requirements. Information about the importance of each requirement and about how well the company's products or services compare to the competition is obtained. Analysis of this information leads to directions of improvement in the design of the product or service. It is fairly typical to step this process down from a high level that begins with the voice of the customer data all the way down to individual process steps and the critical-to-process variables that must be controlled to achieve these results. A typical four-step QFD is shown in Figure 2.7.

Quality Function Deployment is a technique to transform customer requirements into design quality, down to component level and specific elements of the manufacturing system.

DFSS makes specific the recognition that every design decision is a business decision, and that the cost, manufacturability, and performance of the product are determined during design. Once a product is designed and released to manufacturing, it is almost impossible for the manufacturing organization to make it better. Furthermore, overall business improvement cannot be achieved by focusing on reducing variability in manufacturing alone (operational six-sigma), and DFSS is required to focus on customer requirements while simultaneously keeping process capability in mind. Specifically, matching the capability of the production system and the requirements at each stage or level of the design process (refer to Figure 2.8) is essential. When mismatches between process capabilities and design requirements are discovered, either design changes or different production alternatives are considered to resolve the conflicts. Throughout the DFSS process, it is important that the following points be kept in mind:

Every design decision is a business decision.

- Is the product concept well identified?
- Are customers real?

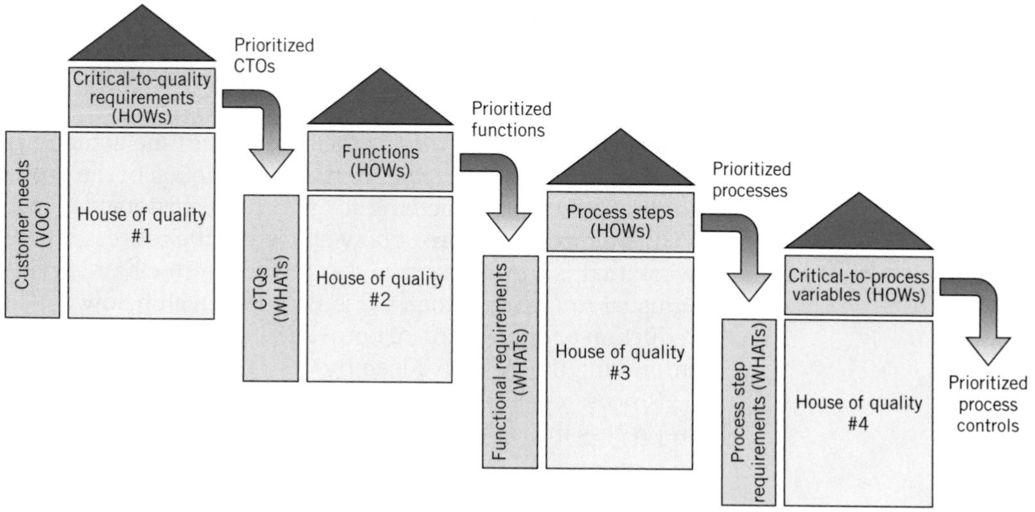

Figure 2.7 **A Four-Step Quality Function Deployment Process**

Figure 2.8 **Matching Product Requirements and Production Capability in DFSS**

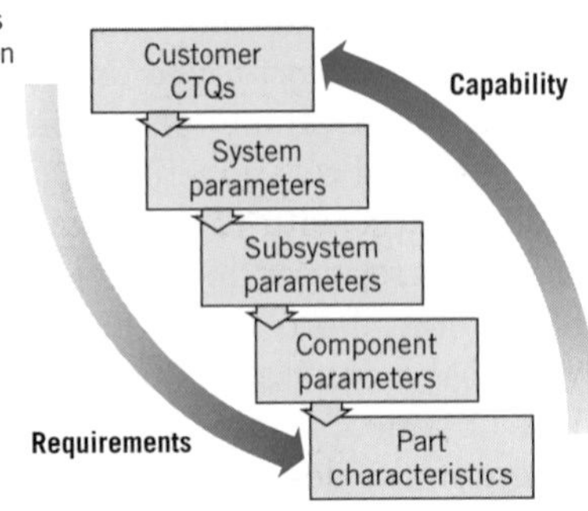

- Will customers buy this product?
- Can the company make this product at competitive cost?
- Are the financial returns acceptable?
- Does this product fit with the overall business strategy?
- Is the risk assessment acceptable?
- Can the company make this product better than the competition?
- Can product reliability and maintainability goals be met?
- Has a plan for transfer to manufacturing been developed and verified?

Lean is a series of practices that focus on the systematic elimination of waste and the promotion of efficiency.

Lean systems are designed to eliminate waste. By waste, we mean unnecessarily long cycle times, or waiting times between value-added work activities. Waste can also include rework of doing something over again to eliminate defects introduced the first time) or scrap. Rework and scrap are often the result of excess variability, so there is an obvious connection between six-sigma and lean. An important metric in lean is the process cycle efficiency PCE), defined as

$$\text{Process cycle efficiency} = \frac{\text{Value-add time}}{\text{Process cycle time}}$$

where the value-add time is the amount of time actually spent in the process that transforms the form, fit, or function of the product or service that results in something for which the customer is willing to pay. PCE is a direct measure of how efficiently the process is converting the work that is in-process into completed products or services. It is not unusual to find that the PCE is very low, well below 25%. Obviously, reduction or elimination of non-value-added activities are important in improving the PCE. In a lean process, the PCE will exceed 25%.

Little's Law

Process cycle time is also related to the amount of work that is in-process through **Little's Law.**

$$\text{Process cycle time} = \frac{\text{Work-in-process}}{\text{Average completion rate}}$$

The average completion rate is a measure of capacity; that is, it is the output of a process over a defined time period. For example, consider a mortgage refinance operation at a bank. If the average completion rate for submitted applications is 100 completions per day, then there are 1,500 applications waiting for processing, the process cycle time is

$$\text{Process cycle time} = \frac{1500}{100} = 15 \text{ days}$$

Often the cycle time can be reduced by eliminating waste and inefficiency in the process, resulting in an increase in the completion rate. Lean takes a diverse set of tools and techniques and integrates them into a system for solving problems where elimination of waste, reduction of cycle time, and increased throughput are the goals.

The following list and summarizes some of the most commonly used lean tools:

• **Value-stream and value-added process mapping**	A graphical approach to describing the important material and information flows in the process. Process mapping is described in detail in Chapter 3.
• **The five Ss**	These principles focus on creating orderliness and discipline in the workplace: Separate, Straighten, Scrub, Standardize, and Systematize.
• **Kanban**	Pull inventory management; that is, don't produce parts until they are needed. In the early days a kanban was a piece of paper or a card that ordered the production of a part. Today it is most likely a computer record.
• **Error-proofing (or Poka-Yoke)**	Designing work or products so that it is nearly impossible to do the work incorrectly. This can include color-coding components so that they are assembled correctly, constructing visual aids for workers so that they can see what the finished item should look like, or designing components so that they can only be assembled one way. Special error-checking or control devices that provide feedback to operators when work is not correctly performed, such as feedback from a computer program when a required field is blank or not filled out in the expected format, are also examples of error-proofing.
• **Set-up time reduction and reduced lot sizes**	A process of identifying waste and inefficiencies in the process of changing over tools and equipment from one product to another and then eliminating this wasted effort. Lengthy changeovers and setups limit the flexibility of an organization to respond quickly to customer needs and also contribute to large lot sizes. Ideally, one should aim for a single minute exchange of dies (SMED). Considerable reduction in setup times can often be achieved by reorganizing the work process surrounding changeovers and by redesigning the tools and dies themselves to facilitate rapid setups. Large lot sizes result in excess inventory (and dollars tied up in inventory), excess handling of materials, larger space requirements, risk of obsolesce, and risk of damage or spoilage. When setup times are short, lot sizes can also be small.

Lean also makes use of many tools of industrial engineering and operations research. One of the most important of these is **discrete-event simulation,** in which a computer model of the system is built

Figure 2.9 Six-Sigma/DMAIC, Lean, and DFSS: How They Fit Together

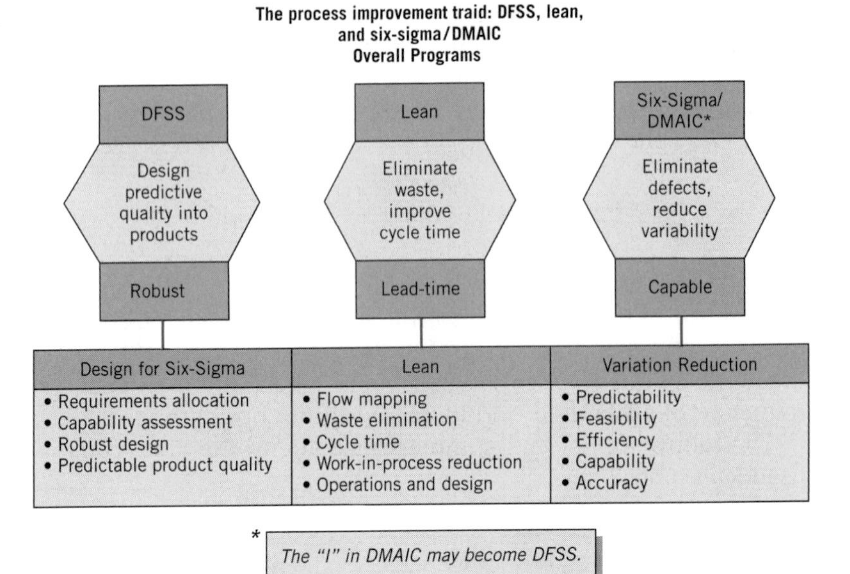

and used to quantify the impact of changes to the system that improve its performance. Simulation models are often very good predictors of the performance of a new or redesigned system. Both manufacturing and service organizations can greatly benefit by using simulation models to study the performance of their processes.

Ideally, six-sigma/DMAIC, DFSS, and lean tools are used simultaneously and harmoniously in an organization to achieve high levels of process performance and significant business improvement. Figure 2.9 highlights many of the important complementary aspects of these three sets of tools.

Six-sigma (often combined with DFSS and lean) has been much more successful than its predecessors, notably TQM. The project-by-project approach and the focus on obtaining improvement in bottom-line business results has been instrumental in obtaining management commitment to six-sigma. Another major component in obtaining success is driving the proper deployment of statistical methods into the right places in the organization. The DMAIC problem-solving framework is an important part of this.

There have been many initiatives devoted to improving the production system. Some of these include the Just-in-Time approach emphasizing in-process inventory reduction, rapid set-up, and a pull-type production system; Poka-Yoke or mistake-proofing of processes; the Toyota production system and other Japanese manufacturing techniques (with once-popular management books by those names); reengineering; theory of constraints; agile manufacturing; and so on. Most of these programs devote far too little attention to variability reduction. It's virtually impossible to reduce the in-process inventory or operate a pull-type or agile production system when a large and unpredictable fraction of the process output is defective and where there are significant uncontrolled sources of variability. These types of activities will not achieve their full potential without a major focus on statistical methods for process improvement and variability reduction to accompany them.

2.7 The DMAIC Process

DMAIC (typically pronounced "doh-MAY-iclc") is a structured five-step problem-solving procedure widely used in quality and process improvement. It is often associated with six-sigma activities, and almost all implementations of six-sigma use the **DMAIC process** for project management and completion. However, DMAIC is not necessarily formally tied to six-sigma, and it can be used regardless of an organization's use of six-sigma. It is a very general procedure. For example, lean projects that focus on cycle time reduction, throughput improvement, and waste elimination can be easily and efficiently conducted using DMAIC.

DMAIC:

- Define
- Measure
- Analyze
- Improve
- Control

The letters DMAIC form an acronym for the five steps: **Define, Measure, Analyze, Improve,** and **Control.** These steps are illustrated in Figure 2.10. Notice that there are **"tollgates"** between each of the major steps in DMAIC. At a tollgate, a project team presents its work to managers and "owners" of the process. In a six-sigma organization, the tollgate participants also would include the project champion, master black belts, and other black belts not working directly on the project. Tollgates are where the project is reviewed to ensure that it is on track; they provide a continuing opportunity to evaluate whether the team can successfully complete the project on schedule. Tollgates also present an opportunity to provide guidance regarding the use of specific technical tools and other information about the problem. Organization problems and other barriers to success—and strategies for dealing with them—also often are identified during tollgate reviews. Tollgates are critical to the overall problem-solving process; It is important that these reviews be conducted very soon after the team completes each step.

The DMAIC structure encourages creative thinking about the problem and its solution within the definition of the original product, process, or service. When the process is operating so badly that it is necessary to abandon the original process and start over, or if it is determined that a new product or service is required, then the improve

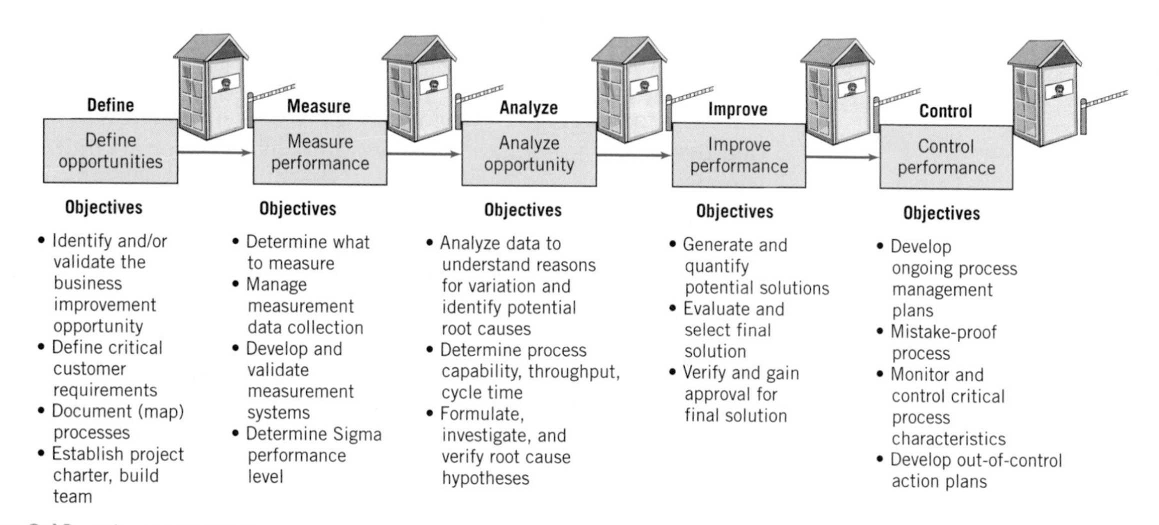

Figure 2.10 The DMAIC Process

Table 2.4

Statistical Tools Used in DMAIC

Tool	Define	Measure	Analyze	Improve	Control
Project charter	Chapter 2				
Process maps and flow charts	Chapter 2	Chapter 3			
Cause-and-effect analysis		Chapter 3			
Process capability analysis		Chapter 5			
Hypothesis tests, confidence intervals			Chapter 4		
Gauge R&R		Chapter 5			
Failure mode and effects analysis			Chapter 2		
Designed experiments			Chapter 8	Chapter 8	
SPC and process control plans		Chapters 5, 6	Chapters 5, 6		Chapters 5, 6

step of DMAIC actually becomes a **design** step. In a six-sigma organization, that probably means that a Design for Six-Sigma (DFSS) effort is required. (See Section 2.2.8 for a discussion of DFSS.)

One of the reasons that DMAIC is so successful is that it focuses on the effective use of a relatively small set of tools. Table 2.4 shows the statistical tools, along with the DMAIC steps where they are most likely to be used, and where the tools are discussed and or illustrated in this textbook. Table 2.5 shows a set of lean tools organized by DMAIC phase. Other tools, or variations of the ones shown here, are used occasionally in DMAIC. Some books on six-sigma give useful overviews of many of these other tools; for example, see George, 2002 and Snee and Hoerl, 2005.

Projects are an essential aspect of quality and process improvement. Projects are an integral component of six-sigma, but quality and business improvement via projects traces its origins back to Juran, who always urged a project-by-project approach to improving quality. Selecting, managing, and completing projects successfully is critical in deploying any systematic business improvement effort, not just six-sigma.

Importance of Financial Results

A project should represent a potential **breakthrough** in the sense that it will result in a major improvement in the product or service. Project impact should be evaluated in terms of its financial benefit to the business, as measured and evaluated by the finance or accounting unit; this helps ensure more objective project evaluations. Obviously, projects with high potential impact are most desirable. This **financial systems integration** is standard practice in six-sigma and should be a part of any DMAIC project, even if the organization isn't currently engaged in a six-sigma deployment.

The **value opportunity** of projects must be clearly identified and projects must be well aligned with corporate business objectives at all

Table 2.5

Lean Tools by DMAIC Phase

Define Opportunities	Measure Performance	Analyze Opportunity	Improve Performance	Control Performance
• Project charter, including business case, problem statement, goal, scope, risk assessment, high-level project plan, stakeholders, team, and multi-generational plan if appropriate • Voice of the customer (VOC) research on lead time, delivery time, cost, etc. • Quality Function Deployment (QFD) • Critical-to-quality requirements (CTQs) • High-level Supplier-Input-Process-Output-Customer (SIPOC) map • High-level process map	• Detailed current-state end-to-end process maps • Current-state value stream map • Activity maps such as spaghetti charts and transportation diagrams for flow of information, product, and people • Data collection plan • Statistical sampling • Measurement systems analysis • Baseline process metrics, including lead or cycle time, rolled-throughput yield, capability and capacity • Pareto charts, histograms, box plots	• Value-add analysis • Waste analysis: overproduction, waiting (product), transportation, excess processing, inventory, motion (person), defects • Brainstorming, cause-and-effect diagrams and matrices, Failure Modes and Effects Analyses (FMEAs) to generate potential causes • Statistical analysis such as Pareto charts, hypothesis testing, regression, and ANOVA to support cause-and-effect relationships	• Brainstorming and benchmarking to develop potential solutions • Designed experiments • Process design, including layout and material and information flow • Standard work • Detailed future-state end-to-end process maps • Process FMEAs • Future-state value stream map • The 5 Ss • Poka-Yoke, mistake-proofing • Preliminary control plan • Simulation and/or pilot of solution • Statistical analysis of simulation/pilot results	• Visual process controls/dashboards • Process control system, including control charts • Total Productive Maintenance • Standard Operating Procedures (SOPs), work instructions • Final Control Plan

levels. At the highest (corporate) level, the stockholders, top executives, members of the board of directors, and business analysts who guide investors are typically interested in return on equity, return on invested capital, stock price, dividends, earnings, earnings per share of stock, growth in operating income, sales growth, generation of new designs, products and patents, and development of future business leaders. At the business unit or operations level, managers and executives are interested in factory metrics such as yield, cycle time and throughput, profit and loss optimization, customer satisfaction, delivery and due-date performance, cost reduction, safety of employees and customers, efficient use of assets, new product introduction, sales and marketing effectiveness, development of people, and supply chain performance (cost, quality, service). Aligning projects with both business-unit goals and corporate-level metrics helps ensure that the best projects are considered for selection.

PROJECT DEFINITION AND SELECTION

Approaching to Project Definition and Selection:

- Opportunistic
- Tied to strategic business objectives

The first types of projects that companies usually undertake are designed to demonstrate the potential success of an overall improvement effort. These projects often focus on the areas of the business that are full of opportunities, but they also often are driven by current problems. Issues that are identified by customers or from customer satisfaction (or dissatisfaction) feedback, such as analysis of field failures and customer returns, sometimes are the source of these projects.

Such initial opportunistic projects often are successful, but they typically are not the basis for long-term success; most easy opportunities soon are exhausted. A different approach to project definition and selection needs to evolve. One widely used approach is basing projects on strategic business objectives. In this approach, defining the key set of critical business processes and the metrics that drive them is the first step toward successful project development. Linking those processes together to form an integrated view of the business then follows. Projects that focus on the key business metrics and strategic objectives, as well as the interfaces among critical business processes, are likely to have significant value to the company. The only risks here are that the projects may be very large, and still may focus only on some narrow aspect of the business, which may reduce the organization's overall exposure to the improvement process and reduce or delay its impact. A good project selection and management system prevents such problems from occurring. Many companies have set up formal project selection committees and conducted regular meetings between customers and the project selection committees to help meet that goal. Ideally, projects are strategic and well aligned with corporate metrics and are not local (tactical). Local projects are often reduced to firefighting, their solutions rarely are broadly implemented in other parts of the business, and too often the solutions aren't permanent; within a year or two, the same old problems reoccur. Some companies use a **dashboard** system—which graphically tracks trends and results—to effectively facilitate the project selection and management process.

Dashboard

Project selection is probably the most important part of any business improvement process. Projects should be able to be completed within a reasonable time frame and should have real impact on key business metrics. This means that a lot of thought must go into defining the organization's key business processes, understanding their interrelationships, and developing appropriate performance measures.

example 2.1 What should be considered when evaluating proposed projects?

Suppose that a company is operating at the 4σ level (that is, about 6,210 ppm defective, assuming the 1.5σ shift in the mean that is customary with six-sigma applications). This is actually reasonably good performance, and many of today's organizations have achieved the 4–4.5σ level of performance for many of their key business processes. The objective is to achieve the 6σ performance level (3.4 ppm). What implications does this have for project selection criteria? Suppose that the criterion is a 25% annual improvement in quality level. Then to reach the six-sigma performance level, it will take x years, where x is the solution to

$$3.4 = 6210(1 - 0.25)^x$$

It turns out that x is about 34 years. Clearly, a goal of improving performance by 25% annually isn't going to work—no organization will wait for 34 years to achieve its goal. Quality improvement is a never-ending process, but no management team that understands how to do the above arithmetic will support such a program.

Raising the annual project goal to 50% helps a lot, reducing x to about 11 years, a much more realistic time frame. If the business objective is to be a six-sigma organization in five years, then the annual project improvement goal should be about 75%.

These calculations are the reasons why many quality-improvement authorities urge organizations to concentrate their efforts on projects that have real impact and high payback to the organization. By that they usually mean projects that achieve at least a 50% annual return in terms of quality improvement.

Is this level of improvement possible? The answer is yes, and many companies have achieved this rate of improvement. For example, Motorola's annual improvement rate exceeded 65% during the first few years of their six-sigma initiative. To do this consistently, however, companies most devote considerable effort to project definition, management, execution, and implementation. It's also why the best possible people in the organization should be involved in these activities.

2.7.1 THE DEFINE STEP

Define

Define opportunities

The objective of the define step of DMAIC is to identify the project opportunity and to verify or validate that it represents legitimate breakthrough potential. A project must be important to customers (voice of the customer) and important to the business. Stakeholders who work in the process and its downstream customers need to agree on the potential usefulness of the project.

Project Charter

One of the first items that must be completed in the define step is a **project charter.** This is a short document (typically about two pages) that contains a description of the project and its scope, the start and the anticipated completion dates, an initial description of both primary

Business case	Opportunity statement
• This project supports the business quality goals, namely (a) reduce customer resolution cycle time by x% and (b) improve customer satisfaction by y%.	• An opportunity exists to close the gap between our customer expectations and our actual performance by reducing the cycle time of the customer return process.
Goal statement	**Project scope**
• Reduce the overall response cycle time for returned product from our customers by x% year to year.	• Overall response cycle time is measured from the receipt of a product return to the time that either the customer has the product replaced or the customer is reimbursed.
Project plan	**Team**
• Activity / Start / End Define 6/04 6/30 Measure 6/18 7/30 Analyze 7/15 8/30 Improve 8/15 9/30 Control 9/15 10/30 Track benefits 11/01	• Team sponsor • Team leader • Team members

Figure 2.11 **A Project Charter for a Customer Returns Process**

and secondary metrics that will be used to measure success and how those metrics align with business unit and corporate goals, the potential benefits to the customer, the potential financial benefits to the organization, milestones that should be accomplished during the project, the team members and their roles, and any additional resources that are likely to be needed to complete the project. Figure 2.11 shows a project charter for a customer product return process. Typically, the project sponsor (or champion in a six-sigma implementation) plays a significant role in developing the project charter and may use a draft charter as a basis for organizing the team and assigning responsibility for project completion. Generally, a team should be able to complete a project charter in two to four working days; if it takes longer, the scope of the project may be too big. The charter should also identify the customer's *critical-to-quality characteristics (CTQs)* that are impacted by the project.

Graphic aids are also useful in the define step; the most common ones used include process maps and flow charts, value stream maps (see Chapter 3), and the SIPOC diagram. Flow charts and value stream maps provide much visual detail and facilitate understanding about what needs to be changed in a process. The **SIPOC diagram** is a high-level map of a process. *SIPOC* is an acronym for *S*uppliers, *I*nput, *P*rocess, *O*utput, and *C*ustomers, defined thus:

SIPOC Diagram

1. The **suppliers** are those who provide the information, material, or other items that are worked on in the process.
2. The **input** is the information or material provided.
3. The **process** is the set of steps actually required to do the work.

Suppliers	Inputs	Process	Output	Customer
Starbucks Purifier Utility company	Ground coffee Water Filter Electricity	Collect materials → Brew coffee → Pour coffee from pot	Hot Taste Correct strength Correct volume	Consumer

Figure 2.12 **A SIPOC Diagram**

4. The **output** is the product, service, or information sent to the customer.
5. The **customer** is either the external customer or the next step in the internal business.

SIPOC diagrams give a simple overview of a process and are useful for understanding and visualizing basic process elements. They are especially useful in the nonmanufacturing setting and in service systems in general, where the idea of a process or process thinking is often hard to understand. That is, people who work in banks, financial institutions, hospitals, accounting firms, e-commerce, government agencies, and most transactional/service organizations don't always see what they do as being part of a process. Constructing a process map can be an eye-opening experience, as it often reveals aspects of the process that people were not aware of or didn't fully understand.

Figure 2.12 is a SIPOC diagram developed by a company for their internal coffee service process. The team was asked to reduce the number of defects and errors in the process and the cycle time to prepare the coffee. The first step they performed was to create the SIPOC diagram to identify the basic elements of the process that they were planning to improve.

The team also will need to prepare an action plan for moving forward to the other DMAIC steps. This will include individual work assignments and tentative completion dates. Particular attention should be paid to the measure step as it will be performed next.

Finally, the team should prepare for the define step tollgate review, which should focus on the items listed on the adjacent display:

Tollgate for Define

1. Does the problem statement focus on symptoms, and not on possible causes or solutions?
2. Are all the key stakeholders identified?
3. What evidence is there to confirm the value opportunity represented by this project?
4. Has the scope of the project been verified to ensure that it is neither too small nor too large?
5. Has a SIPOC diagram or other high-level process map been completed?
6. Have any obvious barriers or obstacles to successful completion of the project been ignored?
7. Is the team's action plan for the measure step of DMAIC reasonable?

2.7.2 THE MEASURE STEP

Measure

Measure performance

The purpose of the measure step is to evaluate and understand the current state of the process. This involves collecting data on measures of quality, cost, and throughput/cycle time. It is important to develop a list of all of the **key process input variables** (sometimes abbreviated **KPIV**) and the **key process output variables (KPOV).** The KPIV and KPOV may have been identified at least tentatively during the define step, but they must be completely defined and measured during the measure step. Important factors may be the time spent to perform various work activities and the time that work spends waiting for additional processing. Deciding what and how much data to collect are important tasks; there must be sufficient data to allow for a thorough

KPIV
KPOV

Gauge or Measurement System Capability

analysis and understanding of current process performance with respect to the key metrics.

Data may be collected by examining historical records, but this may not always be satisfactory, as the history may be incomplete, the methods of record keeping may have changed over time, and, in many cases, the desired information never may have been retained. Consequently, it is often necessary to collect current data through an observational study. This may be done by collecting process data over a continuous period of time (such as every hour for two weeks) or it may be done by sampling from the relevant data streams. When there are many human elements in the system, work sampling may be useful. This form of sampling involves observing workers at random times and classifying their activity at that time into appropriate categories. In transactional and service businesses, it may be necessary to develop appropriate measurements and a measurement system for recording the information that are specific to the organization. This again points out a major difference between manufacturing and services: measurement systems and data on system performance often exist in manufacturing, as the necessity for the data is usually more obvious in manufacturing than in services.

The data that are collected are used as the basis for determining the current state or **baseline performance** of the process. Additionally, the **capability** of the measurement system should be evaluated. This may be done using a formal gauge capability study (called *gauge repeatability* and *reproducibility*, or *gauge R & R*). At this point, it is also a good idea to begin to divide the process cycle time into value-added and non-value-added activities and to calculate estimates of process cycle efficiency and process cycle time, if appropriate.

The data collected during the measure step may be displayed in various ways such as histograms, stem-and-leaf diagrams, run charts, scatted diagrams, and Pareto charts. Chapters 3 and 4 provide information on these techniques.

At the end of the measure step, the team should update the project charter (if necessary), re-examine the project goals and scope, and re-evaluate team makeup. They may consider expanding the team to include members of downstream or upstream business units if the measure activities indicate that these individuals will be valuable in subsequent DMAIC steps. Any issues or concerns that may impact project success need to be documented and shared with the process owner or project sponsor. In some cases, the team may be able to make quick, immediate recommendations for improvement, such as eliminating an obvious non-value-added step or removing a source of unwanted variability.

Finally, it is necessary to prepare for the measure step tollgate review. Issues and expectations that will be addressed during this tollgate include the items shown in the adjacent display:

Tollgate for Measure:

1. There must be a comprehensive process flow chart or value stream map. All major process steps and activities must be identified, along with suppliers and customers. If appropriate, areas where queues and work-in-process accummulate should be identified and queue lengths, waiting times, and work-in-process levels reported.
2. A list of KPIVs and KPOVs must be provided, along with identification of how the KPOVs related to customer satisfaction or the customers CTQs.
3. Measurement systems capability must be documented.
4. Any assumptions that were made during data collection must be noted.
5. The team should be able to respond to requests such as, "Explain where that data came from," and questions such as, "How did you decide what data to collect?" "How valid is your measurement system?" and "Did you collect enough data to provide a reasonable picture of process performance?"

2.7.3 THE ANALYZE STEP

Analyze

Analyze opportunity

In the analyze step, the objective is to use the data from the measure step to begin to determine the cause-and-effect relationships in the process and to understand the different sources of variability. In other words, in the analyze step we want to determine the potential causes of the defects, quality problems, customer issues, cycle time and throughput problems, or waste and inefficiency that motivated the project. It is important to separate the sources of variability into **common causes** and **assignable causes.** We discuss these sources of variability in Chapter 3 but, generally speaking, common causes are sources of variability that are embedded in the system or process itself, while assignable causes usually arise from an external source. Removing a common cause of variability usually means changing the process, while removing an assignable cause usually involves eliminating that specific problem. A common cause of variability might be inadequate training of personnel processing insurance claims, while an assignable cause might be a tool failure on a machine.

There are many tools that are potentially useful in the analyze step. Among these are **control charts,** which are useful in separating common cause variability from assignable cause variability; statistical **hypothesis testing** and **confidence interval** estimation, which can be used to determine if different conditions of operation produce statistically significantly different results and to provide information about the accuracy with which parameters of interest have been estimated; and **regression analysis,** which allows models relating outcome variables of interest to independent input variables to be built.

Discrete-event computer simulation is another powerful tool useful in the analyze step. It is particularly useful in service and transactional businesses, although its use is not confined to those types of operations. For example, there have been many successful applications of discrete-event simulation in studying scheduling problems in factories to improve cycle time and throughput performance. In a discrete-event simulation model, a computer model simulates a process in an organization. For example, a computer model could simulate what happens when a home mortgage loan application enters a bank. The time between arrival of applications, processing times, and even the routing of the applications through the bank's process are random variables. The specific realizations of these random variables influence the backlogs or queues of applications that accummulate at the different processing steps.

Other random variables can be defined to model the effect of incomplete applications, erroneous information and other types of errors and defects, and delays in obtaining information from outside sources, such as credit histories. By running the simulation model for many loans, reliable estimates of cycle time, throughput, and other quantities of interest can be obtained.

Failure modes and effects analysis (FMEA) is another useful tool during the analyze stage. FMEA is used to prioritize the different potential sources of variability, failures, errors, or defects in a product or process relative to three criteria:

1. The likelihood that something will go wrong (ranked on a 1 to 10 scale where 1 = not likely and 10 = almost certain).
2. The ability to detect a failure, defect, or error (ranked on a 1 to 10 scale where 1 = very likely to be detected and 10 = very unlikely to be detected).

Analysis Tools:

- Control Charts
- Hypothesis Testing
- Confidence Interval
- Regression Analysis
- Discrete Event Computer Simulation

Random Variable

A variable where values are unknown but can be described by a probability distribution.

Tollgate for Analyze:

In preparing for the analyze tollgate review, the team should consider the following issues and potential questions:

1. What opportunities are going to be targeted for investigation in the improve step?
2. What data and analysis supports that investigating the targeted opportunities and improving/eliminating them will have the desired outcome on the KPOVs and customer CTQs that were the original focus of the project?
3. Are there other opportunities that are not going to be further evaluated? If so, why?
4. Is the project still on track with respect to time and anticipated outcomes? Are any additional resources required at this time?

Improve

Improve performance

Mistake-Proof Operations

Tollgate for Improve:

1. Adequate documentation of how the problem solution was obtained.
2. Documentation on alternative solutions that were considered.
3. Complete results of the pilot test, including data displays, analysis,

(Continued)

Control

Control performance

3. The severity of a failure, defect, or error (ranked on a 1 to 10 scale where 1 = little impact and 10 = extreme impact, including extreme financial loss, injury, or loss of life).

The three scores for each potential source of variability, failure, error, or defect are multiplied to obtain a **risk priority number (RPN).** Sources of variability or failures with the highest RPNs are the focus for further process improvement or redesign efforts.

The analyze tools are used with historical data or data that was collected in the measure step. This data is often very useful in providing clues about potential causes of the problems that the process is experiencing. Sometimes these clues can lead to breakthroughs and actually identify specific improvements. In most cases, however, the purpose of the analyze step is to explore and understand tentative relationships between and among process variables and to develop insight about potential process improvements. A list of specific opportunities and root causes that are targeted for action in the improve step should be developed. Improvement strategies will be further developed and actually tested in the improve step.

2.7.4 THE IMPROVE STEP

In the measure and analyze steps, the team focused on deciding which KPIVs and KPOVs to study, what data to collect, how to analyze and display the data and identified potential sources of variability and determined how to interpret the data they obtained. In the improve step, they turn to creative thinking about the specific changes that can be made in the process and other things that can be done to have the desired impact on process performance.

A broad range of tools can be used in the improve step. Redesigning the process to improve work flow and reduce bottlenecks and work-in-process will make extensive use of flow charts and/or value stream maps. Sometimes **mistake-proofing** (designing an operation so that it can be done only one way—the right way) an operation will be useful. Designed experiments are probably the most important statistical tool in the improve step. Designed experiments can be applied either to an actual physical process or to a computer simulation model of that process and can be used both for determining which factors influence the outcome of a process and the optimal combination of factor settings.

The objectives of the improve step are to develop a solution to the problem and to **pilot test** the solution. The pilot test is a form of **confirmation experiment:** it evaluates and documents the solution and confirms that the solution attains the project goals. This may be an iterative activity, with the original solution being refined, revised, and improved several times as a result of the pilot test's outcome.

The tollgate review for the improve step should consider the adjacent items:

2.7.5 THE CONTROL STEP

The objectives of the control step are to complete all remaining work on the project and to hand off the improved process to the process owner along with a **process control plan** and other necessary procedures to ensure that the gains from the project will be institutionalized. That is, the goal is to ensure that the gains are of help in the process and, if possible, that the improvements will be implemented in other similar processes in the business.

The process owner should be provided with before and after data on key process metrics, operations and training documents, and updated current process maps. The process control plan should be a system for monitoring the solution that has been implemented, including methods and metrics for periodic auditing. Control charts are an important statistical tool used in the control step of DMAIC; many process control plans involve control charts on critical process metrics.

The transition plan for the process owner should include a validation check several months after project completion. It is important to ensure that the original results are still in place and stable so that the positive financial impact will be sustained. It is not unusual to find that something has gone wrong in the transition to the improved process. The ability to respond rapidly to unanticipated failures should be factored into the plan.

The tollgate review for the control step typically includes the items shown in the adjacent display:

Process Control Plan

(*Continued*)

experiments, and simulation analyses.

4. Plans to implement the pilot test results on a full-scale basis. This should include dealing with any regulatory requirements (FDA, OSHA, legal, for example), personnel concerns (such as additional training requirements), or impact on other business standard practices.
5. Analysis of any risks of implementing the solution, and appropriate risk-management plans.

Tollgate for Control:

1. Data illustrating that the before and after results are in line with the project charter should be available. (Were the original objectives accomplished?)
2. Is the process control plan complete? Are procedures to monitor the process, such as control charts, in place?
3. Is all essential documentation for the process owner complete?
4. A summary of lessons learned from the project should be available.
5. A list of opportunities that were not pursued in the project should be prepared. This can be used to develop future projects; it is very important to maintain an inventory of good potential projects to keep the improvement process going.
6. A list of opportunities to use the results of the project in other parts of the business should be prepared.

2.7.6 EXAMPLES OF DMAIC

example 2.2 Litigation Documents

Litigation usually creates a very large number of documents. These can be internal work papers, consultants' reports, affidavits, court filings, documents obtained via subpoena, and papers from many other sources. In some cases, there can be hundreds of thousands of documents and millions of pages. DMAIC was applied in the corporate legal department of DuPont, led by DuPont lawyer, Julie Mazza, who spoke about the project at an American Society for Quality meeting [Mazza (2000)]. The case is also discussed in Snee and Hoerl (2005). The objective was to

example 2.2 Continued

develop an efficient process to allow timely access to needed documents with minimal errors. Document management is extremely important in litigation; it also can be time-consuming and expensive. The process was usually manual, so it was subject to human error, with lost or incorrect documents fairly common problems. In the specific case presented by Mazza, there was an electronic data base that listed and classified all of the documents, but the documents themselves were in hard copy form.

Define The DuPont legal function and the specific legal team involved in this specific litigation were the customers for this process. Rapid and error-free access to needed documents was essential. For example, if a request for a document could not be answered in 30 days, the legal team would have to file a request for an extension with the court. Such extensions add cost, time, and detract from the credibility of the legal team. A project team consisting of process owners, legal subject-matter experts, clerks, an information systems specialist, and Mazza (who was also a black belt in Dupont's six-sigma program) was formed. The team decided to focus on CTQs involving reduction of cycle time, reduction of errors, elimination of non-value-added process activities, and reduction of costs. They began by mapping the entire document-production process, including defining the steps performed by DuPont legal, outside counsel, and the outside documents-management company. This process map was instrumental in identifying non-value-added activities.

Measure In the measure step, the team formally measured the degree to which the CTQs were being met by reviewing data in the electronic data base; obtaining actual invoices; reviewing copying and other labor charges, the costs of data entry, and the charges for shipping, court fees for filing for extensions; and studying how frequently individual documents in the data base were being handled. It was difficult to accurately measure the frequency of handling. Many of the cost categories contained non-value-added costs because of errors, such as having to copy a different document because the wrong document had been pulled and copied. Another error was allowing a confidential document to be copied.

Analyze The team worked with the data obtained during the measure step and the knowledge of team members to identify many of the underlying causes and cost exposures. A failure modes and effects analysis highlighted many of the most important issues that needed to be addressed to improve the system. The team also interviewed many of the people who worked in the process to better understand how they actually did the work and the problems they encountered. This is often very important in nonmanufacturing and service organizations because these types of operations can have a much greater human factor. Some of the root causes of problems they uncovered were:

1. A high turnover rate for the contractor's clerks
2. Inadequate training
3. Inattention to the job, caused by clerks feeling they had no ownership in the process
4. The large volume of documents

example 2.2 Continued

The team concluded that many of the problems in the system were the result of a manual document-handling system.

Improve To improve the process, the team proposed a digital scanning system for the documents. This solution had been considered previously but always had been discarded because of cost. However, the team had done a very thorough job of identifying the real costs of the manual system and the inability of a manual system to ever really improve the situation. The better information produced during the measure and analyze steps allowed the team to successfully propose a digital scanning system that the company accepted.

The team worked with DuPont's information technology group to identify an appropriate system, get the system in place, and scan all of the documents. They remapped the new process and, on the basis of a pilot study, estimated that the unit cost of processing a page of a document would be reduced by about 50%, which would result in about $1.13 million in savings. About 70% of the non-value-added activities in the process were eliminated. After the new system was implemented, it was proposed for use in all of the DuPont legal functions; the total savings were estimated at about $10 million.

Control The control plan involved designing the new system to automatically track and report the estimated costs per document. The system also tracked performance on other critical CTQs and reported the information to users of the process. Invoices from contactors were also forwarded to the process owners as a mechanism for monitoring ongoing costs. Explanations about how the new system worked and necessary training were provided for all those who used the system. Extremely successful, the new system provided significant cost savings, improvement in cycle time, and reduction of many frequently occurring errors.

example 2.3 Improving On-Time Delivery

A key customer contacted a machine tool manufacturer about poor recent performance they had experienced regarding on-time delivery of the product. On-time deliveries were at 85%, instead of the desired target value of 100%, and the customer could choose to exercise a penalty clause to reduce by up to 15% of the price of each tool, or about a $60,000 loss for the manufacturer. The customer was also concerned about the manufacturer's factory capacity and its capability to meet their production schedule in the future. The customer represented about $8 million of business volume for the immediate future—the manufacturer needed a revised business process to resolve the problem or the customer might consider seeking a second source supplier for the critical tool.

A team was formed to determine the root causes of the delivery problem and implement a solution. One team member was a project engineer who was sent to a supplier factory, with the purpose to work closely with the supplier, to examine all the processes used in manufacturing of the tool, and to identify any gaps in the processes that affected delivery. Some of the supplier's processes might need improvement.

example 2.3 Continued

Define The objective of the project was to achieve 100% on-time delivery. The customer had a concern regarding on-time delivery capability, and a late deliveries penalty clause could be applied to current and future shipments at a cost to the manufacturer. Late deliveries also would jeopardize the customer's production schedule, and without an improved process to eliminate the on-time delivery issue, the customer might consider finding a second source for the tool. The manufacturer could potentially lose as much as half of the business from the customer, in addition to incurring the 15% penalty costs. The manufacturer also would experience a delay in collecting the 80% equipment payment customarily made upon shipment.

The potential savings for meeting the on-time delivery requirement was $300,000 per quarter. Maintaining a satisfied customer also was critical.

Measure The contractual lead time for delivery of the tool was eight weeks. That is, the tool must be ready for shipment eight weeks upon receipt of the purchase order. The CTQ for this process was to meet the target contractual lead time. Figure 2.13 shows the process map for the existing process, from purchase order receipt to shipment. The contractual lead time could be met only when there was no excursion or variation in the process. Some historical data on this process was available, and additional data was collected over approximately a two-month period.

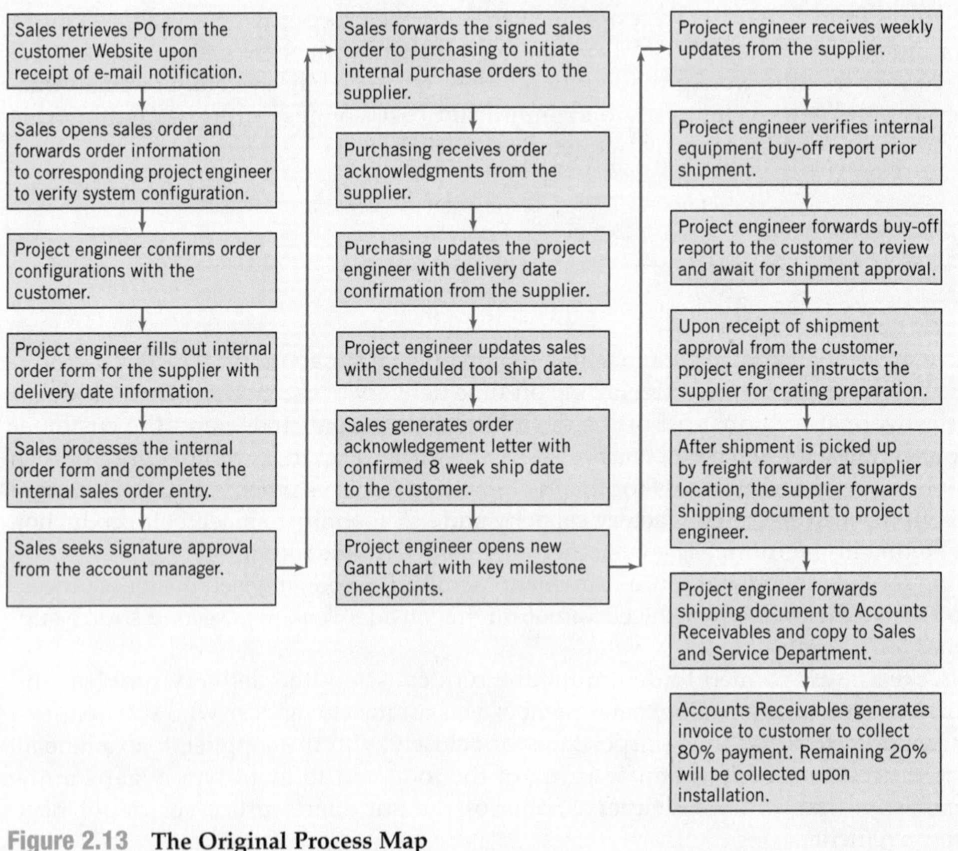

Figure 2.13 The Original Process Map

example 2.3 Continued

Analyze Based on the data collected from the measure step, the team concluded that problem areas came from:

1. Supplier quality issues: Parts failed prematurely. This caused delay in equipment final testing due to trouble shooting or waiting for replacement parts.
2. Purchase order process delay: Purchase orders were not processed promptly, resulting in delayed internal project start dates.
3. Delay in customer confirmation: It took up to three days to confirm the final equipment configuration with the customer. This delayed most of the early manufacturing steps and complicated production scheduling.
4. Incorrect tool configuration orders: There were many processes on the customer side, leading to frequent confusion when the customer placed the order and often resulting in an incorrect tool configuration. This caused rework at the mid-stream of the manufacturing cycle and contributed greatly to the delivery delay problem.

Improve In order to meet the eight-week contractual lead time, the team knew that it was necessary to eliminate any possible process variation, starting from receipt of the purchase order to shipment of the equipment. Three major corrective actions were taken:

1. *Supplier Quality Control and Improvement:* An internal buy-off checklist for the supplier was implemented that contained all required testing of components and subsystems that had to be completed prior to shipment. This action was taken to minimize part failures both in manufacturing and final test as well as in the field. The supplier agreed to provide consigned critical spare parts to the manufacturer so that it could save on shipping time for replacement parts if part failures were encountered during manufacturing and final testing.
2. *Improve the Internal Purchase Order Process:* A common e-mail address was established to receive all purchase order notifications. Three people (a sales support engineer, a project engineer, and an account manager) were to have access to the e-mail account. Previously, only one person checked purchase order status. This step enhanced the transparency of purchase order arrival and allowed the company to act promptly when a new order was received.
3. *Improve the Ordering Process with the Customer:* The team realized that various tool configurations were generated over the years due to new process requirements from the customer. In order to ensure accuracy of tool configurations in a purchase order, a customized spreadsheet was designed together with the customer to identify the key data for the tool on order. The spreadsheet was saved under a purchase order number and stored in a predefined Web location. The tool owner also was to take ownership of what he/she ordered to help to eliminate the confirmation step with the customer and to ensure accuracy in the final order.

Figure 2.14 shows a process map of the new, improved system. The steps in the original process that were eliminated are shown as shaded boxes in this figure.

example 2.3 Continued

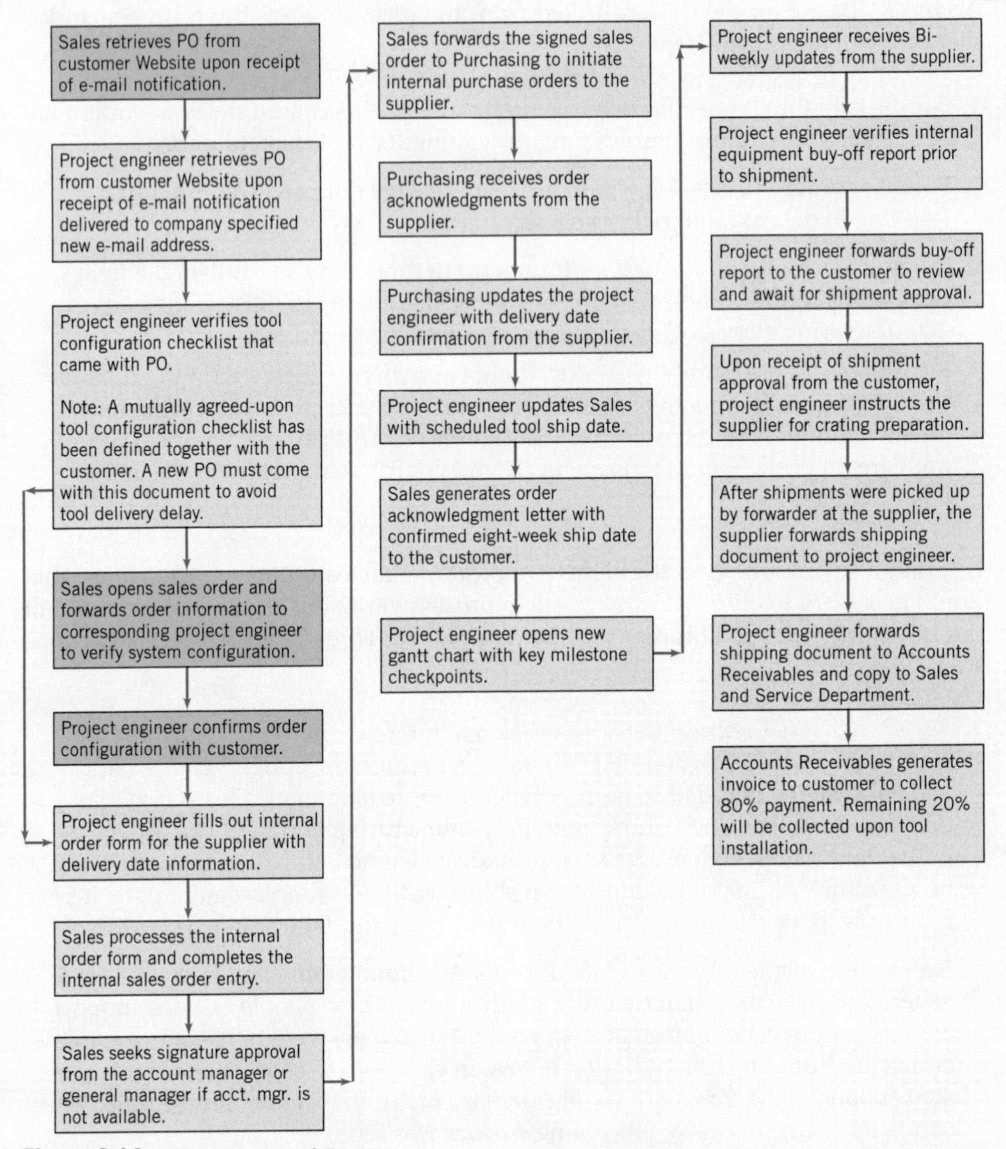

Figure 2.14 The Improved Process

Control To ensure that the new process was in control, the team revised the production tracking spreadsheet with firm milestone dates and provided a more visual format. An updating procedure was provided on a biweekly basis by the factory to reflect up-to-date information. The project engineer would be able to monitor the progress of each tool on order and take action accordingly should any unplanned deviation from the schedule occur.

After implementing the new process, including the new production tracking procedure, the manufacturer was able to ship tools with 100% on-time delivery. The cost savings were more than $300,000 each quarter. Equally important, the customer was satisfied and continued to remain confident in the manufacturer's capability and capacity.

example 2.4 Improving Service Quality in a Bank

Kovach (2007) describes how the DMAIC process can be used to improve service quality for a banking process. During the define and measure phases of this project, the team identified several CTQs to be improved:

1. Speed of service
2. Consistent service
3. An easy-to-use process
4. A pleasant environment
5. Knowledgeable staff

There were many factors that could be investigated to improve these CTQs. The team decided to focus on two areas of improvement: improved teller and customer work stations and new training for the staff. In the improve stage, they decided to use a designed experiment to investigate the effects of these two factors on the CTQs. Four different branches were selected in which to conduct the experiment. Notice that this is a **physical experiment,** not an experiment with a computer simulation model of the branch operations. New teller and customer work stations were designed and installed in two of the branches. The team designed a new training program and delivered it to the staff at two of the branches, one with the new work stations and one without the new facilities. (This was a two-factor factorial experiment, with each of the two factors having two levels. We discuss these types of experiments extensively in this book.)

The team decided to conduct the experiment for 30 working days. Each day was considered to be a block (blocking is a design technique for eliminating the effects of nuisance factors on the experimental results; here the nuisance factors were transaction types, volumes, and different customers at each of the four branches). The response data was obtained by asking customers to complete a survey instrument that registered their degree of satisfaction with the previously identified CTQs.

The results of the experiment demonstrated that there was a statistically significant difference in the CTQs resulting from both the new work stations and the new training, with the best results obtained from the combination of the new work stations and the new training. Implementation of the new stations and training was expected to significantly improve customer satisfaction with the banking process across the bank's branches.

Important Terms and Concepts

Analyze step
Control step
Define step
Design for Six-Sigma (DFSS)
DMAIC
Failure modes and effects analysis (FMEA)
Improve step
Key process input variables (KPIV)
Key process output variables (KPOV)
Measure step
Project charter
SIPOC diagram
Six-Sigma
Tollgate

chapter Two Exercises

2.1 Is Deming's philosophy more or less focused on statistics then Juran?

2.2 What is the Juran Trilogy?

2.3 What is the Malcolm Baldrige National Quality Award? Who is eligible for the award?

2.4 What is a six-sigma process?

2.5 Compare and contrast Deming's and Juran's philosophies of quality.

2.6 What would motivate a company to complete for the MBNQA?

2.7 Hundreds of companies have won the MBNQA. Collect information or two of the winners. What occurs have they had since receiving the award?

2.8 Reconsider the fast-food restaurant out discussed in the chapter.

a. What results do you obtain it the probability of good quality on each meal component was 0.999?
b. What level of quality on each meal component would be required to produce an annual level of quality that was acceptable to you?

2.9 Discuss the similarities between the Shewhart cycle and DMAIC.

2.10 Describe a service system that you use. What are the CTQs that are important to you? How do you think that DMAIC could be applied to this process?

2.11 One of the objectives of the control plan in DMAIC is to "hold the gain." What does this mean?

2.12 Is there a point at which seeking further improvement in quality and productivity isn't economically advisable? Discuss your answer.

2.13 Explain the importance of tollgates in the DMAIC process.

2.14 An important part of a project is to identify the key process input variables (KPIV) and key process output variables (KPOV). Suppose that you are the owner/manager of a small business that provides mailboxes, copy services, and mailing services. Discuss the KPIVs and KPOVs for this business. How do they relate to possible customer CTQs?

2.15 An important part of a project is to identify the key process input variables (KPIV) and key process output variables (KPOV). Suppose that you are in charge of a hospital emergency room. Discuss the KPIVs and KPOVs for this business. How do they relate to possible customer CTQs?

2.16 Why are designed experiments most useful in the improve step of DMAIC?

2.17 Suppose that your business is operating at the three-sigma quality level. If projects have an average improvement rate of 50% annually, how many years will it take to achieve six-sigma quality?

2.18 Suppose that your business is operating at the four-and-a-half-sigma quality level. If projects have an average improvement rate of 50% annually, how many years will it take to achieve six-sigma quality?

2.19 Explain why it is important to separate sources of variability into special or assignable causes and common or chance causes.

2.20 Consider improving service quality in a restaurant. What are the KPIVs and KPOVs that you should consider? How do these relate to likely customer CTQs?

2.21 Suppose that during the analyze phase an obvious solution is discovered. Should that solution be immediately implemented and the remaining steps of DMAIC abandoned? Discuss your answer.

2.22 What information would you have to collect in order to build a discrete-event simulation model of a retail branch-banking operation? Discuss how this model could be used to determine appropriate staffing levels for the bank.

2.23 Suppose that you manage an airline reservation system and want to improve service quality. What are the important CTQs for this process? What are the KPIVs and KPOVs? How do these relate to the customer CTQs that you have identified?

2.24 It has been estimated that safe aircraft carrier landings operate at about the 5s level. What level of ppm defective does this imply?

2.25 Discuss why, in general, determining what to measure and how to make measurements is more difficult in service processes and transactional businesses than in manufacturing.

2.26 Suppose that you want to improve the process of loading passengers onto an airplane. Would a discrete-event simulation model of this process be useful? What data would have to be collected to build this model?

chapter Three Tools and Techniques for Quality Control and Improvement

Chapter Outline

Six-Sigma reduces candy defects for a UK food manufacturer

The European confectionary market is highly competitive. Growth is slow and the market is saturated. To be competitive, companies must find ways to reduce costs while producing the highest-quality goods. Recently, a UK-based producer of cough drops tested the effectiveness of the DMAIC approach for cutting manufacturing costs.

Define:

The team selected cough drop thickness as a key quality characteristic, as it could cause machine downtimes and high levels of scrap/rework. They prepared a *cause-and-effect diagram* to categorize issues causing thickness variances.

Measure:

To measure and assess thickness, the process used a Go/No Go gauge. The team purchased a gauge that could produce an exact measurement of thickness and prepared a chart that showed the average thickness for groups of cough drops over a two-week period (*a control chart*).

Analyze:

A frequency analysis (*histogram*) of thickness found that there were a few very large cough drops but not many small ones. They then prepared a *scatter plot* of size against weight and found that some drops had a low weight compared to their size. Taking a closer look, the team realized that these drops contained air pockets. They then revisited the earlier DMAIC phases, brainstormed, followed the process flow map, and tried to pinpoint where air bubbles were entering the process.

Improve:

Next came a series of *designed experiments* to establish process settings that would minimize air bubbles. The team installed systems to both *identify* and *remove* cough drops containing air bubbles. They also conducted a series of experiments to improve wrapping machine processes.

Control:

Finally, the team fully documented the process changes and trained all personnel on new procedures and expectations. They also implemented *control charts* to detect process changes that could cause air bubbles.

By following these steps, the company saved £290,000 and reduced its scrap rate from 1 in every 5 cough drops to 1 in 10,000 or more. More importantly the organization now has a much better understanding of the impact of variation.

Chapter Overview and Learning Objectives

This chapter has three objectives. The first is to present several basic process improvement tools and to illustrate how these tools form a cohesive, practical framework for quality improvement. These tools form an important basic approach to both reducing variability and monitoring the performance of a process and are widely used in both the analyze and control steps of DMAIC. The second objective is to describe the statistical basis of the Shewhart control chart, one of the most important of the process improvement tools. The reader will see how decisions about sample size, sampling interval, and placement of control limits affect the performance of a control chart. Other key concepts include the idea of rational subgroups, interpretation of control chart signals and patterns, and the average run length as a measure of control chart performance. The third objective is to discuss and illustrate some practical issues in the implementation of control charts and their associated process improvement tools. Collectively, these tools are often called **statistical process control (SPC)**.

After careful study of this chapter you should be able to do the following:

1. Understand chance and assignable causes of variability in a process
2. Explain the statistical basis of the Shewhart control chart
3. Understand the basic process improvement tools of SPC: the histogram or stem-and-leaf plot, the check sheet, the Pareto chart, the cause-and-effect diagram, the defect concentration diagram, the scatter diagram, and the control chart
4. Explain how sensitizing rules and pattern recognition are used in conjunction with control charts

3.1 Introduction

If a product is to meet or exceed customer expectations, generally it should be produced by a process that is stable or repeatable. More precisely, the process must be capable of operating with little variability

Statistical Process Control
A collection of problem-solving tools useful in achieving process stability and improving capability through reduction of variability.

around the target or nominal dimensions of the product's quality characteristics. SPC is a powerful collection of problem-solving tools useful in achieving process stability and improving capability through the reduction of variability.

A process is an organized sequence of activities that produces an output (product or service) that adds value to the organization. All work is performed a process on, and any process can be improved. Thus while we traditionally think of SPC as being applied to manufacturing processes, it can really be applied to any kind of process including service processes.

SPC is one of the greatest technological developments of the twentieth century because it is based on sound underlying principles, is easy to use, has significant impact, and can be applied to any process. Its seven major tools are

The Magnificent Seven

1. Histogram or stem-and-leaf plot
2. Check sheet
3. Pareto chart
4. Cause-and-effect diagram
5. Defect concentration diagram
6. Scatter diagram
7. Control chart

Although these tools, often called **"the magnificent seven,"** are an important part of SPC, they comprise only its technical aspects. The proper deployment of SPC helps create an environment in which all individuals in an organization seek continuous improvement in quality and productivity. This environment is best developed when management becomes involved in the process. Once this environment is established, routine application of the magnificent seven becomes part of the usual manner of doing business, and the organization is well on its way to achieving its quality improvement objectives.

Of the seven tools, the Shewhart control chart is probably the most technically sophisticated. It was developed in the 1920s by Walter A. Shewhart of Bell Telephone Laboratories. To understand the statistical concepts that form the basis of SPC, we must first describe Shewhart's theory of variability.

3.2 Chance and Assignable Causes of Quality Variation

In any production process, regardless of how well designed or carefully maintained it is, a certain amount of inherent or natural variability will always exist. This natural variability or "background noise" is the cumulative effect of many small, essentially unavoidable causes. In the framework of statistical quality control, this natural variability is often called a "stable system of chance causes." A process that is operating with only **chance causes of variation** present is said to be **in**

statistical control. In other words, the chance causes are an inherent part of the process.

Other kinds of variability may occasionally be present in the output of a process. This variability in key quality characteristics usually arises from three sources: improperly adjusted or controlled machines, operator errors, or defective raw material. Such variability is generally large when compared to the background noise, and it usually represents an unacceptable level of process performance. We refer to these sources of variability that are not part of the chance cause pattern as **assignable causes of variation.** A process that is operating in the presence of assignable causes is said to be an **out-of-control process**.

The terminology "chance" and "assignable causes" was developed by Shewhart. Today, some writers use the terminology "common cause" instead of "chance cause" and "special cause" instead of "assignable cause".

These chance and assignable causes of variation are illustrated in Figure 3.1. Until time t_1 the process shown in this figure is in control; that is, only chance causes of variation are present. As a result, both the mean and standard deviation of the process are at their in-control values (say, μ_0 and σ_0). At time t_1 an assignable cause occurs. As shown in Figure 3.1, the effect of this assignable cause is to shift the process mean to a new value $\mu_1 > \mu_0$. At time t_2 another assignable cause occurs, resulting in $\mu = \mu_0$, but now the process standard deviation has shifted to a larger value of $\sigma_1 > \sigma_0$. At time t_3 there is another assignable cause present, resulting in both the process mean and standard deviation taking on out-of-control values. From time t_1 forward, the presence of assignable causes has resulted in an out-of-control process.

Processes will often operate in the in-control state for relatively long periods of time. However, no process is truly stable forever, and, eventually, assignable causes will occur, seemingly at random, resulting in a shift to an out-of-control state where a larger proportion of the process output does not conform to requirements. For example, note from Figure 3.1 that when the process is in control, most of the production will fall between the lower and upper specification limits (LSL and USL, respectively). When the process is out of control, a higher proportion of the process lies outside of these specifications.

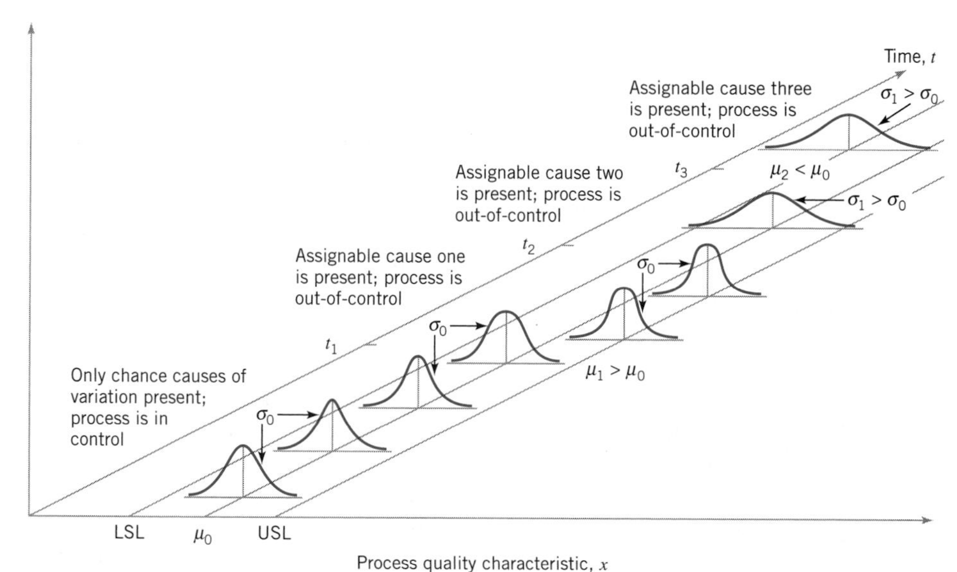

Figure 3.1 Chance and Assignable Causes of Variation

A major objective of statistical process control is to quickly detect the occurrence of assignable causes of process shifts so that investigation of the process and corrective action may be undertaken before many nonconforming units are manufactured. The **control chart** is an on-line process-monitoring technique widely used for this purpose. Control charts may also be used to estimate the parameters of a production process, and, through this information, to determine process capability. The control chart may also provide information useful in improving the process. Finally, remember that the eventual goal of statistical process control is the **elimination of variability in the process.** It may not be possible to completely eliminate variability, but the control chart is an effective tool in reducing variability as much as possible.

We now present the statistical concepts that form the basis of control charts. Chapters 5 and 6 develop the details of construction and use of the most widely used types of control charts.

3.3 The Control Chart

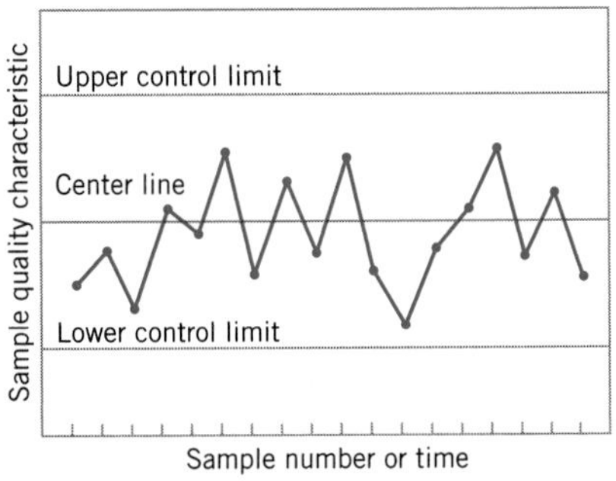

A typical control chart is shown in Figure 3.2. The control chart is a graphical display of a quality characteristic that has been measured or computed from a sample versus the sample number or time. The chart contains a **center line** that represents the average value of the quality characteristic corresponding to the in-control state. (That is, only chance causes are present.) Two other horizontal lines, called the **upper control limit** (UCL) and the **lower control limit** (LCL), are also shown on the chart. These control limits are chosen so that if the process is in control, nearly all of the sample points will fall between them. As long as the points plot within the control limits, the process is assumed to be in control, and no action is necessary. However, a point that plots outside of the control limits is interpreted as evidence that the process is out of control, and investigation and corrective action are required to find and eliminate the assignable cause or causes responsible for this behavior. It is customary to connect the sample points on the control chart with straight-line segments, so that it is easier to visualize how the sequence of points has evolved over time.

Figure 3.2 **A Typical Control Chart**

Even if all the points plot inside the control limits, if they behave in a systematic or non-random manner, then this could be an indication that the process is out of control. For example, if 18 of the last 20 points plotted above the center line but below the upper control limit and only two of these points plotted below the center line but above the lower control limit, we would be very suspicious that something was wrong. If the process is in control, all the plotted points should have an essentially random pattern. Methods for looking for sequences or nonrandom patterns can be applied to control charts as an aid in detecting out-of-control conditions. Usually, there is a reason why a particular nonrandom pattern appears on a control chart, and if it can be found and eliminated, process performance can be improved.

To illustrate the preceding ideas, we give an example of a control chart. In semiconductor manufacturing, an important fabrication step is photolithography, in which a light-sensitive photoresist material is applied to the silicon wafer, the circuit pattern is exposed on the resist (typically through the use of high-intensity UV light), and the unwanted resist material is removed through a developing process. After the resist pattern is defined, the underlying material is removed by either wet chemical or plasma etching. It is fairly typical to follow development with a hard-bake process to increase resist adherence and etch resistance. An important quality characteristic in hard bake is the flow width of the resist, a measure of how much it expands due to the baking process. Suppose that flow width can be controlled at a mean of 1.5 microns, and it is known that the standard deviation of flow width is 0.15 microns. A control chart for the average flow width is shown in Figure 3.3. Every hour, a sample of five wafers is taken, the average flow width ($\bar{x}$) computed, and $\bar{x}$ plotted on the chart. Because this control chart utilizes the sample average $\bar{x}$ to monitor the process mean, it is usually called an $\bar{x}$ control chart. Note that all of the plotted points fall inside the control limits, so the chart indicates that the process is considered to be in statistical control.

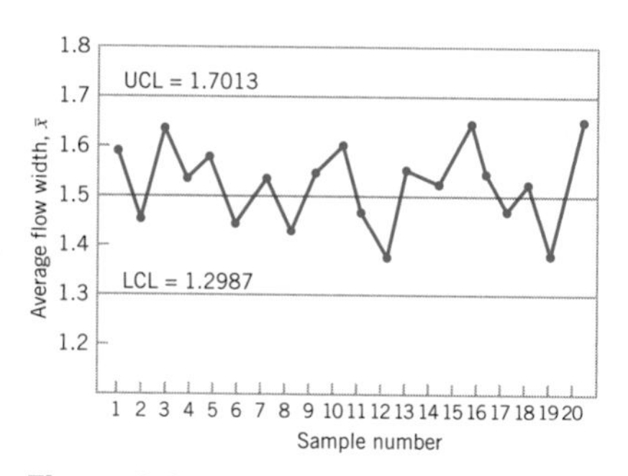

Figure 3.3 **$\bar{x}$ Control Chart for Flow Width**

To assist in understanding the statistical basis of this control chart, consider how the control limits were determined. The process mean is 1.5 microns, and the process standard deviation is $\sigma = 0.15$ microns. Now if samples of size $n = 5$ are taken, the standard deviation of the sample average $\bar{x}$ is

$$\sigma_{\bar{x}} = \frac{\sigma}{\sqrt{n}} = \frac{0.15}{\sqrt{5}} = 0.0671$$

Therefore, if the process is in control with a mean flow width of 1.5 microns, then by using the central limit theorem to assume that $\bar{x}$ is approximately normally distributed, we would expect $100(1 - \alpha)\%$ of the sample means $\bar{x}$ to fall between $1.5 + Z_{\alpha/2}(0.0671)$ and $1.5 - Z_{\alpha/2}(0.0671)$. We will arbitrarily choose the constant $Z_{\alpha/2}$ to be 3, so that the upper and lower control limits become

Three Sigma Control Chart

The "Sigma" in the three sigma control chart refers to the standard deviation of the statistic plotted on the chart (i.e. $\sigma_{\bar{x}}$), not the standard deviation of the quality characteristics.

$$\text{UCL} = 1.5 + 3(0.0671) = 1.7013$$

and

$$\text{LCL} = 1.5 - 3(0.0671) = 1.2987$$

as shown on the control chart. These are typically called three-sigma control limits. The width of the control limits is inversely proportional to the sample size n for a given multiple of sigma. Note that choosing the control limits is equivalent to setting up the critical region for testing the hypothesis

> Note that "sigma" refers to the standard deviation of the statistic plotted on the chart (i.e., $\sigma_{\bar{x}}$), not the standard deviation of the quality characteristic.

$$H_0: \quad \mu = 1.5$$
$$H_1: \quad \mu \neq 1.5$$

where $\sigma = 0.15$ is known. Essentially, the control chart tests this hypothesis repeatedly at different points in time. The situation is illustrated graphically in Figure 3.4.

We may give a general **model** for a control chart. Let w be a sample statistic that measures some quality characteristic of interest, and

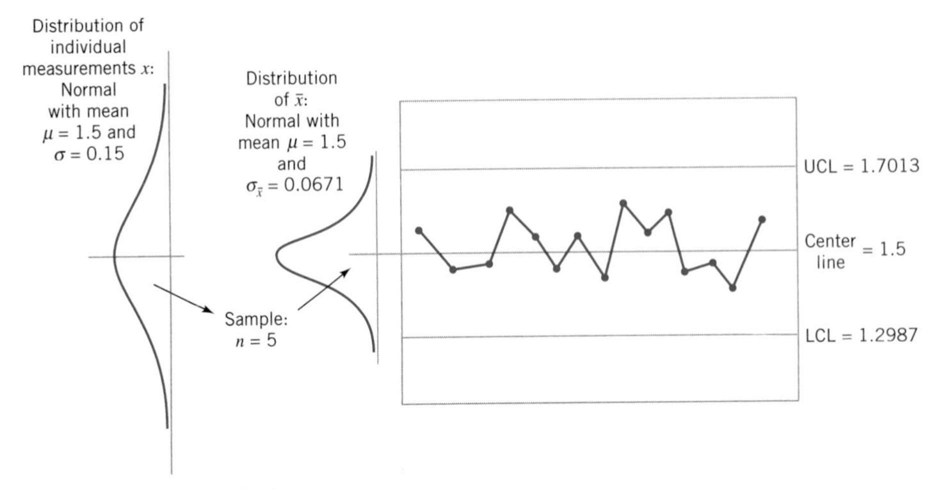

Figure 3.4 How the Control Chart Works

suppose that the mean of w is μ_w and the standard deviation of w is σ_w. Then the center line, the upper control limit, and the lower control limit become

$$\begin{aligned} \text{UCL} &= \mu_w + L\sigma_w \\ \text{Center line} &= \mu_w \\ \text{LCL} &= \mu_w - L\sigma_w \end{aligned} \qquad (3.1)$$

where L is the "distance" of the control limits from the center line, expressed in standard deviation units. This general theory of control charts was first proposed by Walter A. Shewhart, and control charts developed according to these principles are often called **Shewhart control charts.**

The control chart is a device for describing in a precise manner exactly what is meant by statistical control; as such, it may be used in a variety of ways. In many applications, it is used for on-line process surveillance. That is, sample data are collected and used to construct the control chart, and if the sample values of $\bar{x}$ (say) fall within the control limits and do not exhibit any systematic pattern, we say the process is in control at the level indicated by the chart. Note that we may be interested here in determining *both* whether the past data came from a process that was in control and whether future samples from this process indicate statistical control.

The most important use of a control chart is to **improve** the process. We have found that, generally:

1. Most processes do not operate in a state of statistical control.
2. Consequently, the routine and attentive use of control charts will identify assignable causes. If these causes can be eliminated from the process, variability will be reduced and the process will be improved.

This process improvement activity using the control chart is illustrated in Figure 3.5. Note that:

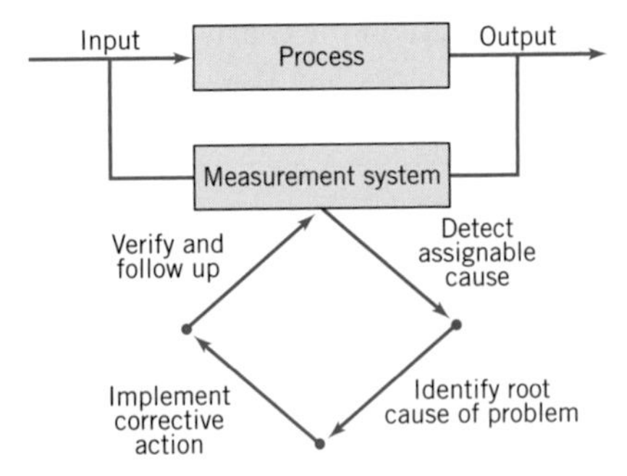

Figure 3.5 Process Improvement Using the Control Chart

3. The control chart will only **detect** assignable causes. Management, operator, and engineering **action** will usually be necessary to eliminate the assignable causes.

In identifying and eliminating assignable causes, it is important to find the **root cause** of the problem and to attack it. A cosmetic solution will not result in any real, long-term process improvement. Developing an effective system for corrective action is an essential component of an effective SPC implementation.

OCAP

Out of Control Action Plan

A very important part of the corrective action process associated with control chart usage is the **out-of-control-action plan (OCAP).** An OCAP is a flow chart or text-based description of the sequence of activities that must take place following the occurrence of an *activating event*. These are usually out-of-control signals from the control chart. The OCAP consists of *checkpoints*, which are potential assignable causes, and *terminators*, which are actions taken to resolve the out-of-control condition, preferably by eliminating the assignable cause. It is very important that the OCAP specify as complete a set as possible of checkpoints and terminators, and that these be arranged in an order that facilitates process diagnostic activities. Often, analysis of prior failure modes of the process and/or product can be helpful in designing this aspect of the OCAP. Furthermore, an OCAP is a *living document* in the sense that it will be modified over time as more knowledge and understanding of the process is gained. Consequently, when a control chart is introduced, an initial OCAP should accompany it. Control charts without an OCAP are not likely to be useful as a process improvement tool.

The OCAP for the hard-bake process is shown in Figure 3.6. This process has two controllable variables, temperature and time. In this process, the mean flow width is monitored with an $\bar{x}$ control chart, and the process variability is monitored with a control chart for the range, or an R chart. Notice that if the R chart exhibits an out-of-control signal, operating personnel are directed to contact process engineering immediately. If the $\bar{x}$ control chart exhibits an out-of-control signal, operators are directed to check process settings and calibration and then make adjustments to temperature in an effort to bring the process back into a state of control. If these adjustments are unsuccessful, process engineering personnel are contacted.

We may also use the control chart as an **estimating device.** That is, from a control chart that exhibits statistical control, we may estimate certain process parameters, such as the mean, standard deviation, fraction nonconforming or fallout, and so forth. These estimates may then be used to determine the **capability** of the process to produce acceptable products. Such **process-capability studies** have considerable impact on many management decision problems that occur over the product cycle, including make or buy decisions, plant and process improvements that reduce process variability, and contractual agreements with customers or vendors regarding product quality.

Two Types of Control Charts

- Variables Control Chart
- Attributes Control Chart

Control charts may be classified into two general types. If the quality characteristic can be measured and expressed as a number on some continuous scale of measurement, it is usually called a **variable.** In such cases, it is convenient to describe the quality characteristic with a measure of central tendency and a measure of variability. Control

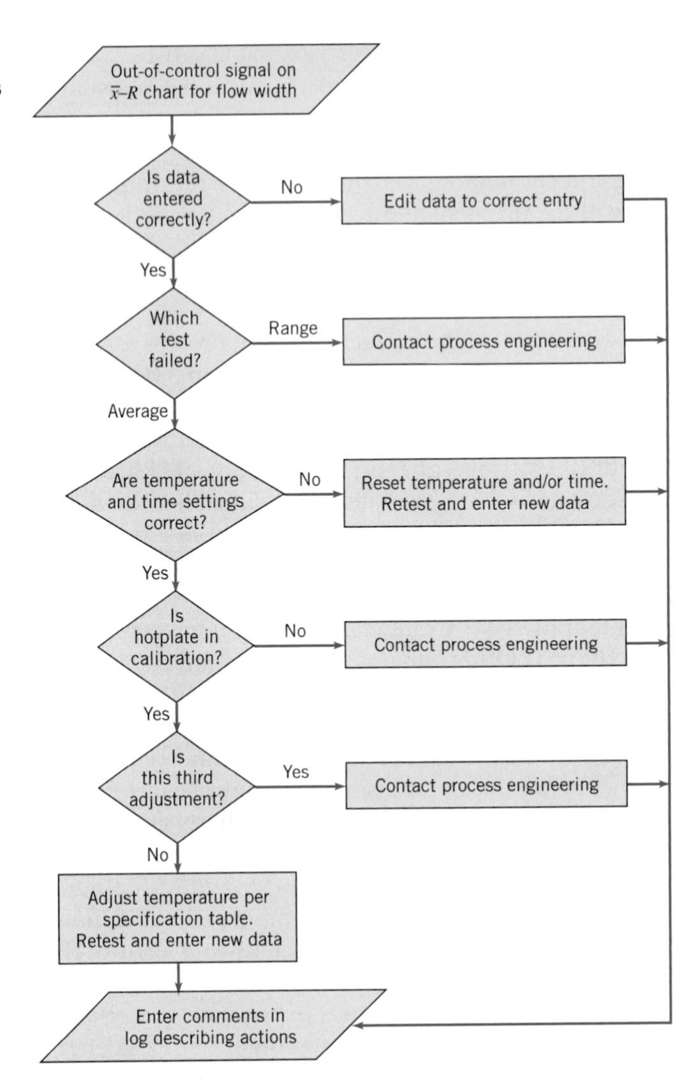

Figure 3.6 **The Out-of-Control-Action Plan (OCAP) for the Hard-Bake Process**

charts for central tendency and variability are collectively called **variables control charts.** The $\bar{x}$ chart is the most widely used chart for controlling central tendency, whereas charts based on either the sample range or the sample standard deviation are used to control process variability. Control charts for variables are discussed in Chapter 5. Many quality characteristics are not measured on a continuous scale or even a quantitative scale. In these cases, we may judge each unit of product as either conforming or nonconforming on the basis of whether or not it possesses certain attributes, or we may count the number of nonconformities (defects) appearing on a unit of product. Control charts for such quality characteristics are called **attributes control charts** and are discussed in Chapter 6.

An important factor in control chart use is the **design of the control chart.** By this we mean the selection of the sample size, control limits, and frequency of sampling. For example, in the $\bar{x}$ chart of Figure 3.3, we specified a sample size of five measurements, three-sigma control limits, and a sampling frequency of every hour. In most quality-control problems, it is customary to design the control chart using primarily statistical considerations. For example, we know that increasing the sample size will decrease the probability error, thus enhancing the chart's ability to detect an out-of-control state, and so forth. The use of statistical criteria such as these along with industrial experience has led to general guidelines and procedures for designing control charts. These procedures usually consider cost factors only in an implicit manner. Recently, however, we have begun to examine control chart design from an **economic** point of view, considering explicitly the cost of sampling, losses from allowing defective product to be produced, and the costs of investigating out-of-control signals that are really false alarms.

Another important consideration in control chart usage is the **type of variability** exhibited by the process. Figure 3.7 presents data from three different processes. Figures 3.7*a* and 3.7*b* illustrate **stationary behavior.** By this we mean that the process data vary around a fixed mean in a stable or predictable manner. This is the type of behavior that Shewhart implied was produced by an **in-control process**.

Even a cursory examination of Figures 3.7*a* and 3.7*b* reveals some important differences. The data in Figure 3.7*a* are **uncorrelated;** that is, the observations give the appearance of having been drawn at random from a stable population, perhaps a normal distribution. This type of data is referred to by time series analysts as **white noise.** (Time-series analysis is a field of statistics devoted exclusively to studying and modeling time-oriented data.) In this type of process, the order in which the data occur does not tell us much that is useful to analyze the process. In other words, the past values of the data are of no help in predicting any of the future values.

Figure 3.7*b* illustrates stationary but **autocorrelated** process data. Notice that successive observations in these data are **dependent;** that is, a value above the mean tends to be followed by another value above the mean, whereas a value below the mean is usually followed by another such value. This produces a data series that has a tendency to move in moderately long "runs" on either side of the mean.

Figure 3.7*c* illustrates **nonstationary** variation. This type of process data occurs frequently in the chemical and process industries. Note that the process is very unstable in that it drifts or "wanders about" without any sense of a stable or fixed mean. In many industrial settings, we stabilize this type of behavior by using **engineering process control** (such as **feedback control**). This approach to process control is required when there are factors that affect the process that cannot be stabilized, such as environmental variables or properties of raw materials. When the control scheme is effective, the process output will *not* look like Figure 3.7*c* but will resemble either Figure 3.7*a* or 3.7*b*.

Shewhart control charts are most effective when the in-control process data look like Figure 3.7*a*. By this we mean that the charts can be designed so that their performance is predictable and reasonable to the user and so that they are effective in reliably detecting out-of-control conditions. Most of our discussion of control charts in this chapter and in Chapters 5 and 6 will assume that the in-control process data are stationary and uncorrelated. With some modifications, Shewhart control charts and other types of control charts can be applied to autocorrelated data. They can also be applied in

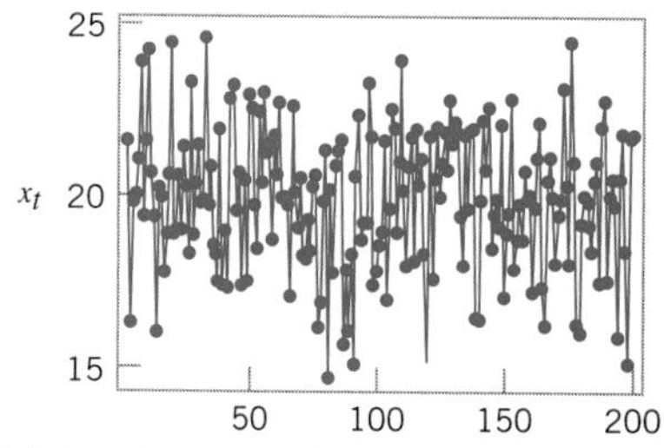

(*a*) Stationary and uncorrelated white noise

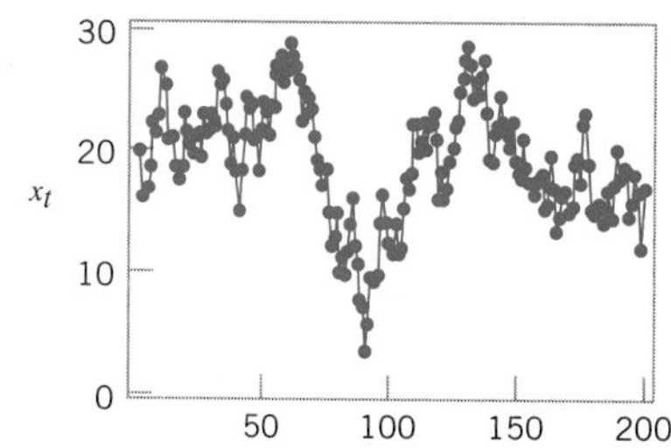

(*b*) Stationary and autocorrelated

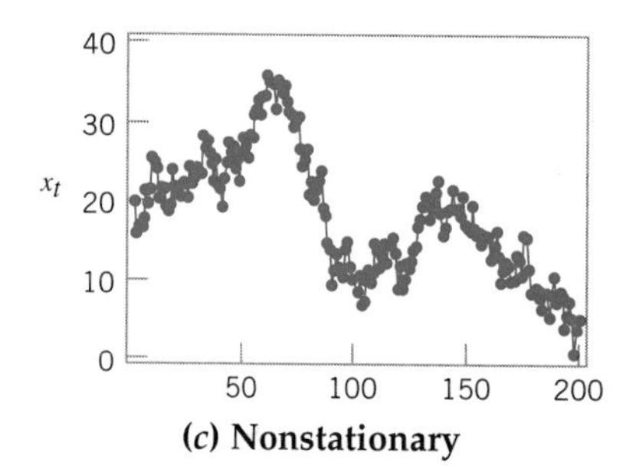

(*c*) Nonstationary

Figure 3.7 **Data from Three Different Processes**

systems where feedback control is employed. See Montgomery (2009) for details.

Control charts are among the most important management control tools; they are as important as cost controls and material controls. Modern computer technology has made it easy to implement control charts in *any* type of process, as data collection and analysis can be performed on a microcomputer or a local area network terminal in real time, on-line at the work center. Some additional guidelines for implementing a control chart program are given at the end of Chapter 6.

3.4 The Rest of the Magnificent Seven

Although the control chart is a very powerful problem-solving and process-improvement tool, it is most effective when its use is fully integrated into a comprehensive SPC program. The seven major SPC problem-solving tools should be widely taught throughout the organization and used routinely to identify improvement opportunities and to assist in reducing variability and eliminating waste. They can be used in several ways throughout the DMAIC problem-solving process. The magnificent seven, introduced in Section 3.1, are listed again here for convenience:

The Magnificent Seven

1. Histogram or stem-and-leaf plot
2. Check sheet
3. Pareto chart
4. Cause-and-effect diagram
5. Defect concentration diagram
6. Scatter diagram
7. Control chart

We introduced the histogram and the stem-and-leaf plot in Chapter 3 and discussed the control chart in Section 3.3. In this section, we illustrate the rest of the tools.

Check Sheet. In the early stages of process improvement, it will often become necessary to collect either historical or current operating data about the process under investigation. This is a common activity in the measure step of DMAIC. A **check sheet** can be very useful in this data collection activity. The check sheet shown in Figure 3.8 was developed by an aerospace firm engineer who was investigating defects that occurred on one of the firm's tanks. The engineer designed the check sheet to help summarize all the historical defect data available on the tanks. Because only a few tanks were manufactured each month, it seemed appropriate to summarize the data monthly and to identify as many different types of defects as possible. The **time-oriented summary** is particularly valuable in looking for **trends** or other meaningful patterns. For example, if many defects occur during the summer, one possible cause that might be the use of temporary workers during a heavy vacation period.

CHECK SHEET
DEFECT DATA FOR 2002–2003 YTD

Part No.: TAX-41
Location: Bellevue
Study Date: 6/5/03
Analyst: TCB

	2002												2003					
Defect	1	2	3	4	5	6	7	8	9	10	11	12	1	2	3	4	5	Total
Parts damaged		1		3	1	2		1		10	3		2	2	7	2		34
Machining problems			3	3				1	8		3		8	3				29
Supplied parts rusted			1	1		2	9											13
Masking insufficient		3	6	4	3	1												17
Misaligned weld	2																	2
Processing out of order	2															2		4
Wrong part issued		1						2										3
Unfinished fairing			3															3
Adhesive failure				1							1		2			1	1	6
Powdery alodine					1													1
Paint out of limits						1								1				2
Paint damaged by etching			1															1
Film on parts						3		1	1									5
Primer cans damaged								1										1
Voids in casting									1	1								2
Delaminated composite										2								2
Incorrect dimensions											13	7	13	1		1	1	36
Improper test procedure										1								1
Salt-spray failure													4					4
TOTAL	4	5	14	12	5	9	9	6	10	14	20	7	29	7	7	6	2	166

Figure 3.8 A Check Sheet to Record Defects on a Tank Used in an Aerospace Application

When designing a check sheet, it is important to clearly specify the type of data to be collected, the part or operation number, the date, the analyst, and any other information useful in diagnosing the cause of poor performance. If the check sheet is the basis for performing further calculations or is used as a worksheet for data entry into a computer, then it is important to be sure that the check sheet will be adequate for this purpose. In some cases, a trial run to validate the check sheet layout and design may be helpful.

Pareto Chart. The **Pareto chart** is simply a frequency distribution (or histogram) of attribute data arranged by category. Pareto charts are often used in both the measure and analyze steps of DMAIC. To illustrate a Pareto chart, consider the tank defect data presented in Figure 3.8. Plotting the total frequency of occurrence of each defect type (the last column of the table in Figure 3.8) against the various defect types will produce Figure 3.9, which is called a Pareto chart. Through

The name *Pareto chart* is derived from Italian economist Vilfredo Pareto (1848–1923), who theorized that in certain economies the majority of the wealth was held by a disproportionately small segment of the population. Quality engineers have observed that defects usually follow a similar Pareto distribution.

Figure 3.9 **Pareto Chart of the Tank Defect Data**

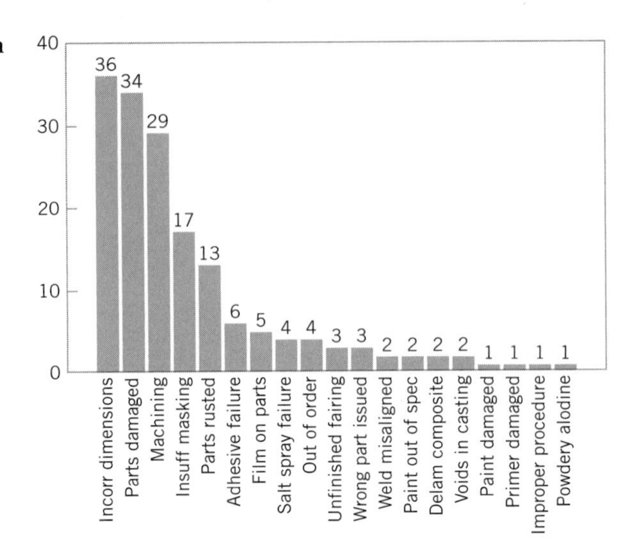

this chart, the user can quickly and visually identify the most frequently occurring types of defects. For example, Figure 3.9 indicates that incorrect dimensions, parts damaged, and machining are the most commonly encountered defects. Thus the causes of these defect types probably should be identified and attacked first.

Note that the Pareto chart does not automatically identify the most *important* defects, but only the most *frequent*. For example, in Figure 3.9 casting voids occur very infrequently (2 of 166 defects, or 1.2%). However, voids could result in scrapping the tank, a potentially large cost exposure—perhaps so large that casting voids should be elevated to a major defect category. When the list of defects contains a mixture of those that might have extremely serious consequences and others of much less importance, one of two methods can be used:

1. Use a weighting scheme to modify the frequency counts. Weighting schemes for defects are discussed in Chapter 6.
2. Accompany the *frequency* **Pareto chart** analysis with a **cost** or *exposure Pareto chart*.

There are many variations of the basic Pareto chart. Figure 3.10*a* shows a Pareto chart applied to an electronics assembly process using surface-mount components. The vertical axis is the percentage of components incorrectly located, and the horizontal axis is the component number, a code that locates the device on the printed circuit board. Note that locations 27 and 39 account for 70% of the errors. This may be the result of the *type* or *size* of components at these locations, or of *where* these locations are on the board layout. Figure 3.10*b* presents another Pareto chart from the electronics industry. The vertical axis is the number of defective components, and the horizontal axis is the component number. Note that each vertical bar has been broken down by supplier to produce a *stacked Pareto chart*. This analysis clearly indicates that supplier A provides a disproportionately large share of the defective components.

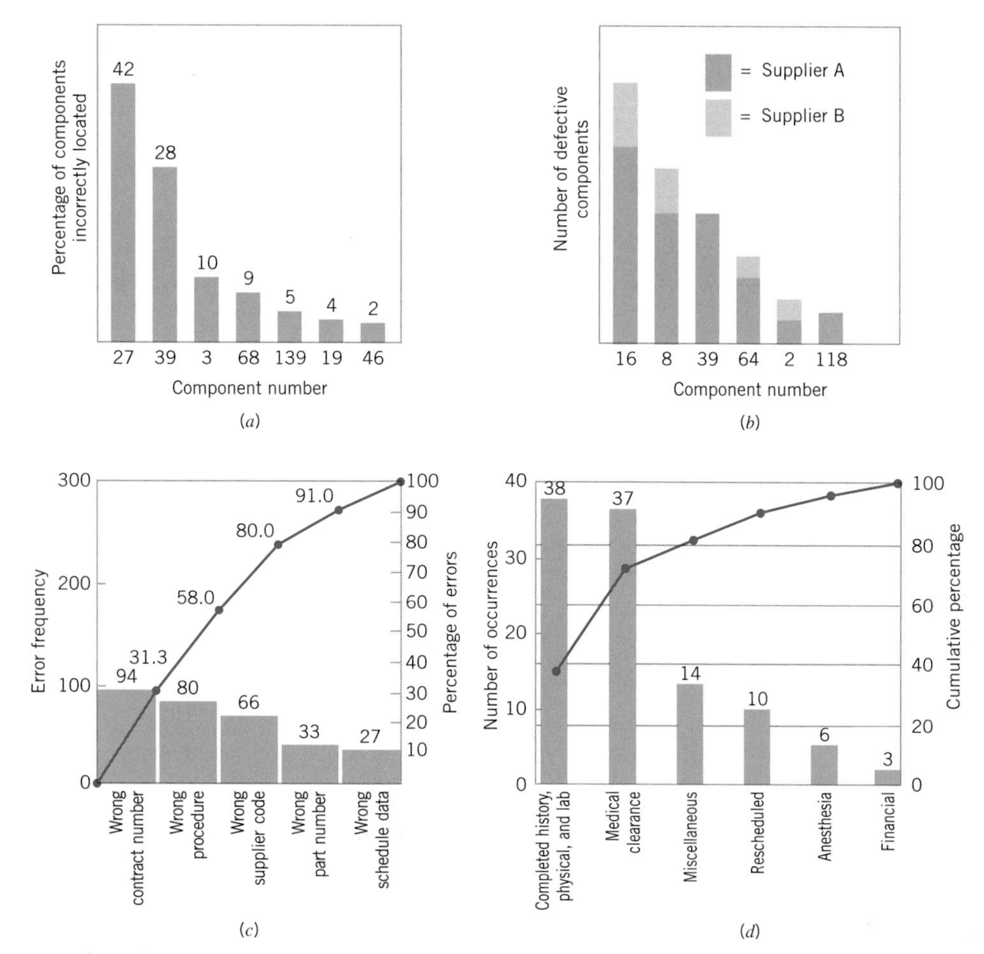

Figure 3.10 Examples of Pareto Charts

Pareto charts are widely used in **nonmanufacturing applications** of quality improvement methods. A Pareto chart used by a quality improvement team in a procurement organization is shown in Figure 3.10*c*. The team was investigating errors on purchase orders in an effort to reduce the organization's number of purchase order changes. (Each change typically cost between \$100 and \$500, and the organization issued several hundred purchase order changes each month.) This Pareto chart has two scales: one for the actual error frequency and another for the percentage of errors. Figure 3.10*d* presents a Pareto chart constructed by a quality improvement team in a hospital to reflect the reasons for cancellation of scheduled outpatient surgery.

In general, the Pareto chart is one of the most useful of the magnificent seven. Its applications to quality improvement are limited only by the ingenuity of the analyst.

Cause-and-Effect (or Ishikawa) Diagram. Once a defect, error, or problem has been identified and isolated for further study, we must begin to analyze potential causes of this undesirable effect. In situations where causes are not obvious (sometimes they are), the **cause-and-effect diagram** is a formal tool frequently useful in unlayering

Figure 3.11 Cause-and-Effect Diagram for the Tank Defect Problem

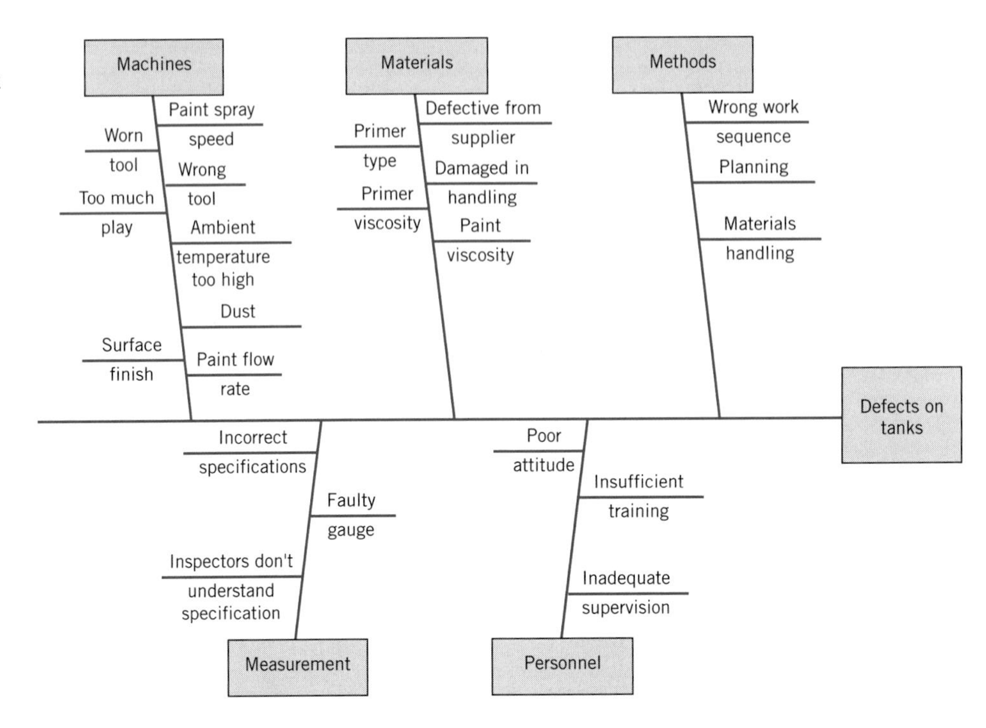

> **How to Construct a Cause-and-Effect Diagram**
>
> 1. Define the problem or effect to be analyzed.
> 2. Form the team to perform the analysis. Often the team will uncover potential causes through brainstorming.
> 3. Draw the effect box and the center line.
> 4. Specify the major potential cause categories and join them as boxes connected to the center line.
> 5. Identify the possible causes and classify them into the categories in step 4. Create new categories, if necessary.
> 6. Rank order the causes to identify those that seem most likely to impact the problem.
> 7. Take corrective action.

potential causes. The cause-and-effect diagram is very useful in the analyze and improve steps of DMAIC. The cause-and-effect diagram constructed by a quality improvement team assigned to identify potential problem areas in the tank manufacturing process mentioned earlier is shown in Figure 3.11.

In analyzing the tank defect problem, the team elected to lay out the major categories of tank defects as machines, materials, methods, personnel, measurement, and environment. A brainstorming session ensued to identify the various subcauses in each of these major categories and to prepare the diagram in Figure 3.11. Then, through discussion and the process of elimination, the group decided that materials and methods contained the most likely cause categories.

Cause-and-effect analysis is an extremely powerful tool. A highly detailed cause-and-effect diagram can serve as an effective troubleshooting aid. Furthermore, the construction of a cause-and-effect diagram as a **team experience** tends to get people involved in attacking a problem rather than in affixing blame.

Defect Concentration Diagram. A **defect concentration diagram** is a picture of the unit showing all relevant views. Then the various types of defects are drawn on the picture, and the diagram is analyzed to determine whether the **location** of the defects on the unit conveys any useful information about the potential causes of the defects. Defect concentration diagrams are very useful in the analyze step of DMAIC.

Figure 3.12 presents a defect concentration diagram for the final assembly stage of a refrigerator manufacturing process. Surface-finish defects are identified by the dark shaded areas on the refrigerator. From inspection of the diagram it seems clear that materials handling is responsible for the majority of these defects. The unit is being moved by securing a belt around the middle, and this belt is either too loose or tight, worn out, made of abrasive material, or too narrow. Furthermore, when the unit is moved, the corners are being damaged. It is possible that worker fatigue is a factor. In any event, proper work methods and improved materials handling will likely improve this process dramatically.

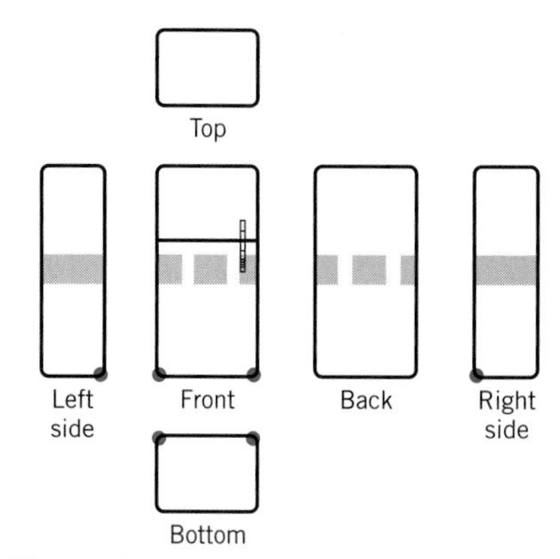

Figure 3.12 Surface-Finish Defects on a Refrigerator

Figure 3.13 shows the defect concentration diagram for the tank problem mentioned earlier. Note that this diagram shows several different broad categories of defects, each identified with a specific code. Often different colors are used to indicate different types of defects.

When defect data are portrayed on a defect concentration diagram over a sufficient number of units, patterns frequently emerge, and the location of these patterns often contains much information about the causes of the defects. We have found defect concentration diagrams to be important problem-solving tools in many industries, including plating, painting and coating, casting and foundry operations, machining, and electronics assembly.

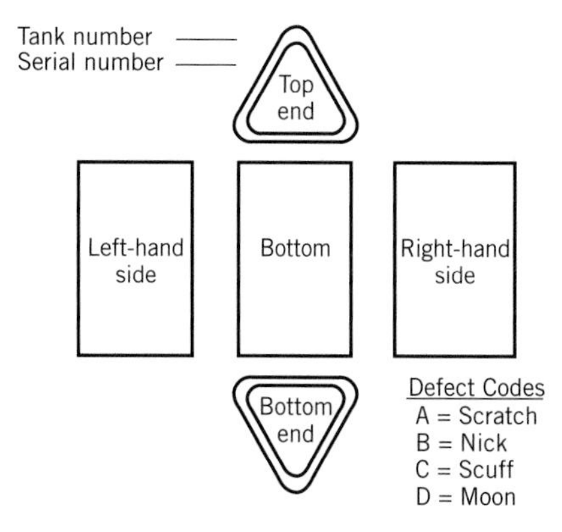

Figure 3.13 Defect Concentration Diagram for the Tank

Scatter Diagram. The **scatter diagram** is a useful plot for identifying a potential relationship between two variables. Data are collected in pairs on the two variables—say, (y_i, x_i)—for $i = 1, 2, \ldots, n$. Then y_i is plotted against the corresponding x_i. The shape of the scatter diagram often indicates what type of relationship may exist between the two variables.

Figure 3.14 shows a scatter diagram relating metal recovery (in percent) from a magnathermic smelting process for magnesium against corresponding values of the amount of reclaim flux added to the crucible. The scatter diagram indicates a strong **positive correlation** between metal recovery and flux amount; that is, as the amount of flux added is increased, the metal recovery also increases. It is tempting to conclude that the relationship is one based on cause and effect: By increasing the amount of reclaim flux used, we can always ensure high metal recovery. This thinking is potentially dangerous, because ***correlation** does not necessarily imply* **causality.** This apparent relationship could be caused by something quite different. For example, both variables could be related to a third one, such as the temperature of the metal prior to the reclaim pouring operation, and this relationship could be responsible for what we see in Figure 3.14. If higher temperatures lead to higher metal recovery and the practice is to add reclaim flux in proportion to temperature, adding more flux when the process is running at low temperatures will do nothing to enhance yield. The scatter diagram is useful for identifying **potential relationships. Designed experiments** [see Montgomery (2009)] **must be used to verify causality.**

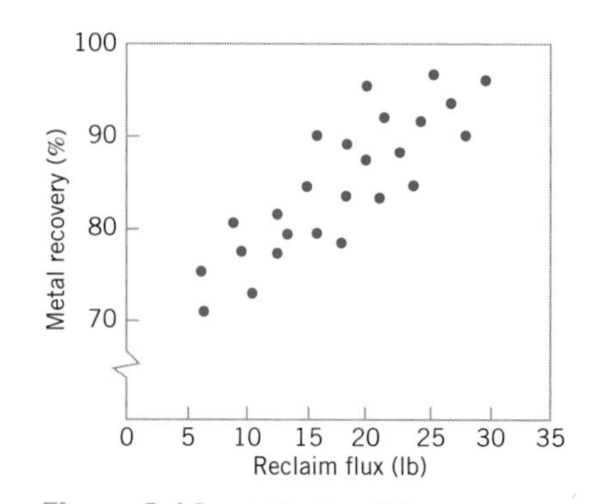

Figure 3.14 A Scatter Diagram

3.5 Implementing SPC in a Quality Improvement Program

The methods of statistical process control can provide significant payback to those companies that can successfully implement them. Although SPC seems to be a collection of statistically based problem-solving tools, there is more to the successful use of SPC than learning and using these tools. SPC is most effective when it is integrated into an overall, companywide quality improvement program. It can be implemented using the DMAIC approach. Indeed, the basic SPC tools are an integral part of DMAIC. Management involvement and commitment to the quality improvement process are the most vital components of SPC's potential success. Management is a role model, and others in the organization look to management for guidance and as an example. A team approach is also important, as it is usually difficult for one person alone to introduce process improvements. Many of the magnificent seven are helpful in building an improvement team, including cause-and-effect diagrams, Pareto charts, and defect concentration diagrams. This team approach also fits well with DMAIC. The basic SPC problem-solving tools must become widely known and widely used throughout the organization. Ongoing education of personnel about SPC and other methods for reducing variability are necessary to achieve this widespread knowledge of the tools.

The objective of an SPC-based variability reduction program is continuous improvement on a weekly, quarterly, and annual basis. SPC is not a one-time program to be applied when the business is in trouble and later abandoned. Quality improvement that is focused on continuous improvement and reduction of variability must become part of the culture of the organization.

The control chart is an important tool for process improvement. Processes do not naturally operate in an in-control state, and the use of control charts is an important step that must be taken early in an SPC program to eliminate assignable causes, reduce process variability, and stabilize process performance. To improve quality and productivity, we must begin to manage with facts and data, and not simply rely on judgment. Control charts are an important part of this change in management approach.

In implementing a companywide effort to reduce variability and improve quality, we have found that several elements are usually present in all successful efforts. These elements are shown on the left:

Elements of a Successful SPC Program

1. Management leadership
2. A team approach, focusing on project-oriented applications
3. Education of employees at all levels
4. Emphasis on reducing variability
5. Measuring success in quantitative (economic) terms
6. A mechanism for communicating successful results throughout the organization

We cannot overemphasize the importance of **management leadership** and the **team/project approach.** Successful quality improvement is a "top-down" management-driven activity. It is also important to measure progress and success in quantitative (economic) terms and to spread knowledge of this success throughout the organization. When successful improvements are communicated throughout the company, this can provide motivation and incentive to improve other processes and to make continuous improvement a usual part of the way of doing business.

3.6 An Application of SPC

In this section, we give an account of how SPC methods were used to improve quality and productivity in a copper plating operation at a printed circuit board fabrication facility. This process was characterized by high levels of defects such as brittle copper and copper voids and by long cycle time. The long cycle time was particularly troublesome, as it had led to an extensive work backlog and was a major contributor to poor conformance to the factory production schedule.

Management chose this process area for an initial implementation of SPC. The DMAIC approach was used. An improvement team was formed, consisting of the plating tank operator, the manufacturing engineer responsible for the process, and a quality engineer. All members of the team had been exposed to DMAIC and the magnificent seven in a company-sponsored seminar. During the define step, it was decided to concentrate on reducing the flow time through the process, as the missed delivery targets were considered to be the most serious obstacle to improving productivity. The team quickly determined (during the measure step) that excessive downtime on the controller that regulated the copper concentration in the plating tank was a major factor in the excessive flow time; controller downtime translated directly into lost production.

As part of the analyze step, the team decided to use a cause-and-effect analysis to begin to isolate the potential causes of controller downtime. Figure 3.15 shows the cause-and-effect diagram that was produced during a brainstorming session. The team was able to quickly identify 11 major potential causes of controller downtime. However, when they examined the equipment logbook to make a more definitive diagnosis of the causes of downtime based on actual process performance, the results were disappointing. The logbook contained little useful information about causes of downtime; instead, it contained only a chronological record of when the machine was up and when it was down.

The team then decided that it would be necessary to collect valid data about the causes of controller downtime. They designed the check sheet

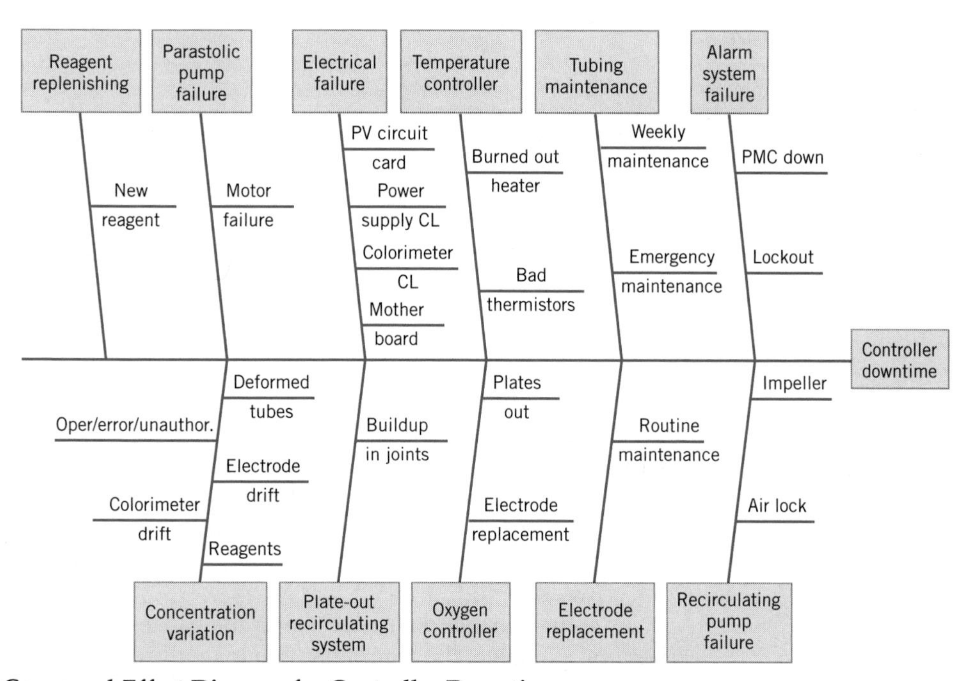

Figure 3.15 **Cause-and-Effect Diagram for Controller Downtime**

WEEKLY TALLY	OPERATOR ____________		
WEEK ENDING ________	ERRORS	DESCRIPTION	ACTION
1. CONCENTRATION VARIATION			
a. Colorimeter drift	________		
b. Electrode failure	________		
c. Reagents	________		
d. Deformed tubes	________		
e. Oper/error/unauthorized	________		
2. ALARM SYSTEM FAILURE			
a. PMC down	________		
b. Lockout	________		
3. RECIRCULATING PUMP FAILURE			
a. Air lock	________		
b. Impeller	________		
4. REAGENT REPLENISHING			
a. New reagent	________		
5. TUBING MAINTENANCE			
a. Weekly maintenance	________		
b. Emergency maintenance	________		
6. ELECTRODE REPLACEMENT			
a. Routine maintenance	________		
7. TEMPERATURE CONTROLLER			
a. Burned out heater	________		
b. Bad thermistors	________		
8. OXYGEN CONTROLLER			
a. Plates out	________		
b. Electrode replacement	________		
9. PARASTOLIC PUMP FAILURE			
a. Motor failure	________		
10. ELECTRICAL FAILURE			
a. PV circuit card	________		
b. Power supply CL	________		
c. Colorimeter CL	________		
d. Motherboard	________		
11. PLATE-OUT RECIRCULATING			
a. Buildup at joints	________		
TOTAL COUNT			

Figure 3.16 Check Sheet for Logbook

shown in Figure 3.16 as a supplemental page for the logbook. The team agreed that whenever the equipment was down, one team member would assume responsibility for filling out the check sheet. Note that the major causes of controller downtime identified on the cause-and-effect diagram were used to structure the headings and subheadings on the check sheet. The team agreed that data would be collected over a four- to six-week period.

As more reliable data concerning the causes of controller downtime became available, the team was able to analyze it using other SPC techniques. Figure 3.17 presents the Pareto analysis of the controller failure

Percent of downtime

Category	Percent of downtime
Concentration variation	36
Tubing maintenance	16
Reagent replenish	14
Rec. pump failure	9
Par. pump failure	6
Plate-out	5
Oxygen controller	5
Alarm failure	5
Electrical failure	2
Replace electrode	2

Figure 3.17 **Pareto Analysis of Controller Failures**

data produced during the six-week study of the process. Note that concentration variation is a major cause of downtime. Actually, the situation is probably more complex than it appears. The third largest category of downtime causes is reagent replenishment. Frequently, the reagent in the colorimeter on the controller is replenished because concentration has varied so far outside the process specifications that reagent replenishment and colorimeter recalibration is the only step that can be used to bring the process back on-line. Therefore, it is possible that up to 50% of the downtime associated with controller failures can be attributed to concentration variation. Figure 3.18 presents a Pareto analysis of only the concentration variation data. From this diagram we know that colorimeter drift and problems with reagents are major causes of concentration variation. This information led the manufacturing engineer on the team to conclude that rebuilding the colorimeter would be an important step in improving the process.

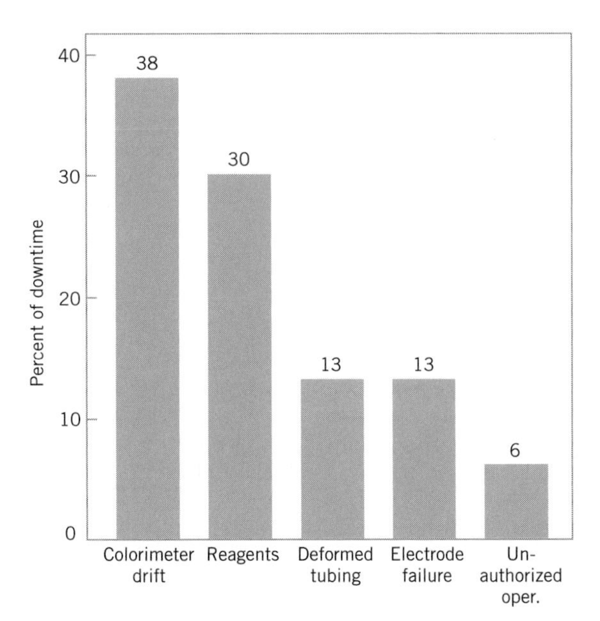

Figure 3.18 **Pareto Analysis of Concentration Variation**

Figure 3.19 $\bar{x}$ **Chart for the Average Daily Copper Concentration**

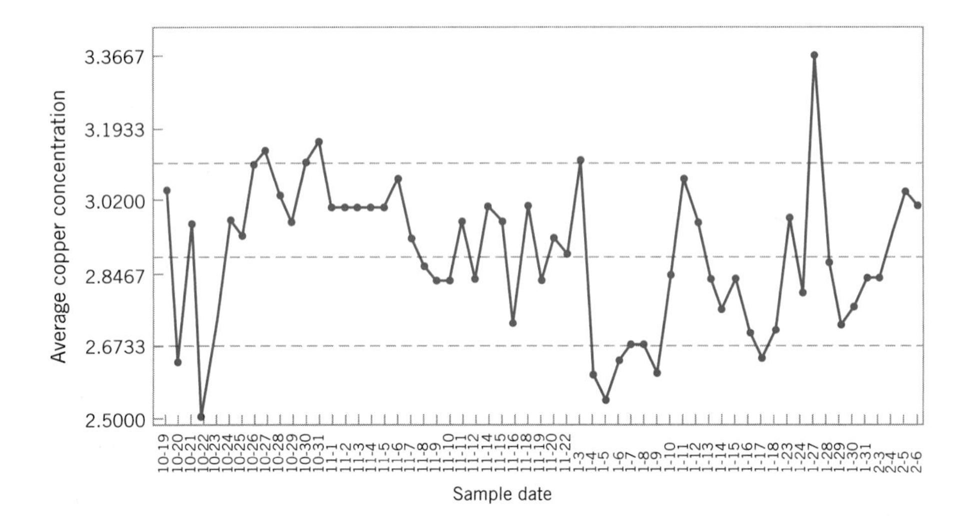

During the time that these process data were collected, the team decided to run statistical control charts on the process. The information collected to this point about process performance was the basis for constructing the initial OCAPs (out-of-control-action plans) for these control charts. These control charts and their OCAP would also be useful in the control step of DMAIC. Copper concentration is measured in this process manually three times per day. Figure 3.19 presents the $\bar{x}$ control chart for average daily copper concentration; that is, each point plotted in the figure is a daily average. The chart shows the center line and three-sigma statistical control limits. (We will discuss the construction of these limits in more detail in the next few chapters.) Note that there are a number of points outside the control limits, indicating that assignable causes are present in the process. Figure 3.20 presents the range or R chart for daily copper concentration. On this chart, R represents the difference between the maximum and minimum copper concentration readings in a day. Note that the R chart also exhibits a lack of statistical control. In particular, the second half of the R chart appears much more unstable than the first half. Examining the dates along the horizontal

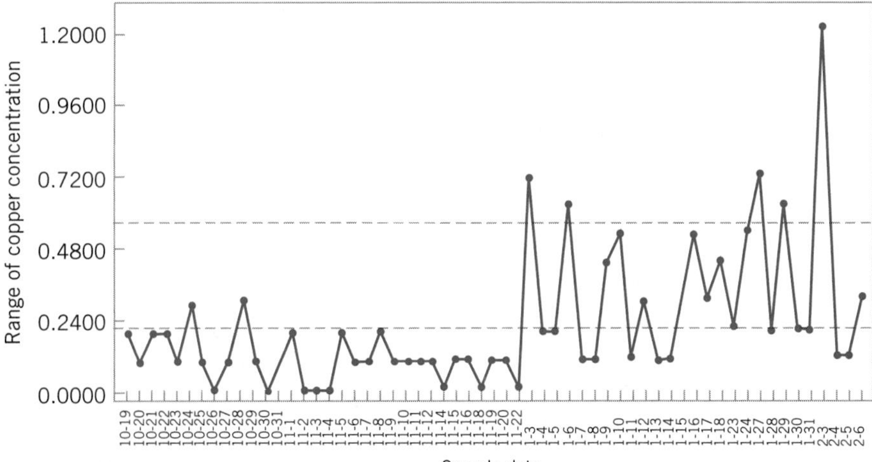

Figure 3.20 ***R*** **Chart for Daily Copper Concentration**

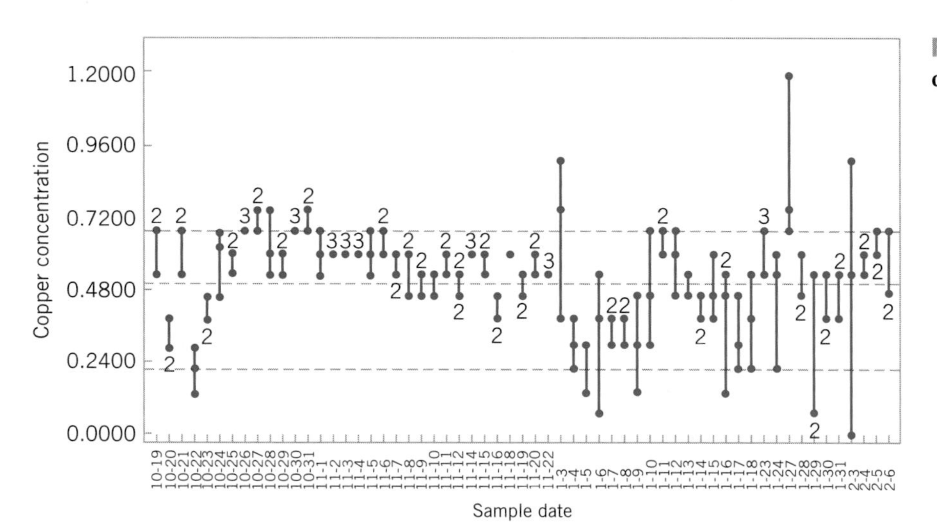

Figure 3.21 Tolerance Diagram of Daily Copper Concentration

axis, the team noted that severe variation in average daily copper concentration only appeared after January 3. The last observations on copper concentration had been taken on November 22. From November 23 until January 3 the process had been in a shutdown mode because of holidays. Apparently, when the process was restarted, substantial deterioration in controller/colorimeter performance had occurred. This hastened engineering's decision to rebuild the colorimeter.

Tolerance Diagram

Figure 3.21 presents a **tolerance diagram** of daily copper concentration readings. In this figure, each day's copper concentration readings are plotted, and the extremes are connected with a vertical line. In some cases, more than one observation is plotted at a single position, so a numeral is used to indicate the number of observations plotted at each particular point. The center line on this chart is the process average over the time period studied, and the upper and lower limits are the specification limits on copper concentration. Every instance in which a point is outside the specification limits would correspond to nonscheduled downtime on the process. Several things are evident from examining the tolerance diagram. First, the process average is significantly different from the nominal specification on copper concentration (the midpoint of the upper and lower tolerance band). This implies that the calibration of the colorimeter may be inadequate. That is, we are literally aiming at the wrong target. Second, we note that there is considerably more variation in the daily copper concentration readings after January 3 than there was prior to shutdown. Finally, if we could reduce variation in the process to a level roughly consistent with that observed prior to shutdown and correct the process centering, many of the points outside specifications would not have occurred, and downtime on the process should be reduced.

To initiate the improve step, the team first decided to rebuild the colorimeter and controller. This was done in early February. The result of this maintenance activity was to restore the variability in daily copper concentration readings to the pre-shutdown level. The rebuilt colorimeter was recalibrated and subsequently was able to hold the correct target. This recentering and recalibration of the process reduced the downtime on the controller from approximately 60% to less than

20%. At this point, the process was capable of meeting the required production rate.

Once this aspect of process performance was improved, the team directed its efforts to reducing the number of defective units produced by the process. Generally, as noted earlier, defects fell into two major categories: brittle copper and copper voids. The team decided that, although control charts and statistical process-control techniques could be applied to this problem, the use of a **designed experiment** might lead to a more rapid solution. As noted previously, the objective of a designed experiment is to generate information that will allow us to understand and model the relationship between the process variables and measures of the process performance.

The designed experiment for the plating process is shown in Table 3.1 and Figure 3.22. The objective of this experiment was to provide information that would be useful in minimizing plating defects. The process

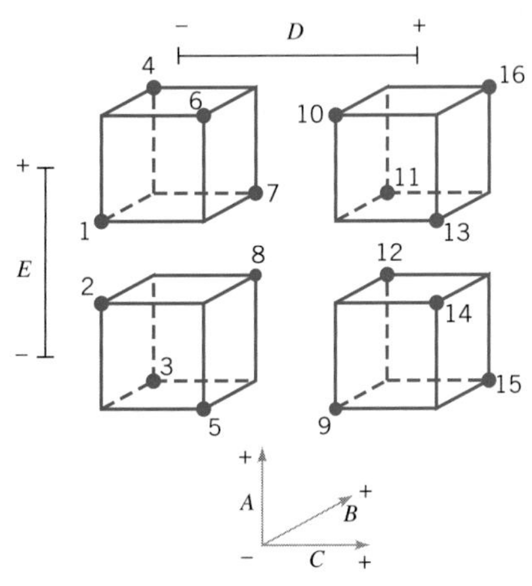

Figure 3.22 A Geometric View of the Fractional Factorial Design for the Plating Process Experiment

Table 3.1

A Designed Experiment for the Plating Process

Objective: Minimize Plating Defects

Process Variables	Low Level	High Level
A = Copper concentration	−	+
B = Sodium hydroxide concentration	−	+
C = Formaldehyde concentration	−	+
D = Temperature	−	+
E = Oxygen	−	+

Experimental Design

Run	Variables A	B	C	D	E	Response (Defects)
1	−	−	−	−	+	
2	+	−	−	−	−	
3	−	+	−	−	−	
4	+	+	−	−	+	
5	−	−	+	−	−	
6	+	−	+	−	+	
7	−	+	+	−	+	
8	+	+	+	−	−	
9	−	−	−	+	−	
10	+	−	−	+	+	
11	−	+	−	+	+	
12	+	+	−	+	−	
13	−	−	+	+	+	
14	+	−	+	+	−	
15	−	+	+	+	−	
16	+	+	+	+	+	

variables considered in the experiment were copper concentration, sodium hydroxide concentration, formaldehyde concentration, temperature, and oxygen. A low level and high level, represented symbolically by the minus and plus signs in Table 3.1, were chosen for each process variable. The team initially considered a **factorial experiment**—that is, an experimental design in which all possible combinations of these factor levels would be run. This design would have required 32 runs—that is, a run at each of the 32 corners of the cubes in Figure 3.22. Since this is too many runs, a **fractional factorial design** that used only 16 runs was ultimately selected. This fractional factorial design is shown in the bottom half of Table 3.1 and geometrically in Figure 3.22. In this experimental design, each row of the table is a run on the process. The combination of minus and plus signs in each column of that row determines the low and high levels of the five process variables to be used during that run. For example, in run 1 copper concentration, sodium hydroxide concentration, formaldehyde concentration, and temperature are run at the low level and oxygen is run at the high level. The process would be run at each of the 16 sets of conditions described by the design (for reasons to be discussed later, the runs would not be made in the order shown in Table 3.1), and a response variable—an observed number of plating defects—would be recorded for each run. Then these data could be analyzed using simple statistical techniques to determine which factors have a significant influence on plating defects, whether or not any of the factors jointly influence the occurrence of defects, and whether it is possible to adjust these variables to new levels that will reduce plating defects below their current level. Although a complete discussion of design of experiments is beyond the scope of this text, we will present examples of designed experiments for improving process performance in Chapter 8.

After the team conducted the experiment shown in Table 3.1 and analyzed the resulting process data, they determined that several of the process variables that they identified for the study were important and had significant impact on the occurrence of plating defects. They were able to adjust these factors to new levels, and as a result, plating defects were reduced by approximately a factor of 10. Therefore, at the conclusion of the team's initial effort at applying SPC to the plating process, it had made substantial improvements in product cycle time through the process and had taken a major step in improving the process capability.

Factorial Experiment
An experimental design in which all possible combinations of these factor levels would be run.

Fractional Factorial Design
An experimental design in which only a portion of all possible combination of factor levels is run.

3.7 Applications of Statistical Process Control and Quality Improvement Tools in Transactional and Service Businesses

This book presents the underlying principles of SPC. Many of the examples used to reinforce these principles are in an industrial, product oriented framework. There have been many successful applications of SPC methods in the manufacturing environment. However, the principles themselves are general; there are many applications of SPC techniques and other quality engineering and statistical tools in nonmanufacturing settings, including transactional and service businesses.

These nonmanufacturing applications do not differ substantially from the more usual industrial applications. As an example, the control chart for fraction nonconforming (which is discussed in Chapter 7) could be applied to reducing billing errors in a bank credit card operation as easily as it could be used to reduce the fraction of nonconforming printed circuit boards produced in an electronics plant. The $\bar{x}$ and R charts discussed in this chapter and applied to the hard-bake process could be used to monitor and control the flow time of accounts payable through a finance function. Transactional and service industry applications of SPC and related methodology sometimes require ingenuity beyond that normally required for the more typical manufacturing applications. There seems to be two primary reasons for this difference:

1. Most transactional and service businesses do not have a natural measurement system that allows the analyst to easily define quality.
2. The system that is to be improved is usually fairly obvious in a manufacturing setting, whereas the observability of the process in a nonmanufacturing setting may be fairly low.

For example, if we are trying to improve the performance of a personal computer assembly line, then it is likely that the line will be contained within one facility and the activities of the system will be readily observable. However, if we are trying to improve the business performance of a financial services organization, then the observability of the process may be low. The actual activities of the process may be performed by a group of people who work in different locations, and the operation steps or workflow sequence may be difficult to observe. Furthermore, the lack of a quantitative and objective measurement system in most nonmanufacturing processes complicates the problem.

The key to applying statistical process-control and other related methods in a nonmanufacturing environment is to focus initial efforts on resolving these two issues. We have found that once the system is adequately defined and a valid measurement system has been developed, most of the SPC tools discussed in this chapter can easily be applied to a wide variety of nonmanufacturing operations including finance, marketing, material and procurement, customer support, field service, engineering development and design, and software development and programming.

Flow charts, operation process charts, and **value stream mapping** are particularly useful in developing process definition and process understanding. A flow chart is simply a chronological sequence of process steps or work flow. Sometimes flow charting is called **process mapping.** Flow charts or process maps must be constructed in sufficient detail to identify **value-added** versus **non-value-added** work activity in the process.

Most nonmanufacturing processes have scrap, rework, and other **non-value-added operations,** such as unnecessary work steps and choke points or bottlenecks. A systematic analysis of these processes can often eliminate many of these non-value-added activities. The flow chart is helpful in visualizing and defining the process so that non-value-added activities can be identified. Some ways to remove non-value-added activities and simplify the process are summarized and listed.

Ways to Eliminate Non-Value-Added Activities

1. Rearrange the sequence of worksteps
2. Rearrange the physical location of the operator in the system
3. Change work methods
4. Change the type of equipment used in the process
5. Redesign forms and documents for more efficient use
6. Improve operator training
7. Improve supervision
8. Identify more clearly the function of the process to all employees
9. Try to eliminate unnecessary steps
10. Try to consolidate process steps

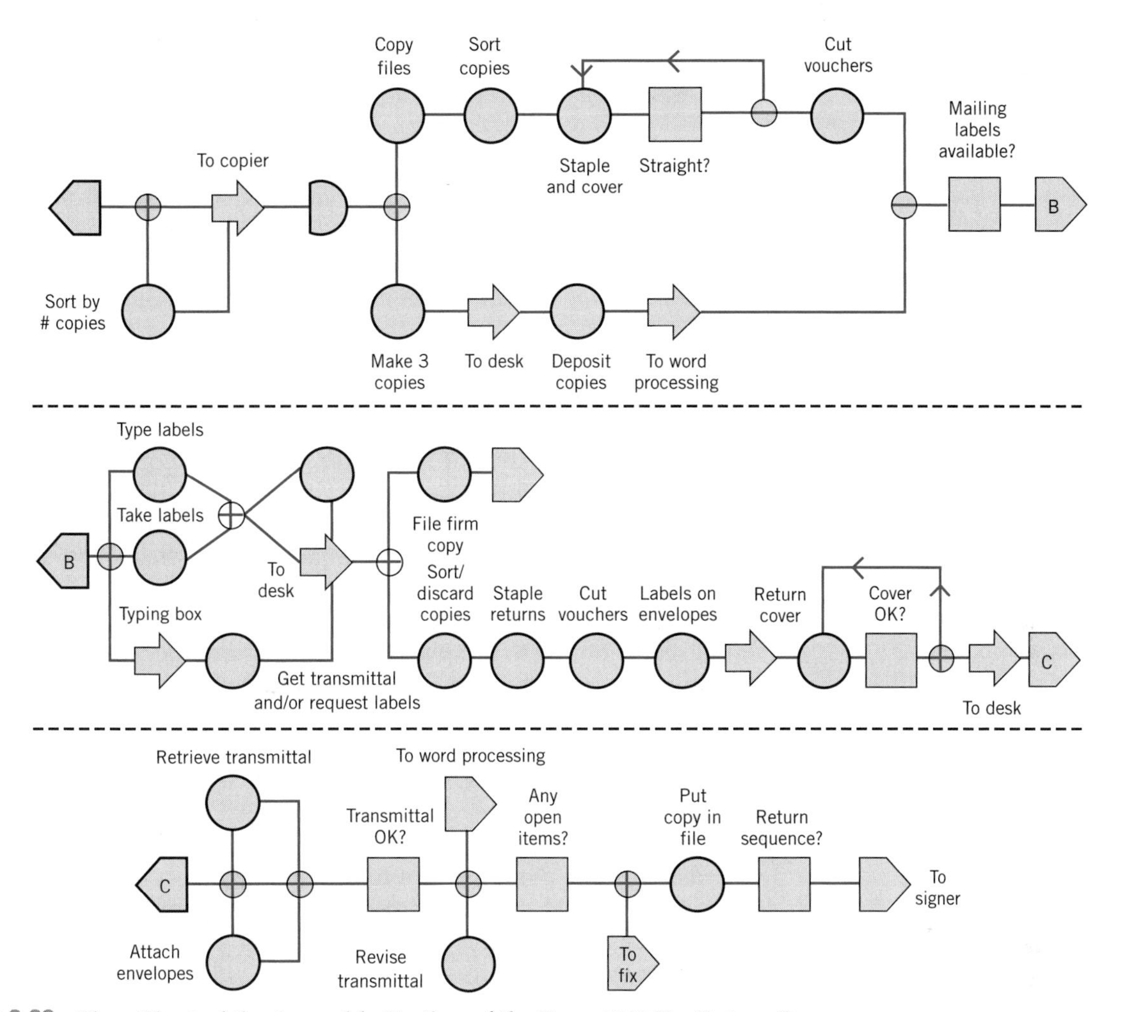

Figure 3.23 Flow Chart of the Assembly Portion of the Form 1040 Tax Return Process

Figure 3.23 is an example of a flow chart for a process in a service industry. It was constructed by a process improvement team in an accounting firm that was studying the process of preparing Form 1040 income tax returns; this particular flow chart documents only one particular subprocess, that of assembling final tax documents. This flow chart was constructed as part of the define step of DMAIC. Note the high level of detail in the flow chart to assist the team find waste or non-value-added activities. In this example, the team used special symbols in their flow chart. Specifically, they used the **operation process chart** symbols shown as follows:

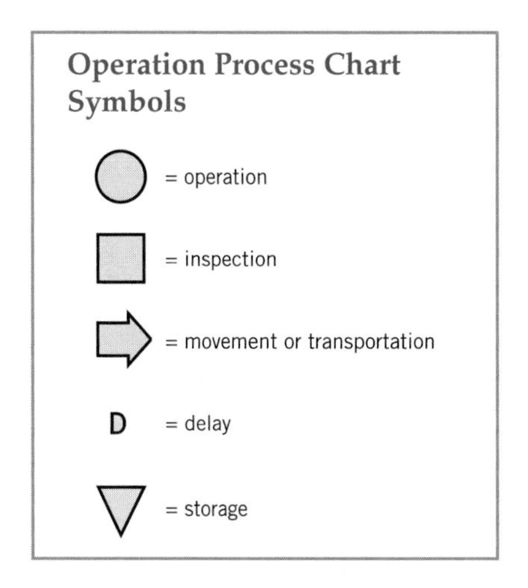

We have found that these symbols are very useful in helping team members identify improvement opportunities. For example, delays, most inspections, and many movements usually represent non-value-added activities. The accounting firm was able to use quality improvement methods and the DMAIC approach successfully in their Form 1040 process, reducing the tax document preparation cycle time (and work content) by about 25%, and reducing the cycle time for preparing the client bill from over 60 days to 0 (that's right, 0!). The client's bill is now included with his or her tax return.

Figure 3.24 **A High-Level Flow Chart of the Planning Process**

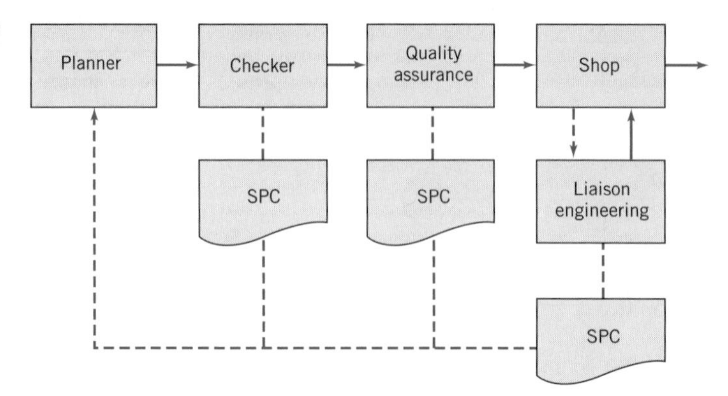

As another illustration, consider an example of applying quality improvement methods in a planning organization. This planning organization, part of a large aerospace manufacturing concern, produces the plans and documents that accompany each job to the factory floor. The plans are quite extensive, often several hundred pages long. Errors in the planning process can have a major impact on the factory floor, contributing to scrap and rework, lost production time, overtime, missed delivery schedules, and many other problems.

Figure 3.24 presents a high-level flow chart of this planning process. After plans are produced, they are sent to a checker who tries to identify obvious errors and defects in the plans. The plans are also reviewed by a quality-assurance organization to ensure that process specifications are being met and that the final product will conform to engineering standards. Then the plans are sent to the shop, where a liaison engineering organization deals with any errors in the plan encountered by manufacturing. This flow chart is useful in presenting an overall picture of the planning system, but it is not particularly helpful in uncovering non-value-added activities, as there is insufficient detail in each of the major blocks. However, each block, such as the planner, checker, and quality-assurance block, could be broken down into a more detailed sequence of work activities and steps. The step-down approach is frequently helpful in constructing flow charts for complex processes. However, even at the relatively high level shown, it is possible to identify at least three areas in which SPC methods could be usefully applied in the planning process.

The management of the planning organization decided to use the reduction of planning errors as a quality improvement project for their organization. A team of managers, planners, and checkers was chosen to begin this implementation. During the measure step, the team decided that each week three plans would be selected at random from the week's output of plans to be analyzed extensively to record all planning errors that could be found. The check sheet shown in Figure 3.25 was used to record the errors found in each plan. These weekly data were summarized monthly, using the summary check sheet presented in Figure 3.26. After several weeks, the team was able to summarize the planning error data obtained using the Pareto analysis in Figure 3.27. The Pareto chart implies that errors in the operations section of the plan are predominant, with 65% of the planning errors in the operations section. Figure 3.28 presents a further Pareto analysis of the operations

DATA SHEET P/N ______			
	ERRORS	DESCRIPTION	ACTION
1. HEADER SECT.			
a. PART NO.			
b. ITEM			
c. MODEL			
2. DWG/DOC SECT.			
3. COMPONENT PART SECT.			
a. PROCUREMENT CODES			
b. STAGING			
c. MOA (#SIGNS)			
4. MOTE SECT.			
5. MATERIAL SECT.			
a. MCC CODE (NON MP&R)			
6. OPERATION SECT.			
a. ISSUE STORE(S)			
b. EQUIPMENT USAGE			
c. OPC FWC MNEMONICS			
d. SEQUENCING			
e. OPER'S OMITTED			
f. PROCESS SPECS			
g. END ROUTE STORE			
h. WELD GRID			
7. TOOL/SHOP AIDS ORDERS			
8. CAR/SHOP STOCK PREP.			
REMARKS: NO. OF OPERATIONS ______		CHECKER ______ DATE ______	

Figure 3.25 **The Check Sheet for the Planning Example**

section errors, showing that omitted operations and process specifications are the major contributors to the problem.

The team decided that many of the operations errors were occurring because planners were not sufficiently familiar with the manufacturing operations and the process specifications that were currently in place. To improve the process, a program was undertaken to refamiliarize planners with the details of factory floor operations and to provide more feedback on the type of planning errors actually experienced. Figure 3.29 presents a run chart of the planning errors per operation for 25 consecutive weeks. Note that there is a general tendency for the planning errors per operation to decline over the first half of the study period. This decline may be due partly to the increased training and supervision activities for the planners and partly to the additional feedback given regarding the types of planning errors that were occurring. The team also recommended that substantial changes be made in the work

Monthly Data Summary					
1. HEADER SECT.					
a. PART NO.					
b. ITEM					
c. MODEL					
2. DWG/DOC SECT.					
3. COMPONENT PART SECT.					
a. PROCUREMENT CODES					
b. STAGING					
c. MOA (#SIGNS)					
4. MOTE SECT.					
5. MATERIAL SECT.					
a. MCC CODE (NON MP&R)					
6. OPERATION SECT.					
a. ISSUE STORE(S)					
b. EQUIPMENT USAGE					
c. OPC FWC MNEMONICS					
d. SEQUENCING					
e. OPER'S OMITTED					
f. PROCESS SPECS					
g. END ROUTE STORE					
h. WELD GRID					
7. TOOL/SHOP AIDS ORDERS					
TOTAL NUMBER ERRORS					
TOTAL OPERATIONS CHECKED					
WEEK ENDING					

Figure 3.26 **The Summary Check Sheet**

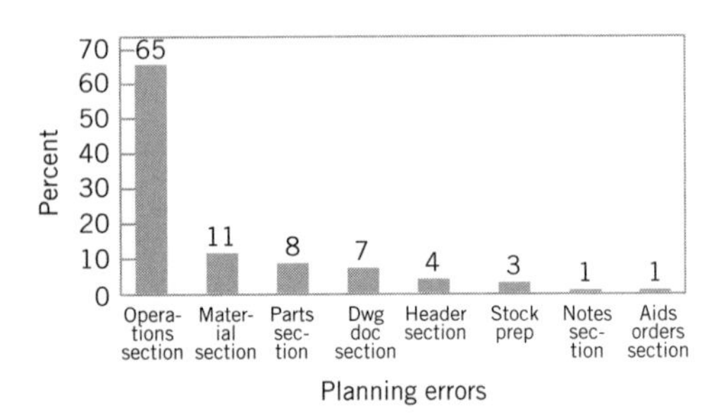

Figure 3.27 **Pareto Analysis of Planning Errors**

methods used to prepare plans. Rather than having an individual planner with overall responsibility for the operations section, it recommended that this task become a team activity so that knowledge and experience regarding the interface between factor and planning operations could be shared in an effort to further improve the process.

The planning organization began to use other SPC tools as part of their quality improvement effort. For example, note that the run chart in Figure 3.29 could be converted to a Shewhart control chart with the addition of a center line and appropriate control limits. Once the planners were exposed to the concepts of SPC, control charts came into use in the organization and proved effective in identifying assignable causes; that is, periods of time in which the error rates produced by the system were higher than those that could be justified by chance cause alone. It is its ability to differentiate between assignable and chance causes that makes the control chart so indispensable.

Management must react differently to an assignable cause than it does to a chance or random cause. Assignable causes are due to phenomena external to the system, and they must be tracked down and their root causes eliminated. Chance or random causes are part of the system itself. They can only be reduced or eliminated by making changes in how the system operates. This may mean changes in work methods and procedures, improved levels of operator training, different types of equipment and facilities, or improved input materials, all of which are the responsibility of management. In the planning process, many of the common causes identified were related to the experience, training, and supervision of the individual planners, as well as poor input information from design and development engineering. These common causes were systematically removed from the process, and the long-term impact of the SPC implementation in this organization was to reduce planning errors to a level of less than one planning error per 1,000 operations.

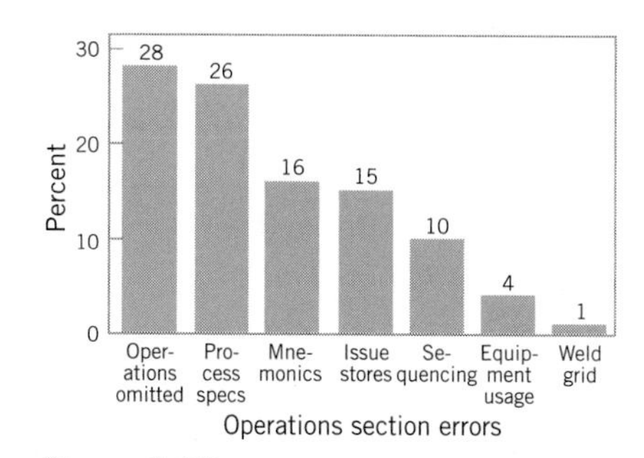

Figure 3.28 **Pareto Analysis of Operations Section Errors**

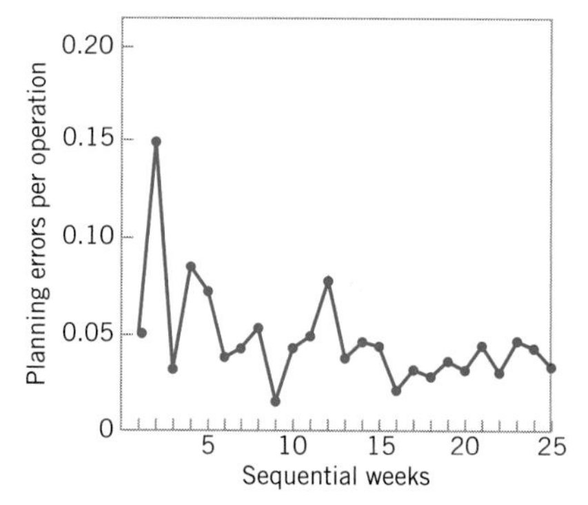

Figure 3.29 **A Run Chart of Planning Errors**

Value stream mapping is another way to see the flow of material and information in a process. A value stream map is much like a flow chart, but it usually incorporates other information about the activities that are occurring at each step in the process and the information that is required or generated. It is a big picture tool that helps an improvement team focus on optimizing the entire process, without focusing too narrowly on only one process activity or step which could lead to sub-optimal solutions.

Like a flow chart or operations process chart, a value stream map usually is constructed using special symbols. The box below presents the symbols usually employed on value stream maps.

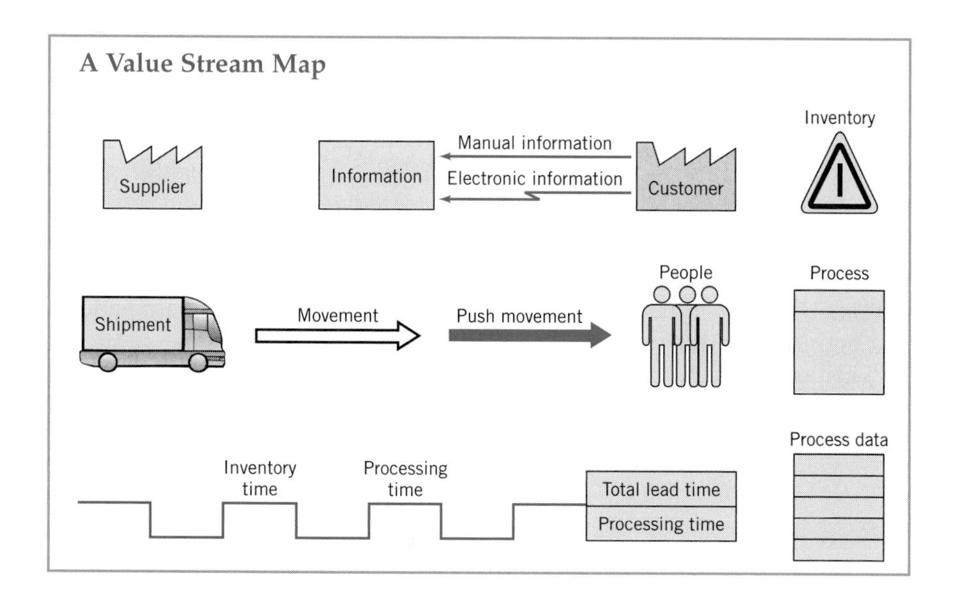

The value stream map presents a picture of the value stream from the product's viewpoint: It is not a flow chart of what people do, but what actually happens to the product. It is necessary to collect process data to construct a value stream map. Some of the data typically collected includes:

1. *Lead time (LT)*	The elapsed time it takes one unit of product to move through the entire value stream from beginning to end.
2. *Processing time (PT)*	The elapsed time from the time the product enters a process until it leaves that process.
3. *Cycle time (CT)*	How often a product is completed by a process. Cycle time is a rate, calculated by dividing the processing time by the number of people or machines doing the work.
4. *Setup time (ST)*	These are activities such as loading/unloading, machine preparation, testing, and trial runs. In other words, all activities that take place between completing a good product until starting to work on the next unit or batch of product.
5. *Available time (AT)*	The time each day that the value stream can operate if there is product to work on.
6. *Uptime (UT)*	The percent of time the process actually operates as compared to the available time or planned operating time.
7. *Pack size*	The quantity of product required by the customer for shipment.
8. *Batch size*	The quantity of product worked on and moved at one time.
9. *Queue time*	The time a product spends waiting for processing.
10. *Work-in-process (WIP)*	Product that is being processed but is not yet complete.
11. *Information flows*	Schedules, forecasts, and other information that tells each process what to do next.

Figure 3.30 shows an example of a value stream map that could be almost anything from a manufactured product (receive parts, preprocess parts, assemble the product, pack and ship the product to the customer) to a transaction (receive information, preprocess information, make calculations and decision, inform customer of decision or results). Notice that in the example we have allocated the setup time on a per-piece basis and included that in the timeline. This is an example of a **current-state value stream map.** That is, it shows what is happening in the process as it is now defined. The DMAIC process can be useful in eliminating waste and inefficiencies in the process, eliminating defects and rework, reducing delays, eliminating non-value-added activities, reducing inventory (WIP, unnecessary backlogs), reducing inspections, and reducing unnecessary product movement. There is a lot of opportunity for improvement in this process, because the process cycle efficiency isn't very good. Specifically,

Process Cycle Efficiency

$$\text{Process cycle efficiency} = \frac{\text{Value-add time}}{\text{Process cycle time}} = \frac{35.5}{575.5} = 0.0617$$

Reducing the amount of work-in-process inventory is one approach that would improve the process cycle efficiency. As a team works on improving a process, often a **future-state value stream map** is constructed to show what a redefined process should look like.

Finally, there are often questions about how the technical quality improvement tools in this book can be applied in service and

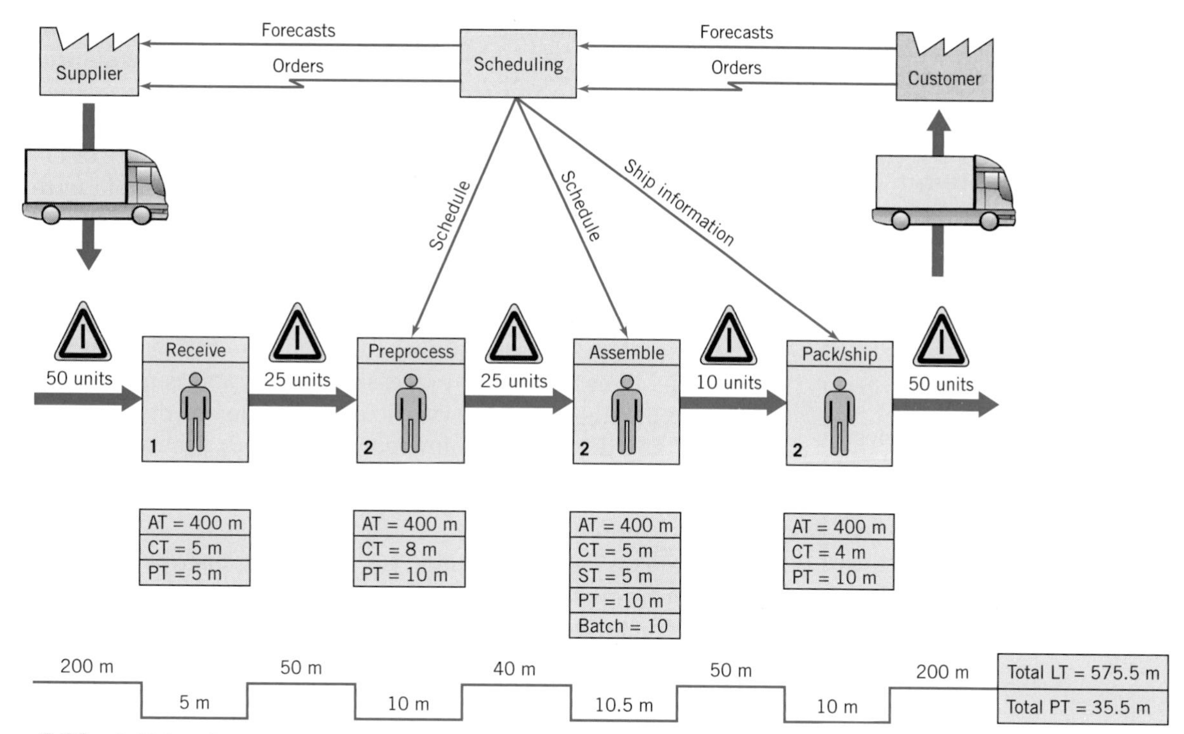

Figure 3.30 **A Value Stream Map**

transactional businesses. In practice, almost all of the techniques translate directly to these types of businesses. For example, designed experiments have been applied in banking, finance, marketing, health care, and many other service/transactional businesses. Designed experiments can be used in any application where we can manipulate the decision variables in the process. Sometimes we will use a **simulation model** of the process to facilitate conducting the experiment. Similarly, control charts have many applications in the service economy, as will be illustrated in this book. It is a big mistake to assume that these techniques are not applicable just because you are not working in a manufacturing environment.

Still, one difference in the service economy is that you are more likely to encounter attribute data. Manufacturing often has lots of continuous measurement data, and it is often safe to assume that these data are at least approximately normally distributed. However, in service and transactional processes, more of the data that you will use in quality improvement projects is either proportion defective, percent good, or counts of errors or defects. In Chapter 6, we discuss control charting procedures for dealing with attribute data. These control charts have many applications in the service economy. However, even some of the continuous data encountered in service and transactional businesses, such as cycle time, may not be normally distributed.

Let's talk about the normality assumption. It turns out that many statistical procedures (such as the *t*-tests and analysis of variance (ANOVA) are very insensitive to the normality assumption. That is, moderate departures from normality have little impact on their effectiveness. There are some procedures that are fairly sensitive to

In extreme cases, there are **nonparametric statistical procedures** that don't have an underlying assumption of normality and can be used as alternatives to procedure such as t-tests and ANOVA. Refer to Montgomery and Runger (2007) for an introduction to many of these techniques. Many computer software packages such as Minitab have nonparametric methods included in their libraries of procedures. There are also special statistical tests for binomial parameters and Poisson parameters.

normality, such as tests on variances, and this book carefully identifies such procedures. One alternative to dealing with moderate to severe non-normality is to **transform** the original data (say, by taking logarithms) to produce a new set of data whose distribution is closer to normal. A disadvantage of this is that nontechnical people often don't understand data transformation and are not comfortable with data presented in an unfamiliar scale. One way to deal with this is to perform the statistical analysis using the transformed data, but to present results (graphs, for example) with the data in the original units.

It also is important to be clear about to what the normality assumption applies. For example, suppose that you are fitting a linear regression model to cycle time to process a claim in an insurance company. The cycle time is y, and the predictors are different descriptors of the customer and the type of claim being processed. The model is

$$y = \beta_0 + \beta_1 x_1 + \beta_2 x_2 + \beta_3 x_3 + \varepsilon$$

The data on y, the cycle time, isn't normally distributed. Part of the reason for this is that the observations on y are impacted by the values of the predictor variables, x_1, x_2, and x_3. It is the errors in this model that need to be approximately normal, not the observations on y. That is why we analyze the residuals from regression and ANOVA models. If the residuals are approximately normal, there are no problems. Transformations are a standard procedure that can often be used successfully when the residuals indicate moderate to severe departures from normality.

There are situations in transactional and service businesses where we are using regression and ANOVA and the response variable y may be an attribute. For example, a bank may want to predict the proportion of mortgage applications that are actually funded. This is a measure of yield in their process. Yield probably follows a binomial distribution. Most likely, yield isn't well approximated by a normal distribution, and a standard linear regression model wouldn't be satisfactory. However, there are modeling techniques based on **generalized linear models** that handle many of these cases. For example, **logistic regression** can be used with binomial data and Poisson regression can be used with many kinds of count data. Montgomery, Peck, and Vining (2006) contains information on applying these techniques. Logistic regression is available in Minitab, and JMP software provides routines for both logistic and Poisson regression.

Important Terms and Concepts

Action limits
Assignable causes of variation
Cause-and-effect diagram
Chance causes of variation
Check sheet
Control chart
Control limits
Defect concentration diagram
Designed experiments
Flow charts, operations process charts, and value stream mapping
Factorial experiment
In-control process
Magnificent seven
Out-of-control-action plan (OCAP)
Out-of-control process
Pareto chart
Patterns on control charts
Scatter diagram
Shewhart control charts
Statistical control of a process
Statistical process control (SPC)
Three-sigma control limits

chapter Three Exercises

3.1 What are chance and assignable causes of variability? What part do they play in the operation and interpretation of a Shewhart control chart?

3.2 Discuss the relationship between a control chart and statistical hypothesis testing.

3.3 What is meant by the statement that a process is in a state of statistical control?

3.4 If a process is in a state of statistical control, does it necessarily follow that all or nearly all of the units of product produced will be within the specification limits?

3.5 Discuss the logic underlying the use of three-sigma limits on Shewhart control charts. How will the chart respond if narrower limits are chosen? How will it respond if wider limits are chosen?

3.6 Consider the control chart shown here. Does the pattern appear random?

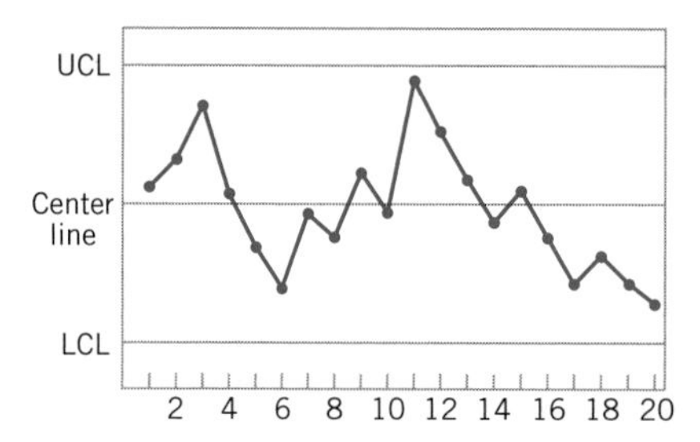

3.7 Consider the control chart shown here. Does the pattern appear random?

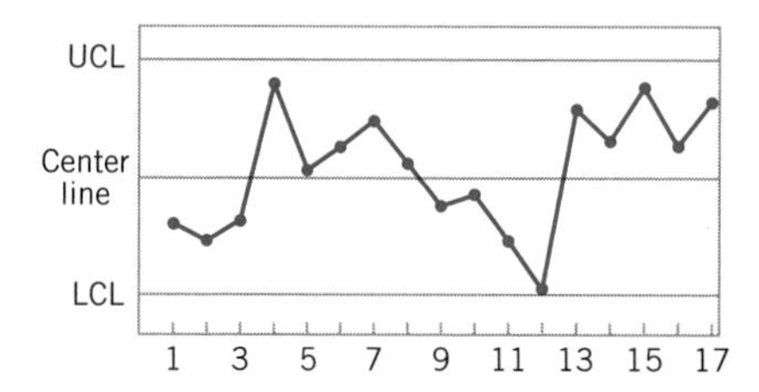

3.8 Consider the control chart shown here. Does the pattern appear random?

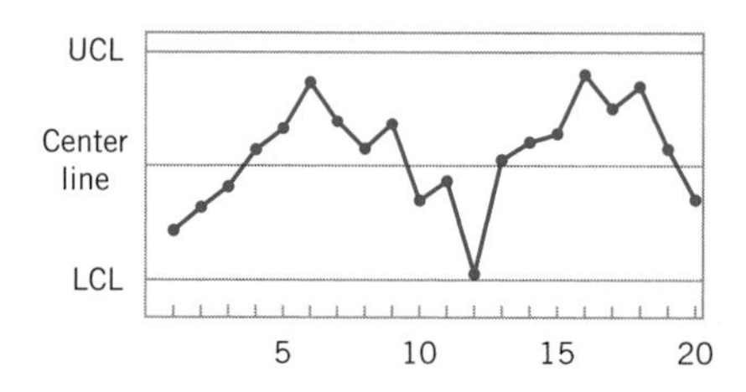

3.9 You consistently arrive at your office about one-half hour later than you would like. Develop a cause-and-effect diagram that identifies and outlines the possible causes of this event.

3.10 A car has gone out of control during a snowstorm and strikes a tree. Construct a cause-and-effect diagram that identifies and outlines the possible causes of the accident.

3.11 Laboratory glassware shipped from the manufacturer to your plant via an overnight package service has arrived damaged. Develop a cause-and-effect diagram that identifies and outlines the possible causes of this event.

3.12 Construct a cause-and-effect diagram that identifies the possible causes of consistently bad coffee from a large-capacity office coffee pot.

3.13 Develop a flow chart for the process that you follow every morning from the time you awake until you arrive at your workplace (or school). Identify the value-added and non-value-added activities.

3.14 Develop a flow chart for the pre-registration process at your university. Identify the value-added and non-value-added activities.

3.15 Many process improvement tools can be used in our personal lives. Develop a check sheet to record "defects" you have in your personal life (such as overeating, being rude, not meeting commitments, missing classes, etc.). Use the check sheet to keep a record of these "defects" for one month. Use a Pareto chart to analyze these data. What are the underlying causes of these "defects"?

chapter Four Statistical Inference about Product and Process Quality

Chapter Outline

Newcrest strikes gold

As Australia's largest gold mine, Newcrest produces approximately 300,000 ounces of gold annually. This is no easy feat, however, as bulk mining procedures and expensive treatment methods are required to harvest the gold from low-grade ore.

The company owns a fleet of large trucks that descend into the mining pit via a narrow ramp, pick up an average 225 tons of ore, and then transport the load out of the mine to a crushing machine. Unfortunately, the trucks can only drive between 5-8 miles per hour on the single-lane ramps. Some trucks move even slower, often causing a large backup of vehicles.

To increase efficiency and improve overall performance, Newcrest sought to reduce such truck speed variations. Over a two-week period, a team observed and recorded data on five trucks and found that *ramp slope* and *fuel injector performance* had the greatest impact on truck speed.

The team then proposed and tested the effectiveness of several solutions. They altered the slope of the uphill ramp and measured the resulting truck speeds. Comparing the before and after speeds with a *t*-test revealed that the slope change had increased truck speed significantly. The team also found a way to identify faulty fuel injectors.

In the end, the team achieved a 2.6 % increase in truck speed and a 7% reduction in truck speed variation. These seemingly small improvements will save the company more than $835,000 in just the first year. Additional improvements are being investigated to further increase these cost savings.

Statistical Inference about Process Quality

You are what you measure Or Worse! (Hauzer & Katz, 1998)

One of the first steps in the DMAIC process is to obtain data to measure the output of a process. This makes sense from a logical perspective because we cannot know if we are improving a process until we can measure its output.

However, all processes have some sort of variability inherent in the system. For example, when coming to campus does it always take you X minutes exactly from start to finish? Probably not. There will always be some inherent variability in the time it takes to come to campus.

The same is true of almost all processes and systems. That is why statistics is so important. It allows us to measure and describe processes that have variability.

This chapter will introduce you to the methods used to describe data. You will learn how to perform simple numerical and graphical analysis to summarize the data in meaningful terms. These methods will become the foundation for all future chapters, so please pay particular attention during this chapter as it will be very easy to get behind without a solid understanding of these techniques.

Chapter Overview and Learning Objectives

The goal of this chapter is to introduce you to basic statistical methods. First, we will show how simple tools can be utilized to describe data visually and numerically. Then, we will introduce the concept of a probability function and demonstrate its usefulness for modeling and describing quality characteristics of a process. Finally, we combine this knowledge to draw conclusions when a sample is taken from a larger population.

After careful study of this chapter you should be able to do the following:

1. Construct and interpret visual data displays, including the stem-and-leaf plot, the histogram, and the box plot
2. Compute and interpret the sample mean, the sample variance, the sample standard deviation, and the sample range
3. Explain the concepts of a random variable and a probability distribution
4. Understand and interpret the mean, variance, and standard deviation of a probability distribution
5. Determine probabilities from probability distributions
6. Construct and interpret confidence intervals on a single mean and on the difference in two means
7. Construct and interpret confidence intervals on a single variance and the ratio of two variances
8. Construct and interpret confidence intervals on a single proportion and on the difference in two proportions
9. Understand how the analysis of variance (ANOVA) is used to compare more than two samples

4.1 Describing Variation

Imagine that you are responsible for shipping operations. Although you have a fully stocked warehouse and underutilized staff, the time to prepare a customer order varies widely. You suspect a problem in this area and would like to collect some data on shipping performance.

The time that it takes from receipt of a customer order to placement on the dock for shipping is recorded in Table 4.1 below (in hours). All orders were of similar size.

Table 4.1

Delivery Data

63	60	59	70	61	58	60	64	57	61
68	53	64	62	60	59	66	62	55	65
53	63	64	64	69	56	71	66	64	62
64	58	63	62	53	62	59	60	62	61
51	59	55	61	55	69	46	58	64	59
71	60	63	55	48	58	58	65	51	67
66	57	65	63	51	63	50	67	59	57
62	50	65	65	60	59	59	59	59	55
69	57	53	61	51	66	63	59	53	56
62	57	62	61	65	71	60	54	57	61

It is difficult to draw conclusions by looking at the raw data. It often is helpful to prepare graphs to describe how the data is distributed. Some common plots are stem and leaf plots, histograms, and box plots. These will be described using the data set above.

Stem and Leaf Plot

A simple text-based graph used to show how frequently values appear in a data set.

4.1.1. STEM AND LEAF PLOT

A stem and leaf plot is a very simple text-based graph used to show how frequently values appear in a data set. As its name implies, it has two main components:

- Stem: The stem is the anchor of the graph. It contains leading digits for the data. In our example above, the minimum value is in the 40s and the largest value is in the 70s. Therefore, we will choose stem values of 3, 4, 5, 6, 7, and 8 (the 3 and 8 were chosen to show no values in the extremities)
- Leaves: The leaves will be the individual data points. However, only the trailing digits will be shown as a leaf. To place a leaf (let's say 51) on the plot, we would place a 1 above the stem position of 5.

example 4.1 Stem & Leaf Plot for Delivery Data

A stem and leaf plot for the delivery data is shown below. We can see that a 6 and 8 were placed above stem position 4 to indicate that values of 46 and 48 were observed.

In reviewing this graph, we see that the data appears to typically be symmetrically dispersed around the value of 60. In addition, we can readily see that the minimum value is 46 and the maximum value is 72. A high percentage of the values appear to be within the range of 55-65.

```
3|
4|  6 8
5|  0 0 1 1 1 1 3 3 3 3 3 4 5 5 5 5 5 6 6 7 7 7 7 7 7 8 8 8 8 8 9 9 9 9 9 9 9 9 9 9 9
6|  0 0 0 0 0 0 0 1 1 1 1 1 1 2 2 2 2 2 2 2 2 2 3 3 3 3 3 3 3 4 4 4 4 4 4 4 5 5 5 5 5 5 6 6 6 6 7 7 8 9 9 9
7|  0 1 1 1 2
8|
```

Figure 4.1 **Stem-and-Leaf Plot**

This version of the stem-and-leaf plot is sometimes called an **ordered stem-and-leaf plot,** because the leaves are arranged in ascending order. This makes it very easy to find **percentiles** of the data. Generally, the Xth percentile is a value such that at least X% of the data values are at or below this value and 100 − X% of the data values are at or above this value.

The **fiftieth percentile** of the data distribution is called the **sample median**. The median can be thought of as the data value that exactly divides the sample in half, with half of the observations smaller than the median and half of them larger.

If the number of data points is odd, then finding the median is fairly straightforward. Consider the case of 7 data points. In this case, the data point in the fourth position would have exactly three data points above this value and three below. Therefore, the value in the fourth position would be the median. (Note: the data must be sorted in ascending order—i.e., smallest to largest.)

We can calculate the position of the median value using the formula below:

$$\text{Median position for odd numbered data sets} = \frac{n-1}{2} + 1 = \frac{n+1}{2} \tag{4.1}$$

where n = number of data points.

If n is even, the median is computed as the average of the average of the $(n/2)$st and $(n/2 + 1)$st ranked observations. In the delivery problem example, $n = 100$ is an even number, therefore, the median is the average of the observations with rank 50 and 51. The 50th value is 60, and the 51st value is 61. Thus the median is 60.5.

Percentile
A value such that X% of data falls below this point and 100 − X% are at or above this point.

50th Percentile
The sample median. This is the data value that divides sample in half. Half of values are above this point and half are below.

First Quartile
The 25th percentile.

Third Quartile
The 75th percentile.

Time Series Plot
A plot of data over time.

Histogram
A column chart that organizes data within intervals.

The **tenth percentile** is the observation with rank $(0.1)(100) + 0.5 = 10.5$ (halfway between the tenth and eleventh observation), or $(53 + 53)/2 = 53$.

The **first quartile** is the observation with rank $(0.25)(100) + 0.5 = 25.5$ (halfway between the twenty-fifth and twenty-sixth observation) or $(57 + 57)/2 = 57$, and the **third quartile** is the observation with rank $(0.75)(100) + 0.5 = 75.5$ (halfway between the seventy-fifth and seventy-sixth observation), or $(64 + 64)/2 = 64$.

The first and third quartiles are occasionally denoted by the symbols Q1 and Q3, respectively, and the interquartile **range** IQR = Q3 − Q1 is occasionally used as a measure of variability. For the delivery data the interquartile range is IQR = Q3 − Q1 = 64 − 57 = 7.

Finally, although the stem-and-leaf display is an excellent way to visually show the variability in data, it does not take the **time order** of the observations into account. Time is often a very important factor that contributes to variability in quality improvement problems. We could, of course, simply plot the data values versus time; such a graph is called a **time series plot** or a **run chart.** This plot is created as a simple line chart in Excel.

Figure 4.2 shows the time series plot of the delivery data. This display clearly indicates that time is an important source of variability in this process. More specifically, it appears that the delivery variability for the first 70 days might be greater than it was during the last 30 days. Something may have changed in the process (or have been deliberately changed by operating personnel) that is responsible for the cycle time improvement.

Later in this book we formally introduce the control chart as a graphical technique for monitoring processes such as this one, and for producing a statistically based signal when a process change occurs.

4.1.2 HISTOGRAM

A **histogram** is a more compact summary of data than a stem-and-leaf plot. To construct a histogram, we divide the range of the data into intervals, which are usually called **class intervals, cells,** or **bins.** If possible, the bins should be of equal width to enhance the visual information in the histogram. Some judgment must be used in selecting the number of bins so that a reasonable display can be developed. The number of bins depends on the number of observations and the amount of scatter or dispersion in the data.

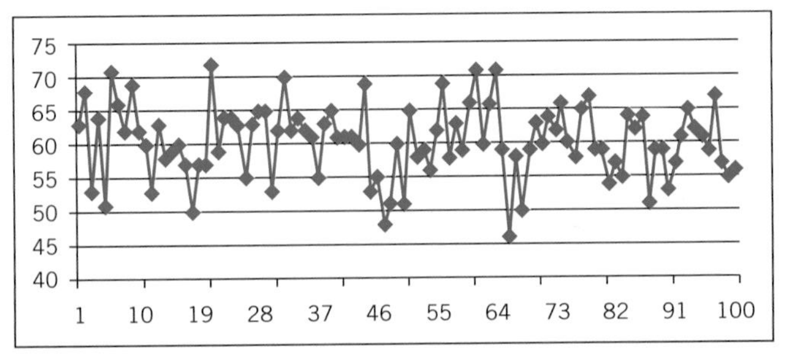

Figure 4.2 Time Series Plot of Delivery Data

Histograms can be relatively sensitive to the choice of the number and width of the bins. For small data sets, histograms may change dramatically in appearance if the number and/or width of the bins changes. For this reason, we prefer to think of the histogram as a technique best suited for **larger data sets** containing, say, 75 to 100 or more observations. A histogram that uses either too few or too many bins will not be informative. In practice, choosing the number of bins approximately equal to the square root of the number of observations often works well.

Histograms for large data sets
Number of bins should be approximately equal to the square root of the number of observations.

Once the number of bins and the lower and upper boundary of each bin have been determined, the data are sorted into the bins and a count is made of the number of observations in each bin. To construct the histogram, use the horizontal axis to represent the measurement scale for the data and the vertical scale to represent the counts, or **frequencies.**

Sometimes the frequencies in each bin are divided by the total number of observations (n). When this is done, the vertical scale of the histogram represents **relative frequencies** or probability of being within that particular bin.

example 4.2 Histogram for Delivery Data

Because the data set contains 100 observations, we suspect that 10 bins will provide a satisfactory histogram. We constructed the histogram using the SPC XL option that allows the user to specify the number of bins.

Looking at the histogram, we can see that the data is distributed fairly symmetrically, though it appears that there is a bias towards lower values. Thus, the data is skewed to the left. In addition, we can see that the average value is between 59.0 and 61.6.

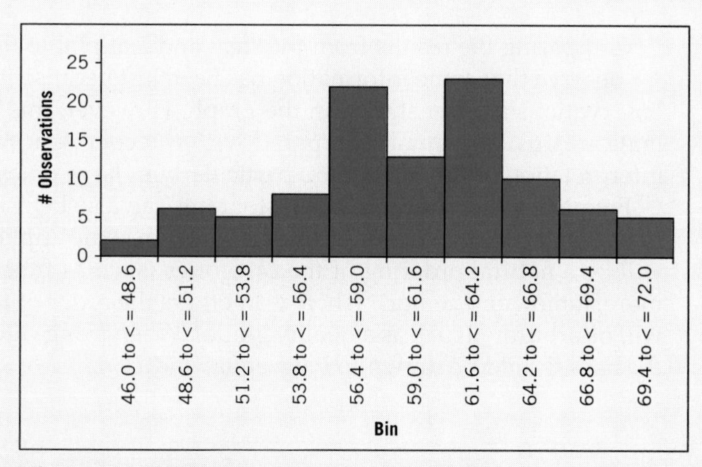

Figure 4.3 Histogram with Appropriate Number of Bins (Generated in SPC XL)

The histogram that follows has 50 bins, far more than the recommended value of 10. In comparing the two plots, we see that many bins have missing values. This makes it more difficult to observe the average value of the data set and determine if the data is skewed in one direction or another.

example 4.2 Continued

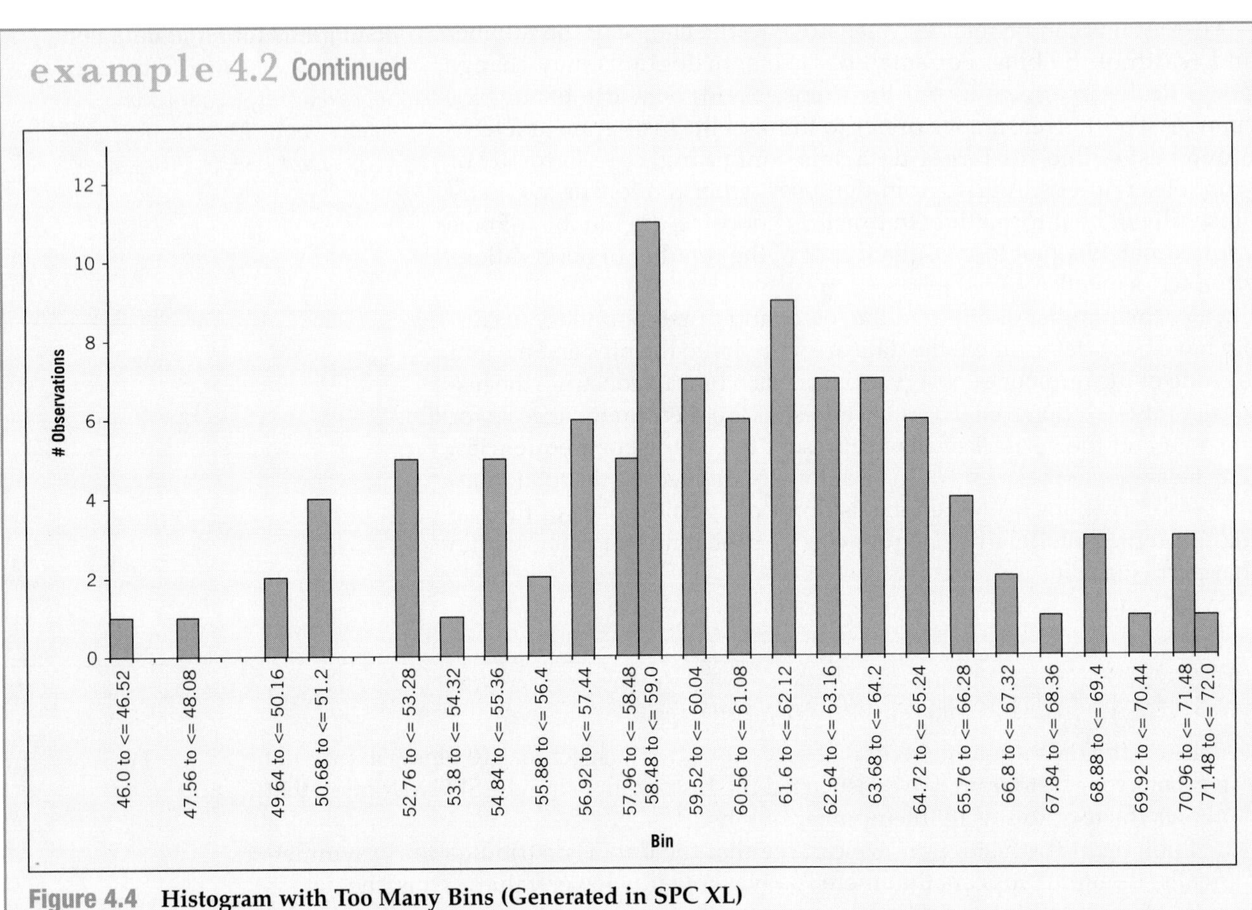

Figure 4.4 Histogram with Too Many Bins (Generated in SPC XL)

Comparing the results from the stem and leaf plot to the histogram, we observe that some information has been lost because the individual observations are not shown in the graph. However, this loss of information is usually small compared with the conciseness and ease of interpretation of the histogram, particularly in large data sets.

Frequency distributions and histograms can also be used with qualitative, categorical, or count (discrete) data. In some applications, there will be a natural ordering of the categories (such as freshman, sophomore, junior, and senior), whereas in others the order of the categories will be arbitrary (such as a quality grade). When using categorical data, the bars should be drawn to have equal width.

example 4.3 Histogram with Categorical Data

There are 200 students in your class: 100 sophomores, 50 juniors, 30 freshmen, and 20 seniors. Prepare a histogram that shows the distribution of students by class level.

Figure 4.5 shows a variation of the histogram available in SPC XL that includes a **cumulative frequency plot**. This blue line shows the total number of students, and the cumulative probabilities associated with various groupings of students.

example 4.3 Continued

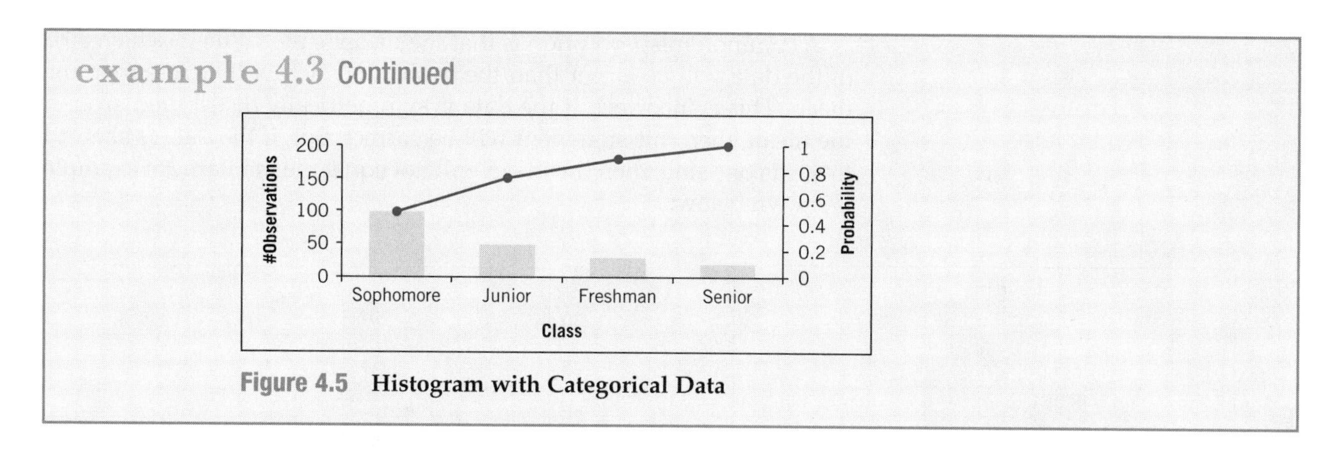

Figure 4.5 Histogram with Categorical Data

Cumulative frequencies are often very useful in data interpretation. Looking at the graph above, we can observe the following:

1. There are 100 sophomores (left vertical axis). This makes up approximately 50% of the total class population.
2. There are 50 juniors. This group is approximately 25% of the class population. Sophomores and juniors clearly make up a majority of the class with a total combined number of 150 students. This represents 75% of the class population.
3. There are 30 freshmen, or 15% of the class population. Sophomores, juniors, and freshman total 180 students, or roughly 90% of the class.
4. There are 20 seniors, or 10% of the class population. Sophomores, juniors, freshman, and seniors make up 100% of the class population (i.e., there are no graduate students).

4.1.3 NUMERICAL SUMMARY OF DATA

The stem-and-leaf plot and the histogram provide a visual display of three properties of sample data: the shape of the distribution of the data, the central tendency in the data, and the scatter or variability in the data. However, it is also helpful to have numerical measures of central tendency and scatter. The two most common metrics are the mean and standard deviation.

The **sample mean**, or **sample average**, is a measure of central tendency. It is essentially the "average" value of the data set. To calculate the sample mean, we add up all of the values in our data sample and divide this by the total number of data points. Mathematically, this is expressed as

Sample Mean
Measure of central tendency
"Average" value of the data set.

$$\bar{x} = \frac{x_1 + x_2 + \cdots + x_n}{n} = \frac{\sum_{i=1}^{n} x_i}{n} \tag{4.2}$$

where $x_1, x_2, \ldots, x_n$ are the observations in a sample and n is the total number of data points.

A common misperception is that the mean is the value in which 50% of the data will be larger than the mean, and 50% will be less than the mean. This is incorrect. If the data is symmetrically distributed around the mean, then this statement will be correct. But, if the data is skewed towards one side, then the mean will not equal the median. An example is shown below:

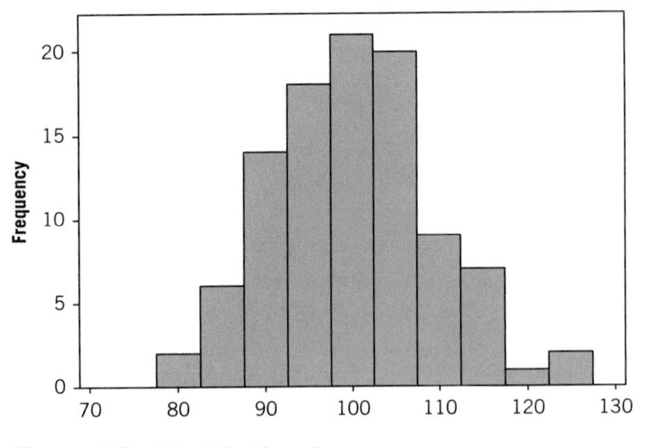

Figure 4.6 **Distribution A**

Figure 4.7 **Distribution B**

Both of the above distributions have a mean of 100. Distribution A is fairly symmetric but Distribution B is not.

Distribution A has an average of 100 and a median of 100. However, Distribution B has an average value of 100 and has a median of 96. In this case, 50% of the data is less than 96, and 50% of the data is above 96. The average is larger than the median due to the higher values that exist within the data set.

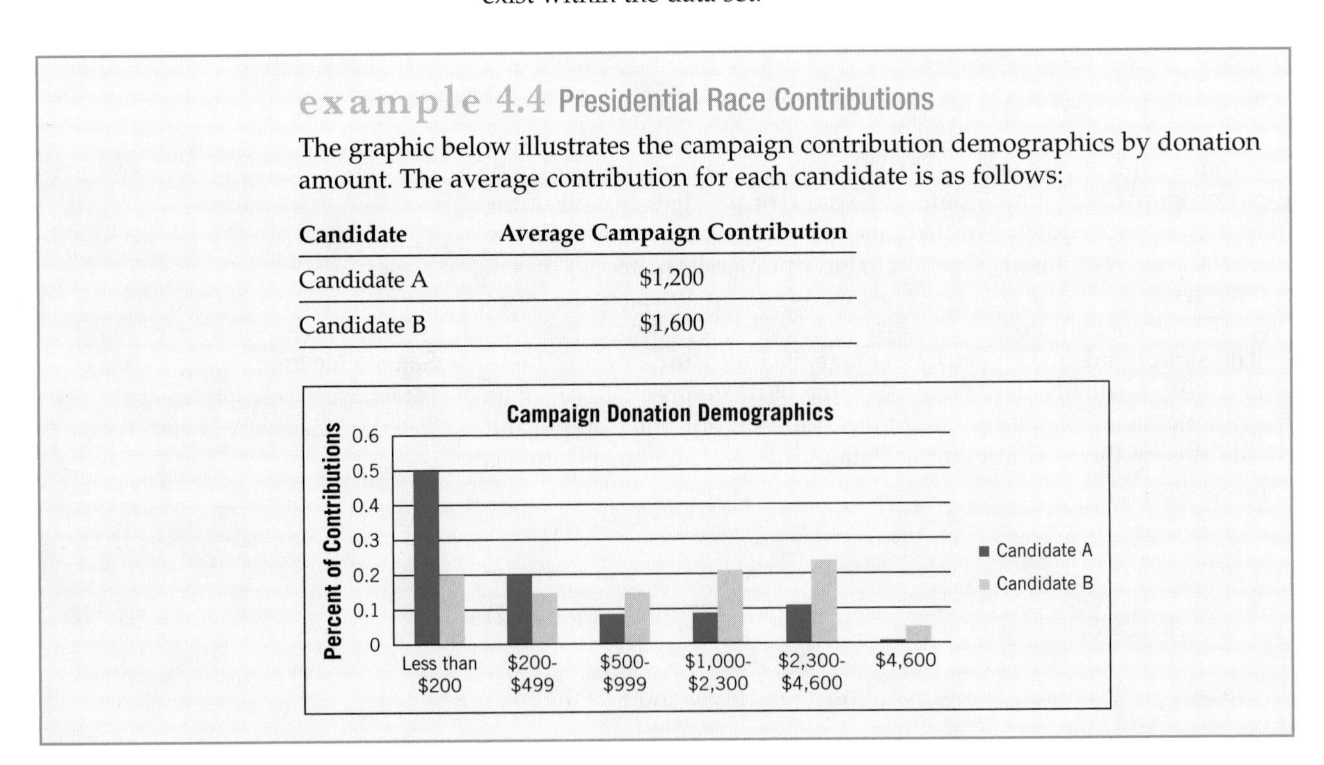

example 4.4 Presidential Race Contributions

The graphic below illustrates the campaign contribution demographics by donation amount. The average contribution for each candidate is as follows:

Candidate	Average Campaign Contribution
Candidate A	$1,200
Candidate B	$1,600

example 4.4 Continued

In looking at the data, it is apparent that the average contribution is heavily skewed by the large donation size. If we look at the 50th percentile values, we see a another picture:

The median value for Candidate A is in the upper end of the less than $200 value category; whereas the median for Candidate B is in the upper end of the $500–$999 category. Though the average donation is similar for both candidates, the types of donors appear to be quite different for the two candidates.

Sample Standard Deviation Describes the spread or variability in the data.

The sample mean and median describe the central tendency of the data, whereas the **sample variance** or **sample standard deviation** describes how much variability these is in the data. A small variance on standard deviation indicates that a majority of the data is near the mean, and a large variance on standard deviation indicates that the data is spread out.

The sample variance is calculated by subtracting each data point from the mean value. This value is then squared to ensure that the deviations are all positive (to avoid positives and negatives canceling each other out) and to add a penalty for large outliers. These values are then summed, divided by the number of data points minus 1 the standard deviation is found by taking the square root of the variance:

$$s = \sqrt{\frac{\sum_{i=1}^{n}(x_i - \bar{x})^2}{n - 1}} \qquad (4.3)$$

To illustrate how the sample standard deviation calculates the dispersion of the data, consider the following three data sets.

	Set 1	Set 2	Set 3
Value 1	10	7	4
Value 2	10	10	10
Value 3	10	13	16
Average Value	10	10	10

Looking at the data, we can see that set 1 has identical measurements: there is no dispersion of the data. Calculating the standard deviation, we find

$$\textit{Set 1 sample standard deviation} = \sqrt{\frac{(10 - 10)^2 + (10 - 10)^2 + (10 - 10)^2}{3 - 1}} = 0$$

Set 2 has some dispersion. Its standard deviation is calculated as

$$\textit{Set 2 sample standard deviation} = \sqrt{\frac{(7 - 10)^2 + (10 - 10)^2 + (13 - 10)^2}{3 - 1}} = 3$$

Finally, set 3 has the greatest amount of variability, or dispersion of the data.

$$\textit{Set 3 sample standard deviation} = \sqrt{\frac{(4-10)^2 + (10-10)^2 + (16-10)^2}{3-1}}$$
$$= 6$$

Note that the standard deviation only describes how spread out the data is around the mean. The standard deviation does not reflect the magnitude of the sample data, only the scatter about the average. To compare how large the dispersion is to the mean, the coefficient of variation metric is often used. It is calculated as the

$$\textit{Coefficient of variation} = \frac{\textit{Standard deviation}}{\textit{Mean}} \qquad \textbf{(4.4)}$$

Consider a case where the standard deviation is 10. If the mean is 100, then the coefficient of variation is 0.1, or 10%. If the mean is 1,000, then the coefficient of variation would be 0.01, or 1%, to indicate that the standard deviation is about 1% of the mean.

Excel Tip

Several Excel functions exist that make calculating descriptive statistics for large data sets relatively straightforward.

AVERAGE(data range) calculates the average of values in a data set.

STDEV(data range) calculates the standard deviation of values in a data set.

MAX(data range) returns the largest value in the data range selected.

MIN (data range) returns the smallest value in the data range selected.

PERCENTILE(Data Range, *k*) returns the value within the data set that represents the *k*th percentile.

4.1.4 THE BOX PLOT

The stem-and-leaf display and the histogram provide a visual impression of a data set, whereas the sample average and standard deviation provide quantitative information about specific features of the data. The **box plot** is a graphical display that simultaneously displays several important features of the data, such as location or central tendency, spread or variability, departure from symmetry, and identification of observations that lie unusually far from the bulk of the data (these observations are often called "outliers").

Box Plot
A graphical display that shows the first quartile, median, and third quartile in a "box" format.

A box plot displays the three quartiles of the data on a rectangular box, aligned either horizontally or vertically. The box encloses the interquartile range with the left (or lower) line at the first quartile Q1 and the right (or upper) line at the third quartile Q3. A line or other marker is drawn through the box at the second quartile (which is the fiftieth percentile or the median) A line at either end extends to the extreme values. These lines are usually called **whiskers.** Some authors refer to the box plot as the **box and whisker plot.** In some computer programs, the whiskers only extend a distance of 1.5 (Q3–Q1) from the ends of the box, at most, and observations beyond these limits are flagged as potential outliers. This variation of the basic procedure is called a **modified box plot.**

A box plot for our delivery data is shown in Figure 4.8 (all 100 data points). Box plots are often used to compare data sets. Early in this chapter, we observed that there may have been a process change that changed delivery times after the 70th data point. A box plot is another way to show differences between data.

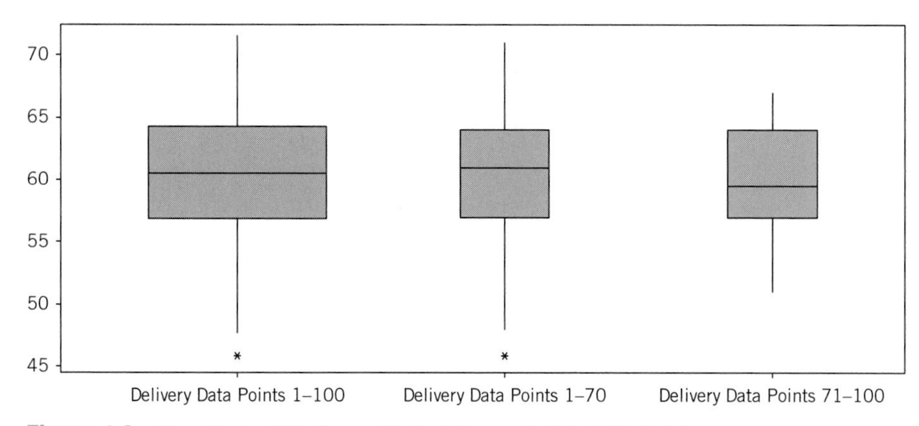

Figure 4.8 **Box Plot of Delivery Data (Generated in Minitab)**

In reviewing the box plots in Figure 4.8, we can learn quite a bit. First, it appears that the median value is slightly lower after data point 70. However, the real difference is the spread of the outliers. The whiskers extend much further for the earlier delivery times.

We can also see this difference numerically.

	Delivery Times (1–70)	Delivery Times (71–100)
Mean	60.5	59.9
Standard deviation	5.9	4.2
Coefficient of variation	0.1	0.1
25th percentile	57.3	57.0
50th percentile (median)	61.0	59.5
75th percentile	64.0	63.5
Min.	46.0	51.0
Max.	72.0	67.0

4.2 Probability Distributions

The histogram (or stem-and-leaf plot, or box plot) is used to describe *sample* data. A **sample** is a collection of measurements selected from some larger source or **population.** For example, the delivery data represent a sample of 100 delivery times selected from the delivery process. There are probably hundreds of thousands of delivery times that could be measured and evaluated.

Another way to look at the data is through a probability distribution. A **probability distribution** is a mathematical model that relates the value of the variable with the probability of occurrence within the population. There are two types of probability distributions.

Examples of discrete and continuous probability distributions are shown in Figures 4.9*a* and 4.9*b*, respectively. The appearance of a discrete distribution is that of a series of vertical "spikes," with the height

Sample: Random Sample
A sample is a subset of the observations in a population and a random sample is one that has been selected from the population in such a way that every possible sample has an equally-likely chance of being sources.

Probability Distribution
A mathematical model that relates the value of the variable with the probability of occurence within the population.

Continuous distributions
When the variable being measured is expressed on a continuous scale, its probability distribution is called a *continuous distribution.* An example of a continuous distribution could be on net present value of a portfolio.

Discrete distributions
When the parameter being measured can only take on certain values, such as the integers 0, 1, 2, . . . , the probability distribution is called a *discrete distribution.* For example, the distribution of the number of nonconformities or defects of a product would be a discrete distribution.

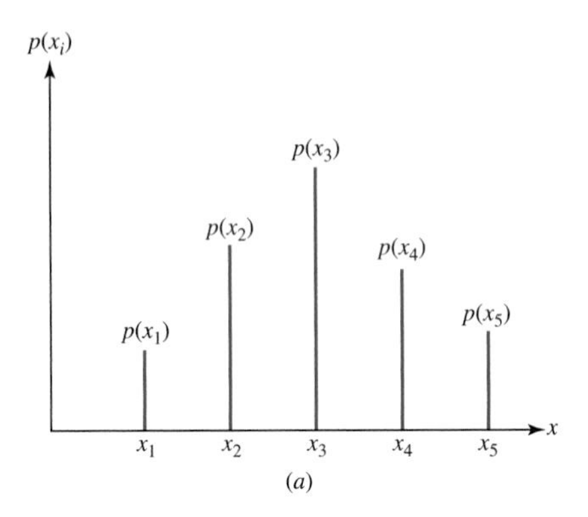

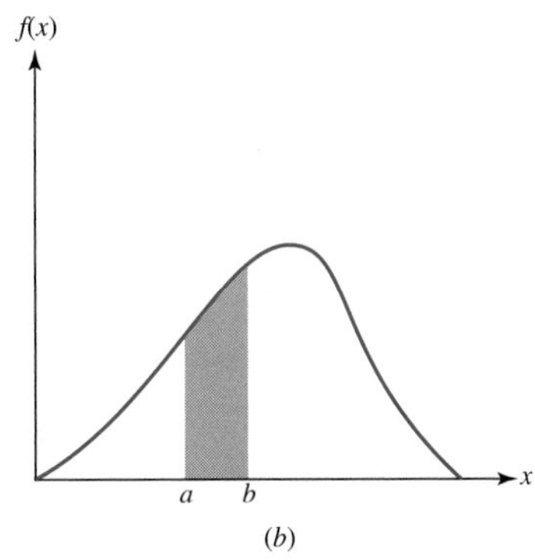

Figure 4.9 Probability Distributions: (a) Discrete Case, (b) Continuous Case

of each spike proportional to the probability. We write the probability that the random variable x takes on the specific value x_i as

$$P\{x = x_i\} = p(x_i)$$

The appearance of a continuous distribution is that of a smooth curve, with the area under the curve equal to probability, so that the probability that x lies in the interval from a to b is written as

$$P\{a \leq x \leq b\} = \int_a^b f(x)dx$$

In both cases, probability functions are defined such that all possible outcomes are captured. Thus, the sum of all possibilities equals 1.

If it is possible to create a mathematical function that represents a probability distribution, then we can calculate probabilities of certain events happening as shown above. We can also directly calculate the mean, μ, variance σ^2, and standard deviation, σ, as well.

$$\mu = \begin{cases} \int_{-\infty}^{\infty} xf(x)\,dx, & x \text{ continuous} \\ \sum_{i=1}^{\infty} x_i p(x_i), & x \text{ discrete} \end{cases}$$

$$\sigma^2 = \begin{cases} \int_{-\infty}^{\infty} (x - \mu)^2\,dx, & x \text{ continuous} \\ \sum_{i=1}^{\infty} (x_i - \mu)^2\,p(x_i) & \text{discrete} \end{cases}$$

The standard deviation $\sigma = \sqrt{\sigma^2}$.

example 4.5 A Discrete Distribution

A manufacturing process produces thousands of semiconductor chips per day. On the average, 5% of these chips do not conform to specifications. Every hour, an inspector selects a random sample of 10 chips and classifies each chip in the sample as conforming or nonconforming. If we let x be the random variable representing the number of nonconforming chips in the sample, then the probability distribution of x is

$$p(x) = \binom{10}{x}(0.05)^x(.95)^{10-x} \quad x = 0, 1, 2, \ldots, 10$$

This is a *discrete* distribution, since the observed number of nonconformance are $x = 0, 1, 2, \ldots, 25$, and is called the **binomial distribution.** A graph of the probability distribution is shown below:

example 4.5 Continued

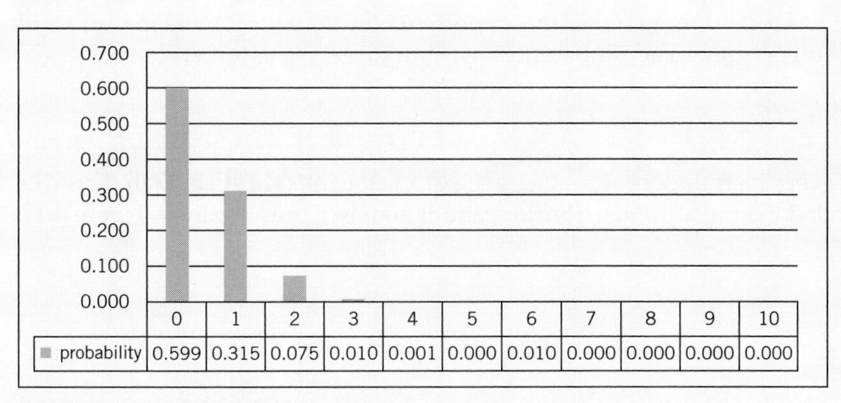

	0	1	2	3	4	5	6	7	8	9	10
probability	0.599	0.315	0.075	0.010	0.001	0.000	0.010	0.000	0.000	0.000	0.000

Figure 4.10 Probability of Obtaining X NonConforming Chips in a Sample of 10 Chips

We can calculate the probability of finding one or fewer nonconforming parts in the sample as

$$
\begin{aligned}
p(x \leq 1) &= P(x = 0) + P(x = 1) \\
&= P(0) + P(1) \\
&= \sum_{x=0}^{x=1} \binom{10}{x} (0.05)^x (0.95)^{10-x} \\
&= \binom{10}{0} (0.05)^0 (0.95)^{10-0} + \binom{10}{1} (0.05)^1 (0.95)^{10-1} \\
&= \frac{10!}{0!10!} (0.05)^0 (0.95)^{10} + \frac{10!}{1!9!} (0.05)^1 (0.95)^9 \\
&= 0.599 + 0.315 = 0.914
\end{aligned}
$$

Conversely, we can find the probability of selecting one or more nonconforming chips in one sample. This is $1 - 0.914 = 0.086$, or 8.6%.

Using this data, we can also calculate the average as:

$$\mu = (0)(0.598) + (1)(0.315) + (2)(0.075) + (3)(0.010) + (4)(0.001) + \cdots + (10)(0.000) = 0.5$$

The standard deviation can be calculated as:

$$\sigma = \sqrt{(0 - 0.5)^2(0.598) + (1 - 0.5)^2(0.3156) + \cdots + (10 - 0.5)^2(0.000)}$$

$$\sigma = 0.68$$

example 4.6 A Continuous Distribution

Suppose that x is a random variable that represents the actual contents in ounces of a 1-lb bag of coffee beans. The probability distribution of x is assumed to be

$$f(x) = \frac{1}{1.5} \quad 15.5 \le x \le 17.0$$

This is a *continuous* distribution, since the range of x is the interval [15.5, 17.0]. This distribution is called the **uniform distribution,** and it is shown graphically in Figure 4.11.

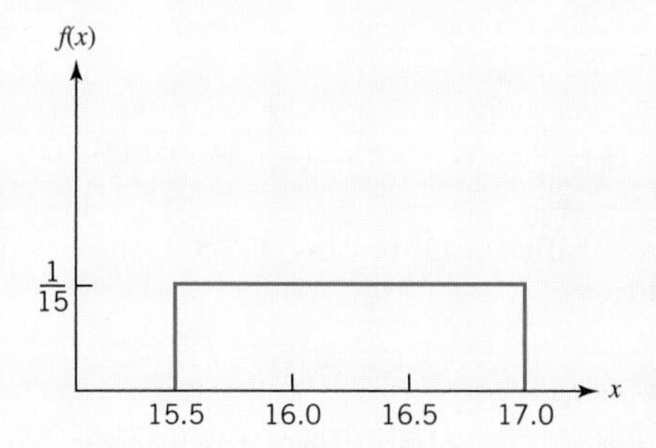

Figure 4.11 Distribution for Example 4.7

Note that the area under the function $f(x)$ corresponds to probability, so that the probability of a bag containing less than 16.0 oz. is

$$P\{x \le 16.0\} = \int_{15.5}^{16.0} f(x)\,dx = \int_{15.5}^{16.0} \frac{1}{15} dx$$

$$= \frac{x}{1.5}\bigg|_{15.5}^{16.0} = \frac{16.0 - 15.5}{1.5} = 0.3333$$

Conversely, the probability of a bag being greater than 16 ounces is 0.6666.

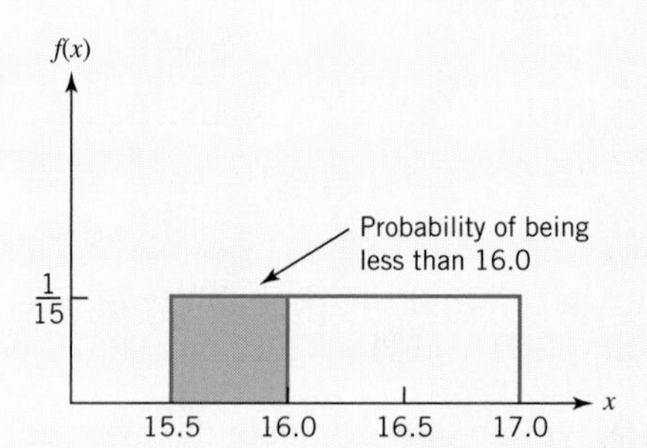

Figure 4.11a

The mean can be calculated as

$$\mu = \int_{-\infty}^{\infty} xf(x)\,dx$$

$$= \int_{15.5}^{17} x * \frac{1}{1.5} dx = \frac{1}{1.5}\int_{15.5}^{17} x\,dx = \frac{1}{1.5} * \frac{1}{2} x^2\Big|_{15.5}^{17}$$

$$= \frac{1}{3.0}(17^2 - 15.5^2) = 16.25$$

example 4.6 Continued

The variance and standard deviation can be calculated as

$$\sigma^2 = \int_{-\infty}^{\infty} (x - \mu)^2 f(x)dx$$

$$= \int_{15.5}^{17.0} (x - 16.25)^2 \frac{1}{15} dx = \frac{1}{1.5}\int_{15.5}^{17.0} (x - 16.25)^2 dx$$

$$= \frac{1}{1.5}\int_{15.5}^{17.0} x^2 - 33.5x + 16.25^2 dx = \frac{1}{1.5} \times \left[\frac{1}{3}x^3 - \frac{33.5}{2}x + 16.25^2 \Big|_{15.5}^{17.0}\right]$$

$$= 0.1875$$

$$\sigma = \sqrt{\sigma^2} = \sqrt{0.1875} = 0.4330$$

Several discrete probability distributions arise frequently in statistical quality control. Tables 4.2 and 4.3 briefly describe these distributions. More information on these distributions may be found in the appendices.

4.3 The Normal Distribution

The normal distribution is probably the most important distribution in both the theory and application of statistics. This is because data is often distributed in a bell curve pattern that is symmetric around the mean.

A rough diagram of a normal distribution is shown in Figure 4.12.

The normal distribution is used so much that we frequently employ a special notation, $x \sim N(\mu, \sigma^2)$ to imply that x is normally distributed with mean and variance σ^2.

There is a simple interpretation of the standard deviation of a normal distribution, which is illustrated in Figure 4.13.

There are three main zones within a normal distribution.

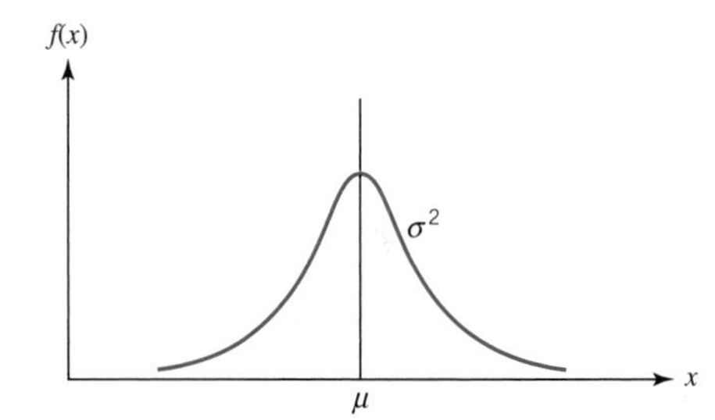

Figure 4.12 Normal Distribution

Zone 1: $\mu \pm 1\sigma$

Zone 1 is where a large percentage of the data are expected. 68.26% of the data are expected to be within one standard deviation of the mean.

Zone 2: $\mu \pm 2\sigma$

Adding one additional standard deviation to zone 1 yields zone 2. 95.46% of the data fall between the limits defined as the mean plus and minus two standard deviations.

Zone 3: $\mu \pm 3\sigma$

97.3% of the data falls within the range of three standard deviations from the mean. This point is often used to sketch rough diagrams of a normal distribution.

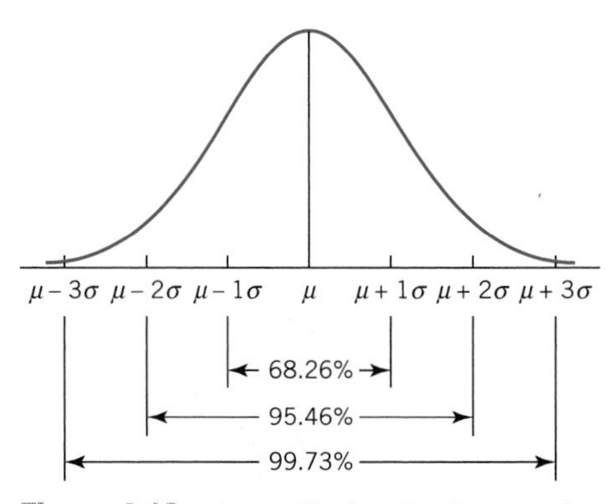

Figure 4.13 Areas Under the Curve of a Normal Distribution

Table 4.2

Discrete Distributions Used in Practice

Name/Use	Probability Density Function	Mean	Standard Deviation
Hypergeometric Sampling from a group without replacement. N: Total number of items D: Number of items in the group with desirable characteristic n: number selected x: number of items that fall within the class of interest	$p(x) = \frac{\binom{D}{x}\binom{N-D}{n-x}}{\binom{N}{n}} \quad x = 0, 1, 2, \ldots, \min(n, N)$	$\mu = \frac{nD}{N}$	$\sigma^2 = \frac{nD}{N}\left(1 - \frac{D}{N}\right)\left(\frac{N-n}{N-1}\right)$
Binomial Describes the number of times an event will occur in a series of events (i.e., if you flip a coin 5 times, how many heads will you receive?) n: number of trials p: probability of success for each trial	$p(x) = \binom{n}{x}p^x(1-p)^{n-x} \quad x = 0, 1, \ldots, n$	$\mu = np$	$\sigma^2 = np(1-p)$
Poisson Describes the number of defects anticipated based upon a defect parameter λ. λ: number of defects per unit (on average)	$p(x) = \frac{e^{-\lambda}\lambda^x}{x!} \quad x = 0, 1, \ldots$	$\mu = \lambda$	$\sigma^2 = \lambda$
Pascal Distribution Used to count occurrences. Two special cases: Negative Binomial ($r > 0$) Similar to binomial but the number of trials are fixed and we observe the number of successes. Geometric ($r = 1$) Determines how many trials until we achieve success.	$p(x) = \binom{x-1}{r-1}p^r(1-p)^{x-r} \quad x = r, r+1, r+2, \ldots$	$\mu = \frac{r}{p}$	$\sigma^2 = \frac{r(1-p)}{p^2}$

Table 4.3

Continuous Distributions Used in Practice

Name/Use	Probability Density Function	Mean	Standard Deviation
Normal Wide variety of applications from product quality assessment, confidence intervals, and SPC. A normal distribution has a mean of μ and a standard deviation of σ. There is no closed form solution to the area under the normal distribution's probability density function, so tables are often utilized to compute probabilities. See section 4.3.	$f(x) = \frac{1}{\sigma\sqrt{2\pi}} e^{-\frac{1}{2}\left(\frac{x-\mu}{\sigma}\right)^2} \quad -\infty < x < \infty$	μ	σ
Exponential Models the time to fail for a component or system. Commonly used in reliability evaluations. λ: failure rate of the system under study	$f(x) = \lambda e^{-\lambda x} \quad x \geq 0$	$\frac{1}{\lambda}$	$\frac{1}{\lambda^2}$

Each of these zones was defined by taking a multiple of a standard deviation and adding/subtracting it from the mean. Thus, the standard deviation is measuring the distance on the horizontal scale associated with the 68.26%, 95.46%, and 99.73% containment limits. It is common practice to round these percentages to 68%, 95%, and 99.7%.

The cumulative normal distribution is defined as the probability that the normal random variable x is less than or equal to some value a, or

$$P\{x \leq a\} = F(a) = \int_{-\infty}^{a} \frac{1}{\sigma\sqrt{2\pi}} e^{-\frac{1}{2}\left(\frac{x-\mu}{\sigma}\right)} dx$$

However, this equation does not have a closed-form solution (i.e. we cannot solve the integration). But, we can take advantage of the knowledge of the above zones to describe the placement within a normal curve through the equation below:

$$z = \left(\frac{x - \mu}{\sigma}\right) \tag{4.5}$$

In this equation, μ is the mean, σ is the standard deviation, and x is a value we would like to evaluate. $x - \mu$ calculates how far we are from the mean. When we divide by σ we are essentially calculating how many standard deviations we are away from the mean. This, with the knowledge of our containment limits, will help us evaluate data from a normal distribution.

example 4.7 Calculating and Interpreting Z

Delivery data is normally distributed with a 60.5 and a standard deviation of 5.9. If a delivery time is 70, how many standard deviations is this away from the mean?

$$z = \frac{70 - 60.5}{5.9} = 1.61$$

This data point is 1.61 standard deviations away from the mean. It is within an area of the curve that is reasonable for a data point to reside.

However, let's consider a new delivery value of 100. How many standard deviations away from the mean is this value?

$$z = \frac{100 - 60.5}{5.9} = 6.69$$

A delivery time of 100 would be 6.69 standard deviations away from the mean. This is well beyond the anticipated range of $\pm 3\sigma$. Therefore, this value would indicate that something has changed from the normal operations.

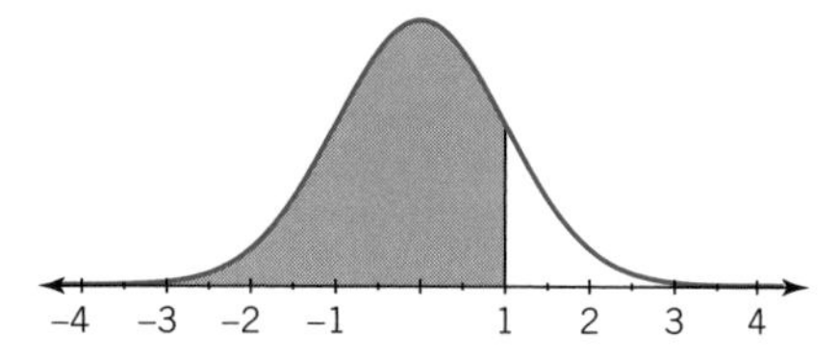

Figure 4.14 Cumulative Normal Distribution

Once we know the value of z, it is relatively easy to determine probabilities from a normal distribution using a table of cumulative probabilities that are a function of the z values. There are two commonly used tables: a standard normal distribution table and a cumulative standard normal distribution table. We choose to present the cumulative standard normal distribution table as it is easier for most students to use and readily aligns with Excel functions which can find these values as well.

A cumulative standard normal distribution is shown in Figure 4.14: The blue area indicates the region in which a desired probability is located. For the cumulative normal curve, the values that are calculated always start from the left side and move towards the right.

Let's calculate the probability of being less than 1 standard deviation above the mean. In this case, we would look up the value of 1 in the cumulative standard normal distribution table in Appendix II (look in the z column). This value is 0.84. Thus there is an 84% chance of observing a value of the standard normal random variable that is less than or equal to 1 standard deviation above the mean (and 16% chance of being greater than 1 standard deviation above the mean).

There are three types of cases to consider in calculating probabilities:

1. The probability of being less than a given value
2. The probability of being greater than a given value
3. The probability of being between two values.

We will illustrate these cases using the example below:

example 4.8 Calculating Probabilities for a Normal Distribution

A just-in-time delivery process has a mean of 60 hours and a standard deviation of 6 hours. The company does not want to deliver the product early as customers may not be ready for the product. They do not want to deliver the product too late as this may negatively impact their customers.

The delivery company has established performance standards for delivery times. The earliest they would like to deliver the product is 50 hours after an order is placed. All products should be delivered within 75 hours. The target range for delivery is between 55 and 65 hours for optimal performance.

A sketch of the normal distribution and its transformation into a standard normal curve via Equation 4.5 is shown below $z = \dfrac{x - \mu}{\sigma}$ function.)

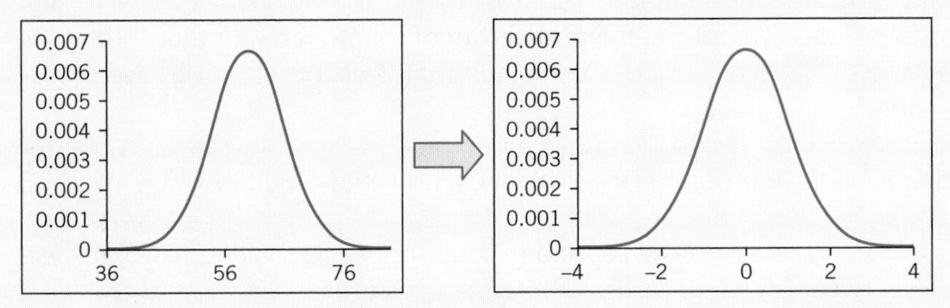

Figure 4.15 Delivery Data Distribution Converted to Standard Normal Curve

What is the probability of delivering product earlier than 50 hours?
The z-value associated with 50 hours is calculated as

$$z = \frac{x - \mu}{\sigma} = \frac{50 - 60}{6} = -1.667$$

The probability of z being less than -1.667 is illustrated in Figure 4.16.

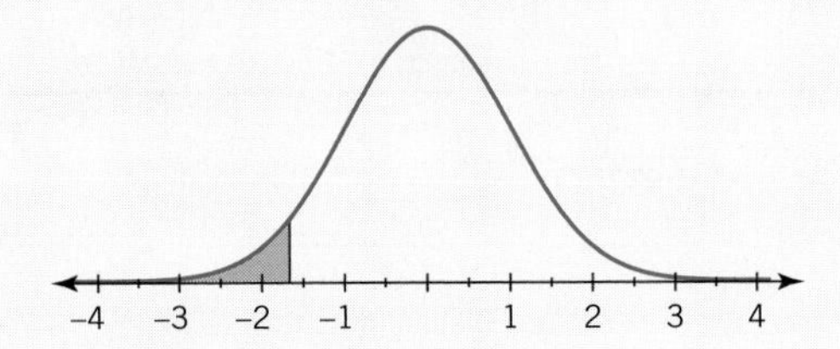

Figure 4.16 Probability of $z < -1.667$

The value of the cumulative normal distribution for $z = -1.667$ is 0.047.
Thus the probability of delivery product earlier than 50 hours is approximately 5%.

What is the probability of delivering product later than 75 hours?
The z-value associated with 75 hours is calculated as

$$z = \frac{x - \mu}{\sigma} = \frac{75 - 60}{6} = 2.5$$

The probability of z greater than 2.5 is illustrated in Figure 4.17.

example 4.8 Continued

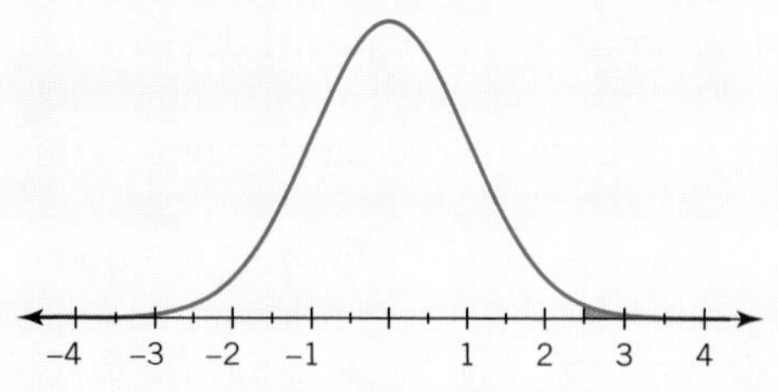

Figure 4.17 Probability of z > −2.5

The probability associated with $z = 2.5$ is 0.99379. This represents the probability within the white area to the left of 2.5. This is not the value we are seeking as we are interested in the blue area to the right of 2.5. To obtain this value, recall that the total probability must equal 1. Therefore, the probability of z being greater than 2.5 can be calculated as $1 - 0.99379 = 0.00621$. This means that 0.6% of deliveries are occurring after 75 hours.

What is the probability of delivering product within the targeted range of 55-65 hours?

The values of 55 and 65 hours correspond to z values of -0.833 and 0.833, respectively. The area between these z values is illustrated in Figure 4.18.

The cumulative probability for $z = 0.8333$ is 0.798 or 79.8%. However, this contains the area below -0.833 that we would like to exclude.

The cumulative probability for $z = -0.833$ is 0.202 or 20.2.

Thus, the probability of being between $-0.833 < z < -0.833$ is 79.8% − 20.2% = 59.5%.

This means that the company is meeting its targeted delivery times 59.5% of the time.

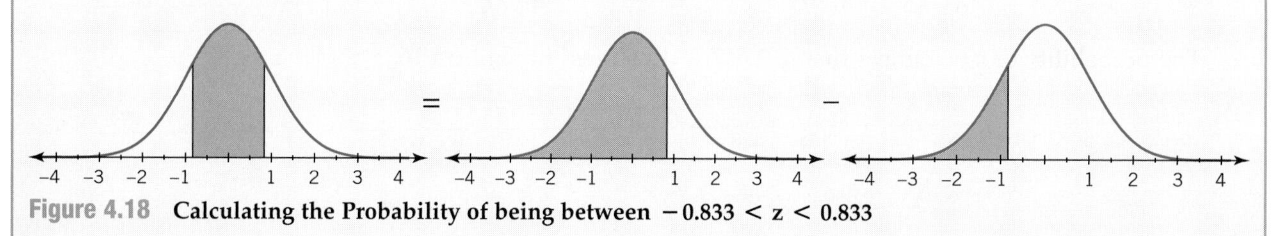

Figure 4.18 Calculating the Probability of being between −0.833 < z < 0.833

4.3.1 THE CENTRAL LIMIT THEOREM

One of the reasons the normal distribution can be used so much in practice is because when samples are taken from distributions that are non-normal and averaged together, they begin to take on a shape that is normal. Consider the following distribution obtained by rolling a fair die 1,000 times.

From the histogram, we can see that this data does not appear to be normally distributed. However, if we average 2, 5, or 10 values together, we would obtain the results in Figure 4.19.

Why does this happen?

When we observe data in practice, if the data is independent of the previous data point, then we would not expect to obtain several high

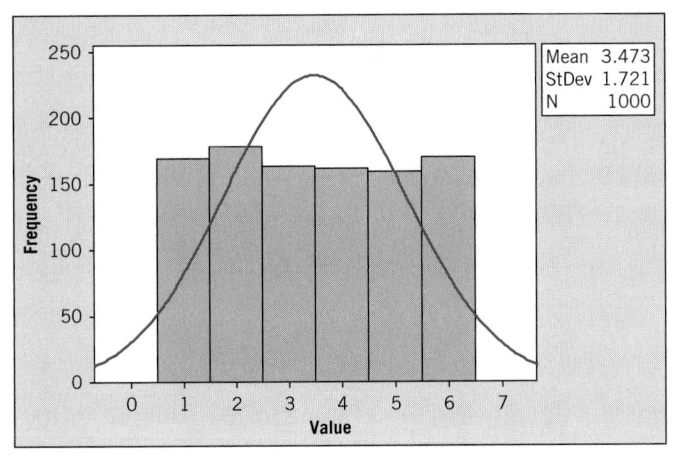

Figure 4.19 **Sample Data from a Non-Normal Distribution**

values or several low values in a row. Rather, we would expect some high values and some low values. For example, if I roll a die, it would be uncommon to receive five 6s in a row. It would be more common to get a few low values, high values, and/or middle values (if it is a fair die). Given this, we would expect the values to "average" out, resulting in a normal distribution centered around the original distribution's mean. However, the more data points that are included in the average, the less variability there will be and the resulting normal distribution will become narrower as shown in Figure 4.20.

Excel Tip

Cumulative normal distribution probabilities can also be calculated directly from Excel using two different functions.

NORMSDIST(Z) should be used if Z has already been calculated. Using the data from the example above, NORMSDIST(0.833) would yield 0.798.

NORMDIST(X, Mean, Standard Deviation, TRUE) goes an extra step and calculates the Z for you.

NORMSDIST(65,60,6,TRUE) would yield 0.798.

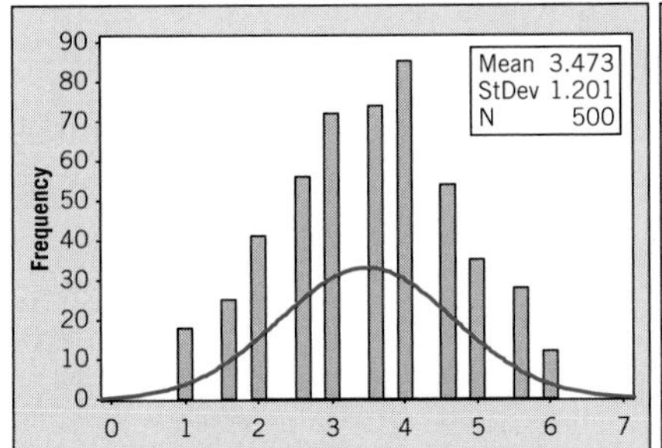

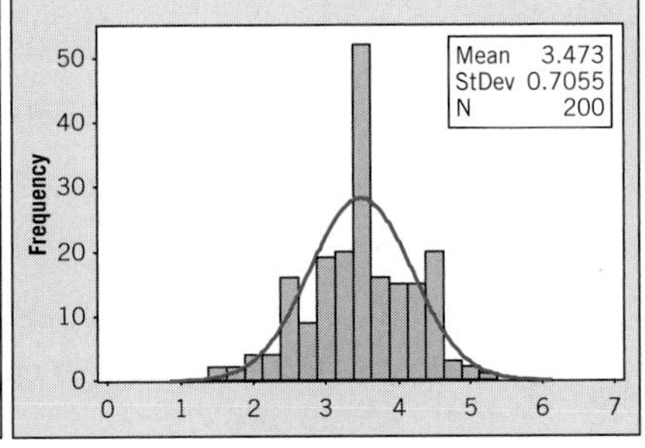

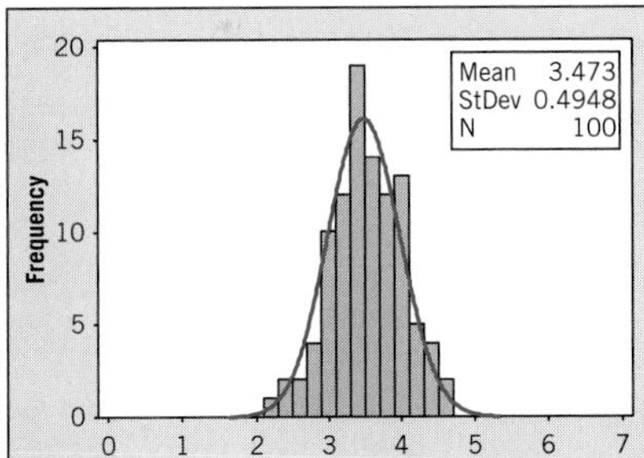

Figure 4.20 **Distribution Created by Averaging 2, 5, and 10 Values Together**

The distribution of averages can be described using the following equations for the mean and standard deviation. The term $\overline{x}$ indicates that the x's are obtained by averaging n values together.

$$E(\overline{x}) = \mu \tag{4.6}$$

$$\sigma_{\overline{x}} = \frac{\sigma}{\sqrt{n}} \tag{4.7}$$

This result is one of the most important in statistics as it allows us to apply many statistical techniques to data that are not normally distributed.

example 4.9 Calculating the Mean and Standard Deviation of the Distribution of Averages

A fair six-sided die has a mean of 3.5 and a standard deviation of 1.7. If we roll the die 1,000 times, but create 250 data points by averaging four rolls of the die together, what would be the mean and standard deviation of the distribution of averages?

The mean of the distribution would remain 3.5. The standard deviation is calculated as

$$\sigma_{\bar{x}} = \frac{\sigma}{\sqrt{n}} = \frac{1.7}{\sqrt{4}} = 0.85$$

If we increased the number of values in the average to 10 (resulting in 100 data points), the mean would again be 3.5, but the standard deviation would be

$$\sigma_{\bar{x}} = \frac{\sigma}{\sqrt{n}} = \frac{1.7}{\sqrt{10}} = 0.54$$

Excel Tip

NORMSINV(Probability) will find a Z value associated with a given probability value.

4.3.2 NORMAL PROBABILITY PLOTS

How do we know whether a data set appears to be normally distributed? We can check to see if the histogram reveals a shape of a normal distribution. However, this is a subjective technique, and the number of bins can alter the results.

Probability plotting is better suited for answering this question and is also a graphical method. It is constructed using the following procedure:

1. Order the data from smallest to largest
2. Calculate the observed cumulative probability for each data point as
   $$\frac{(\text{rank of data point} - 0.5)}{\text{number of data points}}$$
3. Determine the z value for each probability. Note: this means reading the curve backwards. A cumulative probability of 50% will yield a z value of 0.
4. Plot the observed values against the z values.

The result should be a straight line, which indicates that the cumulative probabilities calculated from the data match the anticipated probabilities associated with the normal distribution. An ideal fit is shown below. However, if the plotted points deviate significantly and systematically from a straight line, the normal distribution is not appropriate.

example 4.10 Preparing a Probability Plot

Observations on the octane number of 10 gasoline blends are as follows: 88.9, 87.0, 90.0, 88.2, 87.2, 87.4, 87.8, 89.7, 86.0, and 89.6. We would like to model this data using a normal distribution. Is this a reasonable assumption?

example 4.10 Continued

To create a normal probability plot, we first ordered the data from smallest to largest (the Excel sort function helps here) and calculated their observed cumulative probabilities. For the first point, this was calculated as $(1 - 0.5)/10 = 0.05$.

The function NORMSINV(probability) was then used to convert the cumulative probabilities to *z*-scores and a scatter plot was created.

Rank	Value	Cumulative Probability	Z Value
1	86.0	0.05	−1.64
2	87.0	0.15	−1.04
3	87.2	0.25	−0.67
4	87.4	0.35	−0.39
5	87.8	0.45	−0.13
6	88.2	0.55	0.13
7	88.9	0.65	0.39
8	89.6	0.75	0.67
9	89.7	0.85	1.04
10	90.0	0.95	1.64

The resulting plot is shown in Figure 4.21 for Excel and Minitab. Minitab creates this graphic automatically. The Minitab graph scales the points by probability instead of the *z* score.

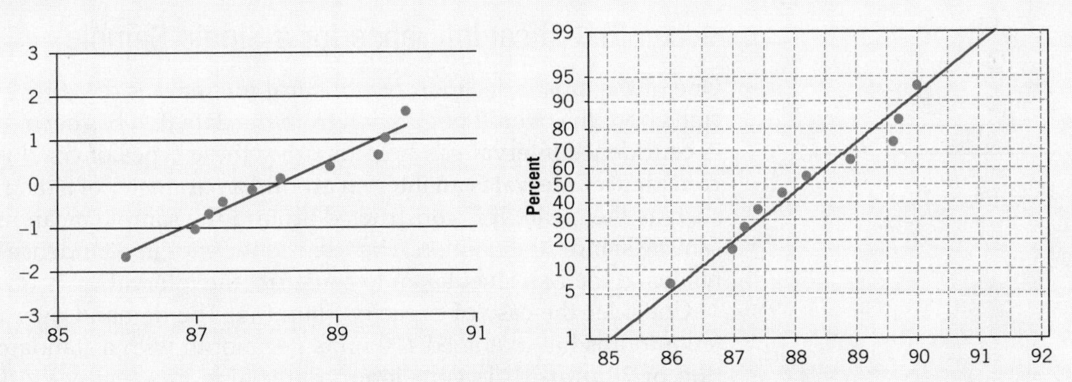

Figure 4.21 Normal Probability Plots Created in Excel and Minitab

The resulting plots show a reasonable straight line that would pass the "fat pencil test". Thus, it is reasonable to use the normal distribution for this data.

4.4 Statistical Inference

So far, we have learned quite a bit. We have learned how to prepare graphics which describe distributions, how to numerically calculate descriptive measures such as a mean, median, and percentile, and we have learned about different types of distributions and how the central

Type I Error
Error of exclusion. Occurs when we conclude that a data set does not meet a criteria even though it does.

Type II Error
Error of inclusion. Occurs when we conclude that a data set meets a criteria even though it does not.

Confidence Interval
An interval constructed around the sample mean using the sample standard devialtion.

limit theorem can result in sample averages from these distributions following a normal distribution.

However, to solve problems, we would also like to be able to confidently draw conclusions on the information collected. This is referred to as statistical inference. Like the criminal justice system, statistical inference techniques are designed to give the benefit of doubt that something is true. We must show, evidence that is not true. It is equivalent to saying that a person is presumed innocent until proven guilty.

When using any statistical method to draw conclusions, we must be aware of two types of error that can be associated with our conclusions: type I and type II errors.

Type I error is an error of exclusion. It occurs when a conclusion is drawn such that we conclude that a data set does not meet a given criteria even though it does. It is the equivalent of saying that a person is guilty even though they are innocent.

Type II error is an error of inclusion. It occurs when a conclusion is drawn such that we conclude that a data set meets a given criteria even though it does not. It is the equivalent of saying that a person is innocent even though he is guilty.

In the criminal justice process, we can never be sure if a person is innocent or guilty (only the person truly knows). Similarly, since we are sampling from a larger population, we can never truly know with 100% certainty that a conclusion will indeed match the true population. However, type I error can be readily controlled through the use of statistical infence technique. Type II error can also be limited by ensuring that adequate sample sizes are chosen for the sampling process.

4.5 Statistical Inference for a Single Sample

One of the most basic tools of statistical inference is drawing a conclusion about the overall population from the data that has been collected. A confidence interval can help us draw these types of conclusions. A **confidence interval** is an interval around a parameter of a distribution, such as the mean. It is constructed around the sample mean using the sample standard deviation. The term confidence in confidence interval relates to the risk level taken to construct the interval.

Consider the case of capacity planning. The demand over the past few months has averaged 100 units per month with a standard deviation of 20 units. It is your job to determine how many workers are needed to cover demand.

One approach is to create a confidence interval that will represent almost all possible demand values. This would be a 99.97% confidence interval and would cover the range of $\pm$ 4 standard deviations around the mean. The confidence interval would be 20–180. This is not very practical from a planning perspective.

Let's try a 90% confidence interval. This means if we maintain similar business conditions then than 90 times we would receive a value within the confidence interval and 10 times we would not. Some interpret this as a 90% chance of having a demand within the confidence interval. This would reduce the range to 67.1–132.9. This is a bit more helpful. An 80% confidence interval would be 74.4–125.6. These ranges are shown in Figure 4.22.

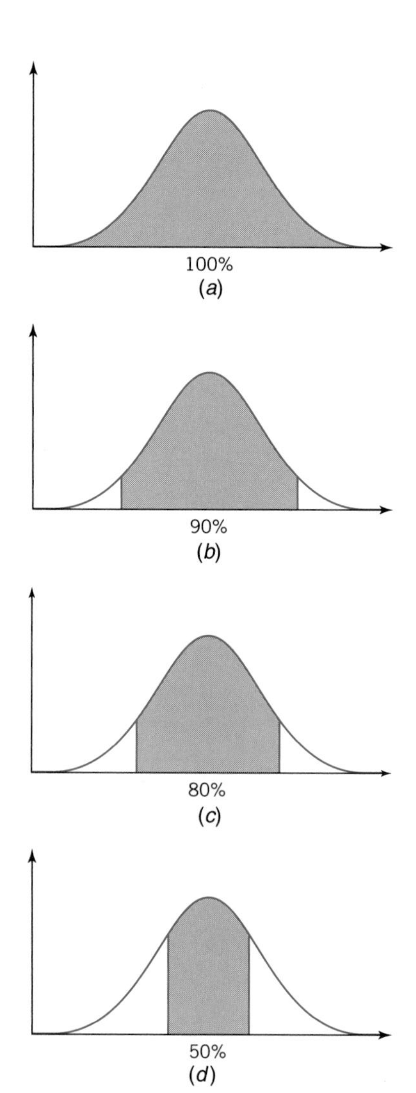

Figure 4.22 Illustration of Confidence Intervals

We can calculate confidence intervals for the overall data spread as shown above. However, it is also possible to construct confidence intervals for estimates of means, standard deviations, proportions (for example 50% prefer our product to our competitors) and variances. Each of these will be described in this section.

4.5.1 CONFIDENCE INTERVAL FOR A MEAN WHEN MORE THAN 30 DATA POINTS ARE AVAILABLE

A confidence interval for the mean expresses a range in which we expect the true population mean to be within. When more than 30 points are available, we can be more certain in our estimates of the mean and standard deviation and use the following formula:

$$\bar{x} - z_{\alpha/2}\frac{\sigma}{\sqrt{n}} \leq \mu \leq \bar{x} + z_{\alpha/2}\frac{\sigma}{\sqrt{n}} \tag{4.8}$$

where $\bar{x}$ is the sample mean, σ is the standard deviation, n is the number of data points used to compute the average, α is the risk level, and $z_{\alpha/2}$ is the corresponding z-value. Note that this formula does not describe where all of the data will be. It just describes an estimate of where the mean will be, as is reflected by the fact that the standard deviation is divided by the square root of the number of data points. This comes directly from the central limit theorem.

This formula assumes that the risk of not being within the confidence interval (α) is dispersed on both sides of the interval. Thus, if we take a risk of 10%, we would put 5% on the low side and 5% on the high side. This is why alpha is divided by 2. It is also possible to create a one-sided confidence interval in which all of the risk is concentrated on one side.

Upper confidence interval: $$\mu \leq \bar{x} + z_{\alpha}\frac{\alpha}{\sqrt{n}} \tag{4.9}$$

Lower confidence interval: $$\bar{x} - z_{\alpha}\frac{\alpha}{\sqrt{n}} \leq \mu \tag{4.10}$$

Interpretation of a Confidence Interval (CI)

A confidence interval is a random interval that covers the parameter with a specific probability, that is, a 95% CI on the mean of a normal distribution is an interval that if we took many repeated samples from the population, 95% of the CIs constructed would include the true value of the mean.

Invalidity of CI

A CI isn't valid unless the samples are random samples.

Two Sided Confidence Interval

Spreads risk equally on both sides. For example, a 90% confidence interval would be designed to exclude the lower 5th percentile and 95th percentile and above (a total of 10%).

One-Sided Confidence Interval

Risk is concentrated in one-side for example, an upper 90% CI would exclude the 90th percentile and above (10%).

example 4.11 Calculating a Confidence Interval for the Mean When More than 30 Points are Available

Joe's coffee company has purchased new equipment to help make its specialty drinks faster. Previously, the average service time per customer was 90 seconds. The system was installed two weeks ago and a sample of 100 customers has shown an average service time of 80 seconds with a standard deviation of 4 seconds. Create a 90% two-sided confidence interval to describe customer service times.

To create a 90% two-sided confidence interval, we distribute 5% of the risk to the low side and 5% to the high side (10%/2). Thus, we need to find the z values for 0.05

example 4.11 Continued

and make it a positive value. This value is 1.65. Thus, the confidence interval can be calculated as

$$\bar{x} - z_{\alpha/2}\frac{\sigma}{\sqrt{n}} \leq \mu \leq \bar{x} + z_{\alpha/2}\frac{\sigma}{\sqrt{n}}$$

$$80 - 1.65\frac{4}{\sqrt{100}} \leq \mu \leq 80 + 1.65\frac{4}{\sqrt{100}}$$

$$79.34 \leq \mu \leq 80.66$$

Using a 90% confidence interval, we conclude that the average service time is between 79.34 and 80.66 seconds. However, we can also make another conclusion. Previously the average service time was 90 seconds per customer. Because our confidence interval does not contain 90 seconds, we can also conclude that the purchase of the new equipment has resulted in an improvement of customer service times.

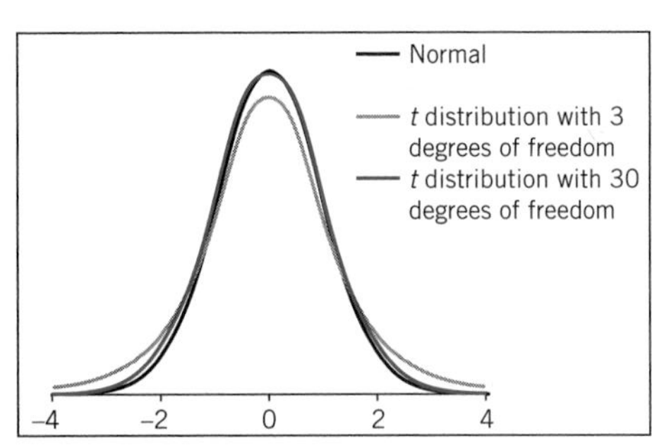

Figure 4.23 Shape of T-Distribution Compared to Normal Distribution

4.5.2 CONFIDENCE INTERVAL FOR A MEAN WHEN LESS THAN 30 DATA POINTS ARE AVAILABLE

When less than 30 data points are available, we need to make a slight adjustment. Recall that the standard deviation calculates the difference between a data point and the mean. If only a few points are available for calculation, it is much more likely that they will be in the center of the distribution than toward the tails. The *t*-distribution accounts for the small number of data points to adjust the multiplier accordingly. This is shown in Figure 4.23.

The *t*-distribution has a parameter, called the degrees of freedom (df). Degrees of freedom are calculated as the number of data points −1. Degrees of freedom will vary depending upon the problem being solved, but in the case of a confidence interval, we calculate it as the number of data points −1.

Note how the *t*-distribution has a larger proportion of data in the tails for smaller data than larger data sets. Also, notice how the *t*-distribution approximates the normal distribution when there are 30 degrees of freedom. This is exactly what we are looking for in our multiplier. We want to create a slightly larger confidence interval when we have smaller data sets than larger data sets.

The formulas to calculate a confidence interval for the mean when there are less than 30 data points are

$$\text{Two-Sided:} \quad \bar{x} - t_{\alpha/2,n-1}\frac{s}{\sqrt{n}} \leq \mu \leq \bar{x} + t_{\alpha/2,n-1}\frac{s}{\sqrt{n}} \qquad (4.11)$$

$$\text{Lower:} \quad \mu \leq \bar{x} + t_{\alpha,n-1}\frac{s}{\sqrt{n}} \qquad (4.12)$$

$$\text{Upper:} \quad \bar{x} - t_{\alpha,n-1}\frac{s}{\sqrt{n}} \leq \mu \qquad (4.13)$$

where $\bar{x}$ is the sample mean, s is the standard deviation, n is the number of data points, α is the risk level, and $t_{\alpha/2,n-1}$ is the corresponding *t*-value.

> **Excel Tip**
>
> TINV(Probability, degrees of freedom) will find a *t* value associated with a given probability value and degrees of freedom.
>
> CAUTION: Excel assumes that you are obtaining *t* for a two-sided test. Therefore, enter the whole probability in the probability value (i.e., not $\alpha/2$). If you are interested in a one-sided confidence interval enter 2α.
>
> Example: Calculation of *t* for service time example. We wish to construct a two-sided 90% confidence interval and have 9 degrees of freedom. Thus, we enter tinv(0.1,9) which yields the result of 1.83.

example 4.12 Calculating a Confidence Interval for the Mean When Less than 30 Points are Available

Suppose we had only taken 10 samples in the Joe's coffee shop example to obtain a sample average of 80 and a sample standard deviation of 4. How would this change the two-sided 90% confidence interval for the mean service time?

To create a 90% two-sided confidence interval, 5% of the risk would be distributed to the low side and 5% would be distributed to the high side (10%/2). Thus, the confidence interval can be calculated as

$$\bar{x} - t_{\frac{\alpha}{2},n-1}\frac{s}{\sqrt{n}} \leq \mu \leq \bar{x} + t_{\frac{\alpha}{2},n-1}\frac{s}{\sqrt{n}}$$

$$80 - 1.83\frac{4}{\sqrt{10}} \leq \mu \leq 80 + 1.83\frac{4}{\sqrt{10}}$$

$$77.7 \leq \mu \leq 82.3$$

Using a 90% confidence interval, we conclude that the average service time is between 77.7 and 82.3 seconds. This is considerably larger than our estimates from the larger population and reflects the smaller number of data points used to form the estimate of the standard deviation. However, since the average service time is between 77.7 and 82.3 seconds, we can still conclude that the purchase of the new equipment has resulted in an improvement of customer service times.

4.5.3 CONFIDENCE INTERVAL FOR PROPORTIONS

Confidence intervals can also be created for a population proportion, p. This comes in very helpful when evaluating market data and product nonconformance data. However, please note that the formula below is only applicable when more than 30 data points are available and the proportion is greater than 10%.

$$\text{Two-Sided:} \quad \hat{p} - z_{\alpha/2}\sqrt{\frac{\hat{p}(1-\hat{p})}{n}} \leq p \leq \hat{p} - z_{\alpha/2}\sqrt{\frac{\hat{p}(1-\hat{p})}{n}} \qquad \textbf{(4.14)}$$

$$\text{Lower:} \quad \hat{p} - z_{\alpha}\sqrt{\frac{\hat{p}(1-\hat{p})}{n}} \leq p \qquad \textbf{(4.15)}$$

$$\text{Upper:} \quad p \leq \hat{p} + z_{\alpha}\sqrt{\frac{\hat{p}(1-\hat{p})}{n}} \qquad \textbf{(4.16)}$$

example 4.13 Calculating a Confidence Interval for Proportion

In a random sample of 80 home mortgage applications processed by an automated decision system. 15 of the applications were not approved. The point estimate of the fraction that was not approved is

$$\hat{p} = \frac{15}{80} = 0.1875$$

example 4.13 Continued

Assuming that the normal approximation to the binomial is appropriate, find a 95% confidence interval on the fraction of nonconforming mortgage applications in the process.

The desired confidence interval is found from Equation 4.14 as

$$0.1875 - 1.96\sqrt{\frac{0.1875(0.8125)}{80}} \leq p \leq 0.1875 + 1.96\sqrt{\frac{0.1875(0.8125)}{80}}$$

which reduces to

$$0.1020 \leq p \leq 0.2730$$

Thus, we expect between 10 and 27 percent of mortgage applications to be denied based upon this data.

4.5.4 CONFIDENCE INTERVAL FOR VARIANCES

Variance

The variance is the square of the standard deviation.

In quality control applications, it is equally important to focus upon the variance of the process as well as the average value. In fact, reducing the variability is sometimes much more important, especially in highly utilized systems. By reducing variability, we can often receive significant performance gains from simply having a more consistent process.

Confidence intervals are often used to describe the variance associated with various processes.

$$\text{Two sided:} \quad \frac{(n-1)s^2}{\chi^2_{\alpha/2,n-1}} \leq \sigma^2 \frac{(n-1)s^2}{\chi^2_{1-\alpha/2,n-1}} \tag{4.17}$$

$$\text{Upper:} \quad \sigma^2 \leq \frac{(n-1)s^2}{\chi^2_{1-\alpha,n-1}} \tag{4.18}$$

$$\text{Lower:} \quad \frac{(n-1)s^2}{\chi^2_{\alpha,n-1}} \leq \sigma^2 \tag{4.19}$$

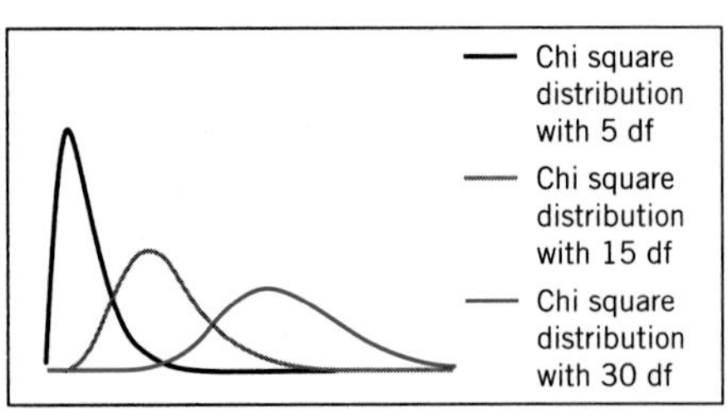

Figure 4.24 Chi Square Distributions

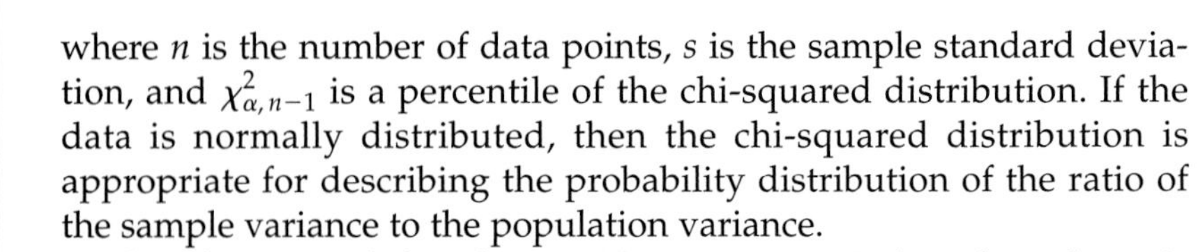

where n is the number of data points, s is the sample standard deviation, and $\chi^2_{\alpha,n-1}$ is a percentile of the chi-squared distribution. If the data is normally distributed, then the chi-squared distribution is appropriate for describing the probability distribution of the ratio of the sample variance to the population variance.

This chi-squared distribution is not symmetrical, and its shape is dependent upon the number of degrees of freedom see Figure 4.24.

example 4.14 Calculating a Confidence Interval for Variances

For the coffee shop example, the raw data is given for the service times of 25 customers. Our previous standard deviation was 5 seconds. Develop 90% confidence intervals on upper side to determine if we have reduced the standard deviation with the new equipment purchase.

81.66	76.84	76.64	82.13	83.35
71.18	80.29	76.75	78.00	79.85
83.71	72.70	84.90	76.23	81.81
80.37	83.21	81.23	80.60	78.09
87.57	81.64	87.82	76.23	77.22

Using this data set, the sample standard deviation is 4.008. There are 25 data points, so there are 24 degrees of freedom. Since we are building a 90% confidence interval, alpha = 10%.

$$\sigma^2 \leq \frac{(n-1)s^2}{\lambda^2_{1-\alpha,n-1}}$$

$$\sigma^2 \leq \frac{(25-1)4.08^2}{\lambda^2_{.9,24}} =$$

$$\sigma^2 \leq \frac{(25-1)4.08^2}{15.56}$$

$$\sigma^2 \leq 25.54$$

$$\sigma \leq 5.05$$

Thus, the 90% upper bound of the standard deviation is less than 5.05. Given this conclusion, we cannot state that we have seen an improvement in the standard deviation. The previous standard deviation was 5 seconds, which is within our confidence interval for the new data. Thus, we cannot conclude that that the new equipment resulted in a decrease of the standard deviation.

4.6 Statistical Inference for Two Samples

The previous section focused upon creating confidence intervals on a single population. However, there are many business situations in which we want to compare two populations to see if there really is a difference. For example, do certain types of acne medicines work better for college students than young teenagers? This is a distinct performance question as well as marketing information. Other examples include the profitability of two different system designs, product portfolios, or marketing campaigns.

When we compare two different items or groups, we would like to determine if they are the same or if they are different. In the case of comparing the mean performance, we could obtain several data points from

Excel Tip

CHIINV(Probability, degrees of freedom) will find a chi-squared value associated with a given probability value and degrees of freedom.

Example: Calculation of chi-squared value for service time example. We wish to construct

(Continued)

a one-sided 90% confidence interval and have 24 degrees of freedom. Thus, we enter chiinv(0.9,24), which yields the result of 15.56.

each group, calculate their means, and see if the difference between the means is 0. If so, we could conclude that there is no difference between the two groups. (E.g., if portfolio A has an average performance of 5 MM and portfolio B has an average performance of 5.05 MM, we might say that the difference of 0.05 is negligible and that they are the same.)

However, due to fact that only a few data points were collected from a much larger population it is wiser to build a confidence interval around difference. This will account for the variability in each population and the number of data points collected. This is illustrated in the graphic below:

The graph on the top in Figure 4.25 has two distributions. Both distributions have a standard deviation of 5. However, the mean of the blue curve is 100 and the mean of the green curve is 101. The means are the same for the graph on the bottom, but the standard deviations are now 0.5. Visually, it is easier to see the difference between the graph on the right and the graph on the left. Confidence intervals numerically account for the variability in the two groups and the fact that the number of data points may be different. These calculations are described below.

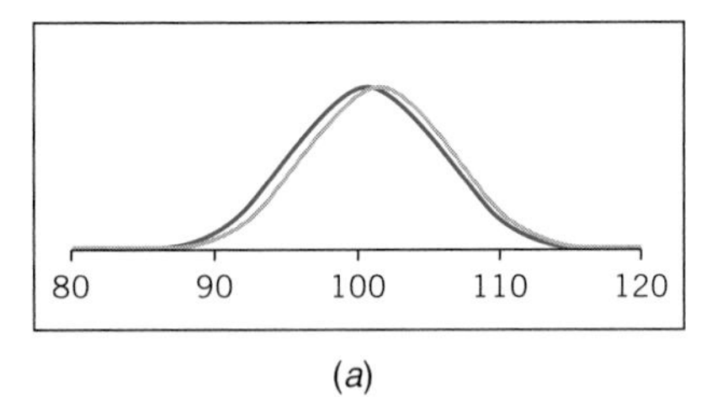

(a)

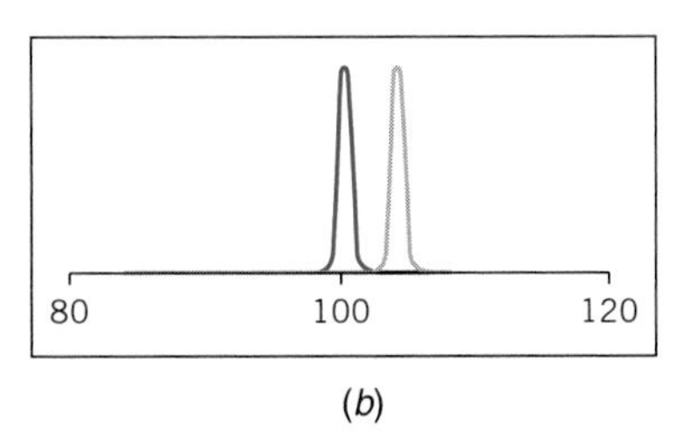

(b)

Figure 4.25 Comparison of the Means of Two Populations

4.6.1 CONFIDENCE INTERVAL FOR DIFFERENCES IN MEANS WHEN MORE THAN 30 DATA POINTS ARE AVAILABLE

When we want to compare two populations and have more than 30 points from each population, then our confidence interval for a mean changes from

Confidence interval for a mean

$$\bar{x} - z_{\alpha/2}\frac{\sigma}{\sqrt{n}} \leq \mu \leq \bar{x} + z_{\alpha/2}\frac{\sigma}{\sqrt{n}} \quad (4.8)$$

to

Confidence interval for difference between two means

$$\bar{x}_1 - \bar{x}_2 - z_{\alpha/2}\sqrt{\frac{\sigma_1^2}{n_1} + \frac{\sigma_2^2}{n_2}} \leq \mu_1 - \mu_2 \leq \bar{x}_1 - \bar{x}_2 + z_{\alpha/2}\sqrt{\frac{\sigma_1^2}{n_1} + \frac{\sigma_2^2}{n_2}} \quad (4.20)$$

These equations are quite similar. Notice that $\bar{x}$ is now replaced with the difference between the mean of population 1 and population 2. Also, $\sigma/\sqrt{n}$ is replaced with a mathematical equivalent of $\sqrt{\sigma^2 n}$ but this is expanded to reflect the sample standard deviations from of the difference between the two means.

example 4.15 Calculating a Confidence Interval for the Difference in Two Means When More than 30 Points are Available

Joe student lives in Ohio and is considering whether or not to pay for a test prep service. The test prep service is relatively new but has been collecting data for marketing purposes.

example 4.15 Continued

Average performance of students who did not take test prep:

Mean = 1050, standard deviation = 148
Number of students sampled = 100

Average performance of students who have used the test prep service:

Mean = 1250, standard deviation = 180
Number of students sampled = 50

The test prep service is concluding that its service is providing a significant benefit to its students. Is this correct?

To answer this question, we must consider the source. How did the test prep service obtain the data? Is it valid and consistent (e.g., did they call students who did not make the honor roll at the local school, did they ensure that the test is limited to first time test takers, etc.).

Assuming that the data is valid, we can begin our analysis using a two-sided 95% confidence interval. We will also denote the test prep service as population 1 so that the difference between the two will be positive (this is a somewhat arbitrary decision).

$$1250 - 1050 - 1.96\sqrt{\frac{180^2}{50} + \frac{148^2}{100}} \le \mu_1 - \mu_2$$

$$\le 1250 - 1050 + 1.96\sqrt{\frac{180^2}{50} + \frac{148^2}{100}}$$

$$142.3 \le \mu_1 - \mu_2 \le 257.7$$

The difference in mean performance for the two groups is between 142.3 and 257.7 using a 95% confidence interval. This indicates that there is a significant difference in performance and that students enrolled in the prep service are scoring on average 142–258 points higher than students who are not.

4.6.2 CONFIDENCE INTERVAL FOR DIFFERENCES IN MEANS WHEN LESS THAN 30 DATA POINTS ARE AVAILABLE

When we have less than 30 data points in each population, we use multipliers from the t-distribution. Our confidence interval for a mean changer from

$$\bar{x} - t_{\alpha/2,n-1}\frac{s}{\sqrt{n}} \le \mu \le \bar{x} + t_{\alpha/2,n-1}\frac{s}{\sqrt{n}} \quad (4.11)$$

Confidence interval for a mean when less than 30 data points are available

to

$$\bar{x}_1 - \bar{x}_2 - t_{\alpha/2,v}\sqrt{\frac{s_1^2}{n_1} + \frac{s_2^2}{n_2}} \le \mu_1 - \mu_2 \le \bar{x}_1 - \bar{x}_2 + t_{\alpha/2,v}\sqrt{\frac{s_1^2}{n_1} + \frac{s_2^2}{n_2}} \quad (4.21)$$

Confidence interval for differences in means when less than 30 data points are available

where ν = pooled degrees of freedom. This is calculated as

$$\nu = \frac{\left(\frac{s_1^2}{n_1} + \frac{s_2^2}{n_2}\right)^2}{\frac{(s_1^2/n_1)^2}{n_1 - 1} + \frac{(s_2^2/n_2)^2}{n_2 - 1}} \quad (4.22)$$

If ν is not a whole number, it is rounded down to the next-smallest whole number.

example 4.16 Calculating a Confidence Interval for the Differences in Two Means When less than 30 Points are Available

Two different arthroscopic procedures are being tested through a clinical trial. Clinical group A has enrolled 10 patients who have experienced a mean recovery time of 60 days with a standard deviation of 10 days. Clinical group B has enrolled 15 patients who have experienced a mean recovery time of 50 days with a standard deviation of 15 days. Using a 99% confidence interval, is there a difference between the two groups?

To begin, we calculate V the degrees of freedom. To simplify the calculations, we first calculate $\frac{s_1^2}{n_1}$ and $\frac{s_2^2}{n_2}$ which reduce to 10 and 15, respectively.

$$\nu = \frac{\left(\frac{s_1^2}{n_1} + \frac{s_2^2}{n_2}\right)^2}{\frac{(s_1^2/n_1)^2}{n_1 - 1} + \frac{(s_2^2/n_2)^2}{n_2 - 1}} = \frac{(10 + 15)^2}{\frac{(10)^2}{9} + \frac{(15)^2}{14}} = 22.97 \text{ which is rounded down to } = 22.0$$

We then find $t_{0.05/2,22} = 1.717$. The confidence interval is then calculated as

$$60 - 50 - 1.717\sqrt{\frac{10^2}{10} + \frac{15^2}{15}} \leq \mu_1 - \mu_2 \leq 60 - 50 + 1.717\sqrt{\frac{10^2}{10} + \frac{15^2}{15}}$$

$$1.42 \leq \mu_1 - \mu_2 \leq 18.59$$

Thus, there appears to be an improvement in recovery time between 1.42 and 18.59 days.

4.6.3 CONFIDENCE INTERVAL FOR THE DIFFERENCES OF PROPORTIONS

Confidence intervals can also be created for differences in population proportions, p. However, please note that the formula below is only applicable when more than 30 data points are available and the proportion is greater than 10%. The confidence interval on the difference in two proportions closely follows the confidence interval on a single proportion.

Confidence interval in a proportion

$$\hat{p} - z_{\alpha/2}\sqrt{\frac{\hat{p}(1-\hat{p})}{n}} \le \hat{p} - z_{\alpha/2}\sqrt{\frac{\hat{p}(1-\hat{p})}{n}} \quad (4.14)$$

> **Note**
>
> When preparing confidence intervals on the differences of means, we assume that the populations are independent. If before and after type testing is performed, a paired test should be utilized. We advise the reader to review Montgomery and Runger (2007) for more information on this topic.

Confidence interval for differences in proportions

$$\hat{p}_1 - \hat{p}_2 - z_{\alpha/2}\sqrt{\frac{\hat{p}_1(1-\hat{p}_1)}{n_1} + \frac{\hat{p}_2(1-\hat{p}_2)}{n_2}} \le p_1 - p_2$$

$$\le \hat{p}_1 - \hat{p}_2 - z_{\alpha/2}\sqrt{\frac{\hat{p}_1(1-\hat{p}_1)}{n_1} + \frac{\hat{p}_2(1-\hat{p}_2)}{n_2}} \quad (4.23)$$

example 4.17 Calculating a Confidence Interval for the Differences in Proportions

A company is nearing final preparations prior to release of its new product. It has asked 100 random consumers in the Midwest and 200 consumers on the West Coast if they were aware of the new product. Although 85% were aware of the product in the Midwest, 90% were aware of the product on the West Coast. Is there a difference between awareness of these two market segments using an 80% confidence interval?

$$\hat{p}_1 - \hat{p}_2 - z_{\alpha/2}\sqrt{\frac{\hat{p}_1(1-\hat{p}_1)}{n_1} + \frac{\hat{p}_2(1-\hat{p}_2)}{n_2}} \le p_1 - p_2$$

$$\le \hat{p}_1 - \hat{p}_2 - z_{\alpha/2}\sqrt{\frac{\hat{p}_1(1-\hat{p}_1)}{n_1} + \frac{\hat{p}_2(1-\hat{p}_2)}{n_2}}$$

$$0.85 - 0.9 - 1.28\sqrt{\frac{0.85(0.15)}{100} + \frac{0.9(0.1)}{200}} \le p_1 - p_2$$

$$\le 0.85 - 0.9 - 1.28\sqrt{\frac{0.85(0.15)}{100} + \frac{0.9(0.1)}{200}}$$

$$-0.10 \le p_1 - p_2 \le 0.003$$

Since the confidence interval on the difference between proportions contains 0, we cannot conclude that there is a difference between awareness in the Midwest and West Coast.

4.6.4 CONFIDENCE INTERVAL FOR THE RATIO OF TWO IN VARIANCES

We have now seen how to create confidence intervals on the differences between means and proportion. Finally, we will discuss how to develop a confidence interval on the differences between variances.

When we have computed differences between means and proportions we evaluated them by calculating the difference. However, when we evaluate the difference in variances, we calculate the difference as a ratio of the variance of population 1 to population 2. If they are equal, then the ratio should equal 1. This allows us to use the F distribution where a random variable F is a ratio of two sample standard deviations.

The confidence interval for the ratio of two sample standard deviations is

$$\frac{s_1^2}{s_2^2}F_{1-\alpha/2,n_2-1,n_1-1} \leq \frac{\sigma_1^2}{\sigma_2^2} \leq \frac{s_1^2}{s_2^2}F_{\alpha/2,n_2-1,n_1-1} \tag{4.24}$$

Excel Tip

FINV(Probability, degrees of freedom 1, degrees of freedom 2) will find a F value associated with a given probability value and degrees of freedoms.

Example: Calculation of F value for DMV example. We utilize the functions FINV(0.05, 49,24) and FINV(.95, 49,24) to find the values of 1.86 and 0.57, respectively.

Given that this is a ratio, we expect this confidence interval to contain the value 1 if there is no difference between the sample standard deviation.

example 4.18 Calculating a Confidence Interval for the Difference in Variances

The DMV has implemented a new procedure for evaluating eyesight. The standard deviation of the time is takes to evaluate eyesight was previously 10 seconds (obtained from a sample of 50 clients). The new standard deviation is 8 seconds (obtained from a second sample of 25 clients). Is there an improvement in variance using a 90% confidence interval?

$$\frac{s_1^2}{s_2^2}F_{1-\alpha/2,n_2-1,n_1-1} \leq \frac{\sigma_1^2}{\sigma_2^2} \leq \frac{s_1^2}{s_2^2}F_{\alpha/2,n_2-1,n_1-1}$$

$$\frac{10^2}{8^2}(0.57) \leq \frac{\sigma_1^2}{\sigma_2^2} \leq \frac{10^2}{8^2}(1.86)$$

$$0.89 \leq \frac{\sigma_1^2}{\sigma_2^2} \leq 2.91$$

Since the confidence interval contains the value 1, we cannot conclude that the sample variances are different. There appears to be no improvement in the variance.

4.6.5 RELATIONSHIP BETWEEN CONFIDENCE INTERVALS AND FORMAL HYPOTHESIS TESTING

In the previous sections, we used confidence intervals to test hypothesis. Specifically, we made conclusions that means, variances, or proportions could be equal to specific values. However, it is also possible to draw such conclusions using formal test statistics and p-values. This will be illustrated through the following example.

example 4.19 Comparision of Confidence Intervals and Formal Hypothesis Testing

The supply chain management department at State University believes that the average off-campus GPA of its upper-class students is 3.1. It would like to test this hypothesis using formal hypothesis testing. This is described below:

1. State the null hypothesis H_o and the alternative hypothesis H_a.
   The null hypothesis states that a parameter is always equal to a target value. The alternate hypothesis is always the opposite of the null hypothesis (not equal to, less than, or greater than).
   For this example, the null hypothesis would be that the average off-campus GPA = 3.1. The alternate hypothesis is that the off-campus GPA ≠ 3.1. This is formally stated as:

$$H_0: \ \mu = 3.1$$
$$H_a: \ \mu \neq 3.1$$

2. Specify the level of significance, α.
   This is the value of type I error that the statistician wishes to use. It is the same value chosen for the confidence interval. For this case, we will use a type 1 error of 5%, so $\alpha = 0.05$.

3. Select the test statistic.
   Test statistics are available for testing means, variances, and proportions, differences in means, differences in variances, and differences in proportions. Test statistics are defined in Appendix 5. The test statistic appropriate for this problem is:

$$z = \frac{x - \mu}{\sigma/\sqrt{n}}$$

4. Collect the sample data and compute the value of the test statistic.
   The data for this problem is included below.

3.30	3.81	2.77	2.78	2.93	3.33	2.85	3.10	3.06	3.44
3.84	3.45	3.17	3.06	3.62	3.11	3.27	3.41	3.44	3.05
3.59	3.50	3.09	3.61	3.48	3.65	3.27	2.92	3.39	3.00
2.82	3.05	3.41	3.33	3.23	3.36	3.39	2.76	2.70	2.59
2.72	3.94	3.56	2.64	3.39	2.95	2.47	3.15	3.18	3.56

   From this raw data, we compute the average GPA as 3.21 and the standard deviation as 0.344. The test statistic is computed as:

$$z = \frac{\overline{x} - \mu}{\sigma/\sqrt{n}} = \frac{3.21 - 3.10}{0.344/\sqrt{50}} = 2.26$$

5. Calculate the p-value using the test statistic value.
   The p-value is a probability associated with the test statistic that expresses the weight of evidence against the null hypothesis. A small p-value implies strong evidence against H_0.
   Our z-value is 2.26. Using Excel, the probability associated with a z value of 2.26 is 0.988. This means there is a probability of 0.012 of obtaining a z value greater

example 4.19 Continued

than 2.26. Since our hypothesis test is two sided (i.e. we are either equal to 3.1 or not), we double this probability to obtain 2 (0.012) = 0.024. Our p-value is thus 0.024 this is strong evidence against the null hypothesis specially, there is only a 2.4% chance that we are wrong when we conclude that $\mu \neq 3.1$.

6. Determine the appropriate conclusion based upon the p-value. This is strong evidence against the null hypothesis specifically
   a. If the p-value is less than alpha, reject H_0.
   b. If the p-value is greater than alpha, fail to reject H_0.
   Since our p-value is less than our type I error of 5%, we reject the null hypothesis that the off-campus GPA is 3.1. There is sufficient evidence to support that the true mean is greater than this value.
   These calculations can be performed in Minitab. The results are shown in Figure 4.26 below:

```
            Test of mu = 3.1 vs not = 3.1
        The assumed standard deviation = 0.344

Variable N    Mean   StDev  SE Mean        95% CI         Z     P
  C1     50 3.2098 0.3440  0.0486 (3.1144, 3.3052) 2.26 0.024
```

Figure 4.26 **Hypothesis Testing in Minitab**

We could also draw the same conclusions using a confidence interval. Since there are 50 points, we calculate the confidence interval for a mean when there are more than 30 data points. This formula is

$$\bar{x} - z_{\alpha/2}\frac{\sigma}{\sqrt{n}} \leq \mu \leq \bar{x} + z_{\alpha/2}\frac{\sigma}{\sqrt{n}}$$

$$3.2098 - 1.95\frac{0.344}{\sqrt{50}} \leq \mu \leq 3.2098 + 1.95\frac{0.344}{\sqrt{50}}$$

$$3.1144 \leq \mu \leq 3.3052$$

Since the value of 3.10 is not contained within the confidence interval, we can conclude that the average GPA of students living off campus is not 3.10. We can also see from the confidence interval that the average GPA is higher than 3.1.

As is shown in the above example, both methods yield the same conclusions. The confidence interval provides a range for where the parameter is likely to be found. This is helpful in many situations. However, the formal hypothesis method provides more structure and slightly more information through the p-value. The use of test statistics and p-values will be used more in the experimental design chapters of this text.

supplement One Reading Cumulative Normal Distribution Table

An abbreviated normal standard distribution table is included below. The *z* column is the *z* that you would like to find a corresponding probability for (i.e., G(*z*)).

Example, find the probability associated with a *z* of 1.64. To find this column, look in the *z* columns until you find 1.64. This probability is 94.5%.

Alternately, you can find *z* for a given probability. For example, find the *z* value associated with a probability of 50%. In this case, look in the G(*z*) columns until you find 0.50. The value of *z* associated with this probability is 0.

Cumulative Standard Normal Distribution

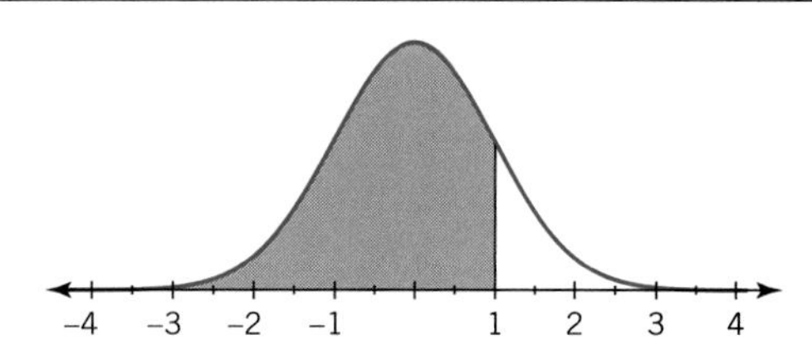

z	G(z)	z	G(z)	z	G(z)	z	G(z)	z	G(z)
−4	0.00003	−2.4	0.00820	−0.8	0.21186	0.8	0.78814	2.4	0.99180
−3.9	0.00005	−2.3	0.01072	−0.7	0.24196	0.9	0.81594	2.5	0.99379
−3.8	0.00007	−2.2	0.01390	−0.6	0.27425	1.0	0.84134	2.6	0.99534
−3.7	0.00011	−2.1	0.01786	−0.5	0.30854	1.1	0.86433	2.7	0.99653
−3.6	0.00016	−2.0	0.02275	−0.4	0.34458	1.2	0.88493	2.8	0.99744
−3.5	0.00023	−1.9	0.02872	−0.3	0.38209	1.3	0.90320	2.9	0.99813
−3.4	0.00034	−1.8	0.03593	−0.2	0.42074	1.4	0.91924	3.0	0.99865
−3.3	0.00048	−1.7	0.04457	−0.1	0.46017	1.5	0.93319	3.1	0.99903
−3.2	0.00069	−1.6	0.05480	0.0	0.50000	1.6	0.94520	3.2	0.99931
−3.1	0.00097	−1.5	0.06681	0.1	0.53983	1.7	0.95543	3.3	0.99952
−3	0.00135	−1.4	0.08076	0.2	0.57926	1.8	0.96407	3.4	0.99966
−2.9	0.00187	−1.3	0.09680	0.3	0.61791	1.9	0.97128	3.5	0.99977
−2.8	0.00256	−1.2	0.11507	0.4	0.65542	2.0	0.97725	3.6	0.99984
−2.7	0.00347	−1.1	0.13567	0.5	0.69146	2.1	0.98214	3.7	0.99989
−2.6	0.00466	−1.0	0.15866	0.6	0.72575	2.2	0.98610	3.8	0.99993
−2.5	0.00621	−0.9	0.18406	0.7	0.75804	2.3	0.98928	3.9	0.99995

supplement Two Reading a *t*-Table

A portion of a *t*-table is included below. The value alpha is the percentage of data that lies beyond the upper tail of the distribution. The value *v* is the degrees of freedom.

Example, find the *t*-value corresponding to 5% and nine degrees of freedom. To find this value, we look in the 0.05 column and find the value that corresponds to nine degrees of freedom. This value is 1.833.

Percentage Points of the *t* Distribution

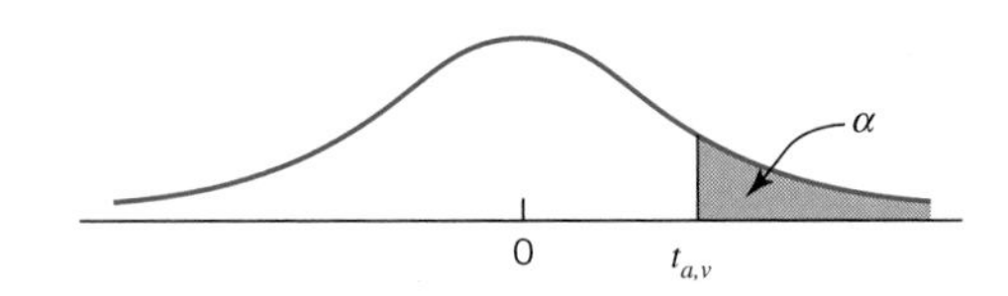

	α									
v	**0.40**	**0.25**	**0.10**	**0.05**	**0.025**	**0.01**	**0.005**	**0.0025**	**0.001**	**0.0005**
1	0.325	1.000	3.078	6.314	12.706	31.821	63.657	127.32	318.31	636.62
2	0.289	0.816	1.886	2.920	4.303	6.965	9.925	14.089	23.326	31.598
3	0.277	0.765	1.638	2.353	3.182	4.541	5.841	7.453	10.213	12.924
4	0.271	0.741	1.533	2.132	2.776	3.747	4.604	5.598	7.173	8.610
5	0.267	0.727	1.476	2.015	2.571	3.365	4.032	4.773	5.893	6.869
6	0.265	0.727	1.440	1.943	2.447	3.143	3.707	4.317	5.208	5.959
7	0.263	0.711	1.415	1.895	2.365	2.998	3.49	4.019	4.785	5.408
8	0.262	0.706	1.397	1.860	2.306	2.896	3.355	3.833	4.501	5.041
9	0.261	0.703	1.383	1.833	2.262	2.821	3.250	3.690	4.297	4.781

supplement Three Reading a Chi-Squared Table

The Chi-Squared table is very similar to a *t*-table. The value alpha is the percentage of data that lies beyond the upper tail of the distribution. The value v is the degrees of freedom.

The chi-squared table has one important point of difference—it is not symmetric. Thus, a chi-squared value for 5% will not be related to the value of 95% (both having 5% in the tails).

Example, find the chi-squared value corresponding to 5% and 95% with nine degrees of freedom. To find this value, we look in the 0.05 column and 0.95 column and find the value that corresponds to nine degrees of freedom. These values are 16.92 and 3.33, respectively.

Percentage Points of the χ^2 Distribution

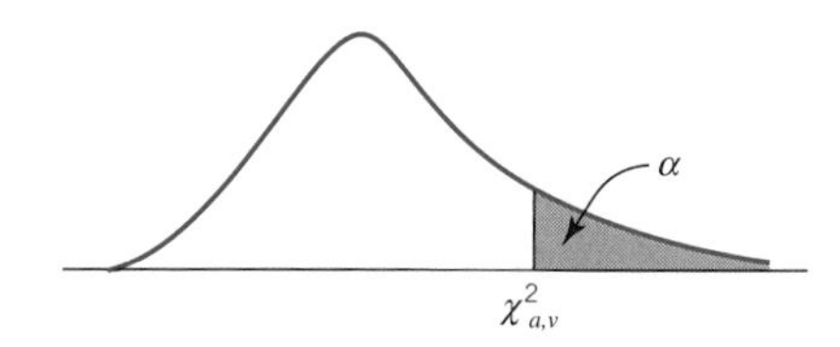

				α					
v	**0.995**	**0.990**	**0.975**	**0.950**	**0.500**	**0.050**	**0.025**	**0.010**	**0.005**
1	0.00+	0.00+	0.00+	0.00+	0.45	3.84	5.02	6.63	7.88
2	0.01	0.02	0.05	0.10	1.39	5.99	7.38	9.21	10.60
3	0.07	0.11	0.22	0.35	2.37	7.81	9.35	11.34	12.84
4	0.21	0.30	0.48	0.71	3.36	9.49	11.14	13.28	14.86
5	0.41	0.55	0.83	1.15	4.35	11.07	12.38	15.09	16.75
6	0.68	0.87	1.24	1.64	5.35	12.59	14.45	16.81	18.55
7	0.99	1.24	1.69	2.17	6.35	14.07	16.01	18.48	20.28
8	1.34	1.65	2.18	2.73	7.34	15.51	17.53	20.09	21.96
9	1.73	2.09	2.70	3.33	8.34	16.92	19.02	21.67	23.59

supplement Four Reading an *F*-Table

The *F*-table requires the knowledge of 3 parameters: alpha, degrees of freedom 1, and degrees of freedom 2. Degrees of freedom 1 and 2 are specified through the statistical test. The *F* value is denoted as $F_{\alpha, df1, df2}$.

There are different *F* tables for different probabilities. The one shown below is for an alpha of 0.25. That is all of the *F* value in the table have probability 0.25 above them. To find, the value of $F_{0.25, 5, 10}$, you would first find the *F*-table for 0.25. Then we look down the column (v_1) associated with five degrees of freedom and find the value associated with 10 degrees of freedom (v_2). This value is 1.59.

Percentage Points of the *F* Distribution

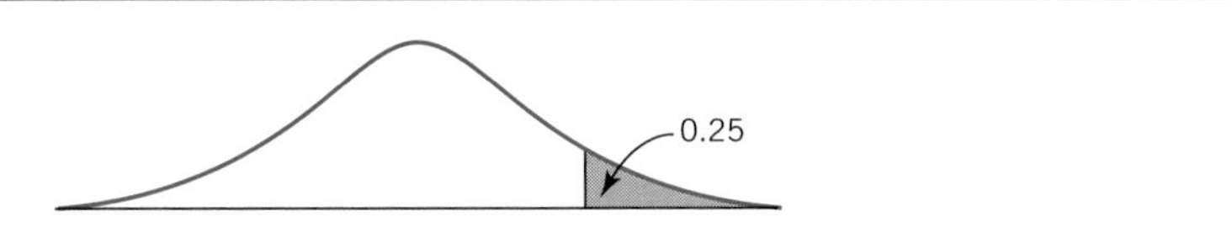

$F_{0.25, v_1, v_2}$

v_2 \ v_1	Degrees of Freedom for the Numerator (v_1)												
Degrees of Freedom for the Denominator (v_2)	1	2	3	4	5	6	7	8	9	10	12	15	20
1	5.83	7.50	8.20	8.58	8.82	8.98	9.10	9.19	9.26	9.32	9.41	9.49	9.5
2	2.57	3.00	3.15	3.28	3.28	3.31	3.34	3.35	3.37	3.38	3.39	3.41	3.4
3	2.02	2.28	2.36	2.39	2.41	2.42	2.43	2.44	2.44	2.44	2.45	2.46	2.4
4	1.81	2.00	2.05	2.06	2.07	2.08	2.08	2.08	2.08	2.08	2.08	2.08	2.0
5	1.69	1.85	1.88	1.89	1.89	1.89	1.89	1.89	1.89	1.89	1.89	1.89	1.8
6	1.62	1.76	1.78	1.79	1.79	1.78	1.78	1.78	1.77	1.77	1.77	1.76	1.7
7	1.57	1.70	1.72	1.72	1.71	1.71	1.70	1.70	1.70	1.69	1.68	1.68	1.6
8	1.54	1.66	1.67	1.66	1.66	1.65	1.64	1.64	1.63	1.63	1.62	1.62	1.6
9	1.51	1.62	1.63	1.63	1.62	1.61	1.60	1.60	1.59	1.59	1.58	1.57	1.5
10	1.49	1.60	1.60	1.59	1.59	1.58	1.57	1.56	1.56	1.55	1.54	1.53	1.5
11	1.47	1.58	1.58	1.57	1.56	1.55	1.54	1.53	1.53	1.52	1.51	1.50	1.4
12	1.46	1.56	1.56	1.55	1.54	1.53	1.52	1.51	1.51	1.50	1.49	1.48	1.4
13	1.45	1.55	1.55	1.53	1.52	1.51	1.50	1.49	1.49	1.48	1.47	1.46	1.4
14	1.44	1.53	1.53	1.52	1.51	1.50	1.49	1.48	1.47	1.46	1.45	1.44	1.4
15	1.43	1.52	1.52	1.51	1.49	1.48	1.47	1.46	1.46	1.45	1.44	1.43	1.4
16	1.42	1.51	1.51	1.50	1.48	1.47	1.46	1.45	1.44	1.44	1.43	1.41	1.4
17	1.42	1.51	1.50	1.40	1.47	1.46	1.45	1.44	1.43	1.43	1.41	1.40	1.3
18	1.41	1.50	1.49	1.48	1.46	1.45	1.44	1.43	1.42	1.42	1.40	1.39	1.3
19	1.41	1.49	1.49	1.47	1.46	1.44	1.43	1.42	1.41	1.41	1.40	1.38	1.3
20	1.40	1.49	1.48	1.47	1.45	1.44	1.43	1.42	1.41	1.40	1.39	1.37	1.3
21	1.40	1.48	1.48	1.46	1.44	1.43	1.42	1.41	1.40	1.39	1.38	1.37	1.3
22	1.40	1.48	1.47	1.45	1.44	1.42	1.41	1.40	1.39	1.39	1.37	1.36	1.3
23	1.39	1.47	1.47	1.45	1.43	1.42	1.41	1.40	1.39	1.38	1.37	1.35	1.3
24	1.39	1.47	1.46	1.44	1.43	1.41	1.40	1.39	1.38	1.38	1.36	1.35	1.3

$F_{0.25}, v_1, p_2$

v_2 \ v_1	Degrees of Freedom for the Numerator (v_1)												
Degrees of Freedom for the Denominator (v_2)	1	2	3	4	5	6	7	8	9	10	12	15	20
25	1.39	1.47	1.46	1.44	1.42	1.41	1.40	1.39	1.38	1.37	1.36	1.34	1.3
26	1.38	1.46	1.45	1.44	1.42	1.41	1.39	1.38	1.37	1.37	1.35	1.34	1.3
27	1.38	1.46	1.45	1.43	1.42	1.40	1.39	1.38	1.37	1.36	1.35	1.33	1.3
28	1.38	1.46	1.45	1.43	1.41	1.40	1.39	1.38	1.37	1.36	1.34	1.33	1.3
29	1.38	1.45	1.45	1.43	1.41	1.40	1.38	1.37	1.36	1.35	1.34	1.32	1.3
30	1.38	1.45	1.44	1.42	1.41	1.39	1.38	1.37	1.36	1.35	1.34	1.32	1.3
40	1.36	1.44	1.42	1.40	1.39	1.37	1.36	1.35	1.34	1.33	1.31	1.30	1.2
60	1.35	1.42	1.41	1.38	1.37	1.35	1.33	1.32	1.31	1.30	1.29	1.27	1.2
120	1.34	1.40	1.39	1.37	1.35	1.33	1.31	1.30	1.29	1.28	1.26	1.24	1.2
∞	1.32	1.39	1.37	1.35	1.33	1.31	1.29	1.28	1.27	1.25	1.21	1.22	1.1

supplement Five Formal Hypothesis Test Statistics

HYPOTHESIS TEST	TEST STATISTIC
Mean, more than 30 data points	$z = \dfrac{\bar{x} - \mu}{\sigma/\sqrt{n}}$
Mean, less than 30 data points	$t = \dfrac{\bar{x} - \mu}{s/\sqrt{n}}$
Variance	$\chi^2 = \dfrac{(n-1)s^2}{\sigma_0^2}$
Proportion	$z = \dfrac{\hat{p} - p_0}{\sqrt{\dfrac{p_0(1-p_0)}{n}}}$
Differences in means, more than 30 data points	$z = \dfrac{\bar{x}_1 - \bar{x}_2 - D_0}{\sqrt{\dfrac{\sigma_1^2}{n_1} + \dfrac{\sigma_2^2}{n_2}}}$
Differences in means, less than 30 data points	$t = \dfrac{\bar{x}_1 - \bar{x}_2 - D_0}{\sqrt{\dfrac{s_1^2}{n_1} + \dfrac{s_2^2}{n_2}}}$ $\nu = \dfrac{\left(\dfrac{s_1^2}{n_1} + \dfrac{s_2^2}{n_2}\right)^2}{\dfrac{(s_1^2/n_1)^2}{n_1 - 1} + \dfrac{(s_2^2/n_2)^2}{n_2 - 1}}$
Differences in variances	$F_0 = \dfrac{s_2^2}{s_1^2}$
Differences in proportions	$z_p = \dfrac{\hat{p}_1 - \hat{p}_2}{\sqrt{\hat{p}(1-\hat{p})\left(\dfrac{1}{n_1} + \dfrac{1}{n_2}\right)}}$

MINITAB

1. Arrange data in columns and name the column by clicking under the CX row heading.
2. Graphics are created using the Graphs menu bar option. Specific options are described below:
   - Stem & Leaf Plot
     Select graph –> stem & leaf plot. Double-click your column name to move it into the graph variables area. Specify an increment (if desired) and select ok to generate the plot.
   - Histogram
     Select graph –> histogram. Choose simple with or without fit. Double click your column name to move it into the graph variables area. Select OK to generate the plot.
   - Box Plot
     Select graph –> box plot. Choose simple or groups to compare multiple data values. Double click your column names to move it into the graph variables area. Select OK to generate the plot.
   - Probability Plot
     Select graph –> probability plot. Choose single or multiple to compare multiple data values. Double click your column names to move it into the graph variables area. Select ok to generate the plot.

SPC XL

1. Arrange data in columns within Excel.
2. Graphics are created using the Analysis Diagrams icon. Specific options are described below:
   - Stem & Leaf Plot
     Not available in SPC XL.
   - Histogram
     Select analysis diagrams –> histogram. SPC XL will recommend a region for the analysis. Confirm this is correct. SPC XL will then recommend a given number of classes and distribution fit. You may alter these or select next to use the recommended settings. You may then customize the axis titles as desired prior to selecting the finish button.
   - Box Plot
     Select analysis diagrams –> box plot. SPC XL will recommend a region for the analysis. Confirm this is correct. SPC XL will then generate a box plot.
   - Probability Plot
     Not available within SPC XL.

supplement Seven Generating Descriptive Statistics with Minitab and SPC XL

MINITAB

1. Arrange data in columns and name the column by clicking under the C# row heading.
2. From the menu bar, select Stat –> Basic Statistics –> Display Descriptive Statistics. Minitab will then bring up a graphical user interface. Double click the appropriate column names to move it into the graph variables area. Click on the graphs icon to specify the generation of a box plot or histogram and then select OK. Click on the statistics icon to specify which descriptive statistics you would like calculated and then select OK. When finished, select the OK button and the statistics and plots will be generated.

SPC XL

1. Arrange data in columns within Excel.
2. Select the analysis diagrams –> Summary Stats (Dot Plot) option. Select the type of analysis you wish to RUN dot plot only, summary stats only, or both. SPC XL will then perform the analysis and display the results.

   Note: Quartiles and coefficients of variation are not presented in this output. Quartiles may be found using the box plot graph, if desired.

supplement Eight Generating Confidence Intervals with Minitab and SPC XL

MINITAB

1. Arrange data in columns and name the column by clicking under the C# row heading.
2. Hypothesis tests and confidence intervals are created using the Stat menu bar. Specific instructions are provided below:
   - Single mean, more than 30 data points
     From the menu bar, select Stat –> Basic Statistics –> 1 Sample Z. Double click the appropriate column names to select for analysis. Enter the hypothesized mean and standard deviation for the data set. Use the options button to change the hypothesis and type I error (default is 5%). Select OK to generate the analysis.
   - Single mean, less than 30 data points
     From the menu bar, select Stat –> Basic Statistics –> 1 Sample t. Double click the appropriate column names to select for analysis and specify the hypothesized mean. Use the options button to change the hypothesis and type I error (default is 5%). Select OK to generate the analysis.
   - Single variance
     From the menu bar, select Stat –> Basic Statistics –> 1 Variance. At the top of the graphical user interface, specify whether you would like the results in a standard deviation or a variance. Then, select the appropriate column for analysis and specify the hypothesized variance. Use the options button to change the hypothesis and type I error (default is 5%). Select OK to generate the analysis.
   - Single proportion
     From the menu bar, select Stat –> Basic Statistics –> 1 Proportion. Note: Proportion data should be entered as two categories only (i.e., male male female male). Select the appropriate column name for the analysis and enter the hypothesized proportion. Use the options button to change the hypothesis and type I error (default is 5%). Select OK to generate the analysis.
   - Differences in means
     From the menu bar, select Stat –> Basic Statistics –> 2 Sample t. Select the radio button that the data are in two separate columns and select them for analysis. Use the options button to change the hypothesis, specify the difference amount, and type I error (default is 5%). Select OK to generate the analysis.
   - Differences in variances
     From the menu bar, select Stat –> Basic Statistics –> 2 Variances. Select the radio button that the data are in two separate columns and select them for analysis. Use the options button to change specify type I error (default is 5%). Select OK to generate the analysis.
   - Differences in proportions.
     From the menu bar, select Stat –> Basic Statistics –> 2 Proportion. Select the radio button that the data are in two separate columns and select them for analysis. Use the options button to change specify type I error (default is 5%). Select OK to generate the analysis.

SPC XL

1. Arrange data in columns within Excel. Calculate the sample mean and standard deviation for your data. If you are analyzing proportions, determine the number of desirable events within your data.

2. Hypothesis tests and confidence intervals are created using the Analysis Tools icon. Specific instructions are provided below:

   - Single mean
     From the menu bar, select Analysis Tools –> Confidence Interval –> Mean. SPC XL will bring up a new worksheet for the analysis. Enter the type of confidence interval you would like to have (e.g., 95%), number of data points, sample average, and sample standard deviation for your data. SPC XL will return the confidence interval. (Note: this method uses the t-distribution.)

   - Single variance
     From the menu bar, select Analysis Tools –> Confidence Interval –> Std Deviation. SPC XL will bring up a new worksheet for the analysis. Enter the type of confidence interval you would like to have (e.g. 95%), number of data points, and sample standard deviation for your data. SPC XL will return the confidence interval.

   - Single proportion
     From the menu bar, select Analysis Tools –> Confidence Interval –> Proportion. SPC XL will bring up a new worksheet for the analysis. Enter the type of confidence interval you would like to have (e.g. 95%), total number of data points, and number of desirable (or undesirable) events observed. SPC XL will return the confidence interval for the proportion of events specified.

   - Differences in means.
     From the menu bar, select Analysis Tools –> T-Test Matrix. SPC XL will then ask you to define where your data is stored. Define this region and select OK. Another window will appear to ask if your data is stored in rows or columns. Indicate that your data is stored in columns and select OK. Finally, indicate which type OK hypothesis you would like to test. SPC XL will then return summary statistics for your data and a p-value for the t-test. The p-value will be colored blue if 5–10% and red if less than 5%.

   - Differences in variances
     From the menu bar, select Analysis Tools –> F-Test Matrix. SPC XL will then ask you to define where your data is stored. Define this region and select OK. Another window will appear to ask if your data is stored in rows or columns. Indicate that your data is stored in columns and select ok. Finally indicate which type of hypothesis you would like to test. SPC XL will then return summary statistics for your data and a p-value for the F-test. The p-value will be colored blue if 5–10% and red if less than 5%.

   - Differences in proportion
     From the menu bar, select Analysis Tools –> Test of Proportions. SPC XL will bring up a new worksheet for the analysis. Enter the total number of data points in each sample as well as the number of desirable (or undesirable) events observed. When presumed indicate which type of hypothesis you would like to test. SPC XL will then return summary statistics for your data and a p-value for the F-test. The p-value will be colored blue if 5–10% and red if less than 5%.

Important Terms and Concepts

Alternative hypothesis
Analysis of variance (ANOVA)
Approximations to probability distributions
Binomial distribution
Box plot
Central limit theorem
Checking assumptions for statistical inference procedures
Chi-square distribution
Confidence interval
Confidence intervals on means, known variance(s)
Confidence intervals on means, unknown variance(s)
Confidence intervals on proportions
Confidence intervals on the variance of a normal distribution
Confidence intervals on the variances of two normal distributions
Continuous distribution
Control limit theorem
Critical region for a test statistic
Descriptive statistics
Discrete distribution
Exponential distribution
F-distribution
Histogram
Hypothesis testing
Interquartile range
Mean of a distribution
Median
Negative binomial distribution
Normal distribution
Normal probability plot
Parameters of a distribution
Percentile
Poisson distribution
Population
Power of a statistical test
Probability distribution
Probability plotting
P-value
Quartile
Random sample
Random variable
Residual analysis
Run chart
Sampling distribution
Standard deviation
Standard normal distribution
Statistic
Statistics
Stem-and-leaf display
t-distribution
Test statistic
Tests of hypotheses on means, known variance(s)
Tests of hypotheses on means, unknown variance(s)
Tests of hypotheses on proportions
Tests of hypotheses on the variance of a normal distribution
Tests of hypotheses on the variances of two normal distributions
Time series plot
Type I error
Type II error
Uniform distribution
Variance of a distribution

chapter Four Exercises

4.1 The fill amount of liquid detergent bottles is being analyzed. Twelve bottles, randomly selected from the process, are measured, and the results are as follows (in fluid ounces): 16.05, 16.03, 16.02, 16.04, 16.05, 16.01, 16.02, 16.02, 16.03, 16.01, 16.00, 16.07.

a. Calculate the sample average.
b. Calculate the sample standard deviation.

4.2 Monthly sales for tissues in the northwest region are (in thousands) 50.001, 50.002, 49.998, 50.006, 50.005, 49.996, 50.003, 50.004.

a. Calculate the sample average.
b. Calculate the sample standard deviation.

4.3 Waiting times for customers in an airline reservation system are (in seconds) 953, 955, 948, 951, 957, 949, 954, 950, 959.

a. Calculate the sample average.
b. Calculate the sample standard deviation.

4.4 Consider the waiting time data in Exercise 4.3.

a. Find the sample median of these data.
b. How much could the largest time increase without changing the sample median?

4.5 The time to complete an order (in seconds) is as follows: 96, 102, 104, 108, 126, 128, 150, 156.

a. Calculate the sample average.
b. Calculate the sample standard deviation.

4.6 The time to failure in hours of an electronic component subjected to an accelerated life test is shown in Table E4.1. To accelerate the failure test, the units were tested at an elevated temperature (read down, then across).

Table 4E.1

Electronic Component Failure Time

127	124	121	118
125	123	136	131
131	120	140	125
124	119	137	133
129	128	125	141
121	133	124	125
142	137	128	140
151	124	129	131
160	142	130	129
125	123	122	126

a. Calculate the sample average and standard deviation.
b. Construct a histogram.
c. Construct a stem-and-leaf plot.
d. Find the sample median and the lower and upper quartiles.

4.7 An article in *Quality Engineering* (Vol. 4, 1992, pp. 487–495) presents viscosity data from a batch chemical process. A sample of these data is presented in Table E4.2 (read down, then across).

a. Construct a stem-and-leaf display for the viscosity data.
b. Construct a frequency distribution and histogram.
c. Convert the stem-and-leaf plot in part (a) into an ordered stem-and-leaf plot. Use this graph to assist in locating the median and the upper and lower quartiles of the viscosity data.
d. What are the ninetieth and tenth percentiles of viscosity?

4.8 Construct and interpret a normal probability plot of the volumes in Exercise 4.1.

4.9 Construct and interpret a normal probability plot of the waiting time measurements in Exercise 4.3.

4.10 Construct a normal probability plot of the failure time data in Exercise 4.6. Does the assumption

Table 4E.2

Viscosity

13.3	14.9	15.8	16.0
14.5	13.7	13.7	14.9
15.3	15.2	15.1	13.6
15.3	14.5	13.4	15.3
14.3	15.3	14.1	14.3
14.8	15.6	14.8	15.6
15.2	15.8	14.3	16.1
14.5	13.3	14.3	13.9
14.6	14.1	16.4	15.2
14.1	15.4	16.9	14.4
14.3	15.2	14.2	14.0
16.1	15.2	16.9	14.4
13.1	15.9	14.9	13.7
15.5	16.5	15.2	13.8
12.6	14.8	14.4	15.6
14.6	15.1	15.2	14.5
14.3	17.0	14.6	12.8
15.4	14.9	16.4	16.1
15.2	14.8	14.2	16.6
16.8	14.0	15.7	15.6

that failure time for this component is well modeled by a normal distribution seem reasonable?

4.11 Construct a normal probability plot of the chemical process yield data in Exercise 4.7. Does the assumption that process yield is well modeled by a normal distribution seem reasonable?

4.12 Consider the yield data Table E4.3. Construct a time-series plot for these data. Interpret the plot.

4.13 Consider the chemical process yield data in Exercise 4.12. Calculate the sample average and standard deviation.

Table 4E.3

Process Yield

94.1	87.3	94.1	92.4	84.6	85.4
93.2	84.1	92.1	90.6	83.6	86.6
90.6	90.1	96.4	89.1	85.4	91.7
91.4	95.2	88.2	88.8	89.7	87.5
88.2	86.1	86.4	86.4	87.6	84.2
86.1	94.3	85.0	85.1	85.1	85.1
95.1	93.2	84.9	84.0	89.6	90.5
90.0	86.7	87.3	93.7	90.0	95.6
92.4	83.0	89.6	87.7	90.1	88.3
87.3	95.3	90.3	90.6	94.3	84.1
86.6	94.1	93.1	89.4	97.3	83.7
91.2	97.8	94.6	88.6	96.8	82.9
86.1	93.1	96.3	84.1	94.4	87.3
90.4	86.4	94.7	82.6	96.1	86.4
89.1	87.6	91.1	83.1	98.0	84.5

4.14 Consider the chemical process yield data in Exercise 4.12. Construct a stem-and-leaf plot and a histogram. Which display provides more information about the process?

4.15 Construct a box plot for the data in Exercise 4.1.

4.16 Construct a box plot for the data in Exercise 4.2.

4.17 Suppose that two fair dice are tossed and the sum of the dice is observed. Determine the probability distribution of x, the sum of the dice.

4.18 Find the mean and standard deviation of x in Exercise 4.17.

4.19 The tensile strength of a metal part is normally distributed with mean 40 lb and standard deviation 5 lb. If 50,000 parts are produced, how many would you expect to fail to meet a minimum specification limit of 35 lb tensile strength? How many would have a tensile strength in excess of 48 lb?

4.20 The output voltage of a power supply is normally distributed with mean 5 V and standard deviation 0.02 V. If the lower and upper specifications for voltage are 4.95 V and 5.05 V, respectively, what is the probability that a power supply selected at random will conform to the specifications on voltage?

4.21 **Continuation of Exercise 4.20.** Reconsider the power supply manufacturing process in Exercise 4-20. Suppose we wanted to improve the process. Can shifting the mean reduce the number of nonconforming units produced? How much would the process variability need to be reduced in order to have all but one out of 1,000 units conform to the specifications?

4.22 The life of an automotive battery is normally distributed with mean 900 days and standard deviation 35 days. What fraction of these batteries would be expected to survive beyond 1,000 days?

4.23 A light bulb has a normally distributed light output with mean 5,000 end foot-candles and standard deviation of 50 end foot-candles. Find a lower specification limit such that only 0.5% of the bulbs will not exceed this limit.

4.24 The specifications on an electronic component in a target-acquisition system are that its life must be between 5,000 and 10,000 h. The life is normally distributed with mean 7,500 h. The manufacturer realizes a price of $10 per unit produced; however, defective units must be replaced at a cost of $5 to the manufacturer.

Two different manufacturing processes can be used, both of which have the same mean life. However, the standard deviation of life for process 1 is 1000 h, whereas for process 2 it is only 500 h. Production costs for process 2 are twice those for process 1. What value of production costs will determine the selection between processes 1 and 2?

4.25 The tensile strength of a fiber used in manufacturing cloth is of interest to the purchaser. Previous experience indicates that the standard deviation of tensile strength is 2 psi. A random sample of eight fiber specimens is selected, and the average tensile strength is found to be 127 psi.

a. Build a 95% lower confidence interval on the mean tensile strength.
b. What can you conclude from this information?

4.26 Payment times of 100 randomly selected customers this month had an average of 35 days. The standard deviation from this group was 2 days.

a. Build a 90% two-sided confidence interval on the mean payment time.
b. Build a 99% two-sided confidence interval on the mean payment time.
c. Is it possible that the average time is 30 days?

4.27 The service life of a battery used in a cardiac pacemaker is assumed to be normally distributed. A random sample of 10 batteries is subjected to an accelerated life test by running them continuously at an elevated temperature until failure, and the following lifetimes (in hours) are obtained: 25.5, 26.1, 26.8, 23.2, 24.2, 28.4, 25.0, 27.8, 27.3, and 25.7. The manufacturer wants to be certain that the mean battery life exceeds 25 hours in accelerated lifetime testing.

a. Construct a 90% two-sided confidence interval on mean life in the accelerated test.
b. Construct a normal probability plot of the battery life data. What conclusions can you draw?

4.28 A local neighborhood has just installed speed bumps to slow traffic. Two weeks after the installation the city recorded the following speeds 500 feet after the last speed bump: 29, 29, 31, 42, 30, 24, 30, 27, 33, 44, 28, 32, 30, 24, 35, 34, 30, 23, 35, 27.

a. Find a 99% one-sided confidence interval on mean speed thickness and assess whether or not the average speed is 25 or less. Assume that the data is normally distributed.
b. Does the normality assumption seem reasonable for these data?

4.29 A company has just purchased a new billboard near the freeway. Sales for the past 10 days have been 483, 532, 444, 510, 467, 461, 450, 444, 540, and 499. Build a 95% two-sided confidence interval on sales. Is there any evidence that the billboard has increased sales from its previous average of 475 per day?

4.30 A machine is used to fill containers with a liquid product. Fill volume can be assumed to be normally distributed. A random sample of 10 containers is selected, and the net contents (oz) are as follows: 12.03, 12.01, 12.04, 12.02, 12.05, 11.98, 11.96, 12.02, 12.05, and 11.99. Suppose that the manufacturer wants to be sure that the mean net contents exceed 12 oz. What conclusions can be drawn from the data using a 95% two-sided confidence interval on the mean fill volume?

4.31 A company is evaluating the quality of aluminum alloy rods received in a recent shipment. Diameters of aluminum alloy rods produced on an extrusion machine are known to have a standard deviation of 0.0001 in. A random sample of 25 rods has an average diameter of 0.5046 in. Test whether or not the mean rod diameter is 0.5025 using a two-sided 95% confidence interval.

4.32 The output voltage of a power supply is assumed to be normally distributed. Sixteen observations taken at random on voltage are as follows: 10.35, 9.30, 10.00, 9.96, 11.65, 12.00, 11.25, 9.58, 11.54, 9.95, 10.28, 8.37, 10.44, 9.25, 9.38, and 10.85.

a. Test the hypothesis that the mean voltage equals 12 V using a 95% two-sided confidence interval.
b. Test the hypothesis that the variance equals 11 V using a 95% two-sided confidence interval.

4.33 Last month, a large national bank's average payment time was 33 days with a standard deviation of 4 days. This month, the average payment time was 33.5 days with a standard deviation of 4 days. They had 1,000 customers both months.

a. Build a 95% confidence interval on the difference between this month's and last month's payment times. Is there evidence that payment times are increasing?

4.34 Two machines are used for filling glass bottles with a soft-drink beverage. The filling processes have known standard deviations $\sigma_1 = 0.010$ liter and $\sigma_2 = 0.015$ liter, respectively. A random sample of $n_1 = 25$ bottles from machine 1 and $n_2 = 20$ bottles from machine 2 results in average net contents of 2.04 liters and 2.07 liters from machine 1 and 2, respectively. Test the hypothesis that both machines fill to the same net contents using a 95% two-sided confidence interval on the differences of fill volume. What are your conclusions?

4.35 A supplier received results on the hardness of metal from two different hardening processes (1) saltwater quenching and (2) oil quenching. The results are shown in Table 4E.4.

a. Construct a 95% confidence interval on the difference in mean hardness.
b. Construct a 95% confidence interval on the ratio of the variances.
c. Does the assumption of normality seem appropriate for this data?
d. What can you conclude?

4.36 A random sample of 200 printed circuit boards contains 18 defective or nonconforming units. Estimate the process fraction nonconforming. Using a 90% two-sided confidence interval, evaluate whether or not it is possible that the true fraction nonconforming in this process is 10%.

Table 4E.4

Saltwater Quench	Oil Quench
145	152
150	150
153	147
148	155
141	140
152	146
146	158
154	152
139	151
148	143

4.37 A random sample of 500 connecting rod pins contains 65 nonconforming units. Estimate the process fraction nonconforming. Construct a 90% upper confidence interval on the true process fraction nonconforming. Is it possible that the true fraction defective is 10%?

4.38 During shipment testing, product was flown from Indianapolis to Seattle and back again to simulate 4 takeoffs and landings which can cause cans to open due to pressure changes. Prototype units of the 100 were shipped and 15 opened. Using a 90% two-sided confidence interval, determine if it is possible that the average failure rate is 11%.

4.39 **Continuation of problem 4.38.** The company has made improvements and has repeated the experiment. In this iteration, 12 opened. Using a 95% two-sided confidence interval on the difference in proportions, is it possible to cite improvement?

4.40 Of 1,000 customers, 200 had payments greater than 30 days last month. This month, there are 1,100 customers, of which 230 had payments greater than 30 days.

a. Estimate the fraction late for last month and this month.

b. Construct a 90% confidence interval on the difference in the percentage of late payments.
c. What can you conclude?

4.41 A new purification unit is installed in a chemical process. Before and after installation data was collected regarding the percentage of impurity:

Before: Sample mean = 9.85 Sample variance = 6.79
Number of samples = 10

After: Sample mean = 8.08 Sample variance = 6.18
Number of samples = 8

a. Can you conclude that the two variances are equal using a two-sided 95% confidence interval?
b. Can you conclude that the new purification device has reduced the mean percentage of impurity using a two-sided 95% confidence interval?

4.42 Two different types of glass bottles are suitable for use by a soft-drink beverage bottler. The internal pressure strength (psi) of the bottle is an important quality characteristic. It is known that $\sigma_1 = \sigma_2 = 3.0$ psi. From a random sample of $n_1 = n_2 = 16$ bottles, the mean pressure strengths are observed to be $\bar{x}_1 = 175.8$ psi and $\bar{x}_2 = 181.3$ psi. The company will not use bottle design 2 unless its pressure strength exceeds that of bottle design 1 by at least 5 psi. Based on the sample data, should they use bottle design 2 if they want no larger than 5% chance of excluding bottle 2 if it meets this target?

4.43 The diameter of a metal rod is measured by 12 inspectors, each using both a micrometer caliper and a vernier caliper. The results are shown in Table 4E.5. Is there a difference between the mean measurements produced by the two types of caliper? Use alpha = 0.01.

Table 4E.5

Measurements Made by the Inspectors

Inspector	Micrometer Caliper	Vernier Caliper
1	0.150	0.151
2	0.151	0.150
3	0.151	0.151
4	0.152	0.150
5	0.151	0.151
6	0.150	0.151
7	0.151	0.153
8	0.153	0.155
9	0.152	0.154
10	0.151	0.151
11	0.151	0.150
12	0.151	0.152

4.44 An experiment was conducted to investigate the filling capability of packaging equipment at a winery in Newberg, Oregon. Twenty bottles of Pinot Gris were randomly selected and the fill volume (in ml) measured. Assume that fill volume has a normal distribution. The data are as follows: 753, 751, 752, 753, 753, 753, 752, 753, 754, 754, 752, 751, 752, 750, 753, 755, 753, 756, 751, and 750.

a. Do the data support the claim that the standard deviation of fill volume is less than 1 ml using a 95% two-sided confidence interval?
b. Does it seem reasonable to assume that fill volume has a normal distribution?

4.45 Rehab Inc. is evaluating patient success results for its Northbrook and Southbrook locations. Each successfully treated 10 patients within the last year following elbow surgery. Total recovery times and % range of motion achieved are listed in Table 4E.6:

a. Build a 95% two-sided confidence intervals for the differences of the average of the treatment times. Is there evidence to support that the facilities are different?
b. Build a 95% two-sided confidence intervals for the differences of the variance of the treatment times. Is there evidence to support that the facilities are different?
c. Build a 95% two-sided confidence intervals for the differences in the differences of percentage of motion restored. Is there evidence to support that the facilities are different?

4.46 An article in Solid State Technology (May 1987) describes an experiment to determine the effect of flow rate on etch uniformity on a silicon wafer used in integrated-circuit manufacturing. Three flow rates are tested, and the resulting uniformity (in percent) is observed for six test units at each flow rate. The data are shown in Table 4E.7.

a. Does flow rate affect etch uniformity? Answer this question by using an analysis of variance.

Table 4E.6

Northbrook		Southbrook	
Recovery Time	% ROM Achieved	Recovery Time	% ROM Achieved
148.81	0.69	135.25	0.98
188.72	0.89	174.99	0.47
186.77	0.65	144.15	0.85
152.72	0.73	161.81	0.71
197.80	0.79	151.35	0.94
162.78	0.81	149.69	0.56
192.18	0.64	136.17	0.57
200.17	0.88	146.25	0.20
181.32	0.67	162.88	0.84
193.03	0.74	183.95	0.62

Table 4E.7

C_2F_6 Flow	Observations					
(SCCM)	1	2	3	4	5	6
125	2.7	2.6	4.6	3.2	3.0	3.8
160	4.6	4.9	5.0	4.2	3.6	4.2
200	4.6	2.9	3.4	3.5	4.1	5.1

b. Construct a box plot of the etch uniformity data. Use this plot, together with the analysis of variance results, to determine which gas flow rate would be best in terms of etch uniformity (a small percentage is best).

c. Plot the residuals versus predicted flow. Interpret this plot.

d. Does the normality assumption seem reasonable in this problem?

4.47 An article in the ACI Materials Journal (Vol. 84, 1987, pp. 213–216) describes several experiments investigating the rodding of concrete to remove entrapped air. A 3-in diameter cylinder was used, and the number of times this rod was used is the design variable. The resulting compressive strength of the concrete specimen is the response. The data are shown in Table 4E.8.

Table 4E.8

Rodding Level	Compressive Strength		
10	1530	1530	1440
15	1610	1650	1500
20	1560	1730	1530
25	1500	1490	1510

a. Is there any difference in compressive strength due to the rodding level? Answer this question by using the analysis of variance with alpha = 0.05.

b. Construct box plots of compressive strength by rodding level. Provide a practical interpretation of these plots.

c. Construct a normal probability plot of the residuals from this experiment. Does the assumption of a normal distribution for compressive strength seem reasonable?

4.48 An article in *Environmental International* (Vol. 18, No. 4, 1992) describes an experiment in which the amount of radon released in showers was investigated. Radon-enriched water was used in the experiment, and six different orifice diameters were tested in shower heads. The data from the experiment are shown in Table 4E.9.

a. Does the size of the orifice affect the mean percentage of radon released? Use the analysis of variance.

b. Analyze the results from this experiment.

Table 4E.9

Orifice Diameter	Radon Released (%)			
0.37	80	83	83	85
0.51	75	75	79	79
0.71	74	73	76	77
1.02	67	72	74	74
1.40	62	62	67	69
1.99	60	61	64	66

chapter Five Control Charts for Variables

Chapter Outline

Coca-Cola's secret is SPC

Travel the world and you will find a variety of sights, sounds, smells, and tastes, but one thing will always remain the same—the taste of Coca-Cola. Coca-Cola has attributed its ongoing success to two primary factors: (1) its secret recipe, locked away in a secured vault, and (2) its unyielding focus on process and product quality, supported by Statistical Process Control (SPC). Let's take a closer look at how SPC fits into the picture.

Bottlers turn the Coca-Cola syrup into a finished product by adding water and carbon dioxide. To ensure their efforts produce that signature "Coca-Cola" taste, bottlers perform frequent taste-testing throughout this mixing process. In addition, SPC ensures that critical parameters—such as the syrup-to-water ratio—are performing as intended. Equally important, SPC also ensures consistent and top-quality packaging of Coca-Cola products, with specific focus on fill heights, package sealing, and labeling.

Recognizing that quality cannot be tested into a product, Coca-Cola always strives to "do it right the first time"—a major tenet of the Coca-Cola company culture. Processes are clearly defined and process inputs rigorously controlled so that bottlers can most accurately predict finished product quality. As a final step, bottlers perform various tests to verify or confirm anticipated results.

Hence, whether being bottled in the United States, India, China, Germany, Africa, Taiwan, or other worldwide locations, Coca-Cola maintains its signature taste due to the company's dedication to a consistent process.

Chapter Overview and Learning Objectives

This chapter has three objectives. The first objective is to describe the statistical basis of the Shewhart control chart. The reader will see how decisions about sample size, sampling interval, and placement of control limits affect the performance of a control chart. Other key concepts include the idea of rational subgroups and interpretation of control chart signals and patterns. The second objective is to discuss and illustrate other control charts that are appropriate to variable data. Examples include dimensions such as length or width, temperature, and volume. The third objective is to introduce the concept of process capability, or how the inherent variability in a process compares with the specifications or requirements for the product.

After careful study of this chapter you should be able to do the following:

1. Understand chance and assignable causes of variability in a process
2. Explain the statistical basis of the Shewhart control chart, including choice of sample size, control limits, and sampling interval
3. Explain the rational subgroup concept
4. Explain how sensitizing rules and pattern recognition are used in conjunction with control charts
5. Know how to design variables control charts
6. Know how to set up and use $\overline{X}$ and R control charts
7. Know how to set up and use $\overline{X}$ and S control charts
8. Know how to set up and use control charts for individual measurements
9. Understand the importance of the normality assumption for individuals control charts and know how to check this assumption
10. Set up and use CUSUM control charts for monitoring the process mean
11. Set up and use EWMA control charts for monitoring the process mean
12. Understand the difference between process capability and process potential
13. Calculate and properly interpret process capability ratios
14. Understand the role of the normal distribution in interpreting most process capability ratios

5.1 Introduction

In any process, regardless of how well-designed or carefully maintained, a certain amount of inherent or natural variability will always exist. This natural variability or "background noise" is the cumulative effect of many small, essentially unavoidable causes. When the background noise in a process is relatively small, we usually consider it an acceptable level of performance. In the framework of statistical quality control, this natural variability is often called a "stable system of chance causes." A process that is operating with only chance causes of

In Control
A process operating with only chance causes of variation.

Assignable Cause
An identifiable source of variation in a process. It is not a part of the normal chance cause of variation of a process e.g. a traffic accident causing significant delay.

Out of Control
A process operating in the presence of assignable causes.

variation present is said to be in statistical control. In other words, the chance causes are a inherent part of the process.

Other kinds of variability may occasionally be present in the process. The variability in key quality characteristics usually arises from three sources: improperly adjusted equipment, operator errors, or defective raw materials. Such variability is generally large when compared to the background noise and it usually represents an unacceptable level of process performance. We refer to these sources of variability that are not part of the chance cause pattern as assignable causes. A process that is operating the presence of assignable causes is said to be out of control. Table 5.1 produces some examples of chances and assignsable causes.

Processes will often operate in the in-control state, producing acceptable product for relatively long periods of time. Occasionally, however, assignable causes will occur, seemingly at random, resulting in a "shift" to an out-of-control state where a large proportion of the output does not conform to requirements. A major objective of statistical process control is to quickly detect the occurrence of assignable causes or process shifts so that the investigation of the process and corrective action may be undertaken before many nonconforming units are processed. The control chart is an online process monitoring technique widely used for this purpose.

Control charts may also be used to estimate the parameters of a process and through this information, determine the capability of a process to conform to specifications. The control chart can also provide information that is useful in improving the process. Finally, remember that the eventual goal of statistical process control is the elimination of variability in the process. Although it may not be possible to eliminate variability completely, the control chart helps reduce it as much as possible.

A typical control chart is shown in Figure 5.1 and plots the value of a quality characteristic over time. Often samples are selected at periodic intervals such as every hour. The chart contains a center line (CL) that represents the average value of the quality characteristic corresponding to the in-control state (that is, only chance causes are present.) Two other horizontal lines, called the upper control limit (UCL) and

Table 5.1

Examples of Chance and Assignable Causes

Process/Quality Characteristics	Chance Causes	Assignable Causes
Hotel Laundry/Sheet Whiteness	Slight variations in laundry detergent, water temperature, load sizes	Fraternity party, different detergent type, different service provider
Utility Company/Number of Outages	Number of storms per year, number of new homes built	Placement of utilities too close to the road (accidents), not trimming trees away from power lines
Production Process/Hole Diameter	Slight variations in metal hardness, measurement system variability	Dull drilling bits, wrong drilling bit

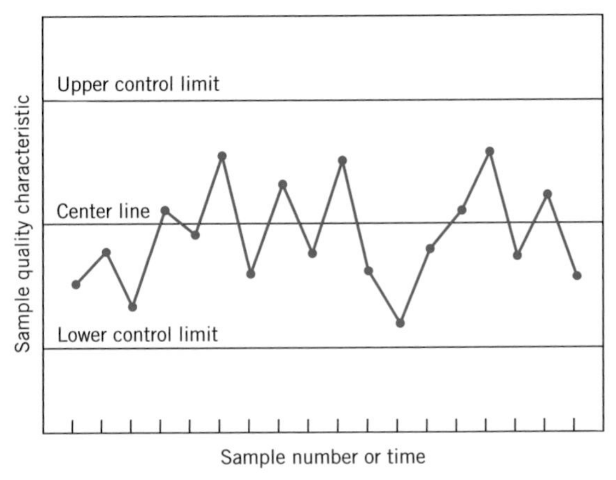

Figure 5.1 A Typical Control Chart

Control Chart
A chart in which observations are plotted over time to indicate whether the process is in control or if assignable causes are present.

the lower control limit (LCL), are also shown on the chart. These control limits are chosen so that if the process is in control nearly all of the sample points will fall between them. In general, as long as the points plot within the control limits, the process is assumed to be in control, and no action is necessary. However, a point that plots outside of the control limits is interpreted as evidence that the process is out of control, and investigation and corrective action are required to find and eliminate the assignable cause for this behavior. The sample points on the control chart are usually connected with straight line segments, so that it is easier to visualize how the sequence of points has evolved over time.

Even if all the points plot inside the control limits, if they behave in a systematic or nonrandom manner, then this could be an indication that the process is out of control. For example, if 18 of the last 20 points plotted above the center line but below the upper control limit, we would be very suspicious that something was wrong. If the process is in control, all the plotted points should have an essentially random pattern with roughly half the points above the center line and half below. Methods for looking for sequences or nonrandom patterns can be applied to control charts as an aid in detecting out-of-control conditions. Usually, there is a reason why a particular nonrandom pattern appears on a control chart, and if it can be found and eliminated, process performance can be improved.

There is a close connection between control charts and hypothesis testing. Essentially, the control chart is a test of the hypothesis that the process is in a state of statistical control. A point plotting within the control limits is equivalent to failing to reject the hypothesis of statistical control, and a point plotting outside the control limits is equivalent to rejecting the hypothesis that the process is in a state of statistical control.

The adjacent display gives a general model for a control chart.

A common choice for the width of the limits as a multiple of sigma is $k = 3$ (because this range contains 99.7% of points within a normal distribution, per chapter 4). This general theory of control charts was first proposed by Dr. Walter A. Shewhart, and control charts developed according to these principles are often called Shewhart control charts.

General Model for a Control Chart

Let W be a sample statistic that measures some quality characteristic of interest, and suppose that the mean of W is μ_W and the standard deviation of W is σ_W. Then the center line, the upper control limit, and the lower control limit become

$$\text{UCL} = \mu_W + k\sigma_W \quad (5.1)$$

$$\text{CL} = \mu_W \quad (5.2)$$

$$\text{LCL} = \mu_W + k\sigma_W \quad (5.3)$$

where k is the "distance" of the control limits away from the center line, expressed in standard deviation units.

Most Important Aspect of a Control Chart

Improving a process

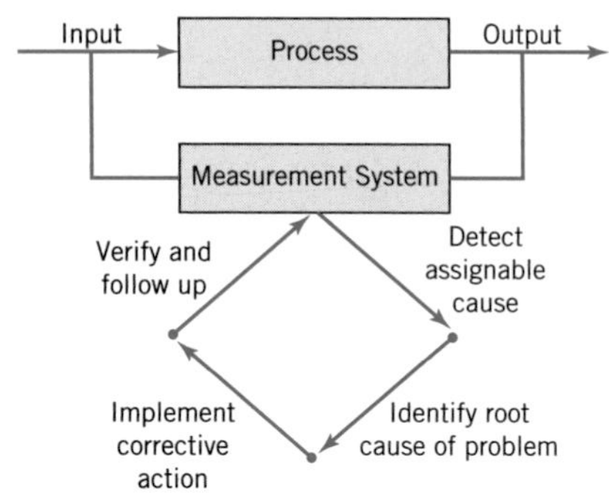

Figure 5.2 Process Improvement Using a Control Chart

The control chart is a device for describing exactly what is meant by statistical control; as such, it may be used in a variety of ways. In many applications, it is used for online process monitoring. Sample data are collected and used to construct the control chart. If the same values fall within the control limits and do not exhibit any systematic pattern, we say the process is in control at the level indicated by the chart. Note that we may be interested here in determining both whether the past data came from a process that was in control and whether future samples from this process indicate statistical control.

This is the most important aspect of a control chart: improving a process. We have found that, generally,

1. Most processes do not operate in a state of statistical control.
2. Consequently, the routine and attentive use of control charts will identify assignable causes. If these causes can be eliminated from the process, variability will be reduced and the process will be improved.

   This process improvement activity using the control chart is illustrated in Figure 5.2.
3. The control chart will only detect assignable causes. Management, operator, and engineering action will usually be necessary to eliminate the assignable causes.

In identifying and eliminating assignable causes, it is important to find the root cause of the problem and to remedy it. A cosmetic solution will not result in any real, long-term process improvement. Developing an effective system for corrective action is an essential component of an effective SPC implementation.

We may also use the control chart as an estimating device. That is, from a control chart that exhibits statistical control, we may estimate certain process parameters, such as the mean, standard deviation, and fraction nonconforming. These estimates may then be used to determine the capability of the process to produce acceptable products. Such process capability studies have considerable impact on many management decision problems that occur over the product cycle, including make-or-buy decisions, process improvements that reduce variability, and contractual agreements with customers or suppliers regarding product quality.

Control charts may be classified into two general types. Many quality characteristics can be measured and expressed as numbers on some continuous scale of measurement. In such cases, it is convenient to describe the quality characteristic with a measure of central tendency and a measure of variability. Control charts for central tendency and variability are collectively called variables control charts. The $\overline{X}$ (pronounced $\overline{X}$) chart is the most widely used chart for monitoring central tendency, whereas charts based either on the sample range or sample standard deviation are used to monitor process variability. Many quality characteristics are not measured on a continuous scale or even a quantitative scale. In these cases, we may judge each unit as either conforming or nonconforming, or we may count the number of defects appearing in a product. Control charts for such quality characteristics are called attributes control charts and will be discussed in chapter 6.

Control charts have had a long history of use in U.S. industries and in many offshore industries as well. There are at least five reasons for their popularity:

Control charts are a proven technique for improving productivity.	A successful control chart program will reduce scrap and rework, which are the primary productivity killers in any operation. If you reduce scrap and rework, then productivity increases, cost decreases, and production capacity (measured in the number of good parts per hour) increases.
Control charts are effective in defect prevention.	The control chart helps keep the process in control, which is consistent with the "do it right the first time" philosophy. It is never cheaper to sort out "good" units from "bad" units later on than it is to build things right initially. If you do not have effective process control, you are paying to make a nonconforming product.
Control charts prevent unnecessary process adjustment.	A control chart can distinguish between background noise and abnormal variation; no other device, including a human operator, is as effective in making this distinction. If process operators adjust the process based on periodic tests unrelated to a control chart program, they will often overreact to the background noise and make unneeded adjustments. Such unnecessary adjustments can actually result in a deterioration of process performance. In other words, the control chart is consistent with the "if it isn't broken, don't fix it" philosophy.
Control charts provide diagnostic information.	Frequently, the pattern of points on the control chart will contain information of diagnostic value to an experienced operator or engineer. This information allows the implementation of a change in the process that improves its performance.
Control charts provide information about process capability.	The control chart provides information about the value of important process parameters and their stability over time. This allows an estimate of process capability to be made. This information is of tremendous use to product and process designers.

Control charts are among the most effective management control tools, and they are as important as cost controls and material controls. Modern computer technology has made it easy to implement control charts in almost any type of process; as data collection and analysis can be performed on a computer or a local area network terminal in real time at a workcenter.

DESIGN OF A CONTROL CHART

Turnaround times for complete blood counts (CBC) are a critical metric within a hospital. The manager of the group has separated urgent requests from routine requests and has generated the control chart in Figure 5.3.

Every day a random sample of five turnaround times is taken, and the average turnaround time ($\overline{x}$) is computed and plotted on the chart. Because this control chart utilizes the sample mean to monitor the

Figure 5.3 Averages Control Chart for CBC Turnaround Time

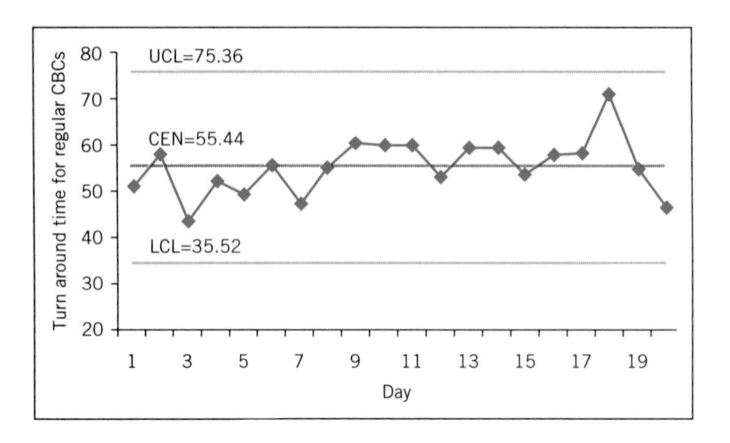

process mean, it is usally called an $\overline{X}$ chart. Note that all points fall within the control limits, so the chart indicates that the CBC turnaround times are in a state of statistical control.

Now let's consider how the control limits were determined. The process average is 55.44 minutes and the process standard deviation is 14.85 minutes. Now, if a sample size of $n = 5$ is taken, the standard deviation of the sample average is

$$\sigma_{\overline{X}} = \frac{\sigma}{\sqrt{n}} = \frac{14.85}{\sqrt{5}} = 6.64$$

Control Limits

3 sigma control limits contain 99.73% of data.

Therefore, if the process is in control with a mean turnaround time of 55.44 minutes, we would expect approximately $100(1 - \alpha)\%$ of the sample turnaround times to fall between $55.44 - z_{\alpha/2}\frac{\sigma}{\sqrt{n}}$ and $55.44 + z_{\alpha/2}\frac{\sigma}{\sqrt{n}}$ (this should look very familiar—if not, check back to review confidence intervals for a mean in Chapter 4).

As discussed, we customarily choose the constant $z_{\alpha/2}$ to be 3 so that the upper and lower control limits become

$$\text{UCL} = 55.44 + 3(6.64) = 75.36$$

$$\text{LCL} = 55.44 - 3(6.64) = 35.52$$

as shown in the control chart. Note that the use of three-sigma limits implies that $\alpha = 0.0027$; that is the probability that a point plots outside the control limits when the process is in control is 0.0027 or 0.27%.

As demonstrated in this example, when we design a control chart, we must specify both the sample size to use (i.e., how many points to average together) as well as the frequency of sampling (i.e., every day). In general, larger samples will make it easier to detect small shifts of the process. When choosing the sample size, we must keep in mind the size of the shift we are trying to detect. If we are interested in detecting a relatively large process shift, then smaller sample sizes would be used than if the shift of interest were relatively small.

We must also determine the frequency of sampling. The most desirable situation from the point of view of detecting shifts would be to take large samples very frequently; however, this is usually not economically feasible. The general problem is one of allocating sampling effort. That is, either we take small samples at short intervals or larger samples at longer intervals. Current industry practice tends to favor smaller, more frequent samples, particularly in high-volume processes or where a great many types of assignable causes can occur. Furthermore, as automatic sensing and measurement technology develops, it is becoming possibly to greatly increase frequencies. Ultimately, every unit can be tested as it is processed. This capability will increase the effectiveness of process control and improve quality.

RATIONAL SUBGROUPS

A fundamental idea in the use of control charts is the collection of sample data according to what Shewhart called the rational subgroup concept. To illustrate this concept, suppose that we are using a control chart to detect changes in the process mean. Then the rational subgroup concept means that *subgroups or samples should be selected so that if assignable causes are present, the chance for differences between subgroups will be maximized, while the chance for differences due to these assignable causes within a subgroup will be minimized*. This is important because the within subgroup variability determines the control limits. If assignable causes occur within a subgroup, this will widen the control limits more than necessary. Ultimately, this will reduce the control chart's capability to detect shifts.

Rational Subgroup

A subgroup or sample selected so that if assignable causes are present, the chance for differences between subgroups will be maximized, while the chance for differences due to the assignable cause within a subgroup will be minimized.

Two approaches to constructing rational subgroups are commonly used. In the first approach, each subgroup consists of units that were produced at the same time (or as closely together as possible). This approach is used when the primary purpose of the control chart is to detect process shifts. It minimizes the chance of variability due to assignable causes within a sample, and it maximizes the chance of variability between samples if assignable causes are present. It also provides a better estimate of the standard deviation of the process in the case of variables control charts. This approach to rational subgrouping essentially gives a snapshot of the process at each point in time where a sample is collected.

Approaches to Constructing Rational Subgroups

1. Subgroup units produced about the same time.
2. Subgroup a random sample of all units produced within a sampling interval.

In the second approach, each sample consists of units of product that are representative of all units that have been produced since the last sample was taken. Essentially, each subgroup is a random sample of all process output over the sampling interval. This method of rational subgrouping is often used when the control chart is employed to make decisions about the acceptance of all units of product that have been produced since the last sample. However, if the process shifts to an out-of-control state and then back in control again between samples, it is sometimes argued that the snapshot method of rational subgrouping will be ineffective against these types of shifts.

When the rational subgroup is a random sample of all units produced over the sampling interval, considerable care must be taken in interpreting the control charts. If the process mean drifts between several levels during the interval between samples, this may cause the range of the observations within the sample to be relatively large, resulting in wider limits on the chart. It is the within-sample variability that determines the width of the control limits on a $\overline{X}$ chart, so this practice will result in wider limits. This makes it harder to detect shifts in the mean. In fact, we can often make any process appear to be in statistical control just by stretching out the interval between observations in the sample. It is also possible for shifts in the process average to cause points on a control chart for the range or standard deviation to plot out of control, even though there has been no shift in process variability.

There are other ways to form rational subgroups. For example, suppose a process consists of several machines that pool their output into a common stream. If we sample from this common stream of output, it will be very difficult to detect whether any of the machines are out of control. A logical approach to rational subgrouping here is to apply control chart techniques to the output for each individual machine. Sometimes this concept needs to be applied to different components on the same machine, different work stations, different operators, and so forth. In many situations, the rational subgroup will consist of a single observation. This situation occurs frequently in the chemical and process industries, where the quality characteristic of the product changes relatively slowly and samples taken very close together in time are virtually identical, apart from measurement or analytical error.

The rational subgroup concept is very important. The proper selection of samples requires careful consideration of the process, with the objective of obtaining as much useful information as possible from the control chart analysis.

ANALYSIS OF PATTERNS ON CONTROL CHARTS

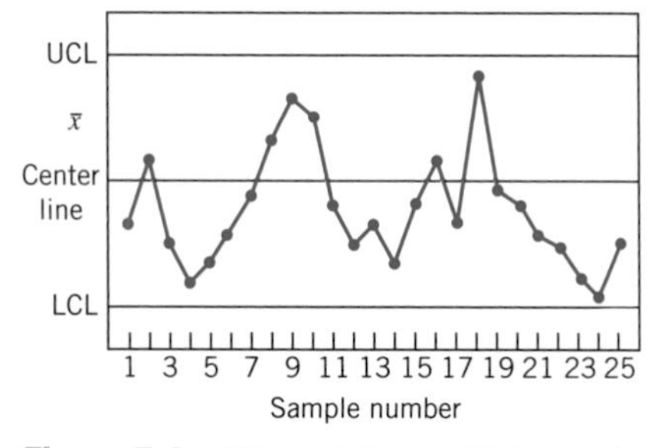

Figure 5.4 "Out of Control" Control Chart

A control chart may indicate an out-of-control condition when one or more points fall beyond the control limits or when the plotted points exhibit some nonrandom pattern of behavior. For example, consider the chart shown in Figure 5.4. Although all 25 points fall within the control limits, the points do not indicate statistical control because their pattern is very nonrandom in appearance. Specifically, we note that 19 of 25 points plot below the center line, while only 6 of them plot above the centerline. If the points truly are random, we should expect a more even distribution above and below the center line. We also observe that following the fourth point, five points in a row increase in magnitude. This arrangement of points is called a *run*. Since the observations are increasing, we could call this a *run up*. Similarly, a sequence of decreasing points is called a *run down*. This control chart has an unusually long run up (beginning with the fourth point) and an unusually long run down (beginning with the eighteenth point). In general, we define a run as a sequence of observations of the same type.

In addition to runs up and runs down, we could define the types of observations as those above and below the center line, respectively, so that two points in a row above the center line would be a run of length 2. A run of length 8 or more points has a very low probability of occurrence in a random sample of points. Consequently, any type of run of length 8 or more is often taken as a signal of an out-of-control condition. For example, eight consecutive points on one side of the center line may indicate that the process is out of control.

Although runs are an important measure of nonrandom behavior on a control chart, other types of patterns may also indicate an out-of-control condition. For example, consider the chart in Figure 5.5. Note that the plotted sample averages exhibit a cyclic behavior, yet they all fall within the control limits. Such a pattern may indicate a problem with the process such as operator fatigue, raw material deliveries, heat or stress buildup, and so forth. Although the process is not really out of control, more product may be produced within the upper and lower specification limits (USL and LSL) by eliminating or reducing the source of variability causing the cycling behavior (see Figure 5.6).

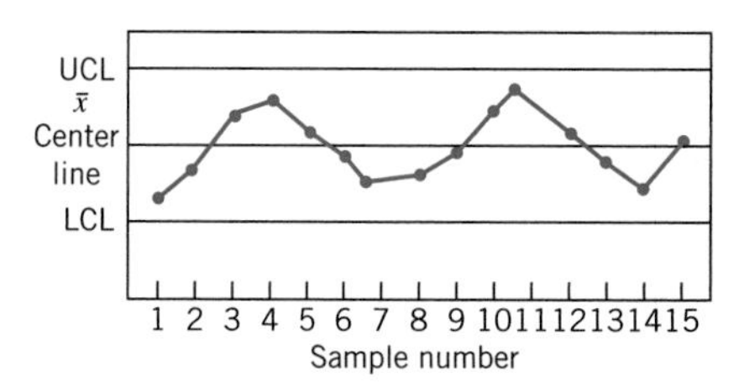

Figure 5.5 A Control Chart with a Cyclic Pattern

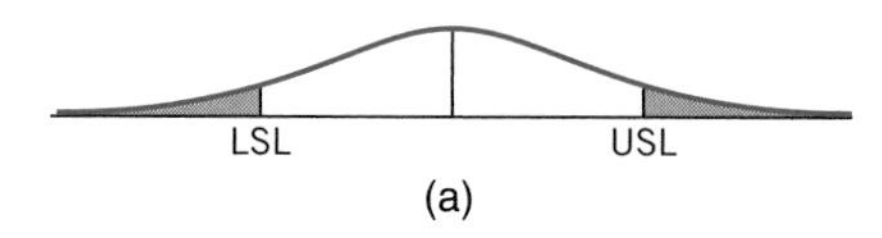

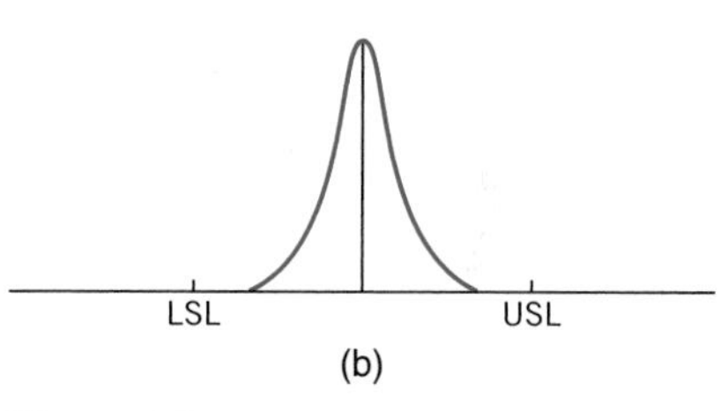

Figure 5.6 (a) Variability with the Cyclic Pattern (b) Variability with the Cyclic Pattern Eliminated

The problem is one of pattern recognition—that is, recognizing systematic or nonrandom patterns on the control chart and identifying the reason for this behavior. The Western Electric Handbook (1956) suggests a set of decision rules for detecting nonrandom patterns on control charts. These rules are as follows:

1. One point plots outside the three-sigma control limits
2. Two out of three consecutive points plot beyond the two-sigma warning limits
3. Four out of five consecutive points plot at a distance of one-sigma or beyond from the center line or
4. Eight consecutive points plot on one side of the center line

These zones are illustrated below in Figure 5.7.

We have found that these rules very effective in practice for enhancing the sensitivity of control charts. Rules 2 and 3 apply to one side of the center line at a time. That is, a point above the upper two-sigma limit followed immediately by a point below the lower two-sigma limit would not signal an out-of-control alarm. Figure 5.8 illustrates these rules in practice for a process with a mean of 100 and a standard deviation of 10.

The Western Electric rules are sometimes used to increase the sensitivity of the control charts to a small process shift so that we may respond more quickly to the assignable cause. When all four rules are applied simultaneously, we often use a graduated response to out-of-control signals. For example, if a point exceeded a control limit, we would immediately begin to search for the assignable cause, but if one or two consecutive points exceeded only the two-sigma warning limit, we might increase the frequency of sampling from every hour to, say, every 10 minutes. This adaptive sampling response might not be as severe as a complete search for an assignable cause, but if the process were really out of control, it would give us a high probability of detecting this situation more quickly than we would by maintaining the longer sampling interval.

However, the use of the four rules simultaneously does have a downside. Recall that when we used the rule of 1 point outside the control limits, we expected a false alarm rate of 0.27%. This corresponds to about 1 every 370 points. Champ and Woodall (1987) investigated the false alarm rate (also called the average run length) for the Shewhart control chart with Western Electric rules and found that the use of all four rules decreased the average run length to 91.25. Thus

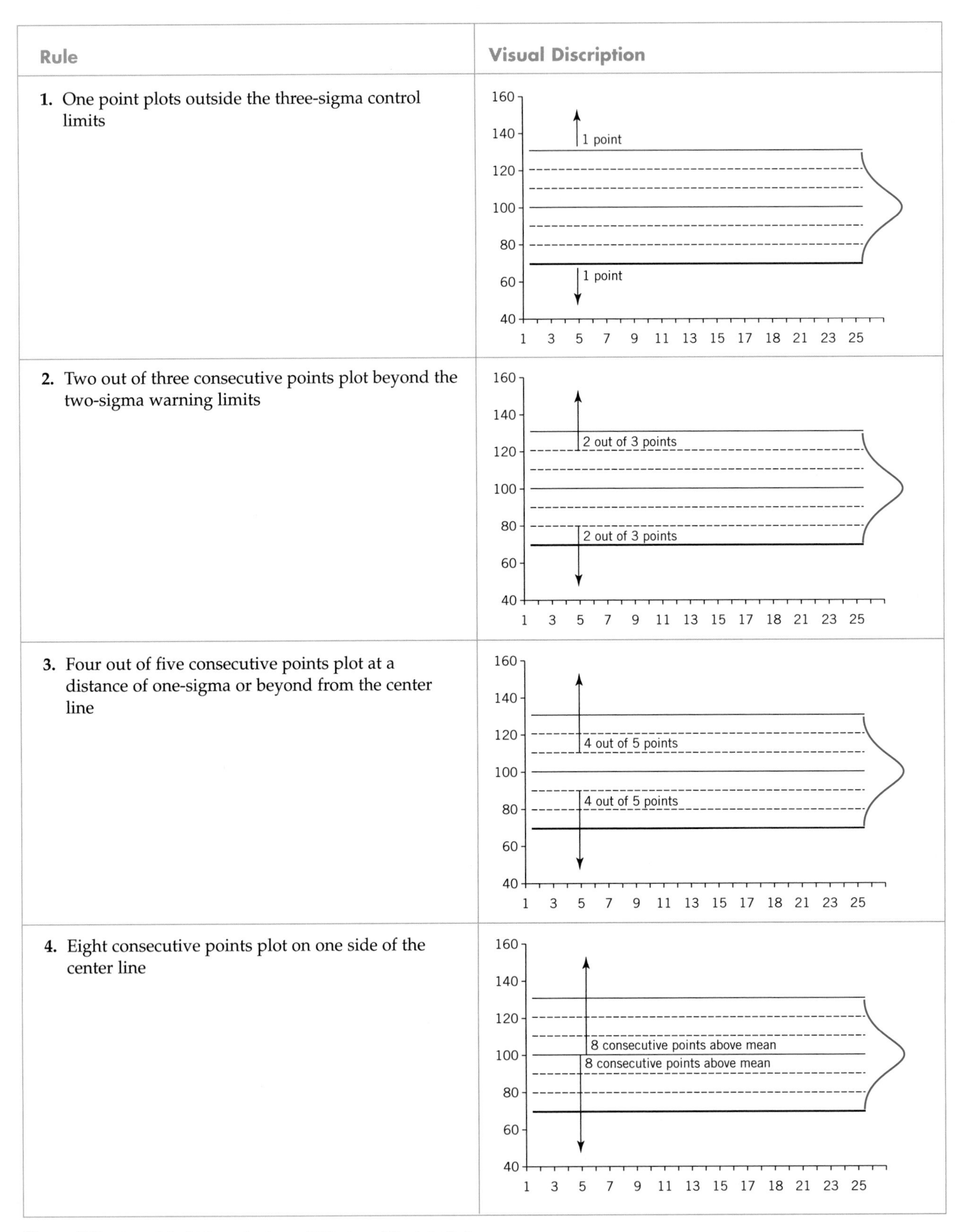

Figure 5.7 Graphical Description of Western Electric Rules

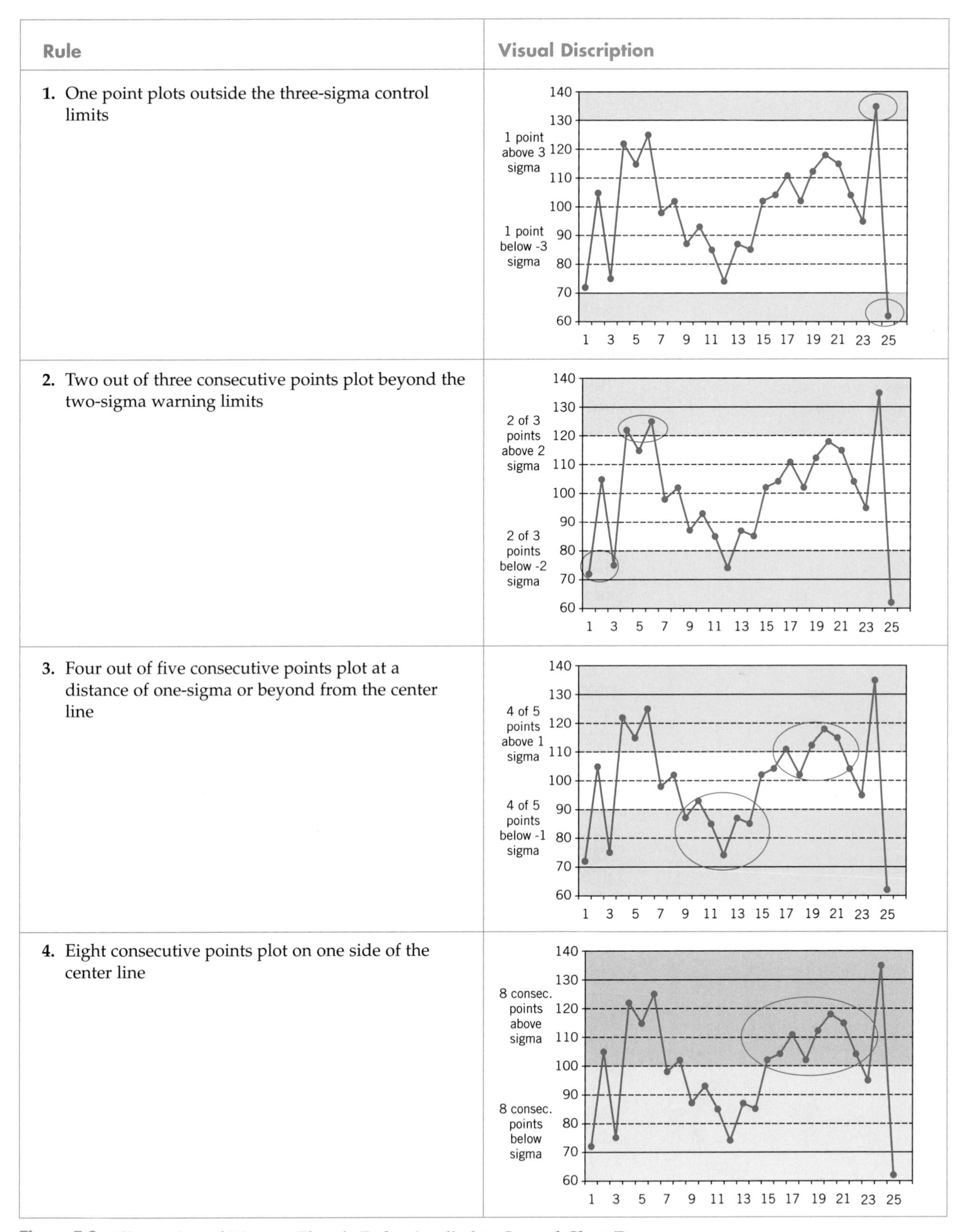

Rule	Visual Discription
1. One point plots outside the three-sigma control limits	1 point above 3 sigma; 1 point below -3 sigma
2. Two out of three consecutive points plot beyond the two-sigma warning limits	2 of 3 points above 2 sigma; 2 of 3 points below -2 sigma
3. Four out of five consecutive points plot at a distance of one-sigma or beyond from the center line	4 of 5 points above 1 sigma; 4 of 5 points below -1 sigma
4. Eight consecutive points plot on one side of the center line	8 consec. points above sigma; 8 consec. points below sigma

Figure 5.8 **Illustration of Western Electric Rules Applied to Control Chart Data**

every 92 points (on average) you should expect to receive a indication that the system is not in a state of statistical control (even though it is possible that it is).

The remainder of this chapter will discuss the creation of control charts for variable data. Chapter 6 will continue the discussion with attribute data.

5.2 $\overline{X}$ and R Charts

$\overline{X}$ Chart

A control chart that depicts the sample mean over time.

R Chart

A control chart that depicts how the range within each subgroup changes over time.

When dealing with a quality characteristic that can be expressed as a measurement, it is customary to monitor both the mean and its variability. The mean is important because it indicates whether or not the process is operating at its target level, and the variability is important because it is an indicator of whether or not the process is operating consistently. An $\overline{X}$ chart is used to monitor the average product quality and a Range Chart is often used to monitor the standard deviation.

If the process mean and standard deviation, μ and σ, are known and we can assume a normal distribution, then we can use μ as the center line for the control chart and place the upper and lower three-sigma limits at $UCL = \mu + 3 * \frac{\sigma}{\sqrt{n}}$ and $LCL = \mu - 3 * \frac{\sigma}{\sqrt{n}}$.

However, this is usually not the case. When the mean and standard deviation are not known, we usually estimate them on the basis of preliminary samples, taken when the process is thought to be in control. We recommend the use of at least 20 to 25 preliminary samples with subgroup size n.

If the chart is being used primarily to detect moderate to large process shifts—say, on the order of 2σ or larger—then relatively small samples of size $n = 4, 5$, or 6 are reasonably effective. On the other hand, if we are trying to detect small shifts, then larger sample sizes of possibly $n = 15$ to 25 are needed. However, when larger samples are used, there is a greater risk of a process shift occurring while a sample is taken. If a shift does occur while a sample is taken, the sample average can obscure this effect. Consequently, this is an argument for using as small a sample size as is consistent with the magnitude of the process shift that one is trying to detect. If you are interested in small shifts, it is best to use the CUSUM or EWMA charts discussed later in this chapter.

The R chart is relatively insensitive to shifts in the process standard deviation for small samples. For example, samples of size $n = 5$ have only about a 40% chance of detecting on the first sample a shift in the process standard deviation from σ to 2σ. Larger samples would seem to be more effective, but we also know that the range method for estimating the standard deviation drops dramatically in efficiency as n increases. Consequently, for moderately large n—say, $n > 10$ or 12—it is probably best to use a control chart for s or s^2 instead of the R chart. Details of the construction of these charts are also discussed in this chapter.

Once we have our initial data, we calculate two items which will be utilized in the control chart construction: the grand mean $\overline{\overline{x}}$ (or the average of the averages) and the average range $\overline{R}$ (which is the average of the subgroup ranges, which are computed as the maximum–minimum data value within each subgroup). The average range is quite important as we can calculate an estimate of the standard deviation using the formula below:

Estimation of Standard Deviation Using Average Range:

$$\hat{\sigma} = \frac{\overline{R}}{d_2} \quad (5.4)$$

Where d_2 is a value from the table Factors for Constructing Variables Control Charts, found in Appendix Six.

example 5.1 Calculation of Grand Mean, Average Range, and Estimate of the Standard Deviation

Three subgroups have been collected, each of size 4. Use the data below to calculate the grand mean and average range.

	Sample number			
Subgroup	**1**	**2**	**3**	**4**
1	11	13	13	8
2	12	11	9	5
3	15	14	12	11

The first step in the calculation is to calculate the subgroup averages and subgroup ranges. This is illustrated below:

Subgroup Averages:	**Subgroup Ranges**
Subgroup 1: $(11 + 13 + 13 + 8)/4 = 11.25$	Subgroup 1: $13 - 8 = 5$
Subgroup 2: $(12 + 11 + 9 + 5)/4 = 9.25$	Subgroup 2: $12 - 5 = 7$
Subgroup 3: $(15 + 14 + 12 + 11)/4 = 13$	Subgroup 3: $15 - 11 = 4$
Grand Average, $\overline{\overline{x}} = (11.25 + 9.25 + 13)/3 = 11.17$	Average range $(\overline{R}) = (5 + 7 + 4)/3 = 5.33$

To calculate the estimate of the standard deviation, we look in Appendix Six for subgroup size of 4 and find that $d_2 = 2.059$. Our estimate of the standard deviation is thus $5.33/2.059 = 2.59$.

Once the grand mean and average range have been calculated, the $\overline{X}$ chart can be calculated using the following equations:

Control Limits for the $\overline{x}$ Chart

$$\text{UCL} = \overline{\overline{x}} + A_2\overline{R} \quad (5.5)$$

$$\text{Center line} = \overline{\overline{x}} \quad (5.6)$$

$$\text{LCL} = \overline{\overline{x}} - A_2\overline{R} \quad (5.7)$$

The constant A_2 is tabulated for various sample sizes in Appendix Six.

Note that the constant $A_2 = \frac{3}{d_2\sqrt{n}}$, so these equations reduce to

$$UCL = \bar{\bar{x}} + A_2\bar{R} = \bar{\bar{x}} + 3*\frac{\bar{R}}{d_2\sqrt{n}} = \bar{\bar{x}} + 3*\frac{\sigma}{\sqrt{n}}$$

$$LCL = \bar{\bar{x}} + A_2\bar{R} = \bar{\bar{x}} - 3*\frac{\bar{R}}{d_2\sqrt{n}} = \bar{\bar{x}} - 3*\frac{\sigma}{\sqrt{n}}$$

The position of the one- and two-sigma lines can be found as follows:

$$\text{One-sigma lines} = \bar{\bar{x}} \pm \frac{A_2\bar{R}}{3}$$

$$\text{Two-sigma lines} = \bar{\bar{x}} \pm 2*\frac{A_2\bar{R}}{3}$$

The range chart is calculated as follows:

Control Limits for the R Chart

$$\text{UCL} = D_4\bar{R} \tag{5.8}$$

$$\text{Center line} = \bar{R} \tag{5.9}$$

$$\text{LCL} = D_3\bar{R} \tag{5.10}$$

The constants D_3 and D_4 are tabulated for various values of n in Appendix Six.

Similarly, we can derive the upper and lower control limits using the simplification coefficients for the range chart given that the standard deviation of the range = d_3 (standard deviation estimate) and the knowledge that $D_4 = 1 + 3(d_3/d_2)$ and $D_3 = 1 - 3(d_3/d_2)$.

The position of the one- and two-sigma lines can be found as follows:

$$\text{one-sigma lines} = \bar{R} \pm \frac{D_4\bar{R} - \bar{R}}{3}$$

$$\text{two-sigma lines} = \bar{R} \pm 2\frac{D_4\bar{R} - \bar{R}}{3}$$

The reader may have noticed that these equations assume that the data is normally distributed. In many situations, we may have reason to doubt the validity of this assumption. For example, we may know that the underlying distribution is not normal, because we have collected extensive data that indicate the normality assumption is inappropriate. Several authors have investigated the effect of departures from normality on control charts and have found that the $\bar{X}$ chart is fairly robust to nonnormal data. Studies indicate that, in most cases, samples of size 4 or 5 are sufficient to ensure reasonable robustness to the normality assumption. However, the R chart is more sensitive to non-normal data.

example 5.2 Development of $\overline{X}$ and R Charts

The IT department at a large corporation is concerned about the time it takes to load a application across the network. They would like to implement an $\overline{X}$ and R chart so that this time can be monitored and improvements can be made to this loading time. Over the past 25 days, they have randomly selected five application loadings and monitored the time from start to completion. This data is shown in the table below:

Table 5.2

Application Loading Time Data

Day	Application Load Time (seconds)					Subgroup Average	Subgroup Range
	1	2	3	4	5		
1	13.2350	14.1280	16.7440	14.5730	16.9140	15.1188	3.6790
2	14.3140	13.5920	16.0750	14.6660	16.1090	14.9512	2.5170
3	14.2840	14.8710	14.9320	14.3240	15.6740	14.8170	1.3900
4	15.0280	16.3520	13.8410	12.8310	15.5070	14.7118	3.5210
5	15.6040	12.7350	15.2650	14.3630	16.4410	14.8816	3.7060
6	15.9550	15.4510	13.5740	13.2810	14.1980	14.4918	2.6740
7	16.2740	15.0640	18.3660	14.1770	15.1440	15.8050	4.1890
8	14.1900	14.3030	16.6370	16.0670	15.5190	15.3432	2.4470
9	13.8840	17.2770	15.3550	15.1760	13.6880	15.0760	3.5890
10	14.0390	16.6970	15.0890	14.6270	15.2200	15.1344	2.6580
11	14.1580	17.6670	14.2780	15.9280	14.1810	15.2424	3.5090
12	15.8210	13.3550	15.7770	13.9800	17.5590	15.2984	4.2040
13	12.8560	14.1060	14.4470	16.3980	11.9280	13.9470	4.4700
14	14.9510	14.0360	15.8930	16.4580	14.9690	15.2614	2.4220
15	13.5890	12.8630	15.9960	12.4970	15.4710	14.0832	3.4990
16	15.7470	15.3010	15.1710	11.8390	18.6620	15.3440	6.8230
17	13.6800	17.2690	13.9570	15.0140	14.4490	14.8738	3.5890
18	14.1630	13.8640	13.0570	16.2100	15.5730	14.5734	3.1530
19	15.7960	14.1850	16.5410	15.1160	17.2470	15.7770	3.0620
20	17.1060	14.4120	12.3610	13.8200	17.6010	15.0600	5.2400
21	14.3710	15.0510	13.4850	15.6700	14.8800	14.6914	2.1850

(Continued)

example 5.2 Continued

Table 5.2

Continued

Day	Application Load Time (seconds)					Subgroup Average	Subgroup Range
	1	2	3	4	5		
22	14.7380	15.9360	16.5830	14.9730	14.7200	15.3900	1.8630
23	15.9170	14.3330	15.5510	15.2950	16.8660	15.5924	2.5330
24	16.3990	15.2430	15.7050	15.5630	15.5300	15.6880	1.1560
25	15.7970	13.6630	16.2400	13.7320	16.8870	15.2638	3.2240
					Grand Average:	15.0567	
					Average Range:		3.2521

When setting up $\overline{X}$ and R charts, it is best to begin with the R chart. Because the control limits on the chart depend on the process variability, unless process variability is in control, these limits will not have much meaning.

Using the data in Table 5.2, we find that the center line for the R chart is 3.2521. For samples of $n = 5$, we find from Appendix Six that $D_3 = 0$ and $D_4 = 2.114$. Therefore, the control limits for the R chart are as follows:

$$\text{UCL} = D_4\overline{R} = (2.114*3.2521) = 6.8749$$

$$\text{CL} = \overline{R} = 3.2521$$

$$\text{LCL} = D_3\overline{R} = (0*2.114) = 0$$

$$\text{One-sigma lines} = \overline{R} \pm \frac{D_4\overline{R} - \overline{R}}{3} = 3.2521 \pm \frac{2.114 + 3.2521 - 3.2521}{3}$$

$$= 2.0445 \text{ and } 4.4597$$

$$\text{Two-sigma lines} = \overline{R} \pm 2\left(\frac{D_4\overline{R} - \overline{R}}{3}\right) = 3.2521 \pm 2\left(\frac{2.114 + 3.2521 - 3.2521}{3}\right)$$

$$= 0.8369 \text{ and } 5.6673$$

The resultant R chart is shown in Fig 5.9 and was created by SPC XL. Reviewing the chart, we find no indication of an out-of-control condition. Since the R chart indicates that process variability is in control, we may now construct the $\overline{X}$ chart.

Prior to calculating the control limits for the $\overline{X}$ chart, we must first find the value of A_2 (our control limit constant). For samples of $n = 5$, we find from Appendix Six that $A_2 = 0.577$. Our control limits are then calculated as:

$$\text{UCL} = \overline{\overline{x}} + A_2\overline{R} = 15.0567 + 0.577(3.2521) = 16.9331$$

$$\text{LCL} = \overline{\overline{x}} - A_2\overline{R} = 15.0567 - 0.577(3.2521) = 13.1802$$

example 5.2 Continued

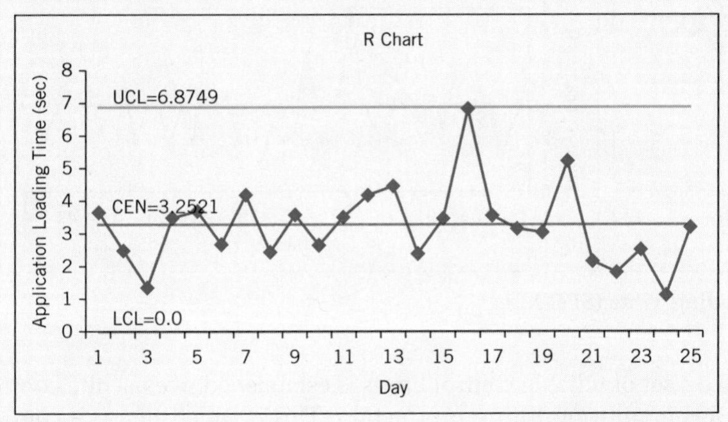

Figure 5.9 ***R* chart for Application Loading Time**

The position of the one- and two-sigma lines can be found as follows:

$$\text{One-sigma lines} = \bar{\bar{x}} \pm \frac{A_2\bar{R}}{3} = 15.0567 \pm \frac{0.577(3.2521)}{3}$$

$$= 14.4312 \text{ and } 15.6822$$

$$\pm \text{ Two-sigma lines} = \bar{\bar{x}} \pm 2\left(\frac{A_2\bar{R}}{3}\right) = 15.0567 \pm 2\left(\frac{0.577(3.2521)}{3}\right)$$

$$= 13.8057 \text{ and } 16.3076$$

The $\overline{X}$ chart is shown in Fig 5.10 and was created by SPC XL. When the preliminary sample averages are plotted on this chart, no indication of an out-of-control condition is observed. Therefore, since both the $\overline{X}$ and R charts exhibit control, we would conclude that the process is in control at the stated levels and adopt the trial control limits for monitoring future production.

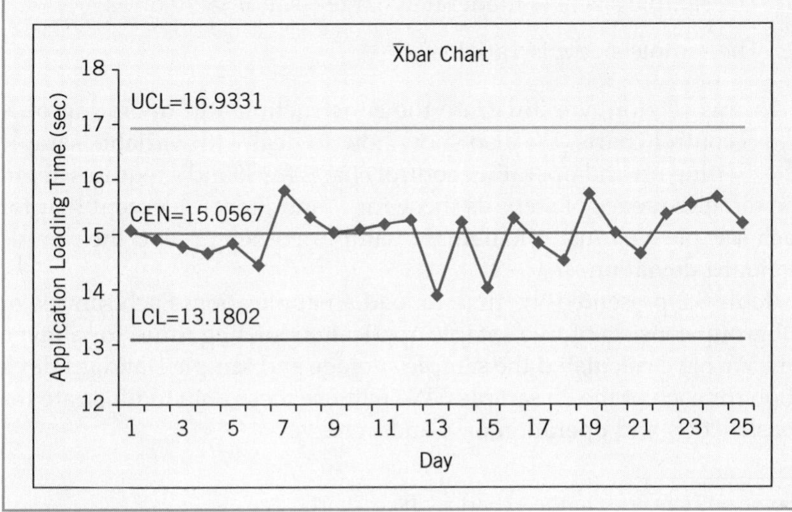

Figure 5.10 **$\overline{X}$ Chart for Application Loading Time**

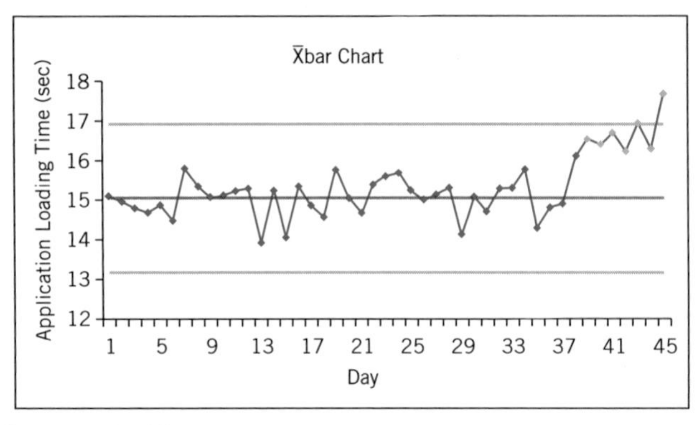

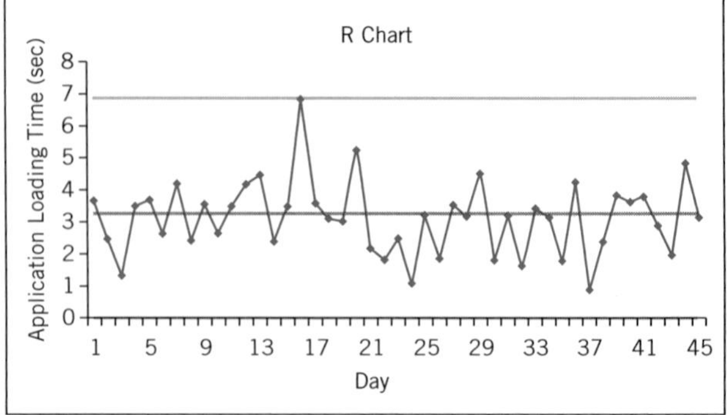

Figure 5.11 $\overline{X}$ and R Charts for Application Loading Time (SPC XL)

Once a set of reliable control limits is established, we use the control chart for monitoring future production. This is called phase II control chart usage. Continuing the previous example, the control limits were established, the IT department continued to monitor the application loading data using a subgroup size of five and a daily sampling frequency. The data was monitored online and the charts were updated daily. The results are shown in Figure 5.11.

The control charts indicate that the process is in control, until the 41st sample is plotted. The general pattern of points on the chart from about subgroup 38 onward is indicative of a shift in the process mean. The company should initiate some type of out of control plan (or OCAP) to find the assignable cause.

5.3 $\overline{X}$ and S Charts

S Chart
A control chart that depicts how the sample standard deviation changes over time.

Although $\overline{X}$ and R charts are widely used, it is occasionally desirable to estimate the process standard deviation directly instead of using the range R. This leads to control charts for σ and S, where S is the sample standard deviation.[1] Generally, $\overline{X}$ and S charts are preferable to their more familiar counterparts, $\overline{X}$ and R charts, when either

1. The sample size n is moderately large—say, $n > 10$ or 12.
2. The sample size n is variable.

In this section, we illustrate the construction and operation of $\overline{X}$ and S control charts. We also show how to deal with variable sample size. Setting up and operating control charts for $\overline{X}$ and s requires about the same sequence of steps as those for $\overline{X}$ and R charts, except that for each sample we must calculate the sample average, $\bar{x}$, and the sample standard deviation, s.

Table 5.2 presented the time to load an application. Each sample or subgroup consists of five sample application loading times for a given day. We have calculated the sample average and sample standard deviation for each of the 25 samples. We will use these data to illustrate the construction and operation of $\overline{X}$ and s charts.

[1] Some authors refer to the s chart as the σ chart.

We will assume that the mean and standard deviation are unknown and will be estimated from initial data. Estimating the mean is performed as before. However, recall that when we sample only a few points within a population, these points tend to be toward the middle of the distribution and not the tails. The more data points we sample, the greater chance we have to obtain a point in the tails. This is the reason why we used the t-distribution to create a confidence interval when we had less than 30 points. We utilize the sample principle in the construction of the s chart. When less than 25 points are obtained within a sample subgroup, a coefficient c_4 is utilized to create an estimate of the standard deviation. Thus, the estimate of the standard deviation = average standard deviation/c_4. Values of c_4 may be found in Appendix Six.

To obtain an $\overline{X}$ chart, we return to our original baseline formula of

$$\text{UCL} = \overline{\overline{x}} + 3\left(\frac{\sigma}{\sqrt{n}}\right)$$

$$\text{CL} = \overline{\overline{x}}$$

$$\text{LCL} = \overline{\overline{x}} - 3\left(\frac{\sigma}{\sqrt{n}}\right)$$

Adjusting these for the estimate of the standard deviation, we obtain

$$\text{UCL} = \overline{\overline{x}} + 3\left(\frac{\overline{s}}{c_4\sqrt{n}}\right)$$

$$\text{CL} = \overline{\overline{x}}$$

$$\text{LCL} = \overline{\overline{x}} - 3\left(\frac{\overline{s}}{c_4\sqrt{n}}\right)$$

If we let $A_2 = \dfrac{3}{c_4\sqrt{n}}$, calculations are simplified and we obtain

> Calculation of Control Limits for $\overline{X}$ Chart using s
>
> $$\text{UCL} = \overline{\overline{x}} + A_3\overline{s} \quad (5.11)$$
>
> $$\text{CL} = \overline{\overline{x}} \quad (5.12)$$
>
> $$\text{LCL} = \overline{\overline{x}} - A_3\overline{s} \quad (5.13)$$
>
> where A_3 is a value from the table Factors for Constructing Variables Control Charts found in Appendix Six.

The position of the one- and two-sigma lines can be found as follows:

$$\text{One-sigma lines} = \overline{x} \pm \frac{A_3\overline{s}}{3}$$

$$\text{Two-sigma lines} = \overline{x} \pm 2\left(\frac{A_3\overline{s}}{3}\right)$$

To obtain an S chart, we calculate the following:

> Calculation of Control Limits for S Chart
>
> $$\text{UCL} = B_4\bar{s} \quad \textbf{(5.14)}$$
> $$\text{CL} = \bar{s} \quad \textbf{(5.15)}$$
> $$\text{LCL} = B_4\bar{s} \quad \textbf{(5.16)}$$
>
> where B_3 and B_4 are values from the table Factors for Constructing Variables Control Charts, found in Appendix Six. These coefficients are used to simplify the calculations.

The position of the one- and two-sigma lines for the S chart can be found as follows:

$$\text{One-sigma lines} = \bar{s} \pm \frac{B_4\bar{s} - \bar{s}}{3}$$

$$\text{Two-sigma lines} = \bar{s} \pm 2\left(\frac{B_4\bar{s} - \bar{s}}{3}\right)$$

example 5.3 Development of $\overline{X}$ and S Charts

Repeat Example 5.2 using an $\overline{X}$, S chart.

Table 5.3

Application Load Time Data ($\overline{X}$, S)

	Application Load Time (seconds)					Subgroup Average	Subgroup Standard Deviation
Day	1	2	3	4	5		
1	13.2350	14.1280	16.7440	14.5730	16.9140	15.1188	1.6350
2	14.3140	13.5920	16.0750	14.6660	16.1090	14.9512	1.1111
3	14.2840	14.8710	14.9320	14.3240	15.6740	14.8170	0.5652
4	15.0280	16.3520	13.8410	12.8310	15.5070	14.7118	1.3891
5	15.6040	12.7350	15.2650	14.3630	16.4410	14.8816	1.4122
6	15.9550	15.4510	13.5740	13.2810	14.1980	14.4918	1.1679
7	16.2740	15.0640	18.3660	14.1770	15.1440	15.8050	1.6136
8	14.1900	14.3030	16.6370	16.0670	15.5190	15.3432	1.0771
9	13.8840	17.2770	15.3550	15.1760	13.6880	15.0760	1.4387

(Continued)

example 5.3 Continued

Table 5.3

Continued

Day	Application Load Time (seconds)					Subgroup Average	Subgroup Standard Deviation
	1	2	3	4	5		
10	14.0390	16.6970	15.0890	14.6270	15.2200	15.1344	0.9885
11	14.1580	17.6670	14.2780	15.9280	14.1810	15.2424	1.5477
12	15.8210	13.3550	15.7770	13.9800	17.5590	15.2984	1.6679
13	12.8560	14.1060	14.4470	16.3980	11.9280	13.9470	1.6992
14	14.9510	14.0360	15.8930	16.4580	14.9690	15.2614	0.9373
15	13.5890	12.8630	15.9960	12.4970	15.4710	14.0832	1.5680
16	15.7470	15.3010	15.1710	11.8390	18.6620	15.3440	2.4232
17	13.6800	17.2690	13.9570	15.0140	14.4490	14.8738	1.4320
18	14.1630	13.8640	13.0570	16.2100	15.5730	14.5734	1.2893
19	15.7960	14.1850	16.5410	15.1160	17.2470	15.7770	1.1954
20	17.1060	14.4120	12.3610	13.8200	17.6010	15.0600	2.2296
21	14.3710	15.0510	13.4850	15.6700	14.8800	14.6914	0.8186
22	14.7380	15.9360	16.5830	14.9730	14.7200	15.3900	0.8321
23	15.9170	14.3330	15.5510	15.2950	16.8660	15.5924	0.9225
24	16.3990	15.2430	15.7050	15.5630	15.5300	15.6880	0.4314
25	15.7970	13.6630	16.2400	13.7320	16.8870	15.2638	1.4816
					Grand Average:	15.0567	
					Average Standard Deviation:		1.3150

The control limits for the s chart are calculated first. Since the subgroup size is 5, we find $B_3 = 0$ and $B_4 = 2.089$ (Appendix Six).

$$\begin{aligned} \text{UCL} &= B_4\bar{s} = 2.089(1.3150) = 2.7470 \\ \text{CL} &= \bar{s} = 1.3150 \\ \text{LCL} &= B_3\bar{s} = 0(1.1350) = 0 \end{aligned}$$

$$\text{One-sigma lines} = 1.3150 \pm \frac{2.7470 - 1.3150}{3} = 0.8376 \text{ and } 1.7923$$

$$\text{Two-sigma lines} = 1.3150 \pm 2\left(\frac{2.7470 - 1.3150}{3}\right) = 0.3603 \text{ and } 2.2697$$

example 5.3 Continued

We then continue with calculation of control limits for the $\overline{X}$ chart. Since the subgroup size is 5, we find $A_2 = 0.577$ (Appendix Six).

$$\text{UCL} = \bar{\bar{x}} + A_3\bar{s} = 15.0567 + 1.427(1.3150) = 16.9331$$
$$\text{CL} = \bar{\bar{x}} = 15.0567$$
$$\text{LCL} = \bar{\bar{x}} + A_3\bar{s} = 15.0567 - 1.427(1.3150) = 13.1802$$

$$\text{One-sigma lines} = \bar{\bar{x}} \pm \frac{A_3\bar{s}}{3} = 15.0567 \pm \frac{1.427(1.3150)}{3}$$
$$= 14.4312 \text{ and } 15.6822$$

$$\text{Two-sigma lines} = \bar{\bar{x}} \pm 2\left(\frac{A_3\bar{s}}{3}\right) = 15.0567 \pm 2\left(\frac{1.427(1.3150)}{3}\right)$$
$$= 13.8057 \text{ and } 16.3077$$

The resulting control charts (prepared in Minitab) are shown as Figure 5.12. Note that there are slight differences in the calculation results due to different numbers of digits being carried through in the calculations.

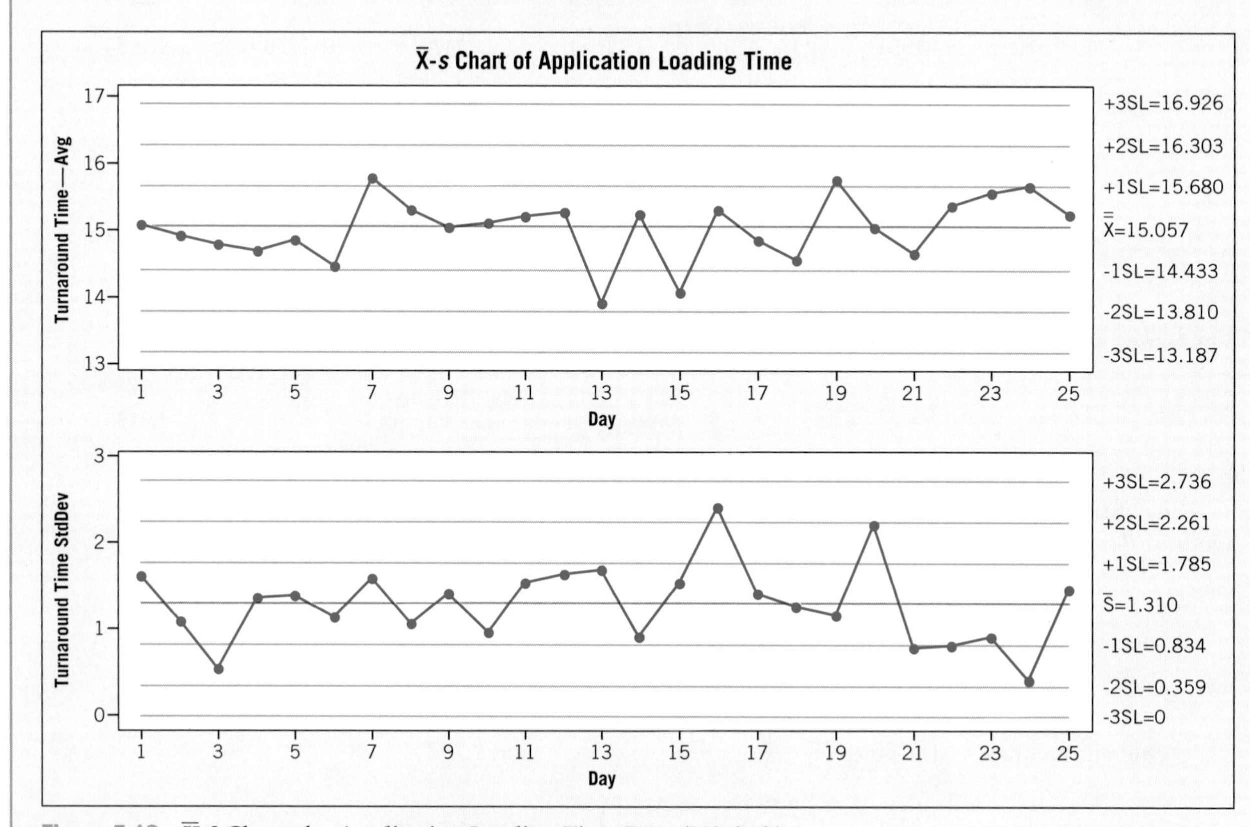

Figure 5.12 $\overline{X}$-S Charts for Application Loading Time Data (Minitab)

It is also possible to adjust the control limits to accommodate for varying sample size. However, this will involve calculating the control limit for each subgroup and using the appropriate multiplier for its subgroup size.

For example, imagine that we have a grand mean of 15.057 and average standard deviation of 1.310 and a subgroup size of 5. In this case, $A_3 = 1.427$ and the upper control limit for the first point would be 16.926. Let's say the second subgroup is missing 2 observations. In this case, there are only 3 data points in the subgroup. A_3 now changes to 1.954 and the upper control limit is now 17.617 (15.057 + 1.954(1.310)).

In this procedure, we adjust the control limits for every subgroup. A sample of an $\overline{X}$ and S chart using variable sample sizes is shown in Figure 5.13. This chart is based upon the application loading time data where some of the data points have been deleted for illustration purposes.

This chart is much more difficult to read and interpret and significantly increases the calculation effort of the control chart calculations. It also is difficult for production and service personnel to understand. Therefore, its use should be avoided.

An alternative to using variable-width control limits is to base the control limit calculations on an average sample size. If the sample

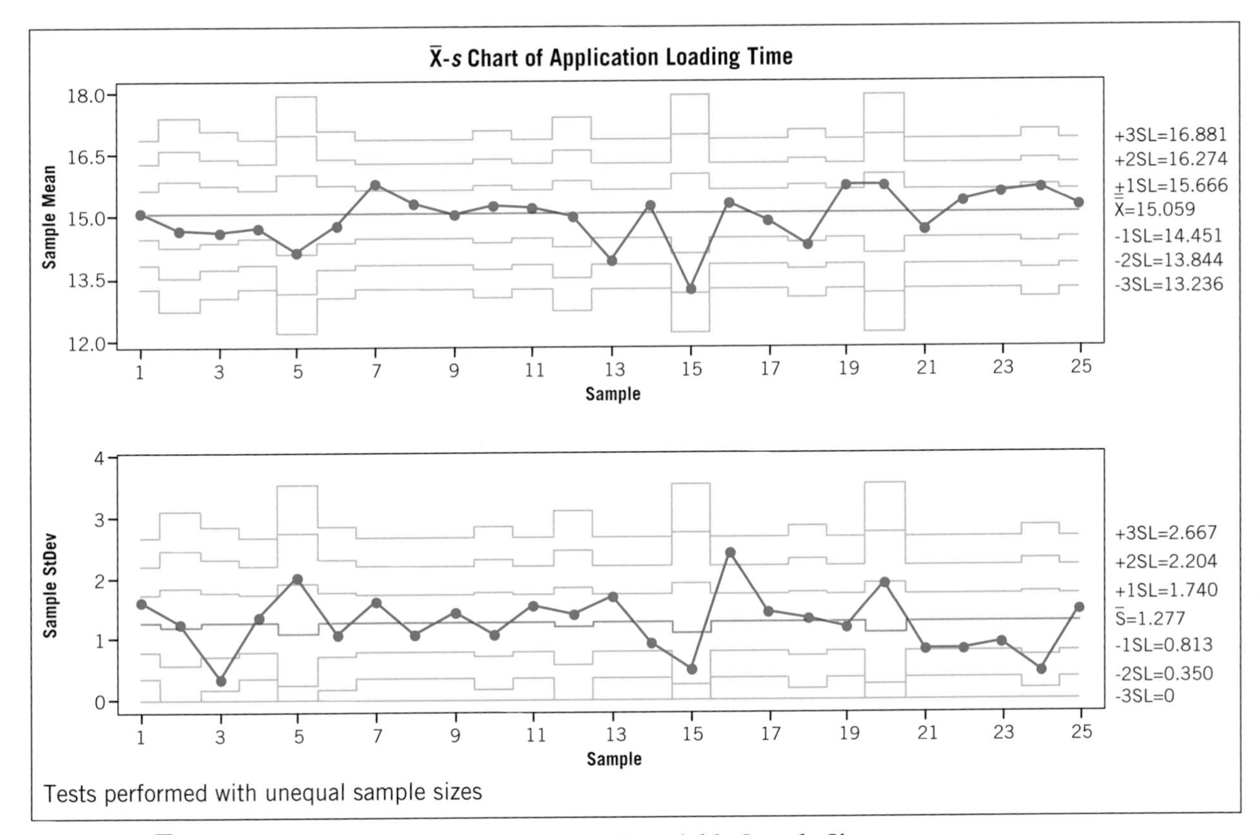

Figure 5.13 $\overline{X}$-*S* **Chart of Application Loading Time with Variable Sample Sizes (Minitab)**

sizes are not very different, this approach may be satisfactory. Since the average sample size may not be an integer, a useful alternative is to base these approximate control limits on the most common sample size.

5.4 Shewart Control Chart for Individual Measurements

There are many situations in which the sample size used for process monitoring is one; that is, the sample consists of an individual unit. Some examples of these situations are as follows:

1. Automated inspection and measurement technology is used, and every unit produced/serviced is analyzed so there is no basis for rational subgrouping.
2. Data comes available relatively slowly, and it is inconvenient to allow sample sizes of $n > 1$ to accumulate before analysis. The long interval between observations will cause problems with rational subgrouping. This occurs frequently in both manufacturing and service situations.
3. The cost of obtaining a measurement is very high.

In such situations, the control chart for individual units is useful. To prepare an individual's control chart, we use a moving range (or difference between two successive points) as the basis of estimating the process variability. It is also possible to establish a moving range control chart. The procedure is illustrated in the following example.

Individuals Chart
A control chart that monitors the process location over time when the subgroup size = 1

Calculation of Control Limits for Individuals Chart

$$\text{UCL} = \bar{x} + 3\frac{\overline{MR}}{d_2} \quad (5.17)$$

$$\text{CL} = \bar{x} \quad (5.18)$$

$$\text{LCL} = \bar{x} - 3\frac{\overline{MR}}{d_2} \quad (5.19)$$

where d_2 is a value from the table Factors for Constructing Variables Control Charts, found in Appendix Six.

The position of the One- and Two-sigma lines can be found as follows:

$$\text{One-sigma lines} = \bar{x} \pm \frac{\overline{MR}}{d_2}$$

$$\text{Two-sigma lines} = \bar{x} \pm 2\frac{\overline{MR}}{d_2}$$

To obtain an MR chart, we calculate the following:

Calculation of Control Limits for MR (Moving Range) Chart

$$UCL = D_4\overline{MR} \quad (5.20)$$

$$CL = \overline{MR} \quad (5.21)$$

$$LCL = D_3\overline{MR} \quad (5.22)$$

where D_3 and D_4 are values from the table Factors for Constructing Variables Control Charts, found in Appendix Six.

Moving Range Chart
A control chart that monitors the variation between successive sub-group over time.

The position of the one- and two-sigma lines can be found as follows:

$$\text{One-sigma lines} = \overline{MR} \pm \frac{D_4\overline{MR} - \overline{MR}}{3}$$

$$\text{Two-sigma lines} = \overline{MR} \pm 2\left(\frac{D_4\overline{MR} - \overline{MR}}{3}\right)$$

example 5.4 Development of Individual Charts

The mortgage loan processing unit of a bank monitors the costs of processing loan applications. The quantity tracked is the average weekly processing costs, obtained by dividing total weekly costs by the number of loans processed during the week. The processing costs for the most recent 20 weeks are shown in Table 5.4. Set up individual and moving range control charts for these data.

Table 5.4
Costs of Processing Mortgage Loan Applications

Weeks	Cost-x	Moving Range MR
1	310	
2	288	22
3	297	9
4	298	1
5	307	9
6	303	4
7	294	9
8	297	3

(Continued)

example 5.4 Continued

Table 5.4
Continued

Weeks	Cost-x	Moving Range MR
9	308	11
10	306	2
11	294	12
12	299	5
13	297	2
14	299	2
15	314	15
16	295	19
17	293	2
18	306	13
19	301	5
20	304	3
	$\bar{x} = 300.5$	$\overline{MR} = 7.79$

To set up the control chart for individual observations, note that the average cost of the 20 observations is 300.5 and that the average of the moving ranges of two observations is 7.79.

To set up the moving range chart, we use $d_2 = 1.128$, $D_3 = 0$ and $D_4 = 3.267$ for $n = 2$. Therefore, the individuals and moving range chart have the following control limits:

Control Limit	**Moving Range**	**Individuals**
UCL	$D_4\overline{MR} = 3.267(7.79) = 25.45$	$\bar{x} + 3\dfrac{\overline{MR}}{d_2} = 300.5 + 3\dfrac{7.79}{1.128} = 321.22$
CL	$\overline{MR} = 7.79$	$CL = 300.5$
LCL	$D_3\overline{MR} = 0(7.79) = 0$	$LCL = \bar{x} - 3\dfrac{\overline{MR}}{d_2} = 300.5 - 3\dfrac{7.79}{1.128} = 279.78$

The control charts (from Minitab) are shown in Figure 5.14. Notice that no points are out of control. The interpretation of the individuals control chart is very similar to the interpretation of the ordinary control chart. However, sometimes a point will plot outside the control limits on both the individuals chart and the moving range

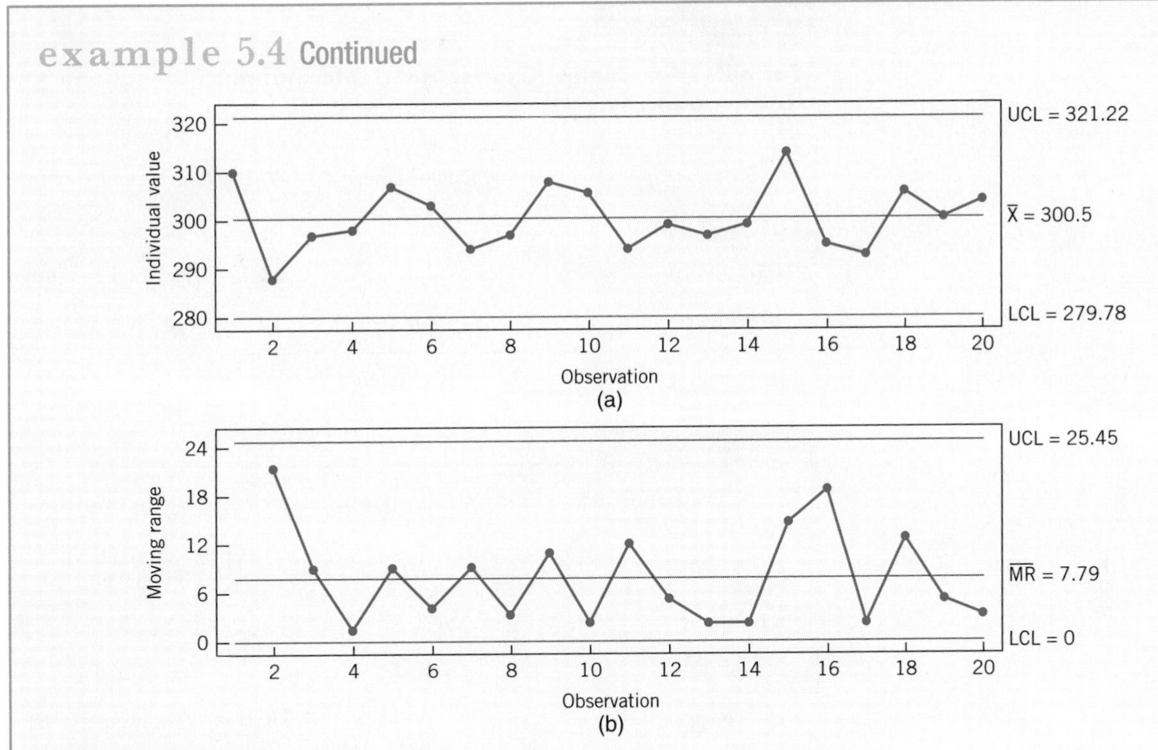

Figure 5.14 Control Charts for (a) Individual Observations on Cost and for (b) the Moving Range

chart. This will often occur because a large value of x will also lead to a large value of the moving range for that sample. This is very typical behavior for the individuals and moving range control charts. It is most likely an indication that the mean is out of control and not an indication that both the mean and the variance of the process are out of control.

Our discussion in this section has made an assumption that the observations follow a normal distribution. Borror, Montgomery, and Runger (1999) have studied the monitoring performance of the Shewhart control chart for individuals when the process data are not normal. They investigated various distributions to represent skewed and symmetric normal-like data. They found that performance of the chart was dramatically impacted for even moderate departures from normality. It is recommended that a normal probability plot be used to check the assumption of normality prior to using this control chart. If departures are noted, a EWMA chart may be more appropriate as it is very insensitive to the normality assumption (this chart is covered later in this chapter).

5.5 Summary of Procedures for $\overline{X}$, *R*, *S*, and Individuals Charts

It is convenient to summarize in one place the various computational formulas for the major types of variables control charts discussed so far. Table 5.5 summarizes the formulas for $\overline{X}$, R, and s charts when the mean

Importance of Normality

The performance of control charts can be significantly impacted for moderate departures in normality.

Table 5.5

Formulas for Control Charts, Mean and Standard Deviation Known and Given

Chart	Center Line	Control Limits
$\bar{x}$ (μ and σ given)	μ	$\mu \pm A\sigma$
R (σ given)	$d_2\sigma$	UCL $= D_2\sigma$, LCL $= D_1\sigma$
s (σ given)	$c_4\sigma$	UCL $= B_6\sigma$, LCL $= B_5\sigma$

Table 5.6

Formulas for Control Charts Given on Estimates of Parameters

Chart	Center Line	Control Limits
$\bar{x}$ (using R)	$\bar{\bar{x}}$	$\bar{\bar{x}} \pm A_2\bar{R}$
$\bar{x}$ (using s)	$\bar{\bar{x}}$	$\bar{\bar{x}} \pm A_3S$
R	$\bar{R}$	UCL $= D_4\bar{R}$, LCL $= D_3\bar{R}$
s	$\bar{s}$	UCL $= B_4\bar{s}$, LCL $= B_3\bar{s}$

and standard deviation are given. Table 5.6 provides the corresponding summary when no standard values are given and trial control limits must be established from analysis of past data. All constants are tabulated for various sample sizes in Appendix Six.

5.6 Example Applications of $\bar{X}$, *R*, *S*, and Individuals Charts

There are many interesting applications of variables control charts. In this section, a few of them will be described to give additional insights into how the control chart works, as well as ideas for further applications.

example 5.5 Using Control Charts to Improve a Supplier's Process

A large aerospace manufacturer purchased an aircraft component from two suppliers. These components frequently exhibited excessive variability on a key dimension that made it impossible to assemble them into the final product. This problem always resulted in expensive rework costs and occasionally caused delays in finishing the assembly of an airplane.

The materials receiving group performed 100% inspection of these parts in an effort to improve the situation. They maintained $\bar{X}$ and *R* charts on the dimension of

example 5.5 Continued

interest for both suppliers. They found that the fraction of nonconforming units was about the same for both suppliers, but for very different reasons. Supplier A could produce parts with mean dimension equal to the required value, but the process was out of statistical control. Supplier B could maintain good statistical control and, in general, produced a part that exhibited considerably less variability than parts from supplier A, but its process was centered so far off the nominal required dimension that many parts were out of specification.

This situation convinced the procurement organization to work with both suppliers, persuading supplier A to implement SPC and to begin working at continuous improvement, and assisting supplier B to find out why his process was consistently centered incorrectly. Supplier B's problem was ultimately tracked to some incorrect code in an NC (numerical-controlled) machine, and the use of SPC at supplier A resulted in considerable reduction in variability over a six-month period. As a result of these actions, the problem with these parts was essentially eliminated.

example 5.6 Using SPC to Purchase Equipment

An article in *Manufacturing Engineering* ("Picking a Marvel at Deere," January 1989, pp. 74–77) describes how the John Deere Company uses SPC methods to help choose production equipment. When a machine tool is purchased, it must go through the company capability demonstration prior to shipment to demonstrate that the tool has the ability to meet or exceed the established performance criteria. The procedure was applied to a programmable controlled bandsaw. The bandsaw supplier cut 45 pieces that were analyzed using $\overline{X}$ and R charts to demonstrate statistical control and to provide the basis for process capability analysis. The saw proved capable, and the supplier learned many useful things about the performance of his equipment. Control and capability tests such as this one are a basic part of the equipment selection and acquisition process in many companies.

example 5.7 SPC Implementation in a Short-Run Job-Shop

One of the more interesting aspects of SPC is the successful implementation of control charts in a job-shop manufacturing environment. Most job-shops are characterized by short production runs, and many of these shops produce parts on production runs of fewer than 50 units. This situation can make the routine use of control charts appear to be somewhat of a challenge, as not enough units are produced in any one batch to establish the control limits.

This problem can usually be easily solved. Since statistical process-control methods are most frequently applied to a characteristic of a product, we can extend SPC to the job-shop environment by focusing on the **process characteristic** in each unit of product. To illustrate, consider a drilling operation in a job-shop. The operator drills holes of various sizes in each part passing through the machine center. Some parts require one hole, and others several holes of different sizes. It is almost impossible to construct an $\overline{X}$ and R chart on hole diameter, since each part is potentially different. The correct approach is to focus on the characteristic of interest in the *process*. In this

example 5.7 Continued

case, the manufacturer is interested in drilling holes that have the correct diameter, and therefore wants to reduce the variability in hole diameter as much as possible. This may be accomplished by control charting the *deviation* of the actual hole diameter from the nominal diameter. Depending on the process production rate and the mix of parts produced, either a control chart for individuals with a moving range control chart or a conventional $\overline{X}$ and R chart can be used. In these applications, it is usually important to mark the start of each lot or to batch carefully on the control chart, so that if changing the size, position, or number of holes drilled on each part affects the process the resulting pattern on the control charts will be easy to interpret.

example 5.8 Using $\overline{X}$ and R Charts in Transactional Service Operations

Variables control charts have found frequent application in both manufacturing and nonmanufacturing settings. A fairly widespread but erroneous notion about these charts is that they do not apply to the nonmanufacturing environment because the "product is different." Actually, if we can make measurements on the product that are reflective of quality, function, or performance, then the *nature* of the product has no bearing on the general applicability of control charts. There are, however, two commonly encountered differences between manufacturing and transactional/service business situations: (1) in the nonmanufacturing environment, specification limits rarely apply to the product, so the notion of process capability is often undefined, and (2) more imagination may be required to select the proper variable or variables for measurement.

One application of $\overline{X}$ and R control charts in a transactional business environment involved the efforts of a finance group to reduce the time required to process its accounts payable. The division of the company in which the problem occurred had recently experienced a considerable increase in business volume, and along with this expansion came a gradual lengthening of the time the finance department needed to process check requests. As a result, many suppliers were being paid beyond the normal 30-day period, and the company was failing to capture the discounts available from its suppliers for prompt payment. The quality-improvement team assigned to this project used the flow time through the finance department as the variable for control chart analysis. Five completed check requests were selected each day, and the average and range of flow time were plotted on $\overline{X}$ and R charts. Although management and operating personnel had addressed this problem before, the use of $\overline{X}$ and R charts was responsible for substantial improvements. Within nine months, the finance department had reduced the percentage of invoices paid late from over 90% to under 3%, resulting in an annual savings of several hundred thousand dollars in realized discounts to the company.

5.7 Cumulative Sum Control Charts

The charts covered in the previous sections are basic SPC methods. They are extremely useful in initial implementations of SPC when the process is likely to be out of control and experiencing assignable causes that result in large shifts in the monitored parameters. They are also very useful in the diagnostic aspects of bringing an unruly process into control, because the patterns on these charts often provide guidance regarding the nature of the assignable cause.

A major disadvantage of these charts is that they use only the information about the process contained in the last sample observation, ignoring information given by the entire sequence of points. This feature makes them relatively insensitive to small process shifts—say, on the order of about 1.5σ or less—unless all four Western Electric Rules are utilized. This may make these charts less useful for monitoring when the process tends to operate in control, reliable estimates of the process mean and standard deviation are available, and assignable causes do not typically result in large process upsets or disturbances.

Two very effective alternatives to the Shewhart control chart may be used when small process shifts are of interest: the cumulative sum (CUSUM) control chart, and the exponentially weighted moving average (EWMA) control chart (covered in the next section)

CUSUM Chart
A control chart capable of detecting small shifts in the process. It works by computing a cumulative sum.

As its name implies, the cumulative sum computes a cumulative sum of the differences of each individual sample point as compared to the process parameter estimate. If the process is operating at random without drift, we would expect that values above the mean and values below the mean would cancel out and result in a cumulative sum of 0 over time. If the cumulative value consistently rises or falls, this would be an indication of a trend.

Consider the following 30 data points in Table 5.7. Values 1–20 are from a distribution with a mean of 10 and a standard deviation of 1. Thus, we expect values between 7 and 13 (given a 3 standard deviation range). Values 21–30 are from a distribution with a mean of 11 and a standard deviation of 1. Thus, we would expect values between 8 and 11. This small shift might be difficult to detect rapidly in a traditional control chart. However, looking at cumulative deviations away from the original mean of 10, we can readily observe the shift.

Table 5.7

Data to Illustrate Cumulative Sum Calculations

Sample	Data Value$_i$	Data Value$_i$ −10	Cumulative Sum (Previous Value + Data Value$_i$ − 10)
1	9.45	−0.55	−0.55
2	7.99	−2.01	−2.56
3	9.29	−0.71	−3.27
4	11.66	1.66	−1.61
5	12.16	2.16	0.55
6	10.18	0.18	0.73
7	8.04	−1.96	−1.23
8	11.46	1.46	0.23

(Continued)

Table 5.7

Continued

Sample	Data Value$_i$	Data Value$_i$ −10	Cumulative Sum (Previous Value + Data Value$_i$ − 10)
9	9.20	−0.80	−0.57
10	10.34	0.34	−0.23
11	9.03	−0.97	−1.20
12	11.47	1.47	0.27
13	10.51	0.51	0.78
14	9.40	−0.60	0.18
15	10.08	0.08	0.26
16	9.37	−0.63	−0.37
17	10.62	0.62	0.25
18	10.31	0.31	0.56
19	8.52	−1.48	−0.92
20	10.84	0.84	−0.08
21	10.90	0.90	0.82
22	9.33	−0.67	0.15
23	12.29	2.29	2.44
24	11.50	1.50	3.94
25	10.60	0.60	4.54
26	11.08	1.08	5.62
27	10.38	0.38	6.00
28	11.62	1.62	7.62
29	11.31	1.31	8.93
30	10.52	0.52	9.45

These results are shown in Figure 5.15. Initially, the cumulative sum has an average of 0 with a random pattern from values 1–20. This indicates that the process is in control. We notice a large departure from this pattern beginning at point 21 (when the mean value shifts from 10–11). An upward trend further indicates that a shift has occurred.

The plot in Figure 5.15 is not a control chart, though it can be expanded into one. The most common implementation of a CUSUM chart (in software packages) is based upon a tabular method that calculates deviations when they occur above the mean as well as below the mean. A slack factor (K) is implemented to provide sensitivity within the chart and is typically chosen to be half the distance between the target value and the shift in the mean that you are trying to detect. In our example, K will be 0.5, as we would like to detect a shift of 1 unit away from the mean (11–10)/2.

Upper deviations will be calculated for deviations above the mean + slack factor (10.5). Cumulative values will only be reported when they are greater than 0. Lower deviations will be calculated for deviations below the mean – slack factor (9.5). They will only be reported when they are less than 0. Whenever multiple deviations are found (upper or lower) they will be counted. This will assist in determining when an out-of-control condition occurs.

Calculations will be described for the first four points.

Sample 1, data value = 9.45.

upper deviation = 9.45 − 10.5 = −1.05. This is reported as 0 (upper deviations are only reported when > 0)

lower deviation = 9.45 − 9.5 = −0.05. This is reported as −0.05 (lower deviations are reported when < 0)

Sample 2, data value = 7.99.

upper deviation = 7.99 − 10.5 = −2.51. The cumulative upper deviation = 0 + −2.51 = −2.51. This is reported as 0.

lower deviation = 7.99 − 9.5 = −1.51. The cumulative lower deviation =−0.05 + −1.51 = −1.56.

Sample 3, data value = 9.29

upper deviation = 9.29 − 10.5 = −1.21. The cumulative upper deviation = 0 + −1.21 = −1.21. This is reported as 0.

lower deviation = 9.29 − 9.5 = −0.21. The cumulative lower deviation = −1.56 + −0.21 = −1.77.

Sample 4, data value = 11.66

upper deviation = 11.66 − 10.5 = 1.16. The cumulative upper deviation = 0 + −1.21 = 1.21.

lower deviation = 11.66 − 9.5 = 2.16. The cumulative lower deviation = −1.77 + 2.16 = 0.39. This is reported as 0.

The system will be considered out of control when the cumulative sum roughly exceeds 5 times the standard deviation of the process. Since the standard deviation is 1, we will consider the process out of control cumulative sum greater than 5 occurs. Finally, we should note that runs tests and other sensitizing rules such as the zone rules cannot be safely applied to the CUSUM.

A full table of calculations is provided in Table 5.8.

Although we have given the development of the tabular CUSUM for the case of individual observations (n = 1), it is readily extended to the case of averages of rational subgroups where the sample size $n > 1$.

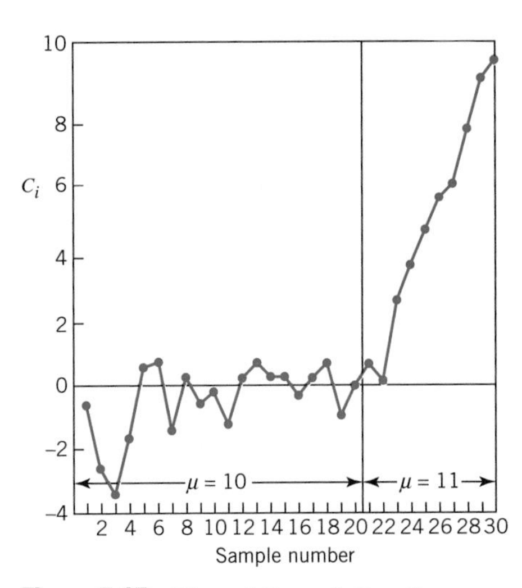

Figure 5.15 **Plot of Cumulative Sums from Table 5.7**

Table 5.8

CUSUM Tabular Calculations

Sample	Data Value	Upper Deviations (10.5 − Sample Value)			Lower Deviations (Sample Value – 9.5)		
		Deviation (X-10.5)	Cumulative Deviation	Count	Deviation (x-9.5)	Cumulative Deviation	Count
1	9.45	−1.05	0		−0.05	−0.05	1
2	7.99	−2.51	0		−1.51	−1.56	2
3	9.29	−1.21	0		−0.21	−1.77	3
4	11.66	1.16	1.16	1	2.16	0	
5	12.16	1.66	2.82	2	2.66	0	
6	10.18	−0.32	2.5	3	0.68	0	
7	8.04	−2.46	0.04	4	−1.46	−1.46	1
8	11.46	0.96	1	5	1.96	0	
9	9.20	−1.30	0		−0.30	−0.3	1
10	10.34	−0.16	0		0.84	0	
11	9.03	−1.47	0		−0.47	−0.47	1
12	11.47	0.97	0.97	1	1.97	0	
13	10.51	0.01	0.98	2	1.01	0	
14	9.40	−1.10	0		−0.10	−0.1	1
15	10.08	−0.42	0		0.58	0	
16	9.37	−1.13	0		−0.13	−0.13	1
17	10.62	0.12	0.12	1	1.12	0	
18	10.31	−0.19	0		0.81	0	
19	8.52	−1.98	0		−0.98	−0.98	1
20	10.84	0.34	0.34	1	1.34	0	
21	10.90	0.40	0.74	2	1.40	0	
22	9.33	−1.17	0		−0.17	−0.17	1
23	12.29	1.79	1.79	1	2.79	0	
24	11.50	1.00	2.79	2	2.00	0	
25	10.60	0.1	2.89	3	1.10	0	
26	11.08	0.58	3.47	4	1.58	0	
27	10.38	−0.12	3.35	5	0.88	0	
28	11.62	1.12	4.47	6	2.12	0	
29	11.31	0.81	5.28	7	1.81	0	
30	10.52	0.02	5.3	8	1.02	0	

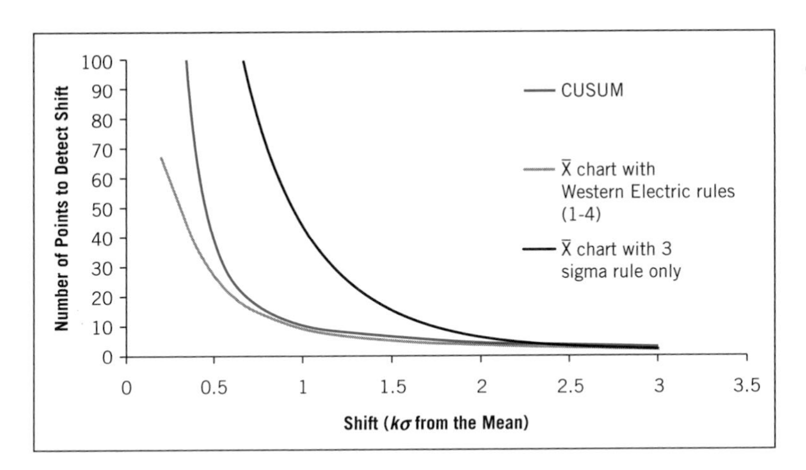

Figure 5.16 Performance of the CUSUM Chart

Simply replace the individual measurements by subgroup averages. When calculating the control chart limits as 5 times the standard deviation, use the standard deviation of the individual measurements/(subgroup size).

However, one must carefully consider this choice. With Shewhart charts, the use of averages of rational subgroups substantially improves control chart performance, but this does not always happen with the CUSUM. For example, a CUSUM chart obtained by taking a sample size of $n = 1$ every half hour performs better than a CUSUM chart obtained through a rational subgroup of size $n = 5$ every 2.5 hours (note that both choices have the same sampling intensity). Only if there is some significant economy of scale or some other valid reason for taking samples of size greater than unity should one consider using $n > 1$ with the CUSUM.

The performance of the CUSUM chart compared to an $\overline{X}$ chart with and without the Western Electric rules is shown in Figure 5.16. Note that the CUSUM chart performs significantly better than the $\overline{X}$ chart without the Western electric rules. Performance of the CUSUM chart is slightly worse than the $\overline{X}$ chart with the Western Electric rules for shifts less than one standard deviation. However, the CUSUM has a siginificantly lower false alarm rate (1 in 465 for the CUSUM versus 1 in 92 for the $\overline{X}$ chart with Western Electric rules)

5.8 Exponentially Weighted Moving Average Control Charts

The exponentially weighted moving average (EWMA) control chart is also a good alternative to the Shewhart control chart when we are interested in detecting small shifts or working with data that is non-normal. The EWMA chart is almost a nonparameteric (distribution free) procedure that can be utilized in a wide variety of situations.

EWMA
A control chart useful for detecting small process shifts or working with non-normal data.

The performance of the EWMA control chart is approximately equivalent to that of the cumulative sum control chart, and in some ways it is easier to set up and operate. As with the CUSUM, the EWMA is typically used with individual observations; however, as with the CUSUM chart, this procedure can be readily modified by replacing an individual measurement x_i with the subgroup average $\overline{x}_i$ and the standard deviation σ as $\sigma/\sqrt{n}$.

The EWMA chart works by weighting each data point such that more recent data has a larger weight than previous data parts. The weight is controlled by a parameter λ which should be between 0.05 and 0.25. In addition, a factor L controls with the width of the control limits. Formulas for the EWMA are described below:

The EWMA Control Chart

$$\text{UCL} = \mu_0 + L\sigma\sqrt{\frac{\lambda}{(2-\lambda)}[1-(1-\lambda)^{2i}]} \quad (5.23)$$

$$\text{Center line} = \mu_0 \quad (5.24)$$

$$\text{LCL} = \mu_0 - L\sigma\sqrt{\frac{\lambda}{(2-\lambda)}[1-(1-\lambda)^{2i}]} \quad (5.25)$$

In general, we have found that values of λ between 0.05 and 0.25 work well in practice, with $\lambda = 0.05$, $\lambda = 0.10$, and $\lambda = 0.20$ being popular choices. A good rule of thumb is to use smaller values of λ to detect smaller shifts. We have also found that $L = 3$ (the usual three-sigma limits) works reasonably well, particularly with the larger value of λ, although when λ is small—say, $\lambda \leq 0.1$—there is an advantage in reducing the width of the limits and using a value of L between about 2.6 and 2.8. This is illustrated in Table 5.9, which details the expected number of points it would take to detect a shift in the mean.

Table 5.9

Number of Points to Detect a Shift in the Mean for an EWMA Chart [Adapted from Lucas and Saccucci (1990)]

Shift in Mean (multiple of sigma)	$L = 3.054$ $\lambda = 0.4$	$L = 2.998$ $\lambda = 0.25$	$L = 2.962$ $\lambda = 0.2$	$L = 2.814$ $\lambda = 0.1$	$L = 2.615$ $\lambda = 0.05$
0.25	224	170	150	106	84.1
0.5	71.2	48.2	41.8	31.3	28.8
0.75	28.4	20.1	18.2	15.9	16.4
1	14.3	11.1	10.5	10.3	11.4
1.5	5.9	5.5	5.5	6.1	7.1
2	3.5	3.6	3.7	4.4	5.2
2.5	2.5	2.7	2.9	3.4	3.5
3	2	2.3	2.4	2.9	3.5
4	1.4	1.7	1.9	2.2	2.7

There is one potential concern about an EWMA with a small value of λ. If the value of the EWMA is on one side of the center line when a shift in the mean in the opposite direction occurs, it could take the EWMA several periods to react to the shift, because the small λ does not weight the new data very heavily. This is called the **inertia effect** and can reduce the effectiveness of the EWMA in shift detection

example 5.9 Development of a EWMA Chart

Set up an EWMA control chart with $\lambda = 0.10$ and $L = 2.7$ using the data in Table 5.7. This combination of L and λ results in a similar detection capability as the CUSUM chart presented earlier in the chapter.

Recall that the target value of the mean is 10 and the standard deviation is 1. The calculations for the EWMA control chart are summarized in Table 5.10, and the control chart (from Minitab) is shown in Figure 5.17.

Table 5.10

Calculation Results for Example 5.9

Sample	Data Value	Plotted Point	LCL	CL	UCL
1	9.45	9.95	9.73	10.00	10.27
2	7.99	9.75	9.64	10.00	10.36
3	9.29	9.70	9.58	10.00	10.42
4	11.66	9.90	9.53	10.00	10.47
5	12.16	10.13	9.50	10.00	10.50
6	10.18	10.13	9.48	10.00	10.52
7	8.04	9.92	9.46	10.00	10.54
8	11.46	10.08	9.44	10.00	10.56
9	9.2	9.99	9.43	10.00	10.57
10	10.34	10.02	9.42	10.00	10.58
11	9.03	9.92	9.41	10.00	10.59
12	11.47	10.08	9.41	10.00	10.59
13	10.51	10.12	9.40	10.00	10.60
14	9.4	10.05	9.40	10.00	10.60
15	10.08	10.05	9.39	10.00	10.61
16	9.37	9.98	9.39	10.00	10.61

(Continued)

example 5.9 Continued

Table 5.10

Continued

Sample	Data Value	Plotted Point	LCL	CL	UCL
17	10.62	10.05	9.39	10.00	10.61
18	10.31	10.07	9.39	10.00	10.61
19	8.52	9.92	9.39	10.00	10.61
20	10.84	10.01	9.39	10.00	10.61
21	10.9	10.10	9.38	10.00	10.62
22	9.33	10.02	9.38	10.00	10.62
23	12.29	10.25	9.38	10.00	10.62
24	11.5	10.37	9.38	10.00	10.62
25	10.6	10.40	9.38	10.00	10.62
26	11.08	10.47	9.38	10.00	10.62
27	10.38	10.46	9.38	10.00	10.62
28	11.62	10.57	9.38	10.00	10.62
29	11.31	10.65	9.38	10.00	10.62
30	10.52	10.63	9.38	10.00	10.62

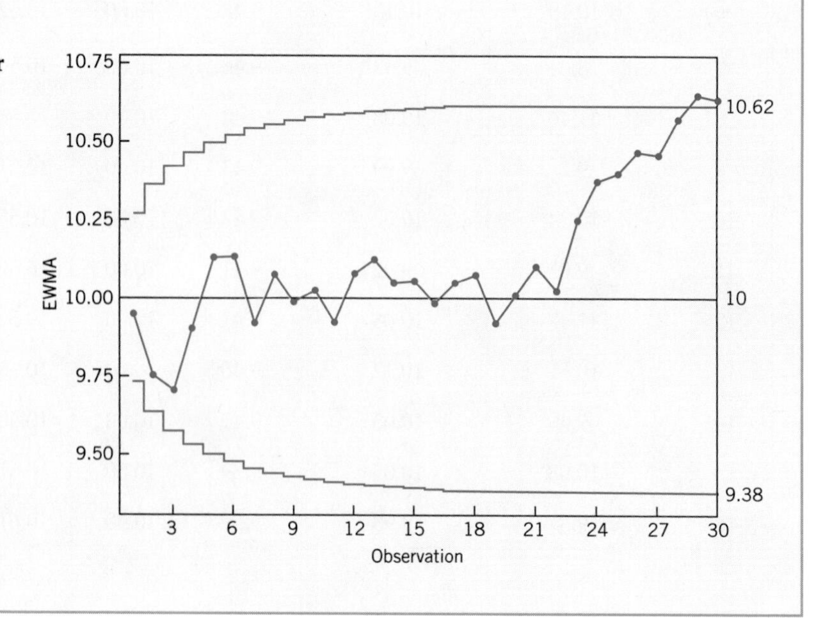

Figure 5.17 EWMA Chart for Example 5.9

example 5.9 Continued

To illustrate, consider the first observation of 9.45. For this first point, we must compute the value of the plotted value, the upper control limit, and the lower control limit. These are defined below:

$$\text{value of plotted point} = \text{weighted average} = (.1)(9.45) + (1 - .1)(10)$$
$$= 9.945.$$

$$\text{upper control limit} = \mu_0 + L\sigma\sqrt{\frac{\lambda}{(2-\lambda)}[(1-\lambda)^{2i}]}$$
$$= 10 + (2.7)(1)\sqrt{\frac{0.1}{(2-0.1)}[1-(1-0.1)^{2*1}]} = 10.27$$

$$\text{lower control limit} = \mu_0 + L\sigma\sqrt{\frac{\lambda}{(2-\lambda)}[(1-\lambda)^{2i}]}$$
$$= 10 - (2.7)(1)\sqrt{\frac{0.1}{(2-0.1)}[1-(1-0.1)^{2*1}]} = 9.73$$

The second observation is 7.99. The values for the plotted value, upper control limit, and lower control limit are as follows:

$$\text{value of plotted point} = (.1)(9.945) + (.9)(7.99) = 9.7495$$

$$\text{upper control limit} = \mu_0 + L\sigma\sqrt{\frac{\lambda}{(2-\lambda)}[(1-\lambda)^{2i}]}$$
$$= 10 + (2.7)(1)\sqrt{\frac{0.1}{(2-0.1)}[1-(1-0.1)^{2*2}]} = 10.36$$

$$\text{lower control limit} = \mu_0 + L\sigma\sqrt{\frac{\lambda}{(2-\lambda)}[(1-\lambda)^{2i}]}$$
$$= 10 - (2.7)(1)\sqrt{\frac{0.1}{(2-0.1)}[1-(1-0.1)^{2*2}]} = 9.64$$

Note that the control limits gradually increase during the startup of the chart but stabilize after about 20 points. This is because the term $[1-(1-\lambda)^{2i}]$ goes to 1 as i increases.

5.9 Process Capability Analysis Using Control Charts

It is usually necessary to obtain information about the capability of a process—that is, the performance of the process when it is operating in control. Two graphical tools, the tolerance chart and the histogram, are helpful in assessing process capability.

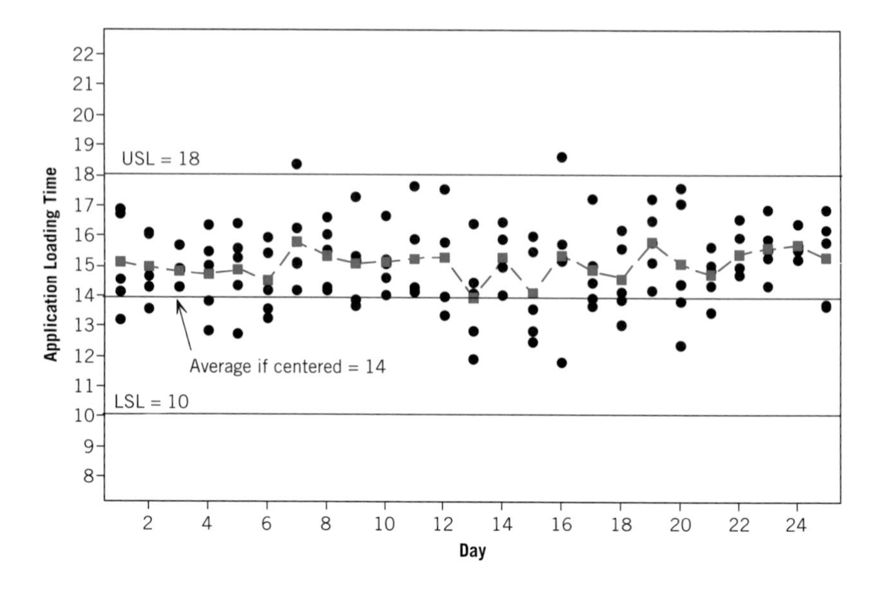

Figure 5.18 Tolerance Chart for Application Loading Time Data

A tolerance chart for the first 25 samples of application loading time (Example 5.2) is shown in Figure 5.18 and includes the upper and lower specification limits for application loading time. These specification limits define necessary business conditions for successful operating conditions and have no relationship to the control limits. In this case, business units experience a loss in productivity and customer service when the application loading time exceeds 18 seconds, and critical business processes can be impacted when an application is loaded faster than 10 seconds. The times of 18 and 10 seconds are referred to as the upper and lower specification limits (USL and LSL). Ideally, the process would have a mean "centered" between these two values. In this case, the process would ideally be centered around 14 seconds.

The tolerance chart is useful in revealing patterns over time in the individual measurements or may show that an unusual shift in the average or range was produced by one or two unusual observations in the sample. Note that it is also appropriate to plot the specification limits on the tolerance chart, since it is a chart of individual measurements. It is never appropriate to plot specification limits on a control chart or to use the specifications in determining the control limits. Specification limits and control limits are unrelated (but ideally the control limits should be within the specification limits to avoid unnecessary reprocessing efforts and poor customer service).

A more common approach is to use a histogram to describe process capability. A histogram for the application loading data is shown in Figure 5.19. The general impression from this histogram is that the process is loading slower than desired (>18 seconds) but if this distribution should be shifted so that it is centered within the specification limits it would be more capable of meeting business requirements.

Process Capability Ratios, C_p

The ratio of the spread between the specification limit to the spread of the process values.

It is frequently convenient to have a simple, quantitative way to express process capability. One way to do so is through the process capability ratio (PCR), C_p, which for a quality characteristic with

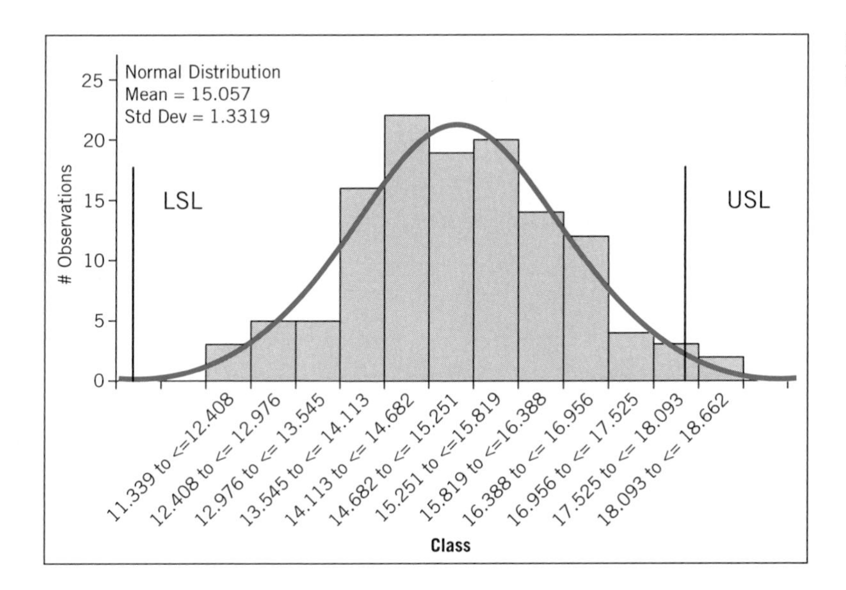

Figure 5.19 **Histogram for Application Loading Data**

both upper and lower specification limits (USL and LSL, respectively) is

$$C_p = \frac{\text{USL} - \text{LSL}}{6\sigma} \tag{5.26}$$

Note that the 6σ spread of the process is the basic definition of process capability. Since σ is usually unknown, we must replace it with an estimate. We frequently use $\hat{\sigma} = \frac{\overline{R}}{d_2}$ as an estimate of σ, resulting in an estimate $\hat{C}_{\overline{p}}$ of C_p.

For the application loading process, the USL = 18, LSL=10, $\overline{R} = 3.251$, $d_2 = 3.326$. The estimate of the standard deviation is thus $3.251/2.326 = 1.398$. Therefore, an estimate for the process capability ratio, C_p, is

$$C_p = \frac{18 - 10}{6(1.398)} = 0.95$$

If C_p is less than 1, the distribution does not fit within the specification limits. In this case, C_p is relatively close to 1, which indicates that some applications will not load within desired business requirements. The PCR C_p may be interpreted another way. The quantity

$$P = \left(\frac{1}{C_p}\right)100\% \tag{5.27}$$

is the percentage of the specification range that the process uses up. For the application loading process, an estimate of P is $(1/0.95)*100 = 105$. That is, the process uses up about 105% of the specification band. Since this number is greater than 100%, it implies that we need to reduce the variability within the process to ensure that applications are loaded within the desired time frame.

Figure 5.20 illustrates three cases of interest relative to the PCR C_p and process specifications. In all cases, the mean of the process is 100 and the standard deviation is 10.

Note that all the cases in Figure 5.20 assume that the process is centered at the midpoint of the specification range. While this is generally

Specification limits are 50 and 150. Therefore the C_p is greater than one (it actually equals 1.67). This means that the process uses up much less than 100% of the specification range. Consequently, relatively few nonconforming units will be produced by this process.

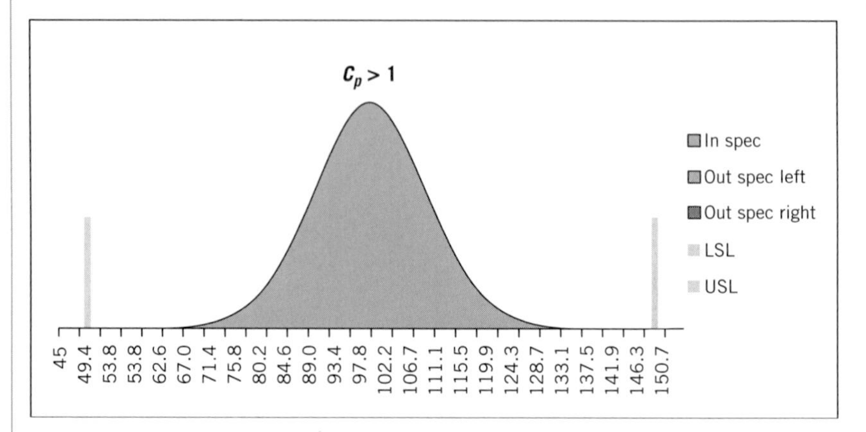

Specification limits are 70 and 130. Therefore, C_p is equal to 1. This means the process uses up all the specification range. For a normal distribution this would imply about 0.27% (or 2700 parts per million—ppm) nonconforming units.

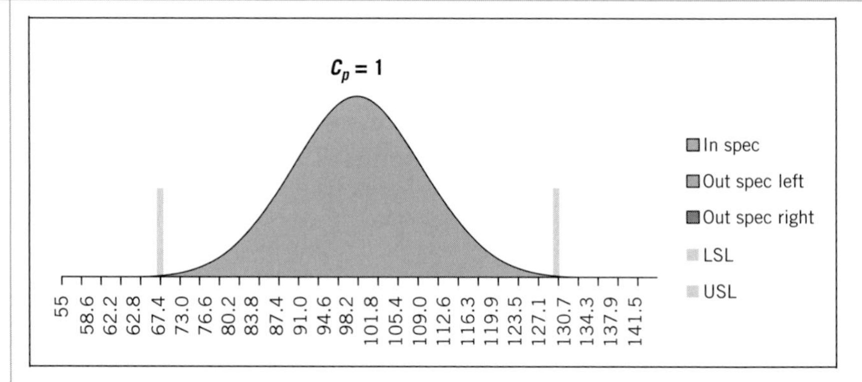

Specification limits are 80 and 120. Therefore, C_p is less than 1 (it actually equals 0.67). This means process uses up more than 100% of the specification range. In this case, a large number of nonconforming units will be produced.

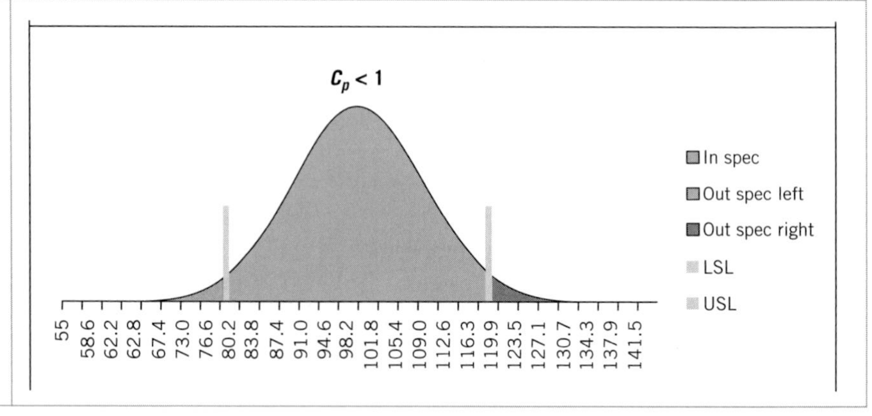

Figure 5.20 **Illustration of C_p**

the ideal situation, it is not always possible to achieve this in practice. Therefore, we also compute one-sided capability ratios to reflect how well the process fits within each specification limit.

$$C_{pu} = \frac{\text{USL} - \mu}{3\sigma} \quad \text{PCR for upper specification limit} \tag{5.28}$$

$$C_{pl} = \frac{\mu - \text{LSL}}{3\sigma} \quad \text{PCR for lower specification limit} \tag{5.29}$$

$$C_{pk} = \min(C_{pl}, C_{pk}) \text{ worst-case performance} \tag{5.30}$$

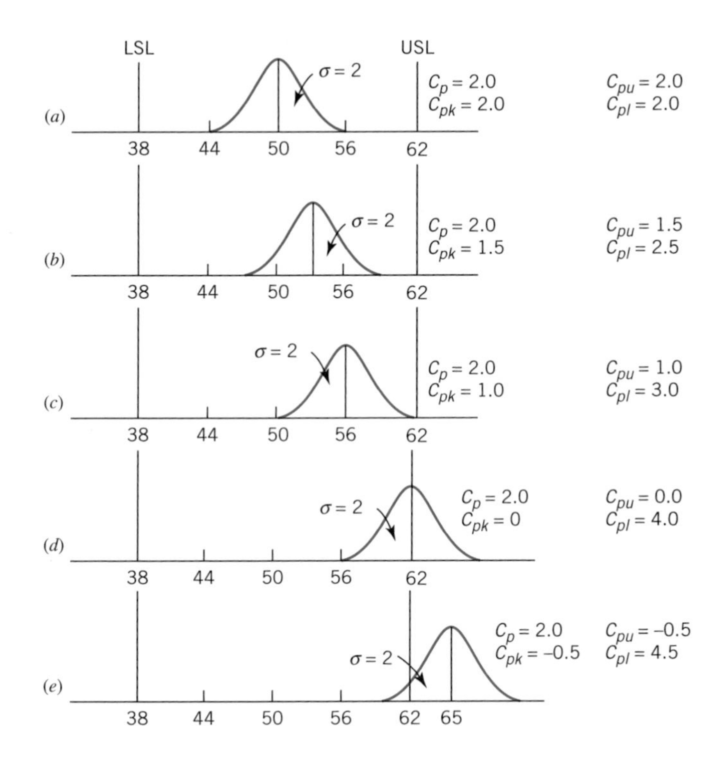

Figure 5.21 **Relationship of C_p and C_{pk}**

As with C_p, estimates of $\hat{C}_{pu}$ and $\hat{C}_{pl}$ can be obtained by replacing σ with an estimate of sigma (such as $\hat{\sigma} = \overline{R}/d_2$).

When the process is perfectly centered, C_{pu} will equal C_{pl}. However, when the process is not centered, an offset will occur. The magnitude of C_{pk} relative to C_p is a direct measure of how off-center the process is operating. Several commonly encountered cases are illustrated in Figure 5.21.

Table 5.11 presents several values of the PCR C_p along with the associated values of process fallout, expressed in defective parts or nonconforming units of product per million (ppm). When the process is centered for a two-sided specification, the column for two-sided specifications is used. For example, a $C_p = 1.00$ implies a fallout rate of 2,700 ppm, whereas a PCR of $C_p = 1.50$ for this process implies a fallout rate of 7 ppm. When the process is not centered, C_{pu} and C_{pl} are utilized with the one sided specification limits to determine the overall fall out rate. For example, if $C_{pu} = 1.5$ and $C_{pl} = 2.5$, then we obtain 4 ppm for the upper specification limit and 0 ppm for the lower specification limit. Adding them together results in an overall fallout rate of 4 ppm.

The ppm quantities were calculated using the following important assumptions:

1. The quality characteristic has a normal distribution.
2. The process is in statistical control.
3. In the case of two-sided specifications, the process mean is centered between the lower and upper specification limits.

These assumptions are absolutely critical to the accuracy and validity of the reported numbers, and if they are not valid, then the

Table 5.11

Values of the Process Capability Ratio (C_p) and Associated Process Fallout for a Normally Distributed Process (in Defective ppm) for a Process that is in Statistical Control

	Process Fallout (in defective ppm)	
PCR	One-Sided Specifications	Two-Sided Specifications
0.25	226,628	453,255
0.50	66,807	133,614
0.60	35,931	71,861
0.70	17,865	35,729
0.80	8,198	16,395
0.90	3,467	6,934
1.00	1,350	2,700
1.10	484	967
1.20	159	318
1.30	48	96
1.40	14	27
1.50	4	7
1.60	1	2
1.70	0.17	0.34
1.80	0.03	0.06
2.00	0.0009	0.0018

reported quantities may be seriously in error. For example, even if a distribution is symmetric but has longer tails (such as the *t*-distribution versus the normal distribution) substantial errors can result. Therefore, it is recommended that a normality plot be utilized to confirm this assumption.

Stability or statistical control of the process is also essential to the correct interpretation of any PCR. Unfortunately, it is fairly common practice to compute a PCR from a sample of historical process data without any consideration of whether or not the process is in statistical control. If the process is not in control, then of course its parameters are unstable, and the value of these parameters in the future is uncertain. Thus the predictive aspects of the PCR regarding process ppm performance are lost.

Table 5.12 presents some recommended guidelines for minimum values of the process capability indices. We point out that the values in this table are only minimums. In recent years, many companies have adopted criteria for evaluating their processes that include process capability objectives that are more stringent. For example, a six-sigma company would require that when the process mean is in control, it will no be closer than six standard deviations from the nearest specification limit. This, in effect, requires that the minimum acceptable value of the process capability ratio will be at least 2.0.

Table 5.12

Recommended Minimum Values of the Process Capability Ratio

	Minimum Process Capability Index (C_{pk}) for Processes with Upper and Lower Specification Limits	Minimum Process Capability (C_{pu} or C_{pl}) for Processes with One Specification (Upper or Lower)
Existing process	1.33	1.25
New processes	1.5	1.45
Safety, strength, or critical parameter, existing process	1.5	1.45
Safety, strength, or critical parameter, new process	1.67	1.60

example 5.10 Process Capability Study

Joe's Salsa Company is considering using a new container for its salsa. According to their research, the glass containers for the salsa should have a strength between 185 and 325 psi. Joe's vendor has provided the following data for evaluation. Is there evidence that the process is in control? Is it capable of producing jars with a breaking strength between 185 and 325 psi?

Table 5.13

Glass Breaking Strength

Sample			Data			$\bar{X}$	R
1	265	205	263	307	220	252.0	102
2	268	260	234	299	215	255.2	84
3	197	286	274	243	231	246.2	89
4	267	281	265	214	318	269.0	104
5	346	317	242	258	276	287.8	104
6	300	208	187	264	271	246.0	113
7	280	242	260	321	228	266.2	93
8	250	299	258	267	293	273.4	49

(Continued)

example 5.10 Continued

Table 5.13

Continued

Sample			Data			$\overline{X}$	R
9	265	254	281	294	223	263.4	71
10	260	308	235	283	277	272.6	73
11	200	235	246	328	296	261.0	128
12	276	264	269	235	290	266.8	55
13	221	176	248	263	231	227.8	87
14	334	280	265	272	283	286.8	69
15	265	262	271	245	301	268.8	56
16	280	274	253	287	258	270.4	34
17	261	248	260	274	337	276.0	89
18	250	278	254	274	275	266.2	28
19	278	250	265	270	298	272.2	48
20	257	210	280	269	251	253.4	70
						$\overline{\overline{x}} = 264.06$	$\overline{R} = 77.3$

Since there are five jars in each subgroup, $A_2 = 0.577$, $D_4 = 2.115$, $D_3 = 0$, and $d_2 = 2.326$. This yields the following:

$\overline{X}$ chart:

$$\text{Center line} = \overline{\overline{x}} = 264.06$$
$$\text{UCL} = \overline{\overline{x}} + A_2\overline{R} = 264.06 + (0.577)(77.3) = 308.66$$
$$\text{LCL} = \overline{\overline{x}} - A_2\overline{R} = 264.0 - (0.577)(77.3) = 219.46$$

R chart:

$$\text{Center line} = \overline{R} = 77.3$$
$$\text{UCL} = D_4\overline{R} = (2.115)(77.3) = 163.49$$
$$\text{LCL} = D_3\overline{R} = (0)(77.3) = 0$$

Figure 5.22 presents the $\overline{X}$ and R charts for glass strength data. Both charts exhibit statistical control. The process parameters may be estimated from the control chart as

$$\hat{\mu} = \overline{\overline{x}} = 264.06$$
$$\hat{\sigma} = \frac{\overline{R}}{d_2} = \frac{77.3}{2.326} = 33.23$$

example 5.10 Continued

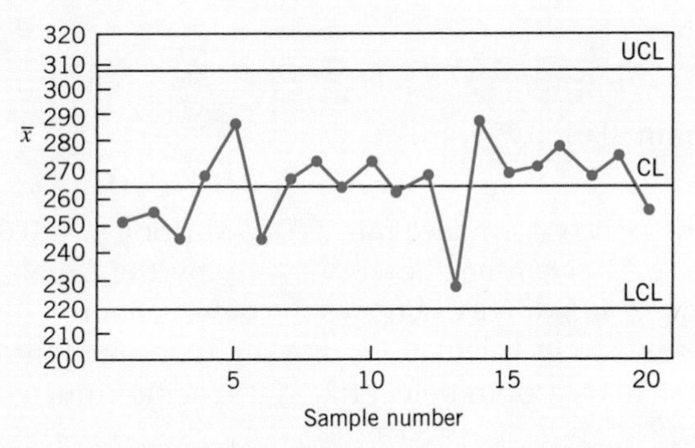

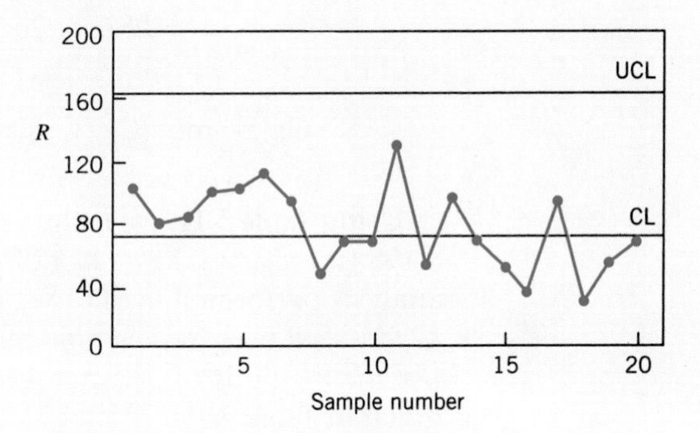

Figure 5.22 $\overline{X}$ **and R Control Charts for Glass Strength**

Using the estimates of the mean and standard deviation, capability indices are calculated as

$$C_p = \frac{\text{USL} - \text{LSL}}{6\hat{\sigma}} = \frac{325 - 185}{6(33.23)} = 0.70$$

$$C_{pu} = \frac{\text{USL} - \hat{\mu}}{3\hat{\sigma}} = \frac{325 - 264.06}{3(33.23)} = 0.61$$

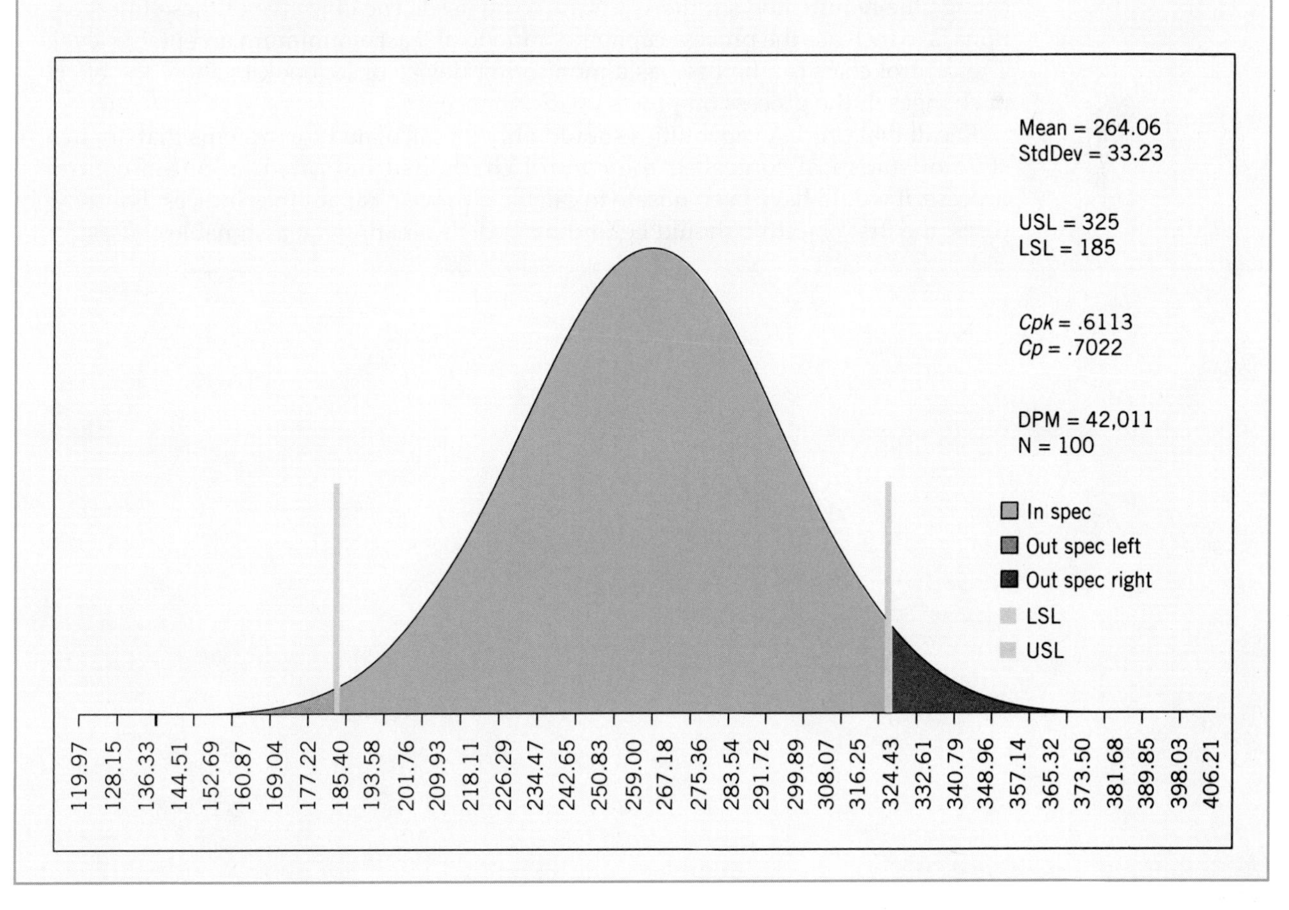

example 5.10 Continued

$$C_{p1} = \frac{\hat{\mu} - LSL}{3\hat{\sigma}} = \frac{264.06 - 185}{3(33.23)} = 0.79$$

$$C_{pk} = \min(C_{pu}, C_{pk}) = \min(0.61, 0.79) = 0.61$$

Using Table 5.11, we estimate the nonconformance rate as 35,931 (for $C_{pu} = 0.6$) + 8.198 (for $C_{pl} = 0.7$) = 44,129 ppm. This can be shown visually by the capability analysis performed in SPC XL. Note that SPC XL calculates the defects per million at 42,011, as it uses more significant digits in its calculations of ppm. If desired, one can use the methods of Chapter 4 to calculate a ppm defect rate if a specific value is not given in Table 5.11.

This example highlights a case of a process that is operating in control but at an unacceptable level of quality. In addition, it illustrates a common mistake of looking at the control limits and comparing them to the specification limits. Since the control limits are based upon subgroup averages, they are based upon tighter limits than individual units (per the central limit theorem). Thus, it appears as if the process is capable of producing good quality at an individual unit basis when it is not (recall that if five units had an average of 325, roughly $\frac{1}{2}$ of those units would have values above 325 and be beyond the specification limits).

In this case, there is no evidence of assignable cause which may reduce the variability within the system. Invention will be required to either improve the process or change the requirements if these jars are to be used. The objective of these interventions is to increase the process capability ratio to at least a minimum acceptable level. The control chart can be used as a monitoring device or logbook to show the effect of changes in the process on process performance.

Recall that process capabilities should only be calculated for systems that are in a state of statistical control. If the control charts had indicated an out-of-control process, it would have been unsafe to calculate process capabilities indices. For these cases, the first objective should be finding and eliminating the assignable causes.

MINITAB

1. Arrange data in columns such that subgroup observations are in the same row, but in different columns. For example, subgroup 1 contains 1, 2, 3, 4, and 5 and subgroup 2 contains 6, 7, 8, 9, and 10. This would be formatted as

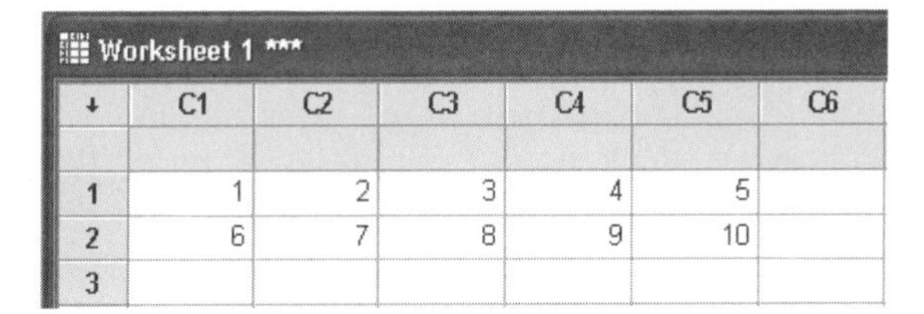

2. From the menu bar, select Stat –> Control Charts –> Variable Charts for Subgroups –> $\overline{X}$-R. Minitab will then bring up a graphical user interface. At the top, use the drop down box to select all observations for a subgroup are in one row of columns and then double click the appropriate column names to move it into the graph variables area.

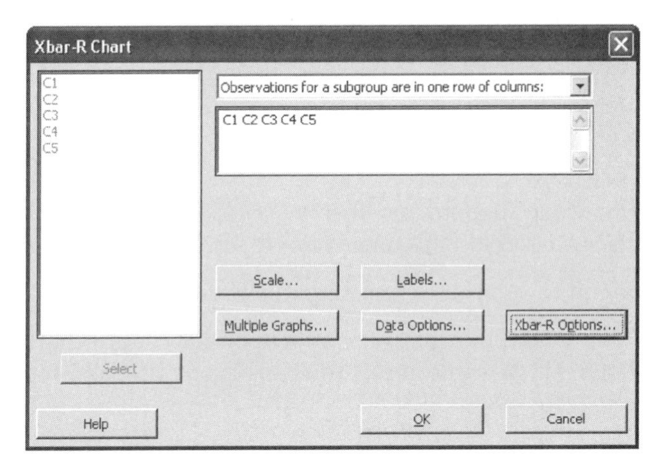

3. Click on the $\overline{X}$-R options to select various options for the $\overline{X}$–R chart. Common options to be considered are detailed below:

   a. Parameters tab: allows you to specify a mean and standard deviation to be used for control limit calculations.
   b. S-limits tab: allows you to specify zones within the control limits. To show Zones A, B, and C specify −3 −2 −1 1 2 3 within the text box "Display Control Limits at These Multiples of the Standard Deviation".

c. Tests tab: allows you to specify the use of the Western Electric rules. The four rules are detailed in the graphic below:

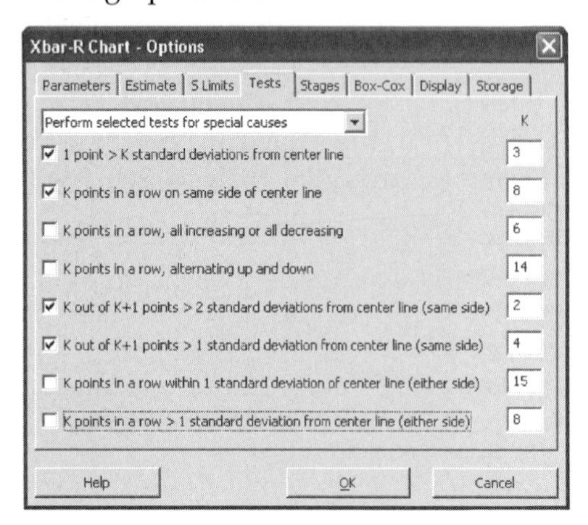

4. Once all options are defined, select OK to generate the chart.

SPC XL

1. Arrange data in columns such that subgroup observations are in the same row, but in different columns. For example, subgroup 1 contains 1, 2, 3, 4, and 5 and subgroup 2 contains 6, 7, 8, 9, and 10. This would be formatted as

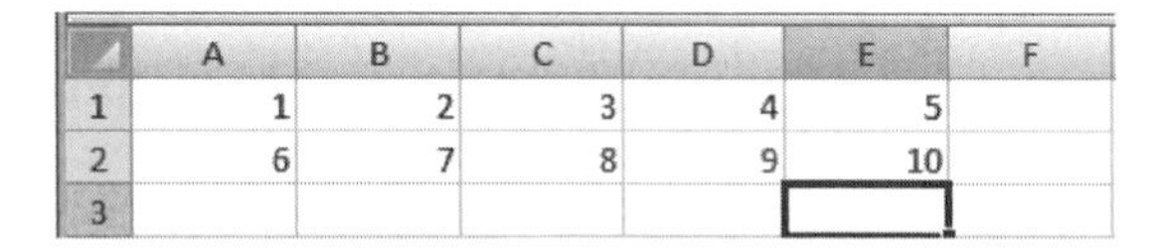

	A	B	C	D	E	F
1	1	2	3	4	5	
2	6	7	8	9	10	
3						

2. Select the control charts icon from the Sigma Zone Ribbon, and then select $\overline{X}$-R chart. You will be prompted to select a range for where your data resides. Select this range prior to selecting the next button.
3. On the next screen, specify that your data are in rows prior to selecting the next button.
4. On the next screen, specify whether or not you would like the zones shown on the control chart and if you would like both the $\overline{X}$ and R chart shown on the same sheet. You can also request a histogram and process capability analysis.

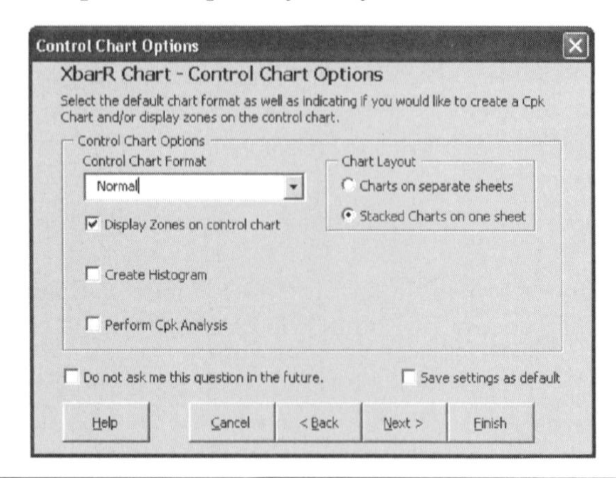

5. Select Next to generate the control chart.

   Note: Default control chart rules are shown below and are accessible from the options icon in the SPC XL Ribbon. These apply to all control charts created.

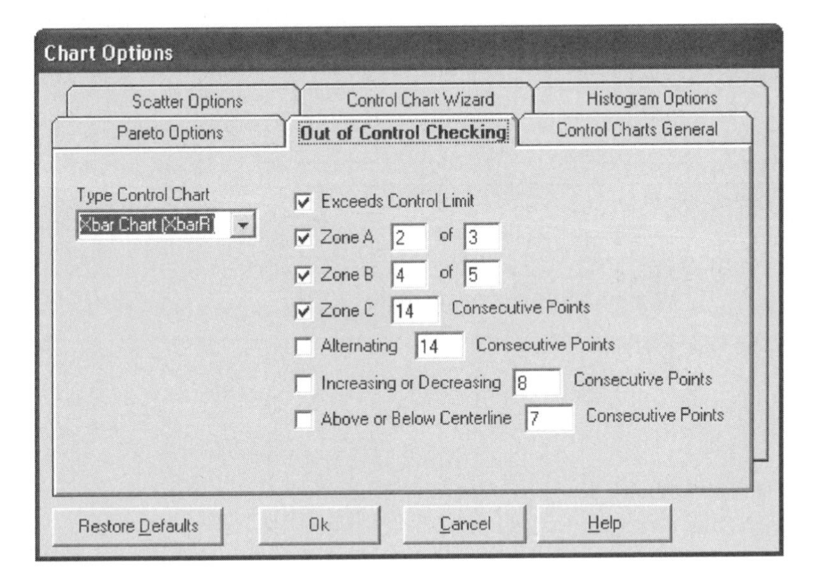

supplement Two Creating an $\overline{X}$–S Chart with SPC XL and Minitab

MINITAB

1. Follow the instructions in Supplement 1, but follow the path of Stat –> Control Charts –> Variable Charts for Subgroups –> $\overline{X}$-s instead.

SPC XL

1. Follow the instructions Supplement 1, but select the $\overline{X}$-s chart instead.

supplement Three Creating an Individuals Moving Range Chart with SPC XL and Minitab

MINITAB

1. Arrange data in a single column.
2. From the menu bar, select Stat –> Control Charts –> Variable Charts for Individuals –> I-MR. Minitab will then bring up a graphical user interface. Select the column of interest for the control charts.

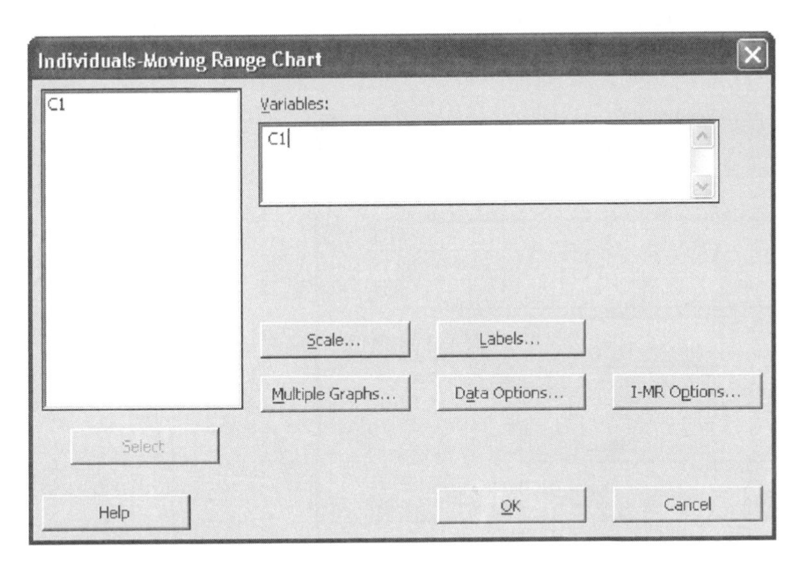

3. Click on the I-MR options to select various options for the I-MR chart. Common settings are defined in Supplement One.
4. Once all options are defined, select OK to generate the chart.

SPC XL

1. Follow the instructions in Supplement One, but select the IndivMovR chart skipping step 3 instead.

MINITAB

1. If individual data are being utilized, format the worksheet per Supplement Three; otherwise, format per Supplement One.

2. From the menu bar, select Stat –> Control Charts –> Time Weighted Charts –> CUSUM. Minitab will then bring up a graphical user interface. From the dropdown box, select "All observations for a chart are in one column" if individual measurements are used. Otherwise, select "All observations for a subgroup are in one row of columns". Then select the columns of interest and specify the subgroup size, and target value.

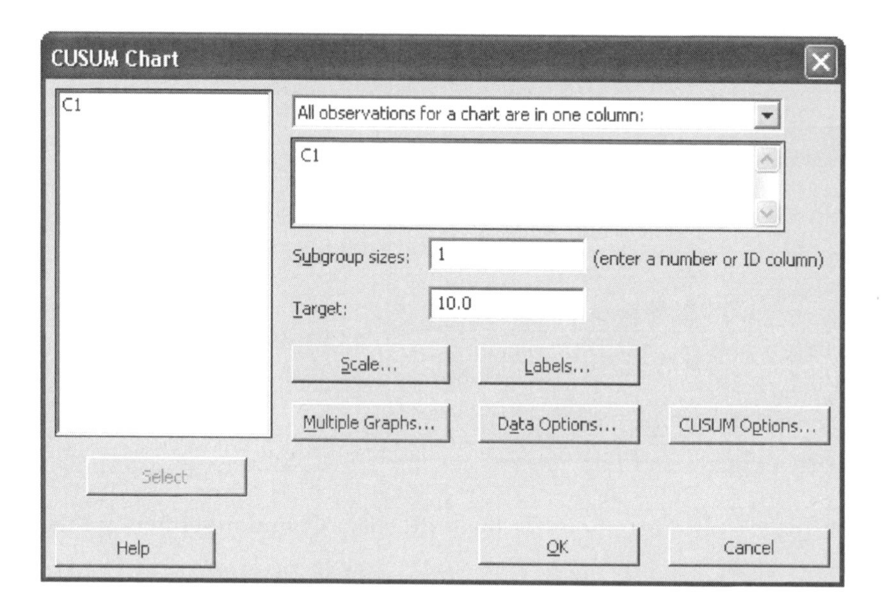

3. Select the CUSUM options icon. Within the plan/type tab, specify h and k.

4. Once all options are defined, select OK to generate the chart.

SPC XL

1. SPC XL does not have the capability to create CUSUM charts.

MINITAB

1. If individual data are being utilized, format the worksheet per Supplement Three; otherwise, format per Supplement One.
2. From the menu bar, select Stat –> Control Charts –> Time Weighted Charts –> EWMA. Minitab will then bring up a graphical user interface. From the dropdown box, select "All observations for a chart are in one column" if individual measurements are used. Otherwise, select "All observations for a subgroup are in one row of columns". Then select the columns of interest and specify the subgroup size, and lambda.

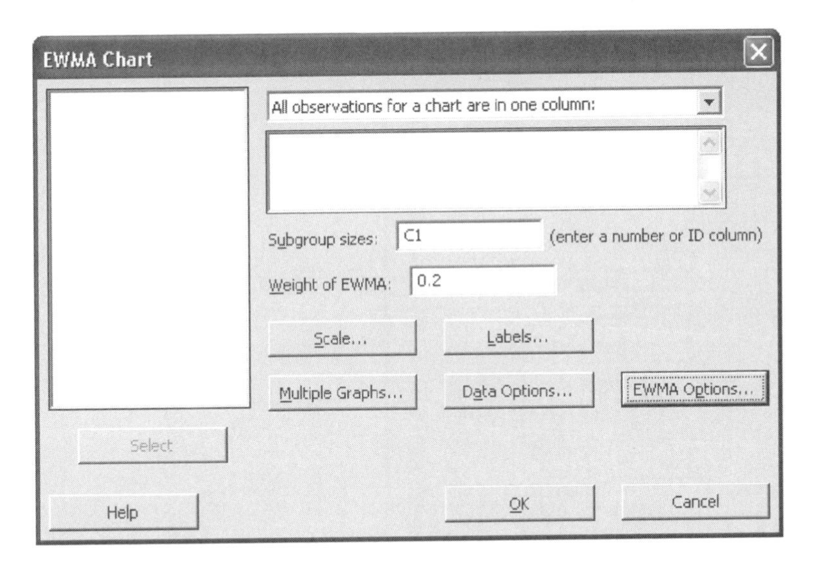

3. Click on EWMA options to select various options. Common settings are defined in Supplement One. Select OK to generate the chart.

SPC XL

1. SPC XL does not have the capability to create EWMA charts.

supplement Six Performing Process Capability Analysis with SPC XL and Minitab

MINITAB

1. If individual data are being utilized, format per Supplement Three. If subgroup sizes greater than 1 are utilized, format the worksheet per Supplement One.

2. From the menu bar select Stat –> Quality Tools –> Capability Analysis –> Normal. The following interface will appear. If individual data is being utilized, select the single column radio button, select the column of interest, and specify a subgroup size of 1. Otherwise, select the radio button for subgroups across rows and select the columns of interest.

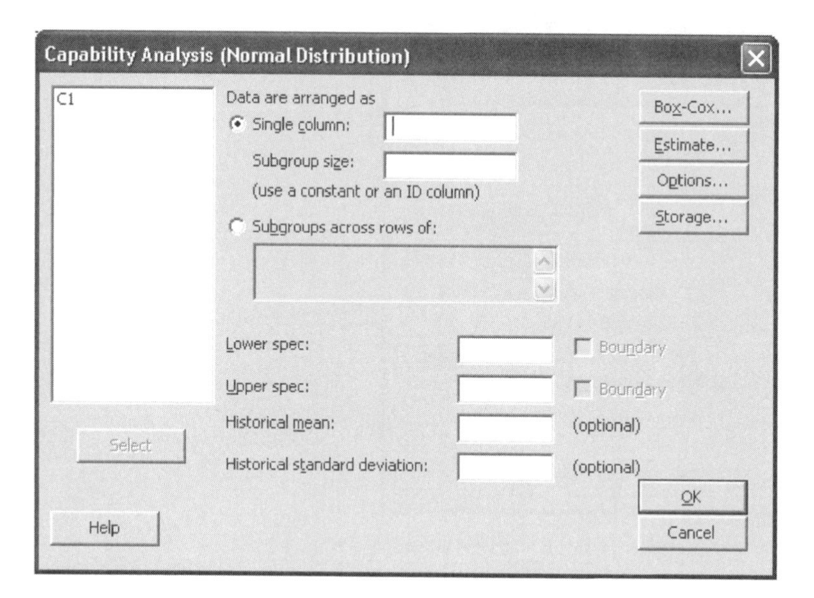

3. Define the upper and lower specifications and select OK.

SPC XL

1. From the SPC XL ribbon, select Analysis Diagrams, then Cpk Analysis. The following screen will appear. If performing a capability analysis from data, select the top radio button and select ok. If creating a capability analysis from long-term process data, select the bottom radio button, click OK, and go to step 3.

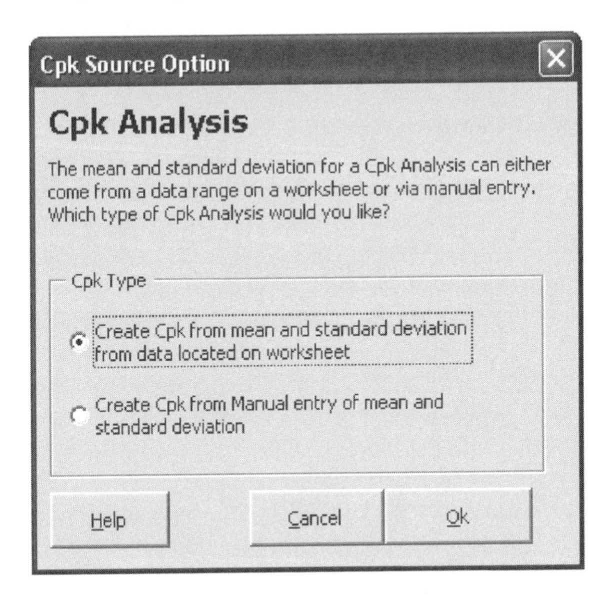

2. Select the data range of interest.

3. Specify the upper and lower specification limits (and mean and standard deviation if not using sample data). If the specification limits do not show on the graph, select the Recompute Scaling for X-axis icon. Then select OK to complete the analysis.

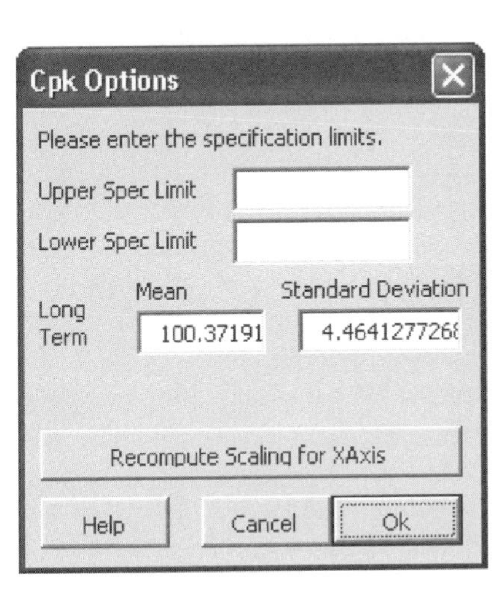

Important Terms and Concepts

Assignable causes of variation
Control chart
Control limits
CUSUM control chart
EWMA control chart
In-control process
Individuals control chart
One-sided process-capability ratios
Out-of-control process
Out-of-control-action plan (OCAP)
Pareto chart
PCR C_p
PCR C_{pk}
PCR C_{pm}
Process capability
R control chart
Rational subgroups
s control chart
Sampling frequency for control charts
Shewhart control charts
Signal resistance of a control chart
Specification limits
Statistical control of a process
Statistical process control (SPC)
Three-sigma control limits
Variable sample size on control charts
Variables control charts
$\bar{x}$ control chart

chapter Five Exercises

5.1 What are chance and assignable causes of variability? What part do they play in the operation and interpretation of a Shewhart control chart?

5.2 Discuss the relationship between a control chart and statistical hypothesis testing.

5.3 Discuss type I and type II errors relative to the control chart. What practical implication in terms of process operation do these two types of errors have?

5.4 What is meant by the statement that a process is in a state of statistical control?

5.5 If a process is in a state of statistical control, does it necessarily follow that all or nearly all of the units of product produced will be within the specification limits?

5.6 Discuss the logic underlying the use of three-sigma limits on Shewhart control charts. How will the chart respond if narrower limits are chosen? How will it respond if wider limits are chosen?

5.7 Discuss the rational subgroup concept. What part does it play in control chart analysis?

5.8 When taking samples or subgroups from a process, do you want assignable causes occurring within the subgroups or between them? Fully explain your answer.

5.9 A molding process uses a five-cavity mold for a part used in an automotive assembly. The wall thickness of the part is the critical quality characteristic. It has been suggested to use $\overline{X}$ and R charts to monitor this process, and to use as the subgroup or sample all five parts that result from a single "shot" of the machine. What do you think of this sampling strategy? What impact does it have on the ability of the charts to detect assignable causes?

5.10 A manufacturing process produces 500 parts per hour. A sample part is selected about every half-hour, and after five parts are obtained, the average of these five measurements is plotted on a control chart. (a) Is this an appropriate sampling scheme if the assignable cause in the process results in an instantaneous upward shift in the mean that is of very short duration? (b) If your answer is no, propose an alternative procedure.

5.11 Consider the sampling scheme proposed in Exercise 5.10. Is this scheme appropriate if the assignable cause in the process results in a slow, prolonged upward shift? If your answer is no, propose an alternative procedure.

5.12 If the time order of production has not been recorded in a set of data from a process, is it possible to detect the presence of assignable causes?

5.13 How do the costs of sampling, the costs of producing an excessive number of defective units, and the costs of searching for assignable causes impact the choice of parameters of a control chart?

5.14 Consider the control chart shown here. Does the pattern appear random?

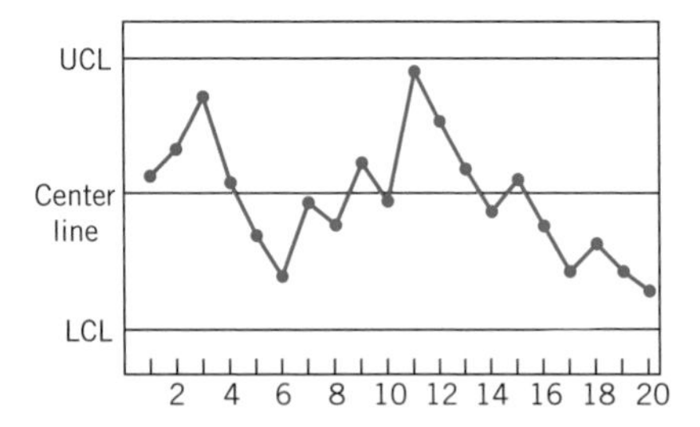

5.15 Consider the control chart shown here. Does the pattern appear random?

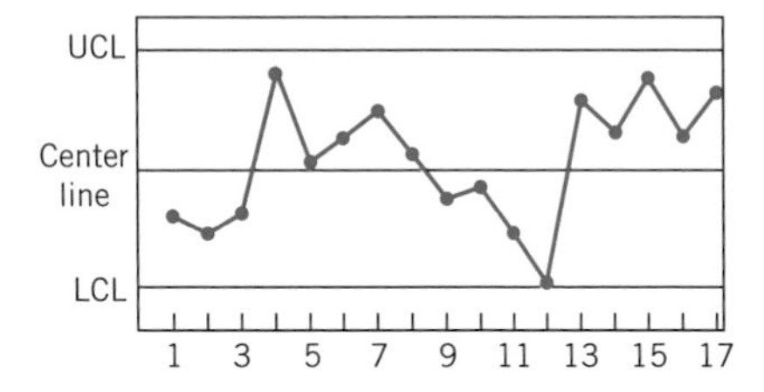

5.16 Consider the control chart shown here. Does the pattern appear random?

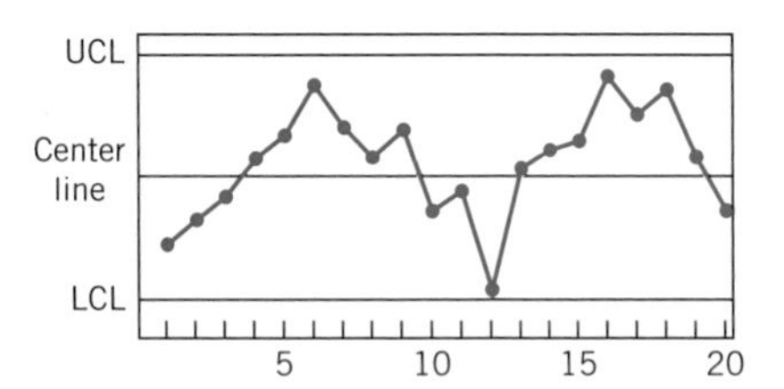

5.17 Apply the Western Electric rules to the control chart in Exercise 5.14. Are any of the criteria for declaring the process out of control satisfied?

5.18 Apply the Western Electric rules to the control chart presented in Exercise 5.16. Would these rules result in any out-of-control signals?

5.19 Consider the time-varying process behavior shown below. Match each of these several patterns of process performance to the corresponding and R charts shown in Figures (a) to (e) below.

5.20 The thickness of a printed circuit board is an important quality parameter. Data on board thickness (in inches) are given in Table 5E.1 for 25 samples of three boards each.

a. Set up $\overline{X}$ and R control charts. Is the process in statistical control?
b. Estimate the process standard deviation.
c. What are the limits that you would expect to contain nearly all the process measurements?

5.21 The net weight of a soft drink is to be monitored by $\overline{X}$ and R control charts using a sample size

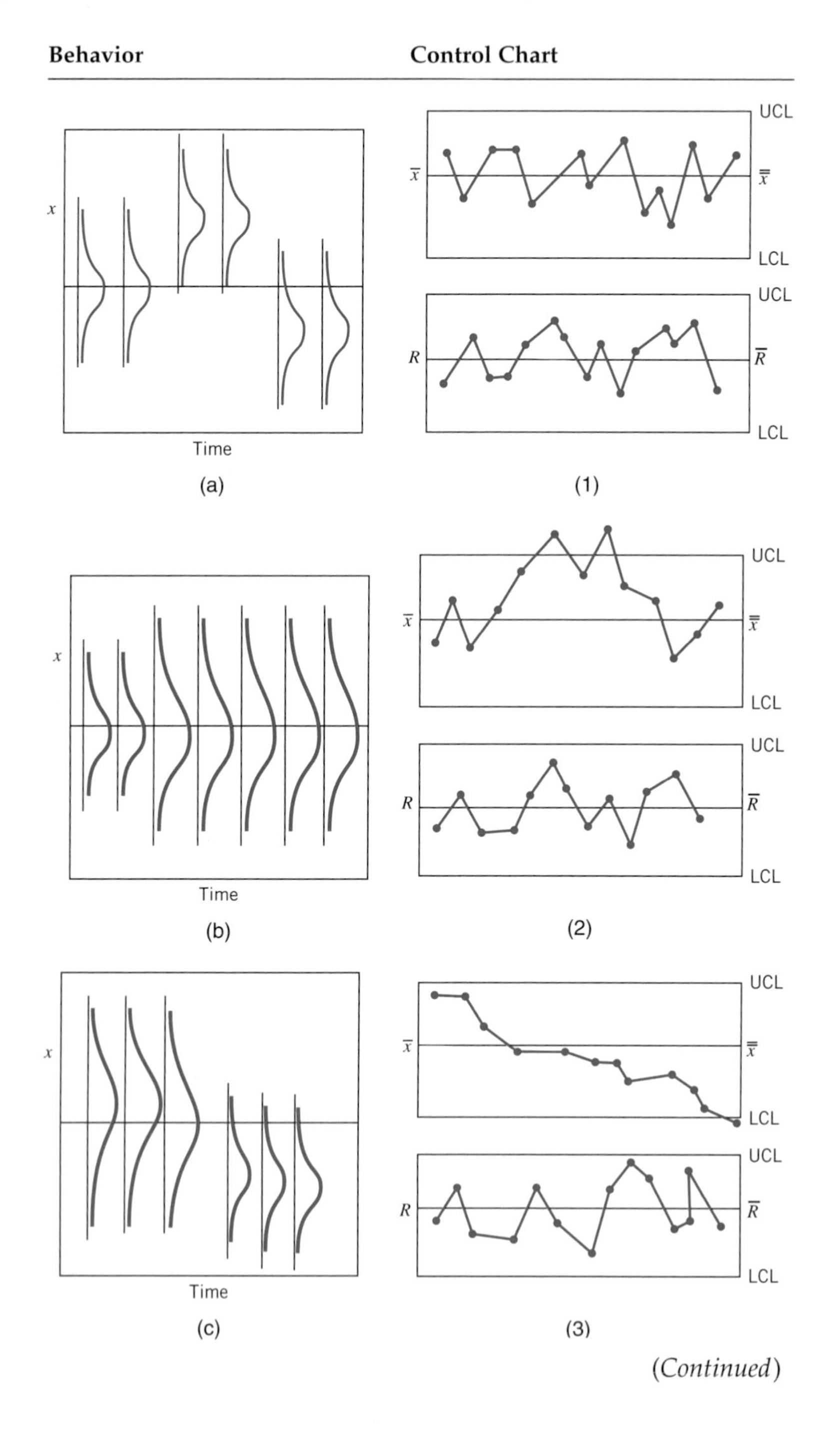

(Continued)

Behavior | **Control Chart**

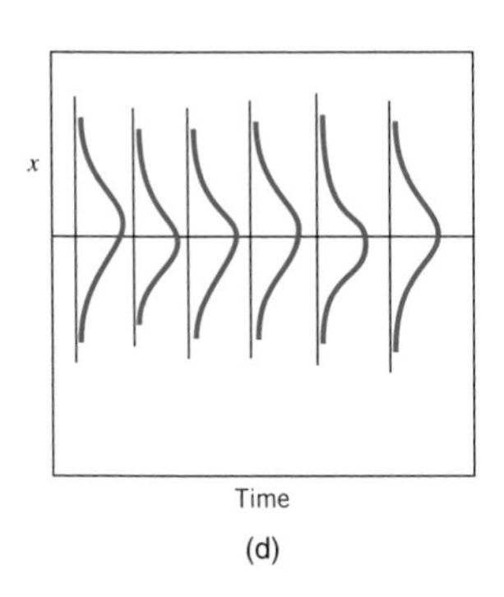

(d)

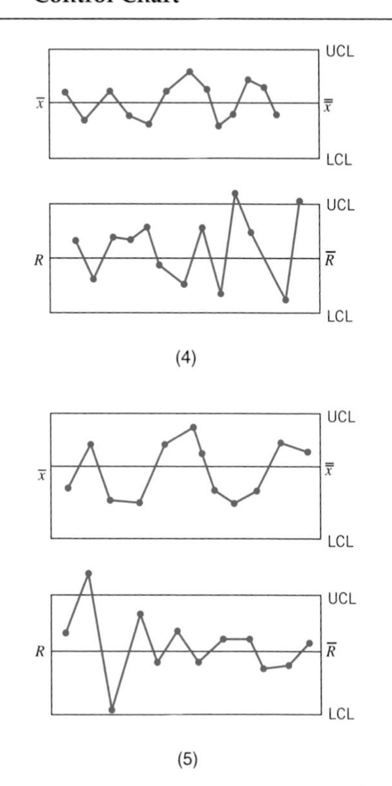

(4)

(5)

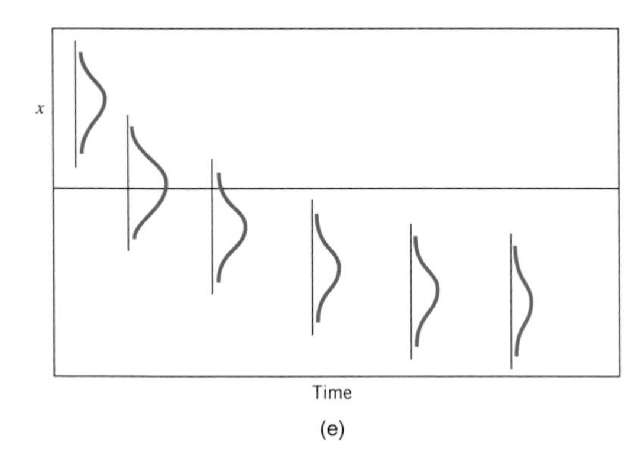

(e)

Table 5E.1

Printed Circuit Board Thickness for Exercise 5.20

Sample Number	x_1	x_2	x_3
1	0.0629	0.0636	0.0640
2	0.0630	0.0631	0.0622
3	0.0628	0.0631	0.0633
4	0.0634	0.0630	0.0631
5	0.0619	0.0628	0.0630
6	0.0613	0.0629	0.0634
7	0.0630	0.0639	0.0625
8	0.0628	0.0627	0.0622
9	0.0623	0.0626	0.0633
10	0.0631	0.0631	0.0633
11	0.0635	0.0630	0.0638
12	0.0623	0.0630	0.0630
13	0.0635	0.0631	0.0630

Table 5E.1

Continued

Sample Number	x_1	x_2	x_3
14	0.0645	0.0640	0.0631
15	0.0619	0.0644	0.0632
16	0.0631	0.0627	0.0630
17	0.0616	0.0623	0.0631
18	0.0630	0.0630	0.0626
19	0.0636	0.0631	0.0629
20	0.0640	0.0635	0.0629
21	0.0628	0.0625	0.0616
22	0.0615	0.0625	0.0619
23	0.0630	0.0632	0.0630
24	0.0635	0.0629	0.0635
25	0.0623	0.0629	0.0630

of $n = 5$. Data for 20 preliminary samples are shown in Table 5E.2.

a. Set up $\overline{X}$ and R control charts using these data. Does the process exhibit statistical control?
b. Estimate the process mean and standard deviation.
c. Does fill weight seem to follow a normal distribution?

Table 5E.2

Soft Drink Data for Problem 5.21

Sample Number	x_1	x_2	x_3	x_4	x_5
1	15.8	16.3	16.2	16.1	16.6
2	16.3	15.9	15.9	16.2	16.4
3	16.1	16.2	16.5	16.4	16.3
4	16.3	16.2	15.9	16.4	16.2
5	16.1	16.1	16.4	16.5	16.0
6	16.1	15.8	16.7	16.6	16.4
7	16.1	16.3	16.5	16.1	16.5
8	16.2	16.1	16.2	16.1	16.3
9	16.3	16.2	16.4	16.3	16.5
10	16.6	16.3	16.4	16.1	16.5
11	16.2	16.4	15.9	16.3	16.4
12	15.9	16.6	16.7	16.2	16.5
13	16.4	16.1	16.6	16.4	16.1
14	16.5	16.3	16.2	16.3	16.4
15	16.4	16.1	16.3	16.2	16.2
16	16.0	16.2	16.3	16.3	16.2
17	16.4	16.2	16.4	16.3	16.2
18	16.0	16.2	16.4	16.5	16.1
19	16.4	16.0	16.3	16.4	16.4
20	16.4	16.4	16.5	16.0	15.8

5.22 Rework Exercise 5.20 using an $\overline{X}$-S chart.

5.23 Rework Exercise 5.21 using an $\overline{X}$-S chart.

5.24 Samples of six items are taken from a service process at regular intervals. A quality characteristic is measured and $\overline{x}$ and R values are calculated for each sample. After 50 groups of size six have been taken we have $\overline{\overline{x}} = 40$ and $\overline{R} = 4$. The data is normally distributed.

a. Compute control limits for the $\overline{X}$ and R control charts. Do all points fall within the control limits?
b. Estimate the mean and standard deviation of the process. What are the $\pm$ 3 standard deviation limits for the individual data?
c. If the specification limits are 41 ± 5, do you think the process is capable of producing within these specifications?

5.25 Table 5E.3 presents 20 subgroups of five measurements on the time it takes to service a customer.

a. Set up $\overline{X}$ and R control charts for this process and verify that it is in statistical control.
b. Following establishing of the control charts in part (a), 10 new samples have been provided in Table 5E.4. Plot the new $\overline{X}$ and R values using the control chart limits you established in part (a) and draw conclusions.
c. Suppose that the assignable cause responsible for the action signals generated in part (b) has been identified and adjustments made to the process to correct its performance. Plot the $\overline{X}$ and R values from the new subgroups shown in Table 5E.5 which were taken following the adjustment against the control chart limits established in part (a). What are your conclusions?

Table 5E.3

Service Time Data for Problem 5.25

Sample Number	x_1	x_2	x_3	x_4	x_5	$\overline{x}$	R
1	138.1	110.8	138.7	137.4	125.4	130.1	27.9
2	149.3	142.1	105.0	134.0	92.3	124.5	57.0
3	115.9	135.6	124.2	155.0	117.4	129.6	39.1
4	118.5	116.5	130.2	122.6	100.2	117.6	30.0
5	108.2	123.8	117.1	142.4	150.9	128.5	42.7
6	102.8	112.0	135.0	135.0	145.8	126.1	43.0
7	120.4	84.3	112.8	118.5	119.3	111.0	36.1

(Continued)

Table 5E.3

Continued

Sample Number	x_1	x_2	x_3	x_4	x_5	$\bar{x}$	R
8	132.7	151.1	124.0	123.9	105.1	127.4	46.0
9	136.4	126.2	154.7	127.1	173.2	143.5	46.9
10	135.0	115.4	149.1	138.3	130.4	133.6	33.7
11	139.6	127.9	151.1	143.7	110.5	134.6	40.6
12	125.3	160.2	130.4	152.4	165.1	146.7	39.8
13	145.7	101.8	149.5	113.3	151.8	132.4	50.0
14	138.6	139.0	131.9	140.2	141.1	138.1	9.2
15	110.1	114.6	165.1	113.8	139.6	128.7	54.8
16	145.2	101.0	154.6	120.2	117.3	127.6	53.3
17	125.9	135.3	121.5	147.9	105.0	127.1	42.9
18	129.7	97.3	130.5	109.0	150.5	123.4	53.2
19	123.4	150.0	161.6	148.4	154.2	147.5	38.3
20	144.8	138.3	119.6	151.8	142.7	139.4	32.2

Table 5E.4

New Service Data for Problem 5.25

Sample Number	x_1	x_2	x_3	x_4	x_5	$\bar{x}$	R
1	131.0	184.8	182.2	143.3	212.8	170.8	81.8
2	181.3	193.2	180.7	169.1	174.3	179.7	24.0
3	154.8	170.2	168.4	202.7	174.4	174.1	48.0
4	157.5	154.2	169.1	142.2	161.9	157.0	26.9
5	216.3	174.3	166.2	155.5	184.3	179.3	60.8
6	186.9	180.2	149.2	175.2	185.0	175.3	37.8
7	167.8	143.9	157.5	171.8	194.9	167.2	51.0
8	178.2	186.7	142.4	159.4	167.6	166.9	44.2
9	162.6	143.6	132.8	168.9	177.2	157.0	44.5
10	172.1	191.7	203.4	150.4	196.3	182.8	53.0

Table 5E.5

Final Service Data for Problem 5.25

Sample Number	x_1	x_2	x_3	x_4	x_5	$\bar{x}$	R
1	131.5	143.1	118.5	103.2	121.6	123.6	39.8
2	111.0	127.3	110.4	91.0	143.9	116.7	52.8
3	129.8	98.3	134.0	105.1	133.1	120.1	35.7
4	145.2	132.8	106.1	131.0	99.2	122.8	46.0
5	114.6	111.0	108.8	177.5	121.6	126.7	68.7
6	125.2	86.4	64.4	137.1	117.5	106.1	72.6
7	145.9	109.5	84.9	129.8	110.6	116.1	61.0
8	123.6	114.0	135.4	83.2	107.6	112.8	52.2
9	85.8	156.3	119.7	96.2	153.0	122.2	70.6
10	107.4	148.7	127.4	125.0	127.5	127.2	41.3

5.26 Parts manufactured by an injection modeling process are subjected to a compressive strength test. Twenty samples of part parts each are collected, and the compressive strengths (in psi) are shown in Table 5E.6.

a. Establish $\overline{X}$ and R control charts for compressive strength using these data. Is the process in statistical control?

b. After establishing the control charts in part (a), 15 new subgroups were collected; the compressive strengths are shown in Table 5E.7. Plot the $\overline{X}$ and R values against the control limits from part (a) and draw conclusions.

Table 5E.6

Part Strength Data for Problem 5.26

Sample Number	x_1	x_2	x_3	x_4	x_5	$\bar{x}$	R
1	83.0	81.2	78.7	75.7	77.0	79.1	7.3
2	88.6	78.3	78.8	71.0	84.2	80.2	17.6
3	85.7	75.8	84.3	75.2	81.0	80.4	10.4
4	80.8	74.4	82.5	74.1	75.7	77.5	8.4
5	83.4	78.4	82.6	78.2	78.9	80.3	5.2

Table 5E.6

Continued

Sample Number	x_1	x_2	x_3	x_4	x_5	$\bar{x}$	R
6	75.3	79.9	87.3	89.7	81.8	82.8	14.5
7	74.5	78.0	80.8	73.4	79.7	77.3	7.4
8	79.2	84.4	81.5	86.0	74.5	81.1	11.4
9	80.5	86.2	76.2	64.1	80.2	81.4	9.9
10	75.7	75.2	71.1	82.1	74.3	75.7	10.9
11	80.0	81.5	78.4	73.8	78.1	78.4	7.7
12	80.6	81.8	79.3	73.8	81.7	79.4	8.0
13	82.7	81.3	79.1	82.0	79.5	80.9	3.6
14	79.2	74.9	78.6	77.7	75.3	77.1	4.3
15	85.5	82.1	82.8	73.4	71.7	79.1	13.8
16	78.8	79.6	80.2	79.1	80.8	79.7	2.0
17	82.1	78.2	75.5	78.2	82.1	79.2	6.6
18	84.5	76.9	83.5	81.2	79.2	81.1	7.6
19	79.0	77.8	81.2	84.4	81.6	80.8	6.6
20	84.5	73.1	78.6	78.7	80.6	79.1	11.4

Table 5E.7

Additional Data for Problem 5.26

Sample Number	x_1	x_2	x_3	x_4	x_5	$\bar{x}$	R
1	68.9	81.5	78.2	80.8	81.5	78.2	12.6
2	69.8	68.6	80.4	84.3	83.9	77.4	15.7
3	78.5	85.2	78.4	80.3	81.7	80.8	6.8
4	76.9	86.1	86.9	94.4	83.9	85.6	17.5
5	93.6	81.6	87.8	79.6	71.0	82.7	22.5
6	65.5	86.8	72.4	82.6	71.4	75.9	21.3
7	78.1	65.7	83.7	93.7	93.4	82.9	27.9
8	74.9	72.6	81.6	87.2	72.7	77.8	14.6
9	78.1	77.1	67.0	75.7	76.8	74.9	11.0

Table 5E.7

Continued

Sample Number	x_1	x_2	x_3	x_4	x_5	$\bar{x}$	R
10	78.7	85.4	77.7	90.7	76.7	81.9	14.0
11	85.0	60.2	68.5	71.1	82.4	73.4	24.9
12	86.4	79.2	79.8	86.0	75.4	81.3	10.9
13	78.5	99.0	78.3	71.4	81.8	81.7	27.6
14	68.8	62.0	82.0	77.5	76.1	73.3	19.9
15	83.0	83.7	73.1	82.2	95.3	83.5	22.2

5.27 Reconsider the data presented in Exercise 5.26. (a) Rework both parts (a) and (b) of Exercise 5.26 using the $\overline{X}$ and S charts. (b) Does the s chart detect the shift in process variability more quickly than the R chart did originally in part (b) of Exercise 5.26?

5.28 One-pound coffee cans are filled by a machine, sealed, and then weighed by a local coffee store. After adjusting for the weight of the can, any package that weighs less than 16 oz is cut out of the conveyor. The weights of 25 successive cans are shown in Table 5E.8. Set up a moving range control chart and a control chart for individuals. Estimate the mean and standard deviation of the amount of coffee packed in each can. Is it reasonable to assume that can weight is normally distributed? If the process remains in

Table 5E.8

Coffee Weight Data for Exercise 5.28

Can Number	Weight	Can Number	Weight
1	16.11	14	16.12
2	16.08	15	16.10
3	16.12	16	16.08
4	16.10	17	16.13
5	16.10	18	16.15
6	16.11	19	16.12

Table 5E.8

Continued

Can Number	Weight	Can Number	Weight
7	16.12	20	16.10
8	16.09	21	16.08
9	16.12	22	16.07
10	16.10	23	16.11
11	16.09	24	16.13
12	16.07	25	16.10
13	16.13		

control at this level, what percentage of cans will be underfilled?

5.29 Fifteen successive heats of a steel alloy are tested for hardness. The resulting data are shown in Table 5E.9. Set up a control chart for the moving range and a control chart for individual hardness measurements. Is it reasonable to assume that hardness is normally distributed?

Table 5E.9

Hardness Data for Exercise 5.29

Heat	Hardness (coded)	Heat	Hardness (coded)
1	52	9	58
2	51	10	51
3	54	11	54
4	55	12	59
5	50	13	53
6	52	14	54
7	50	15	55
8	51		

5.30 The viscosity of a polymer is measured hourly. Measurements for the last 20 hours are shown in Table 5E.10.

Table 5E.10

Viscosity Data for Exercise 5.30

Test	Viscosity	Test	Viscosity
1	2838	11	3174
2	2785	12	3102
3	3058	13	2762
4	3064	14	2975
5	2996	15	2719
6	2882	16	2861
7	2878	17	2797
8	2920	18	3078
9	3050	19	2964
10	2870	20	2805

a. Does viscosity follow a normal distribution?
b. Set up a control chart on viscosity and a moving range chart. Does the process exhibit statistical control?
c. Estimate the process mean and standard deviation.

5.31 **Continuation of Exercise 5.30.** The next five measurements on viscosity are 3163, 3199, 3054, 3147, and 3156. Do these measurements indicate that the process is in statistical control?

5.32 A machine is used to fill cans with an energy drink. A single sample can is selected every hour and the weight of the can is obtained. Since the filling process is automated, it has very stable variability, and long-term experience indicates that $\sigma = 0.05$ oz. The individual observations for 24 hours of operation are shown in Table 5E.11.

a. Assuming that the process target is 8.02 oz, create a CUSUM chart for this process. Design the chart using the standardized values $h = 4.77$ and $k = 0.5$.
b. Does the value of $\sigma = 0.05$ seem reasonable for this process?

5.33 Rework Exercise 5.32 using the standardized CUSUM parameters of $h = 8.01$ and $k = 0.25$. Compare the results with those obtained previously in Exercise 5.32. What can you say about the theoretical performance of those two CUSUM schemes?

Table 5E.11

Fill Data for Exercise 5.32

Sample Number	x	Sample Number	x
1	8.00	13	8.05
2	8.01	14	8.04
3	8.02	15	8.03
4	8.01	16	8.05
5	8.00	17	8.06
6	8.01	18	8.04
7	8.06	19	8.05
8	8.07	20	8.06
9	8.01	21	8.04
10	8.04	22	8.02
11	8.02	23	8.03
12	8.01	24	8.05

5.34 The data in Table 5E.12 are the times it takes for a local payroll company to process checks (in minutes). The target value for the turnaround time is $\mu_0 = 950$ minutes (two working days).

a. Estimate the process standard deviation.

b. Create a CUSUM chart for this process, using standardized values $h = 5$ and $k = 0.5$. Interpret this chart.

Table 5E.12

Payroll Processing Times for Problem 5.34

953	985	949	937	959	948	958	952
945	973	941	946	939	937	955	931
972	955	966	954	948	955	947	928
945	950	966	935	958	927	941	937
975	948	934	941	963	940	938	950
970	957	937	933	973	962	945	970
959	940	946	960	949	963	963	933

Table 5E.12

Continued

973	933	952	968	942	943	967	960
940	965	935	959	965	950	969	934
936	973	941	956	962	938	981	927

5.35 Calcium hardness is measured hourly for a public swimming pool. Data (in ppm) for the last 32 hours are shown in Table 5E.13 (read down from left). The process target is mo = 175 ppm.

a. Estimate the process standard deviation.

b. Construct a CUSUM chart for this process using standardized values of $h = 5$ and $k = 1/2$.

Table 5E.13

Pool Testing Data for Problem 5.35

160	186	190	206
158	195	189	210
150	179	185	216
151	184	182	212
153	175	181	211
154	192	180	202
158	186	183	205
162	197	186	197

5.36 Reconsider the data in Exercise 5.32. Set up an EWMA control chart with $\lambda = 0.2$ and $L = 3$ for this process. Interpret the results.

5.37 Reconstruct the control chart in Exercise 5.36 using $\lambda = 0.1$ and $L = 2.7$. Compare this chart with the one constructed in Exercise 5.36.

5.38 Reconsider the data in Exercise 5.34. Apply an EWMA control chart to these data using $\lambda = 0.1$ and $L = 2.7$.

5.39 Reconstruct the control chart in Exercise 5.34 using $\lambda = 0.4$ and $L = 3$. Compare this chart to the one constructed in Exercise 5.38.

5.40 Reconsider the data in Exercise 5.35. Set up and apply an EWMA control chart to these data using $\lambda = 0.05$ and $L = 2.6$.

5.41 A process is in control with $\bar{\bar{x}} = 100$, $\bar{s} = 1.05$, and $n = 5$. The process specifications are at 95 ± 10. The quality characteristic has a normal distribution.

a. Estimate the potential capability.
b. Estimate the actual capability.
c. How much could the fallout in the process be reduced if the process were corrected to operate at the nominal specification?

5.42 A process is in statistical control with $\bar{\bar{x}} = 199$ and $\bar{R} = 3.5$. The control chart uses a sample size of $n = 4$. Specifications are at 200 ± 8. The quality characteristic is normally distributed.

a. Estimate the potential capability of the process.
b. Estimate the actual process capability.
c. How much improvement could be made in process performance if the mean could be centered at the nominal value?

5.43 A process is in statistical control with $\bar{\bar{x}} = 39.7$ and $\bar{R} = 2.5$. The control chart uses a sample size of $n = 2$. Specifications are at 40 ± 5. The quality characteristic is normally distributed.

a. Estimate the potential capability of the process.
b. Estimate the actual process capability.
c. How much improvement could be made in process performance if the mean could be centered at the nominal value?

5.44 A process is in control with $\bar{\bar{x}} = 75$ and $\bar{s} = 2$. The process specifications are at 80 ± 8. The sample size $n = 5$.

a. Estimate the potential capability.
b. Estimate the actual capability.
c. How much could process fallout be reduced by shifting the mean to the nominal dimension? Assume that the quality characteristic is normally distributed.

5.45 The weights of nominal 1-kg containers of a concentrated chemical ingredient are shown in Table 5E.14. Prepare a normal probability plot of the data and estimate process capability.

Table 5E.14

Weights Data for Problem 5.45

0.9475	0.9775	0.9965	1.0075	1.0180
0.9705	0.9860	0.9975	1.0100	1.0200
0.9770	0.9960	1.0050	1.0175	1.0250

5.46 Consider the package weight data in Exercise 5.45. Suppose there is a lower specification at 0.985 kg. Calculate an appropriate process capability ratio for this material. What percentage of the packages produced by this process is estimated to be below the specification limit?

5.47 The height of the disk used in a computer disk drive assembly is a critical quality characteristic. Table 5E.15 gives the heights (in mm) of 25 disks randomly selected from the manufacturing process. Prepare a normal probability plot of the disk height data and estimate process capability.

Table 5E.15

Height Data for Exercise 5.47

20.0106	20.0090	20.0067	19.9772	20.0001
19.9940	19.9876	20.0042	19.9986	19.9958
20.0075	20.0018	20.0059	19.9975	20.0089
20.0045	19.9891	19.9956	19.9884	20.0154
20.0056	19.9831	20.0040	20.0006	20.0047

5.48 The length of time required to reimburse employee expense claims is a characteristic that can be used to describe the performance of the process. Table 5E.16 gives the cycle times (in days) of 30 randomly selected employee expense claims. Estimate the capability of this process.

Table 5E.16

Reimbursement Data for Exercise 5.48

5	5	16	17	14	12
8	13	6	12	11	10
18	18	13	12	19	14
17	16	11	22	13	16
10	18	12	12	12	14

5.49 An electric utility tracks the response time to customer reported outages. The data in Table 5E.17 are a random sample of 40 of the response times (in minutes) for one operating division of this utility during a single month.

a. Estimate the capability of the utility's process for responding to customer-reported outages.
b. The utility wants to achieve a 90% response rate in under two hours, as response to emergency outages is an important measure of customer satisfaction. What is the capability of the process with respect to this objective?

Table 5E.17

Utility Data for Exercise 5.49

80	102	86	94	86	106	105	110	127	97
110	104	97	128	98	84	97	87	99	94
105	104	84	77	125	85	80	104	103	109
115	89	100	96	96	87	106	100	102	93

5.50 The failure time in hours of 10 memory devices follows: 1210, 1275, 1400, 1695, 1900, 2105, 2230, 2250, 2500, and 2625. Plot the data on normal probability paper and, if appropriate, estimate process capability. Is it safe to estimate the proportion of circuits that fail below 1200 h?

chapter Six Control Charts for Attributes

Chapter Outline

A Medical Group's Implementation of SPC

The medical industry has changed rapidly over the past decade. Doctors are pressured by insurance companies to continually lower costs and increase profits. One medical practice sought to implement SPC charts in tandem with its profit and loss (P&L) statements to help provide a more complete picture of the medical group's performance.

The metrics the company created tied back to five major factors: productivity, growth, affordability, efficiency, and contribution. A team tracked these factors at the individual practice level and then rolled them up into major categories, such as internal medicine, pediatrics, and family medicine.

The team then benchmarked the SPC indicators across each of the primary care practices. This exercise provided the practice with both short- and long-term goals and provided operations managers with a tool for investigating root causes of performance variation.

The benefits of linking SPC to the P&L statements have been significant. The company reports a sense of order, priority, and proactive management. When the company notices unfavorable outliers, they can correct the issues before larger problems occur. Likewise, when positive outliers are found, they can analyze and broadly proliferate the conditions and factors leading to success.

The company has done a great job of explaining the concept of "normal variation" to its staff. As a result, practice managers no longer waste time analyzing what are actually normal variations. Thus, there is less tampering with company processes and procedures, leaving more time for focusing on what's most important—the patient.

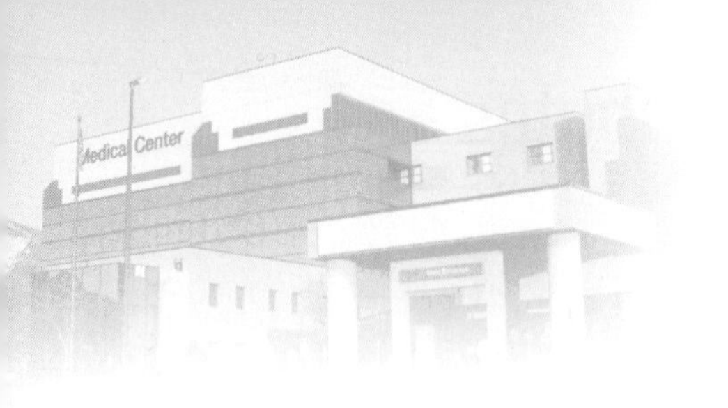

Chapter Overview and Learning Objectives

Many quality characteristics cannot be conveniently represented numerically. In such cases, we usually classify each item inspected as either conforming or nonconforming to the specifications on that quality characteristic. The terms "defective" and "nondefective" are often used to identify these two classifications of product. More recently, the terms "conforming" and "nonconforming" have become popular.

Quality characteristics of this type are called attributes. Some examples of quality characteristics that are attributes are the proportion of warped components in a day's production, the number of nonfunctional semiconductor chips on a wafer, the number of errors or mistakes made in completing a loan application, and the number of medical errors made in a hospital.

This chapter presents three widely used attributes control charts. The first of these relates to the fraction of nonconforming or defective product produced by a process, and is called the control chart for fraction nonconforming, or p chart. In some situations it is more convenient to deal with the number of defects or nonconformities observed rather than the fraction nonconforming. The second type of control chart that we study, called the control chart for nonconformities, or the c chart, is designed to deal with this case. Finally, we present a control chart for nonconformities per unit, or the u chart, which is useful in situations where the average number of nonconformities per unit is a more convenient basis for process control. The chapter concludes with some guidelines for implementing control charts.

After careful study of this chapter you should be able to do the following:

1. Understand the statistical basis of attributes control charts
2. Know how to design attributes control charts
3. Know how to set up and use the p chart for fraction nonconforming
4. Know how to set up and use the np control chart for the number of nonconforming items
5. Know how to set up and use the c control chart for defects
6. Know how to set up and use the u control chart for defects per unit
7. Use attributes control charts with variable sample sizes
8. Understand the advantages and disadvantages of attributes versus variables control charts
9. Understand the rational subgroup concept for attributes control charts

6.1 Introduction

In Chapter 5, we introduced control charts for quality characteristics that are expressed as variables. Although these control charts enjoy widespread application, they are not universally applicable, because not all quality characteristics can be expressed with variable data. For example, consider a glass container for a liquid product. Suppose we examine a container and classify it into one of the two categories called conforming or nonconforming, depending on whether the container meets the requirements on one or more quality characteristics. This is an example of attribute data, and a control chart for the fraction of nonconforming containers could be established (we show how to do this in Section 6.2). Alternatively, in some processes we may examine a unit of product and count defects or nonconformities on the unit. In Section 6.3 we show how to establish control charts for counts or for the average number of counts per unit.

Attributes charts are generally not as informative as variables charts, because there is typically more information in a numerical measurement than in merely classifying a unit as conforming or nonconforming. However, attributes charts do have important applications. They are particularly useful in service industries and in nonmanufacturing quality-improvement efforts because so many of the quality characteristics found in these environments are not easily measured on a numerical scale.

Attribute charts do not provide as much information as variable charts.

6.2 The Control Chart for Fraction Nonconforming

The control chart for fraction nonconforming is widely used in transactional and service industry applications of statistical process control. In the nonmanufacturing environment, examples would include the number of employee paychecks that are in error or distributed late during a pay period, the number of check requests that are not paid within the standard accounting cycle, and the number of deliveries made by a supplier that are not on time.

The fraction nonconforming is defined as the ratio of the number of nonconforming items in a population to the total number of items in that population. The items may have several quality characteristics that are examined simultaneously by the inspector. If the item does not conform to standard for one or more of these characteristics, it is classified as nonconforming. We usually express the fraction nonconforming as a decimal, although occasionally the percent nonconforming (which is simply 100% times the fraction nonconforming) is used. When demonstrating or displaying the control chart to production personnel or presenting results to management, the percent nonconforming is often used; however, we could also analyze the fraction conforming just as easily.

If we assume that the probability of obtaining a nonconforming unit is constant and successive units of production are independent, then we can use the binomial distribution as a model and estimate the mean and standard deviation of this fraction using the formulas below. Note: Additional information for the binomial distribution may be found in Appendix One.

Key Parameters for Fraction Nonconforming

$$\text{Estimate of fraction nonconforming, } \hat{p} = \frac{D}{n} \tag{6.1}$$

(Continued)

(*Continued*)

$$\text{Mean of fraction nonconforming, } \mu_{\hat{p}} = p \tag{6.2}$$

$$\text{Standard deviation of fraction nonconforming} = \sqrt{\frac{p(1-p)}{n}} \tag{6.3}$$

where D = number of defectives found and n = total number of units inspected.

General Model for a Control Chart

$$\text{UCL} = \mu_W + k\sigma_W$$
$$\text{CL} = \mu_W$$
$$\text{LCL} = \mu_W - k\sigma_W$$

In Chapter 5, we introduced the general model for a control chart, which defined the upper control limit, center line, and lower control limit as shown in the adjacent display. In this model, k, normally equals 3.

If there is a known historical percentage nonconforming, we can apply the above formulas to create a fraction nonconforming, or p chart.

p Chart
A control chart for the fraction nonconforming.

Fraction Nonconforming Control Chart Formulas—p is Known

$$\text{UCL} = p + 3\sqrt{\frac{p(1-p)}{n}} \tag{6.4}$$

$$\text{CL} = p \tag{6.5}$$

$$\text{LCL} = p - 3\sqrt{\frac{p(1-p)}{n}} \tag{6.6}$$

Depending on the values of p and n, sometimes the lower control limit, LCL, is less than zero. In these cases, we customarily set LCL = 0 and assume that the control chart only has an upper control limit.

The actual operation of this chart would consist of taking subsequent samples of n units, computing the sample fraction nonconforming $\hat{p}$, and plotting the statistic $\hat{p}$ on the chart. As long as $\hat{p}$ remains within the control limits and the sequence of plotted points does not exhibit any systematic nonrandom pattern, we can conclude that the process is in control at the level p. If a point plots outside of the control limits, or if a nonrandom pattern in the plotted points is observed, we can conclude that the process fraction nonconforming has most likely shifted to a new level and the process is out of control.

When the process fraction nonconforming is not known, then it must be estimated from observed data. The usual procedure is to select 20 to 25 preliminary samples, each of size n. The proportion nonconforming is computed for each sample, and these are averaged together to create a grand average $\bar{p}$.

The center line and control limits of the control chart for fraction nonconforming are computed as follows:

Fraction Nonconforming Control chart Formulas—
***p* is Unknown**

$$\text{UCL} = \bar{p} + 3\sqrt{\frac{\bar{p}(1-\bar{p})}{n}} \tag{6.7}$$

$$\text{CL} = \bar{p} \tag{6.8}$$

$$\text{LCL} = \bar{p} - 3\sqrt{\frac{\bar{p}(1-\bar{p})}{n}} \tag{6.9}$$

The control limits using $\bar{p}$ should be regarded as trial control limits and the sample values used to calculate $\bar{p}$ should be plotted against the trial limits to determine whether the process was in control when the data was collected. Any points that exceed the trial control limits should be investigated. If assignable causes for these points are discovered, they should be discarded and new trial control limits determined.

If the control chart is based on a known or standard value for the fraction nonconforming p, then the calculation of trial control limits is generally unnecessary. However, one should be cautious when working with a standard value for p. In practice, the true value of p is rarely known with certainty and we would usually be given a standard value of p that represents a desired or target value for the process fraction nonconforming. If this is the case and future samples indicate an out-of-control condition, we must determine whether the process is out of control at the target p but in control at some other value of p.

Phase I Control Charts
Control charts used to test whether the process was in control when the initial subgroups were obtained.

Phase II Control Charts
Control charts used to test whether the process is in control when future samples are drawn.

example 6.1 Construction of a *p* Chart

Frozen orange juice concentrate is packed in 6 oz cardboard cans. These cans are formed on a machine by spinning them from cardboard stock and attaching a metal bottom panel. By inspection of a can, we may determine whether, when filled, it could possibly leak either on the side seam or around the bottom joint. Such a nonconforming can has an improper seal on either the side seam or the bottom panel. Set up a control chart to improve the fraction of nonconforming cans produced by this machine.

To establish the control chart, 30 samples of 50 cans each were selected at half-hour intervals over a three-shift period in which the machine was in continuous operation. The data are shown in Table 6.1. We construct a phase I control chart using this preliminary data to determine if the process was in control when these data were collected.

$$\text{UCL} = \bar{p} + 3\sqrt{\frac{\bar{p}(1-\bar{p})}{n}} = 0.2313 + 3\sqrt{\frac{0.2313(1-0.2313)}{50}} = 0.4102$$

$$\text{CL} = \bar{p} = 0.2313$$

$$\text{LCL} = \bar{p} - 3\sqrt{\frac{\bar{p}(1-\bar{p})}{n}} = 0.2313 - 3\sqrt{\frac{0.2313(1-0.2313)}{50}} = 0.0524$$

example 6.1 Continued

Table 6.1

Sample Number	Number of Nonconforming Cans, D_i	Sample Fraction Nonconforming, $\hat{p}_i$	Sample Number	Number of Nonconforming Cans, D_i	Sample Fraction Nonconforming, $\hat{p}_i$
1	12	0.24	16	8	0.16
2	15	0.30	17	10	0.20
3	8	0.16	18	5	0.10
4	10	0.20	19	13	0.26
5	4	0.08	20	11	0.22
6	7	0.14	21	20	0.40
7	16	0.32	22	18	0.36
8	9	0.18	23	24	0.48
9	14	0.28	24	15	0.30
10	10	0.20	25	9	0.18
11	5	0.10	26	12	0.24
12	6	0.12	27	7	0.14
13	17	0.34	28	13	0.26
14	12	0.24	29	9	0.18
15	22	0.44	30	6	0.12
				347	$\overline{p} = 0.2313$

The control chart is a **Phase I or Trial Control Chart** and provides a center line and upper and lower control limits as shown in Figure 6.1. The sample fraction nonconforming from each preliminary sample is plotted on this chart. We note that two points, samples 15 and 23, plot above the upper control limit, so the process is not in control. These points must be investigated to see whether an assignable cause can be determined.

Analysis of the data from sample 15 indicates that a new batch of cardboard stock was put into production during that half-hour period. The introduction of new batches of raw material sometimes causes irregular production performance, and it is reasonable to believe that this has occurred here. Furthermore, during the half-hour period in which sample 23 was obtained, a relatively inexperienced operator had been temporarily assigned to the machine, and this could account for the high fraction nonconforming obtained from that sample. Consequently, samples 15 and

example 6.3 Construction of a *np* Chart

Set up an np control chart for the orange juice concentrate can process in Table 6.1 of Example 6.1.

In Example 6.1, we found that $n = 50$ and $\bar{p} = 0.2313$.

Therefore, the control limits for the np chart are

$$\text{UCL} = n\bar{p} + 3\sqrt{n\bar{p}(1 - \bar{p})} = (50)(0.2313) + 3\sqrt{(50)(0.2313)(1 - 0.2313)} = 20.51$$

$$\text{CL} = n\bar{p} = (50)(0.2313) = 11.565$$

$$\text{LCL} = n\bar{p} - 3\sqrt{n\bar{p}(1 - \bar{p})} = (50)(0.2313) + 3\sqrt{(50)(0.2313)(1 - 0.2313)} = 2.620$$

The results of this chart are shown in Figure 6.7.

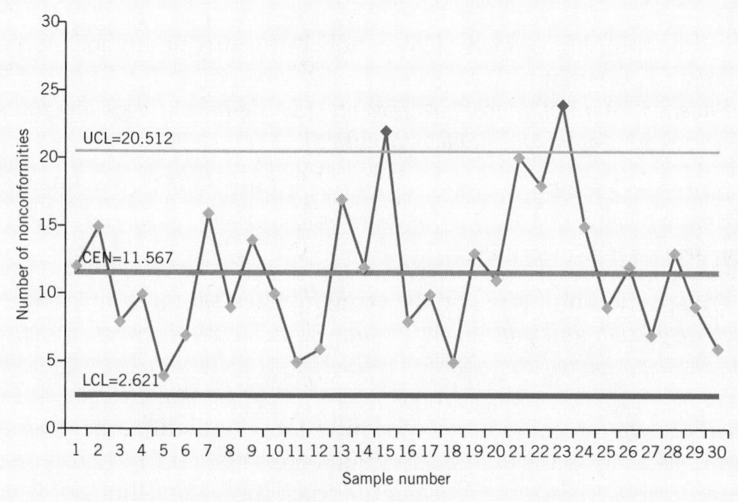

Figure 6.7 ***np* Chart for Data in Table 6.1**

As with the p chart, this chart correctly identifies points 15 and 23 as out-of-control points. However, note that because this chart is based upon the number of nonconforming units, the numbers plotted will always be integers. Some practitioners prefer to use integer values for control limits on the np chart instead of their decimal fraction counterparts.

6.3 Control Charts for Nonconformities (Defects)

A nonconforming item is a unit of product that does not satisfy one or more of the specifications for that product. Each specific point at which a specification is not satisfied results in a defect or nonconformity. Consequently, a nonconforming item will contain at least one nonconformity. However, depending on their nature and severity, it is quite possible for a unit to contain several nonconformities and not be classified as nonconforming.

As an example, suppose we are manufacturing personal computers. Each unit could have one or more very minor flaws in the cabinet finish, and since these flaws do not seriously affect the unit's functional operation, it could be classified as conforming. However, if there are too many of these flaws, the personal computer should be classified as nonconforming, since the flaws would be very noticeable to the customer and might affect the sale of the unit.

There are many practical situations in which we prefer to work directly with the number of defects or nonconformities rather than the fraction nonconforming. These include the number of blemishes in a roll of paper, the number of defects in an electronic device, the number of typographical errors in a report, etc.

c Chart
A control c chart for the total number of nonconformities in a sample.

Control charts can be created for either the total number of nonconformities in a sample (c chart) or the average number of nonconformities per unit (u chart). These charts are based upon the Poisson distribution and assume that

- The probability of occurrence of a nonconformity at any location is small and constant.
  The average count of nonconformities must be significantly smaller than the total count of possible nonconformities.
- The probability of occurrence of a nonconformity is independent.
  The occurrence of one does not increase or decrease the chance of the next item being a nonconformity.
- The inspection unit must be the same for each sample.
  Each inspection unit must always represent an identical area of opportunity for the occurrence of nonconformities.

In addition, we can count nonconformities of several different types on one unit, as long as the above conditions are satisfied for each type of nonconformity. In most practical situations, these conditions will not be satisfied exactly. However, as long as these departures from the assumptions are not severe, the Poisson model will usually work reasonably well.

6.3.1 THE c CHART

The c chart describes the total number of nonconformities in a sample. It relies on the variable c, which reflects this total count. Formulas for the c chart are described below:

Control chart for the number of nonconformities (c chart):

$$\text{UCL} = \bar{c} + 3\sqrt{\bar{c}} \quad (6.14)$$

$$\text{CL} = \bar{c} \quad (6.15)$$

$$\text{LCL} = \bar{c} - 3\sqrt{\bar{c}} \quad (6.16)$$

where $\bar{c}$ is an estimate the average number of nonconformities found per sample. When there is a standard, c replaces $\bar{c}$.

example 6.4 Construction of a c Chart

Table 6.4 presents the number of nonconformities observed in 26 successive samples of 100 circuit boards. Set up a c chart for this data.

The 26 samples contain 516 total nonconformities, so we estimate $\bar{c}$ as

$$\bar{c} = \frac{516}{26} = 19.85$$

Therefore, the Phase I control limits are

$$\text{UCL} = \bar{c} + 3\sqrt{\bar{c}} = 19.85 + 3\sqrt{19.85} = 33.22$$
$$\text{CL} = \bar{c} = 19.85$$
$$\text{LCL} = \bar{c} - 3\sqrt{\bar{c}} = 19.85 - 3\sqrt{19.85} = 6.48$$

The control chart is shown in Figure 6.9 and shows the number of observed nonconformities from the 26 samples. Two points plot outside the control limits, samples 6 and 20.

Investigation of sample 6 revealed that a new inspector had examined the boards in this sample and that he did not recognize several types of nonconformities that

Table 6.4

Data for Example 6.4

Sample Number	Number of Nonconformities	Sample Number	Number of Nonconformities
1	21	14	19
2	24	15	10
3	16	16	17
4	12	17	13
5	15	18	22
6	5	19	18
7	28	20	39
8	20	21	30
9	31	22	24
10	25	23	16
11	20	24	19
12	24	25	17
13	16	26	15

example 6.4 Continued

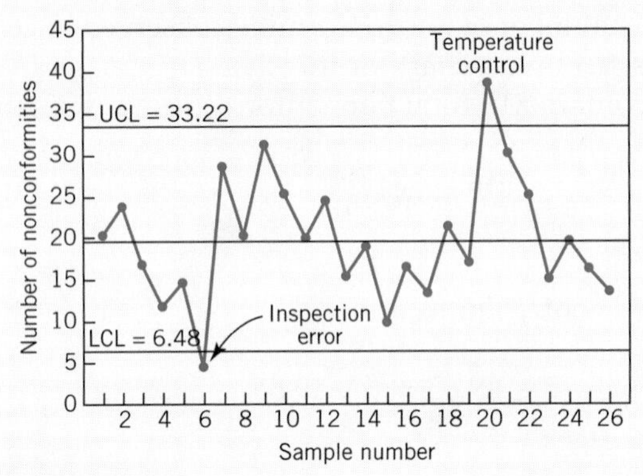

Figure 6.8 ***c* Chart Example 6.4**

could have been present. Furthermore, the unusually large number of nonconformities in sample 20 resulted from a temperature control problem in the wave soldering machine, which was subsequently repaired. Therefore, it seems reasonable to exclude these two samples and revise the trial control limits.

When data points 6 and 20 are excluded, the estimate of $\bar{c}$ is now computed as $472/24 = 19.67$. As such, the Phase II control limits become

$$\text{UCL} = \bar{c} + 3\sqrt{\bar{c}} = 19.67 + 3\sqrt{19.67} = 32.98$$

$$\text{CL} = \bar{c} = 19.67$$

$$\text{LCL} = \bar{c} - 3\sqrt{\bar{c}} = 19.67 - 3\sqrt{19.67} = 6.36$$

These become the standard values against which future production can be compared. Twenty new samples, each consisting of one inspection unit (i.e., 100 boards), are collected and the total number of nonconformities is recorded in Table 6.5 and

Table 6.5

Additional Data for Example 6.4

Sample Number	Number of Nonconformities	Sample Number	Number of Nonconformities
27	16	37	18
28	18	38	21
29	12	39	16
30	15	40	22
31	24	41	19

example 6.4 Continued

Table 6.5

Continued

Sample Number	Number of Nonconformities	Sample Number	Number of Nonconformities
32	21	42	12
33	28	43	14
34	20	44	9
35	25	45	16
36	19	46	21

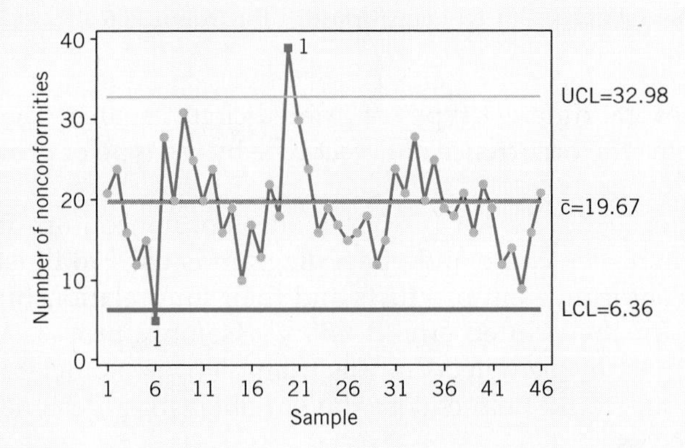

Figure 6.9 **Continuation of Control Chart for Example 6.4**

plotted in the control chart in Figure 6.9. No lack of control is indicated; however, the number of nonconformities per board is still unacceptably high. Further action is necessary to improve the process.

In terms of finding improvements, defect or nonconformity data is typically more informative than fraction nonconforming because there will usually be several different types of nonconformities. By analyzing the nonconformities by type, we can often gain considerable insight into their cause. This can be of considerable assistance in developing the out-of-control-action plans (OCAPs) that must accompany control charts.

For example, in the printed circuit board process, there are sixteen different types of defects. Defect data for 500 boards are plotted on a Pareto chart in Figure 6.10. Note that over 60% of the total number of defects is due to two defect types: solder insufficiency and solder cold joints. This points to further problems with the wave soldering process. If these problems can be isolated and eliminated, there will be a dramatic increase in process yield.

Notice that the nonconformities follow the Pareto distribution; that is, most of the defects are attributable to a few (in this case, two) defect types. This process

example 6.4 Continued

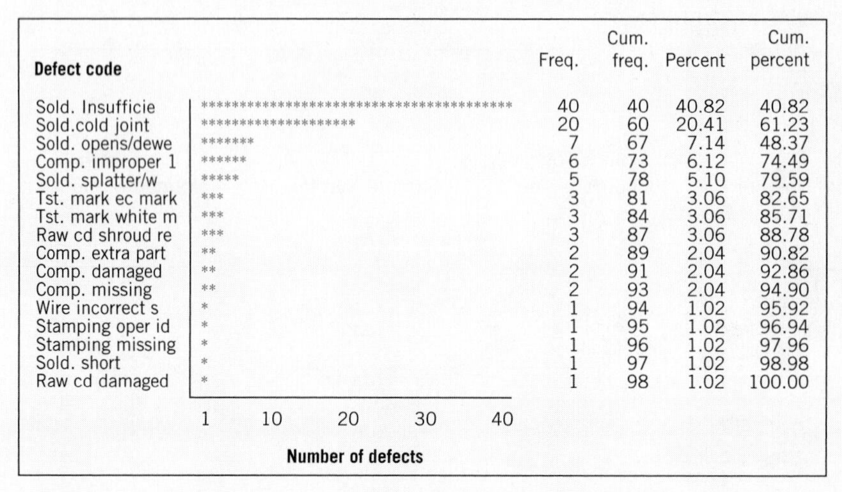

Defect code		Freq.	Cum. freq.	Percent	Cum. percent
Sold. Insufficie	**	40	40	40.82	40.82
Sold.cold joint	********************	20	60	20.41	61.23
Sold. opens/dewe	*******	7	67	7.14	48.37
Comp. improper 1	******	6	73	6.12	74.49
Sold. splatter/w	*****	5	78	5.10	79.59
Tst. mark ec mark	***	3	81	3.06	82.65
Tst. mark white m	***	3	84	3.06	85.71
Raw cd shroud re	***	3	87	3.06	88.78
Comp. extra part	**	2	89	2.04	90.82
Comp. damaged	**	2	91	2.04	92.86
Comp. missing	**	2	93	2.04	94.90
Wire incorrect s	*	1	94	1.02	95.92
Stamping oper id	*	1	95	1.02	96.94
Stamping missing	*	1	96	1.02	97.96
Sold. short	*	1	97	1.02	98.98
Raw cd damaged	*	1	98	1.02	100.00

Figure 6.10 **Pareto Analysis of Nonconformities for Example 6.4**

manufactures several different types of printed circuit boards. Therefore, it may be helpful to examine the occurrence of defect type by the type of printed circuit board (part number).

Another useful technique for further analysis of nonconformities is the cause-and-effect diagram. The cause-and-effect diagram is used to illustrate the various sources of nonconformities in products and their interrelationships. A cause-and-effect diagram for the printed circuit board assembly process is shown in Figure 6.11. Since most of the defects in this example were solder-related, the cause-and-effect diagram could help choose the variables for a designed experiment to improve the process.

Note the distinction. A control chart in an established process can help us monitor the process to detect changes or shifts. However, to make significant improvements in an established process, improvements must often be made to the process itself.

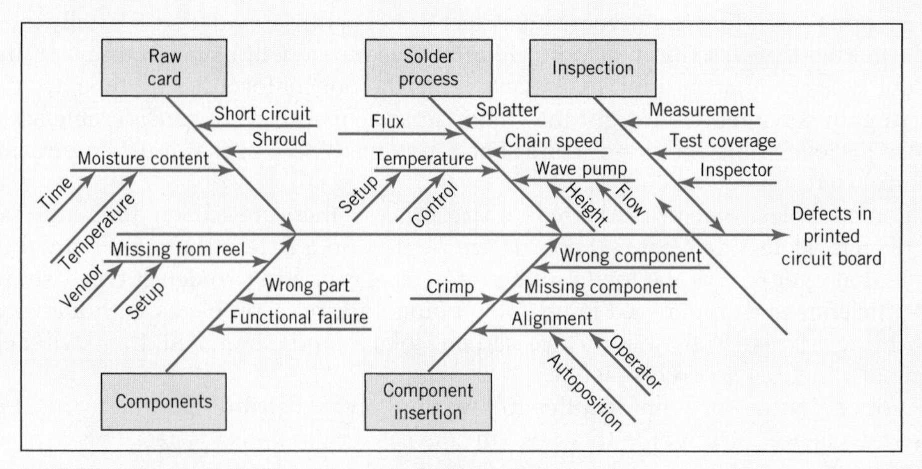

Figure 6.11 **Cause-and-Effect Diagram for Example 6.4**

6.3.2 THE *u* CHART

u-Chart
A control chart for the average number of nonconformities per unit.

Example 6.4 illustrates a control chart for nonconformities with the sample size exactly equal to one inspection unit. The inspection unit is chosen for operational or data-collection simplicity. However, there is no reason why the sample size must be restricted to one inspection unit. In fact, we would often prefer to use several inspection units in the sample, thereby increasing the area of opportunity for the occurrence of nonconformities. The sample size should be chosen according to statistical considerations, such as specifying a sample size large enough to obtain a particular probability of detecting a process shift. Alternatively, economic factors could enter into sample size determination.

Suppose we decide to base the control chart on a sample size of n inspection units. Note that n does not have to be an integer. To illustrate this, suppose that in Example 6.4 we were to specify a subgroup size of $n = 2.5$ inspection units. Then the sample size becomes $(2.5)(100) = 250$ boards. There are two general approaches to constructing the revised chart once a new sample size has been selected.

One approach is simply to redefine a new inspection unit that is equal to n times the old inspection unit. In this case, the center line on the new control chart is $n\bar{c}$ and the control limits are located at $n\bar{c} \pm 3\sqrt{n\bar{c}}$, where $\bar{c}$ is the observed mean number of nonconformities in the original inspection unit. Suppose that in Example 6.3, after revising the trial control limits, we decided to use a sample size of $n = 2.5$ inspection units. Then the center line would have been located at $(2.5)(19.67) = 49.18$ and the control limits would have been $49.18 \pm 3\sqrt{49.18}$ or LCL = 28.14 and UCL = 70.22.

The second approach involves setting up a control chart based on the average number of nonconformities per inspection unit. If we find x total nonconformities in a sample of n inspection units, then the average number of nonconformities per inspection unit is

$$\bar{u} = \frac{\text{Total nonconformities, } x}{\text{Number of inspection units, } n}$$

Consequently, the parameters of the control chart for the average number of nonconformities per unit are as follows:

Control chart for the average number of nonconformities per unit (u chart):

$$\text{UCL} = \bar{u} + 3\sqrt{\frac{\bar{u}}{n}} \qquad (6.17)$$

$$\text{CL} = \bar{u} \qquad (6.18)$$

$$\text{LCL} = \bar{u} - 3\sqrt{\frac{\bar{u}}{n}} \qquad (6.19)$$

where $\bar{u}$ = the average number of nonconformities per inspection unit. When there is a standard, u replaces $\bar{u}$.

example 6.5 Constructing a u-Chart

A supply chain engineering group monitors shipments of materials through the distribution network. Errors on either the delivered material or the accompanying documentation are tracked on a weekly basis. Fifty randomly selected shipments are examined and the errors recorded. Data for 20 weeks are shown in Table 6.6. Set up a u control chart to monitor this process.

Table 6.6

Data for Example 6.5

Sample Number (week), i	Sample Size, n	Total Number of Errors (Nonconformities), x_i
1	50	2
2	50	3
3	50	8
4	50	1
5	50	1
6	50	4
7	50	1
8	50	4
9	50	5
10	50	1
11	50	8
12	50	2
13	50	4
14	50	3
15	50	4
16	50	1
17	50	8
18	50	3
19	50	7
20	50	4
		74

example 6.5 Continued

From the data in Table 6.6, we estimate the number of errors (nonconformities) per unit (shipment) to be

$$\bar{u} = \frac{74 \text{ nonconformities}}{(20)\ 50 \text{ shipments}} = 0.074 \text{ nonconformities per shipment}$$

Therefore, the control limits are

$$\text{UCL} = \bar{u} + 3\sqrt{\frac{\bar{u}}{n}} = 0.074 + 3\sqrt{\frac{0.074}{50}} = 0.1894$$

$$\text{CL} = \bar{u} = 0.074$$

$$\text{LCL} = \bar{u} - 3\sqrt{\frac{\bar{u}}{n}} = 0.074 - 3\sqrt{\frac{0.074}{50}} = -0.0414 = 0$$

Since the LCL < 0, we would set LCL = 0 for the u chart. The control chart is shown in Figure 6.12. The preliminary data do not exhibit lack of statistical control; therefore, the trial control limits given here would be adopted for monitoring of future operations. Once again, note that, although the process is in control, the average number of errors per shipment is high. Action should be taken to improve the supply chain system.

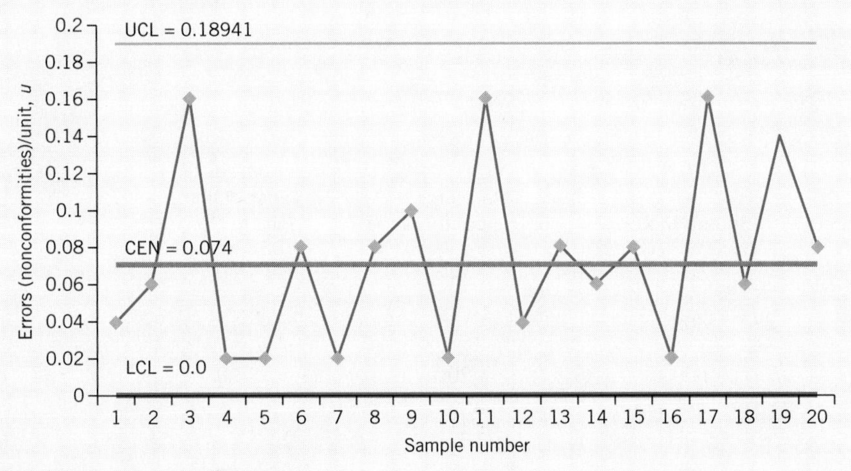

Figure 6.12 ***u*** **Chart for Example 6.5**

6.3.3 VARIABLE SAMPLE SIZES

Control charts for nonconformities are occasionally formed using 100% inspection of the product. When this method of sampling is used, the number of inspection units in a sample will usually not be constant. For example, the inspection of rolls of cloth or paper often leads to a situation in which the size of the sample varies, because not all rolls are of exactly the same length or width. If a control chart for nonconformities (c chart) is used in this situation, both the center line and the control limits will vary with the sample size. Such a control chart would be very difficult to interpret. The correct procedure is to use a control chart for nonconformities per unit (u chart) with variable sample sizes. This chart will have a constant center line; however, the control limits will vary inversely with the square root of the sample size n.

This is illustrated in Example 6.6.

example 6.6 *u* Charts with Varying Sample Sizes

In a textile finishing plant, dyed cloth is inspected for the occurrence of defects per 50 square meters. The data on the rolls of cloth are shown in Table 6.7. Use these data to set up a control chart for nonconformities per unit.

The center line of the chart is the average number of nonconformities per inspection unit—that is, the average number of nonconformities per 50 square meters is

$$\bar{u} = \frac{\text{Total number of defects found}}{\text{Total number of inspection units}} = \frac{153}{107.5} = 1.42$$

The control limits for the chart are computed using $\bar{u}$ with n replaced by the number of inspection units for each roll. The calculations are displayed in Table 6.7 and the control chart is shown in Figure 6.13.

Table 6.7

Occurrence of Nonconformities in Dyed Cloth

Roll Number	Square Meters	Total Number of Nonconformities	Number of Inspection Units in Roll, n
1	500	14	10.0
2	400	12	8.0
3	650	20	13.0
4	500	11	10.0
5	475	7	9.5
6	500	10	10.0
7	600	21	12.0
8	525	16	10.5
9	600	19	12.0
10	625	23	12.5
		153	107.50

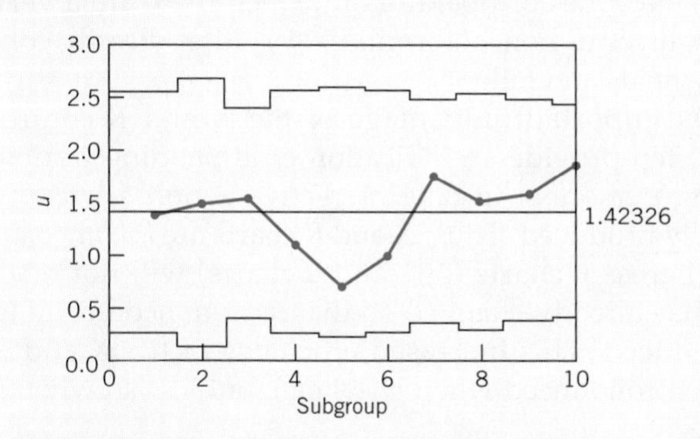

Figure 6.13 *u* **Chart for Example 6.6 Using Variable Sample Sizes**

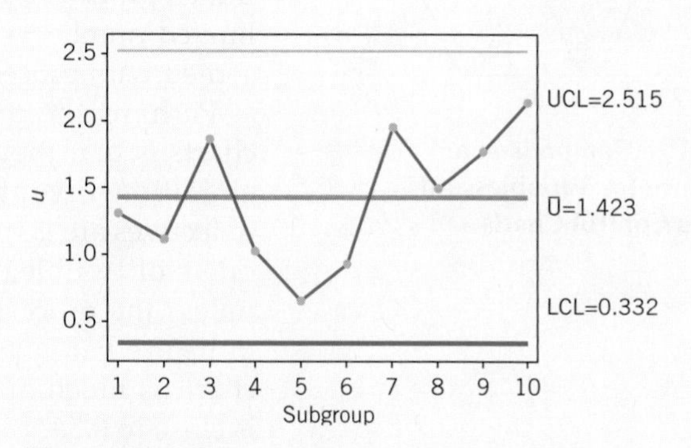

Figure 6.14 *u* **Chart for Example 6.6 Using Average Subgroup Size**

Alternatively, we could have used the average inspection size since the values are close in size (like Example 6.2). In this case, the average inspection unit is 10.75. This would have resulted in Figure 6.14.

6.4 Choice between Attributes and Variables Control Charts

In many applications, the analyst will have to choose between using a variables control chart, such as the $\overline{X}$ and R charts, and an attributes control chart, such as the p chart. In some cases, the choice will be clear-cut. For example, if the quality characteristic is the color of the item, such as might be the case in carpet or cloth production, then attributes inspection would often be preferred over an attempt to quantify the quality characteristic "color." In other cases, the choice will not be obvious, and the analyst must take several factors into account in choosing between attributes and variables control charts.

Attributes control charts have the advantage whereby several quality characteristics can be considered jointly and the unit classified as nonconforming if it fails to meet the specification on any one characteristic. On the other hand, if the several quality characteristics are treated as variables, then each one must be measured, and either a separate $\overline{X}$ and R chart must be maintained on each or some multivariate process control scheme employed. There is an obvious simplicity associated with the attributes chart in this case. Furthermore, expensive and time-consuming measurements may sometimes be avoided by attributes inspection.

Variables control charts, in contrast, provide much more useful information about process performance than an attributes control chart. Specific information about the process mean and variability is obtained directly. In addition, when points plot out of control on variables control charts, usually much more information is provided relative to the potential *cause* of that out-of-control signal. For a process capability

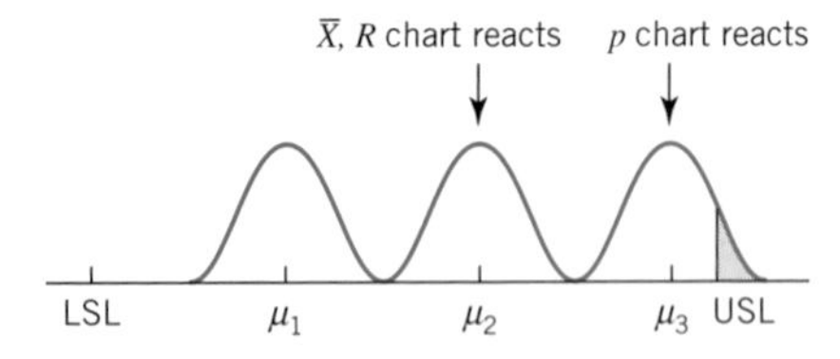

Figure 6.15 Comparison of Performance for Variables and Attributes Control Charts

study, variables control charts are almost always preferable to attributes control charts. The exceptions to this are studies relative to nonconformities produced by machines or operators in which there are a very limited number of sources of nonconformities, or studies directly concerned with process yields and fallouts.

Perhaps the most important advantage of the $\overline{X}$ and R control charts is that they often provide an indication of impending trouble and allow operating personnel to take corrective action *before* any defectives are actually produced. Thus, $\overline{X}$ and R charts are *leading indicators* of trouble, whereas p charts (or c and u charts) will not react unless the process has already changed so that *more* nonconforming units are being produced. This increased efficiency of the $\overline{X}$ and R charts is much more pronounced when p is small, but less so when p is close to 0.5.

Variables control charts can be leading indications of poor process performance

To illustrate, consider the production process depicted in Figure 6.15. When the process mean is at μ_1, few nonconforming units are produced. Suppose the process mean begins to shift upward. By the time it has reached μ_2, the $\overline{X}$ and R charts will have reacted to the change in the mean by generating a strong nonrandom pattern and possibly several out-of-control points. However, a p chart would not react until the mean had shifted all the way to μ_3, or until the actual number of nonconforming units produced had increased. Thus, the $\overline{X}$ and R charts are more powerful control tools than the p chart.

Variables control charts usually require a smaller sample size for a specific level of protection

For a specified level of protection against process shifts, variables control charts usually require a much smaller sample size than the corresponding attributes control chart. Thus, although variables-type inspection is usually more expensive and time-consuming on a per-unit basis than attributes inspection, fewer units must be examined. This is an important consideration, particularly in cases where inspection is destructive (such as opening a can to measure the volume of product within or to test chemical properties of the product). The following example demonstrates the economic advantage of variables control charts.

example 6.7 Economic Advantages of Variable Control Charts

Consider the case of a process that has a lower specification limit of 44 and an upper specification limit of 56. The process is operating at an average value of 50 and a standard deviation of 2.

If we would like to detect a shift of the mean to 52, what is the appropriate sample size for an $\overline{X}$ and p chart?

The sample size on the $\overline{X}$ chart must be large enough so that the upper three-size limit is less than 52. This implies that

$$50 + \frac{3(2)}{\sqrt{n}} = 52$$

Solving this equation, we obtain $n = 9$. If a p chart is used, then we use Duncan's method to calculate the sample size. To use this method, we need to better understand the impact of the shift upon the fraction nonconforming. When the process is

example 6.7 Continued

in control using the standard three-sigma control limits, the fraction nonconforming is 0.0027. When the process shifts to 52, the fraction nonconforming produced is 0.0228. Therefore, the amount of the shift, d, that we would like to detect is (0.0228 − 0.0027) = 0.0201. These values can be used in Duncan's equation:

$$n = \left(\frac{3}{\delta}\right)^2 p(1 - p) = \left(\frac{3}{0.0201}\right)^2 0.0027(1 - 0.0027) = 59.98 \text{ or } 60$$

Thus, 60 units would be required for each subgroup within the p chart. This is seven times as much as the $\overline{X}$ chart. Therefore, the $\overline{X}$ chart is less expensive to operate.

Regardless of the choice of chart, it is important not to exclude the use of attributes in a process control setting. Example 6.8 highlights a case study showing the organizational impact of using these charts in practice.

example 6.8 Impact of Control Chart on an Organization

A paper by T. Moses, A. Stahelski, and G. Knapp ("Effects of Attribute Charts on Organizational Performance," *Journal of Organizational Behavior Management*, Vol. 20, No. 1, 2000) illustrates the impact that attributes control charts can have on an organization. This case study is briefly summarized below.

In early 1990, a small manufacturing company that produces seamless tubing for the nuclear and aerospace industries began to implement TQM. They implemented process improvement teams and trained all employees in math skills, basic probability, process improvement, and construction of control charts. They implemented $\overline{X}$ and R charts first and had very good early successes. One chart alone yielded $200,000 per year savings ($1,990).

However, metal defects such as tearing pits, cuts, and dents occurred frequently and could not be readily evaluated using variable control charts. This company operated three shifts per day, and employees in all three shifts were authorized to evaluate the defects and determine if a product was conforming or not. However, these decisions were rarely made in the second or third shifts. The company wanted to remedy this and create a structure in which these decisions could be made in the presence of data to support the decisions.

A team was created to implement an attribute control chart for this type of data. In addition to creating control charts, they also created a detailed checklist so that inspectors would know exactly what they were evaluating and how a defect should be scored. An out of control action plan (OCAP) was also developed to require inspectors to make decisions about how to eliminate the sources of defects when an out-of-control event occurred.

Within 15 days of implementation, the results were significant. All three shifts were now making decisions of whether a product was conforming or nonconforming. This reduced rework significantly. By detecting and correcting out-of-control conditions quickly, the company reversed a downward trend in yield and improved their yield by 1.8% in just 15 days. Assuming no further improvement, this one

example 6.8 Continued

control chart would yield a $343,000 cost savings in one year. However, it is likely that more assignable causes will be found and additional cost savings realized.

We highlight this story as it may be tempting to choose attribute data as one of the last implemented control charts. However, these charts can often provide considerable costs savings and organizational behavior. In the case of this company, it was a study on important was not necessarily the specific chart that was created (variables or attributes) for the specific type of data, but the organizational structure and importance in reducing variability and addressing nonconformances.

6.5 Guidelines for Implementing Control Charts

Almost any process will benefit from SPC, including the use of control charts. In this section, we present some general guidelines helpful in implementing control charts.

Specifically, we deal with the following:

1. Determining which process characteristics to control
2. Determining where the charts should be implemented in the process
3. Choosing the proper type of control charts
4. Taking actions to improve processes as the result of SPC/control chart analysis
5. Selecting data-collection systems and computer software

The guidelines are applicable to both variables and attributes control charts. Remember, control charts are not only for process surveillance; they should be used as an active, on-line method for reducing process variability.

6.5.1 DETERMINING WHICH CHARACTERISTICS TO CONTROL AND WHERE TO PUT THE CONTROL CHARTS

At the start of a control chart program, it is usually difficult to determine which product or process characteristics should be controlled and at which point in the process to apply control charts. Some useful guidelines follow.

1. At the beginning, control charts should be applied to any product characteristics or manufacturing operations believed to be important. The charts will provide immediate feedback as to whether they are actually needed.
2. The control charts found to be unnecessary should be removed, and others that engineering and operator judgment indicates may be required should be added. More control charts will usually be employed until the process has stabilized.

(Continued)

3. Information on the number and types of control charts on the process should be kept current. It is best to keep separate records on the variables and attributes charts. In general, after the control charts are first installed, the number of control charts tends to increase rather steadily. After that, it will usually decrease. When the process stabilizes, we typically find that it has the same number of charts from one year to the next. However, they are not necessarily the same charts.

4. If control charts are being used effectively and if new knowledge is being gained about the key process variables, we should find that the number of $\overline{X}$ and R charts increases and the number of attributes control charts decreases.

5. At the beginning of a control chart program there will usually be more attributes control charts applied to semi-finished or finished units near the end of the manufacturing process. As we learn more about the process, these charts will be replaced with $\overline{X}$ and R charts applied earlier in the process to the critical parameters and operations that result in nonconformities in the finished product. Generally, the earlier process control can be established the better. In a complex assembly process, this may imply that process controls need to be implemented at the vendor or supplier level.

6. Control charts are an on-line, process-monitoring procedure. They should be implemented and maintained as close to the work center as possible so that feedback will be rapid. Furthermore, the process operators and process engineering should have direct responsibility for collecting the process data, maintaining the charts, and interpreting the results. The operators and engineers have the detailed knowledge of the process required to correct process upsets and use the control chart to improve process performance. Computers can speed up feedback and should be an integral part of any modern, on-line, process control procedure.

7. The out-of-control-action plan (OCAP) is a vital part of the control chart. Operating and engineering personnel should strive to keep OCAPs up to date and valid.

6.5.2 CHOOSING THE PROPER TYPE OF CONTROL CHART

A. $\overline{X}$ and R (or $\overline{X}$ and S) Charts

Consider using variables control charts in these situations:

1. A new process is coming on stream, or a new product is being manufactured by an existing process.
2. The process has been in operation for some time, but it is chronically in trouble or unable to hold the specified tolerances.
3. The process is in trouble, and the control chart can be useful for diagnostic purposes (troubleshooting).
4. Destructive testing (or other expensive testing procedures) is required.
5. It is desirable to reduce acceptance-sampling or other downstream testing to a minimum when the process can be operated in control.
6. Attributes control charts have been used, but the process is either out of control or in control but with unacceptable yield.
7. There are very tight specifications, overlapping assembly tolerances, or other difficult manufacturing problems.
8. The operator must decide whether or not to adjust the process, or when a setup must be evaluated.
9. A change in product specifications is desired.
10. Process stability and capability must be continually demonstrated, such as in regulated industries.

B. **Attributes Charts (p Charts, c Charts, and u Charts)**

Consider using attributes control charts in these situations:

1. Operators control the assignable causes, and it is necessary to reduce process fallout.
2. The process is a complex assembly operation, and product quality is measured in terms of the occurrence of nonconformities, successful or unsuccessful product function, and so forth. (Examples include computers, office automation equipment, automobiles, and the major subsystems of these products.)
3. Process control is necessary, but measurement data cannot be obtained.
4. A historical summary of process performance is necessary. Attributes control charts, such as p charts, c charts, and u charts, are very effective for summarizing information about the process for management review.
5. Remember that attributes charts are generally inferior to charts for variables. Always use R or s charts whenever possible.

C. **Control Charts for Individuals**

Consider using the control chart for individuals in conjunction with a moving-range chart in these situations:

1. It is inconvenient or impossible to obtain more than one measurement per sample, or repeat measurements will only differ by laboratory or analysis error. Examples often occur in chemical processes.
2. Automated testing and inspection technology allow measurement of every unit produced. In these cases, also consider the cumulative sum control chart and the exponentially weighted moving average control chart. The data become available very slowly, and waiting for a larger sample will be impractical or make the control procedure too slow to react to problems. This often happens in nonproduct situations; for example, accounting data may become available only monthly.
3. Generally, once we are in phase II, individuals charts perform poorly in shift detection and can be very sensitive to departures from normality. Always use the EWMA and CUSUM charts in process monitoring instead of individuals charts whenever possible.

6.5.3 ACTIONS TAKEN TO IMPROVE THE PROCESS

Process improvement is the primary objective of statistical process control. The application of control charts will give information on two key aspects of the process: statistical control and capability.

Technically speaking, the capability of a process cannot be adequately assessed until statistical control has been established, but we will use a less precise definition of capability which is simply a qualitative assessment of whether or not the level of nonconforming units produced is low enough to warrant no immediate additional effort to further improve the process.

Figure 6.16 shows the possible states in which the process may exist with respect to these two issues.

Figure 6.16 gives the answers to two questions: "Is the process in control?" and "Is the process capable?" (in the sense of the earlier paragraph).

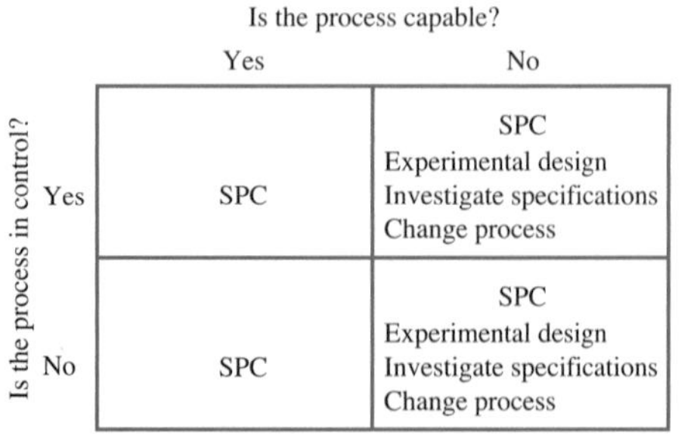

Figure 6.16 Actions Taken to Improve a Process

Each of the four cells in the figure contains some recommended courses of action depending on the answers to these two questions.

The box in the northwest corner is the ideal state: the process is in statistical control and exhibits adequate capability for present business objectives. In this case, SPC methods are valuable for process monitoring and for warning against the occurrence of any new assignable causes that could cause slippage in performance.

The northeast corner implies that the process exhibits statistical control but has poor capability. Perhaps the PCR is lower than the value required by the customer, or there is sufficient variability remaining to result in excessive scrap or rework. In this case, SPC methods may be useful for process diagnosis and improvement, primarily through the recognition of patterns on the control chart, but the control charts will not produce many out-of-control signals. It will usually be necessary to intervene actively in the process to improve it. Experimental design methods are helpful in this regard. Usually, it is also helpful to reconsider the specifications: they may have been set at levels tighter than necessary to achieve function or performance from the part. As a last resort, we may have to consider changing the process; that is, investigating or developing new technology that has less variability with respect to this quality characteristic than the existing process.

The lower two boxes in Figure 6.16 deal with the case of an out-of-control process. The southeast corner presents the case of a process that is out of control and not capable. (Remember our nontechnical use of the term "capability".) The actions recommended here are identical to those for the box in the northeast corner, except that SPC would be expected to yield fairly rapid results now, because the control charts should be identifying the presence of assignable causes. The other methods of attack will warrant consideration and use in many cases, however.

Finally, the southwest corner treats the case of a process that exhibits lack of statistical control but does not produce a meaningful number of defectives because the specifications are very wide. SPC methods should still be used to establish control and reduce variability in this case, for the following reasons:

1. Specifications can change without notice.
2. The customer may require both control and capability.
3. The fact that the process experiences assignable causes implies that unknown forces are at work; these unknown forces could result in poor capability in the near future.

6.5.4 SELECTION OF DATA-COLLECTION SYSTEMS AND COMPUTER SOFTWARE

The past few years have produced an explosion of quality control software and electronic data-collection devices. Some SPC consultants recommended against using the computer, noting that it is unnecessary, since most applications of SPC in Japan emphasized the manual use of control charts. If the Japanese were successful in the 1960s and 1970s

Manual control charting is useful, but computer-generated control charts are best in the long-run.

using manual control charting methods, then does the computer truly have a useful role in SPC?

The answer to this question is yes, for several reasons:

1. Although it can be helpful to begin with manual methods of control charting at the start of an SPC implementation, it is necessary to move successful applications to the computer very soon. The computer is a great productivity improvement device. We don't drive cars with the passenger safety systems of the 1960s, and we don't fly airplanes with 1960s avionics technology. We shouldn't use 1960s technology with control charts, either.
2. The computer will enable the SPC data to become part of the companywide manufacturing database, and in that form, the data will be useful to (and hence more likely to be used by) everyone—management, engineering, marketing, and so on, not only manufacturing and quality.
3. A computer-based SPC system can provide more information than any manual system. It permits the user to monitor many quality characteristics and to provide automatic signaling of assignable causes.

What type of software should be used? That is a difficult question to answer, because all applications have unique requirements and the capability of software is constantly changing. However, several features are necessary for successful results:

1. The software should be capable of stand-alone operation on a personal computer, on a multiterminal local area network, or on a multiterminal minicomputer-host system. SPC packages that are exclusively tied to a large mainframe system are frequently not very useful, because they usually cannot produce control charts and other routine reports in a timely manner.
2. The system must be user-friendly. If operating personnel are to use the system, it must have limited options, be easy to use, provide adequate error correction opportunities, and contain many on-line help features. It should ideally be possible to tailor or customize the system for each application, although this installation activity may have to be carried out by engineering/technical personnel.
3. The system should provide display of control charts for at least the last 25 samples. Ideally, the length of record displayed should be controlled by the user. Printed output should be immediately available on either a line printer or a plotter.
4. File storage should be sufficient to accommodate a reasonable amount of process history. Editing and updating files should be straightforward. Provisions to transfer data to other storage media or to transfer the data to a master manufacturing database is critical.

5. The system should be able to handle multiple files simultaneously. Only rarely does a process have only one quality characteristic that needs to be examined.
6. The user should be able to calculate control limits from any subset of the data on the file. The user should have the capability to input center lines and control limits directly.
7. The system should be able to accept a variety of inputs, including manual data entry, input from an electronic instrument, or input from another computer or instrument controller. It is important to have the capability for real-time process monitoring, or to be able to transfer data from a real-time data acquisition system.
8. The system should support other statistical applications, including as a minimum histograms and computation of process capability indices.
9. Service and support from the software supplier after purchase is always an important factor in deciding which software package to use.

The purchase price of commercially available software varies widely. Obviously, the total cost of software is very different from the purchase price. In many cases, a $500 SPC package is really a $10,000 package when we take into account the total costs of making the package work correctly in the intended application. It is also relatively easy to establish control charts with most of the popular spreadsheet software packages. However, it may be difficult to integrate those spreadsheet control charts into the overall manufacturing database or other business systems.

supplement One Constructing a *p* Chart with SPC XL and Minitab

MINITAB

1. Arrange the data so that the number of defects is in one column. If sample sizes differ for each subgroup, create a second column for this data.

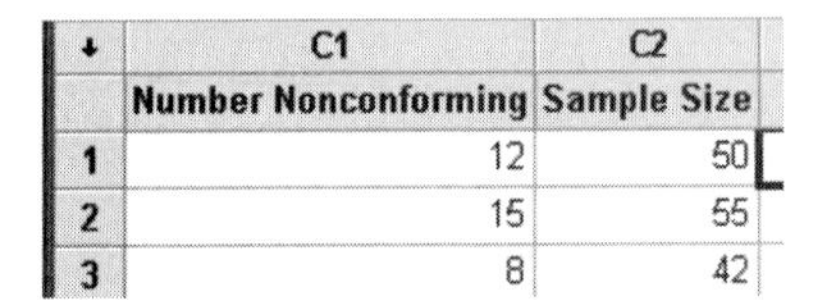

↓	C1	C2
	Number Nonconforming	Sample Size
1	12	50
2	15	55
3	8	42

2. From the menu bar, select Stat –> Control Charts –> Attributes Charts –> *p*. Minitab will then bring up a graphical user interface. Enter the column that defines the number nonconforming in the variables area. In the subgroup size, either type the constant sample size or select a column that defines the sample size.

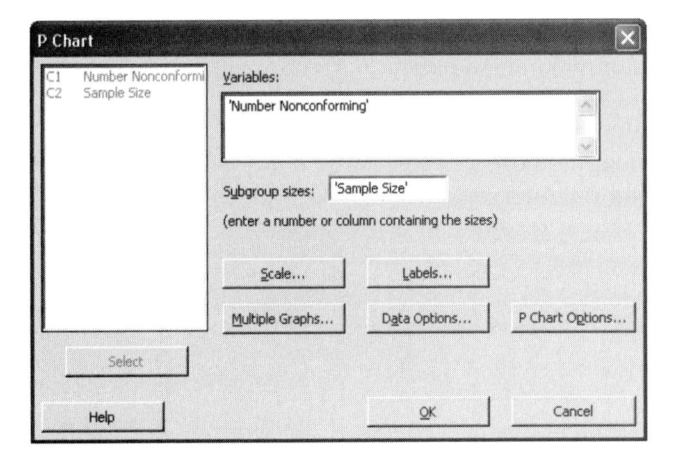

3. Click on the *p* chart options to select various options for the chart. Common options to be considered are detailed below:

   a. Parameters tab: allows you to specify a population proportion to be used for control limit calculations.
   b. *S*-limits tab: allows you to specify zones within the control limits. Note: this is not recommended for *p* charts with varying subgroup sizes.
   c. Tests tab: allows you to specify the use runs tests within the chart.

4. Once all options are defined, select OK to generate the chart.

SPC XL

1. Arrange the data so that the number of defects is in one column. Create a second column that contains information on the sample size of each subgroup (even if it is constant).

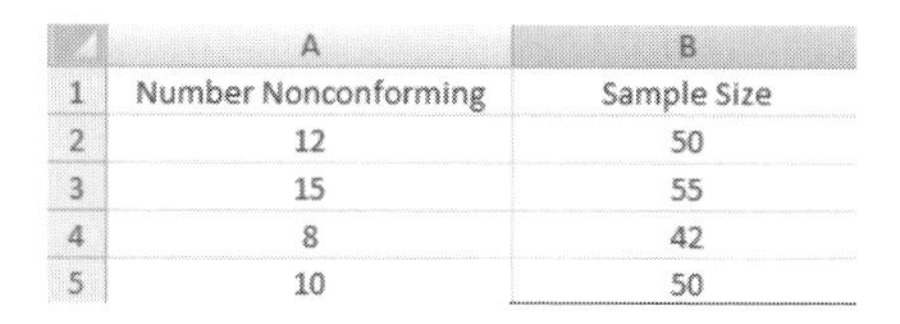

	A	B
1	Number Nonconforming	Sample Size
2	12	50
3	15	55
4	8	42
5	10	50

2. Select the control charts icon from the Sigma Zone Ribbon, then select p chart. You will be prompted to select a range for where your data resides. Then select next.

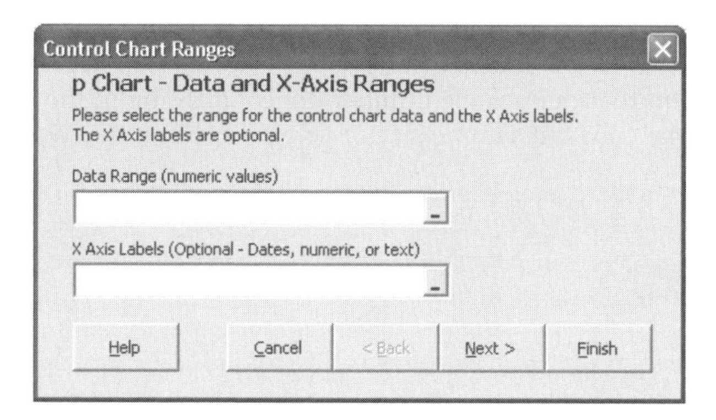

3. On the next screen, specify whether or not you would like the zones shown on the control chart. You can also request a histogram and process capability analysis.

4. Select Next to generate the control chart.

supplement Two Constructing an *np* Chart with SPC XL and Minitab

MINITAB

1. Arrange the data so that the number of defects is in one column. If sample sizes differ for each subgroup, create a second column for this data.

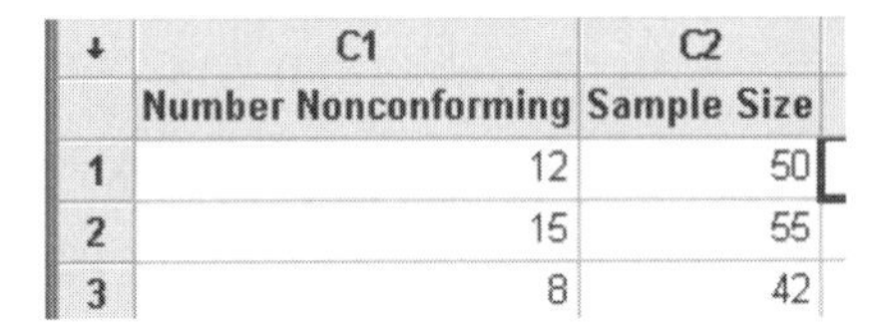

↓	C1	C2
	Number Nonconforming	Sample Size
1	12	50
2	15	55
3	8	42

2. From the menu bar, select Stat –> Control Charts –> Attributes Charts –> *np*. Minitab will then bring up a graphical user interface. Enter the column that defines the number nonconforming in the variables area. In the subgroup size, either type the constant sample size or select a column that defines the sample size.

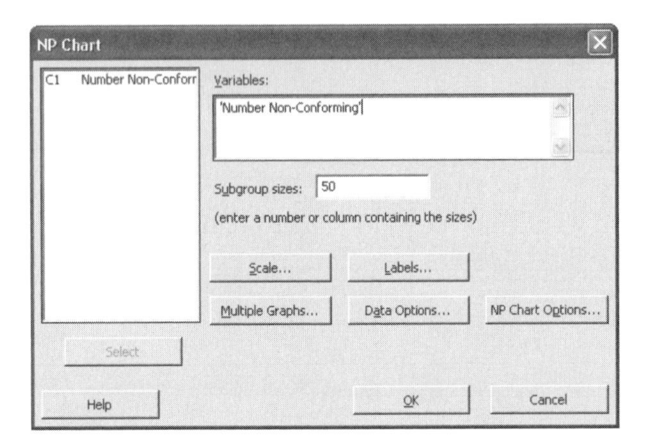

3. If desired, click on the *np* chart options to select various options for the chart. (See Appendix One for more details.)

4. Once all options are defined, select OK to generate the chart.

SPC XL

1. Arrange the data so that the number of defects is in one column.

	A
1	Number Nonconforming
2	12
3	15
4	8
5	10

2. Select the control charts icon from the Sigma Zone Ribbon, then select *np* chart. You will be prompted to select a range for where your data resides. Select this range and enter the subgroup size (variable subgroup sizes are not an option).

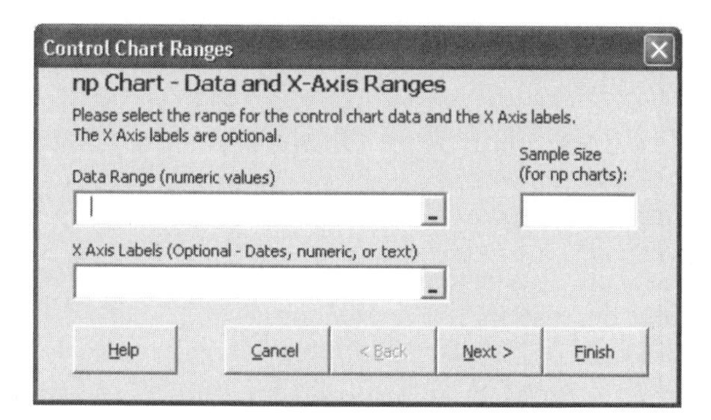

3. On the next screen, specify whether you would like the zones shown on the control chart. You can also request a histogram and process capability analysis.

4. Select Next to generate the control chart.

supplement Three Constructing a *c* Chart with SPC XL and Minitab

MINITAB

1. Arrange the data so that the number of defects is in one column.

2. From the menu bar, select Stat –> Control Charts –> Attributes Charts –> *c*. Minitab will then bring up a graphical user interface. Enter the column that defines the number nonconforming in the variables area. In the subgroup size, either type the constant sample size or select a column that defines the sample size.

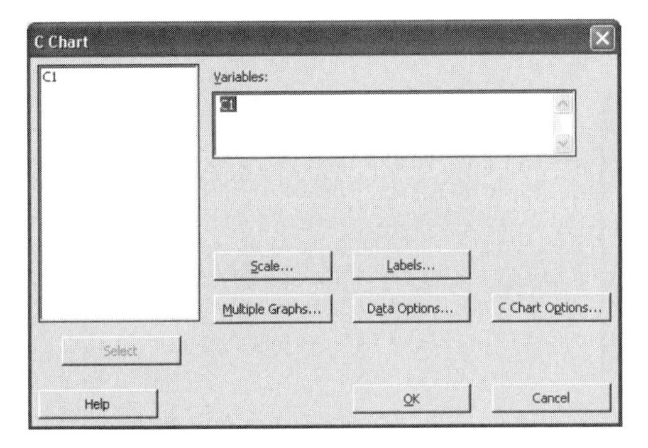

3. Click on the *c* chart options to select various options for the chart. Common options to be considered are detailed below are defined in Appendix One.

4. Once all options are defined, select OK to generate the chart.

SPC XL

1. Arrange the data so that the number of defects is in one column.

2. Select the control charts icon from the Sigma Zone Ribbon, then select *c* chart. You will be prompted to select a range for where your data resides. Then select next.

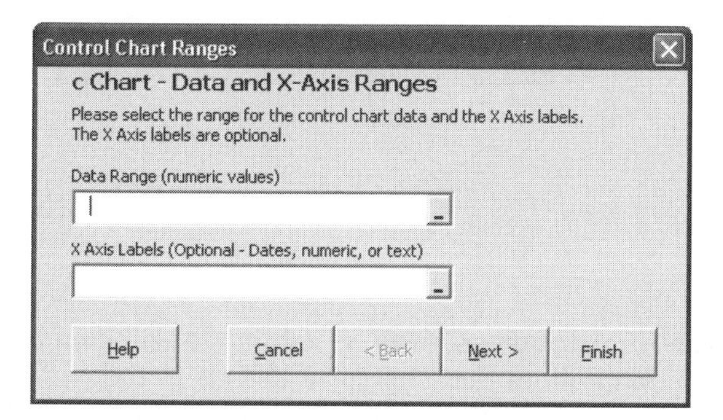

3. On the next screen, specify whether or not you would like the zones shown on the control chart. You can also request a histogram and process capability analysis.

4. Select Next to generate the control chart.

MINITAB

1. Arrange the data so that the number of defects is in one column and the number of units inspected in a separate column.

↓	C1	C2
	Number of Defects	Number of Units
1	14	10.0
2	12	8.0
3	20	13.0

2. From the menu bar, select Stat –> Control Charts –> Attributes Charts –> *u*. Minitab will then bring up a graphical user interface. Enter the column that defines the number of defects in the variables area. In the subgroup size, either type the constant sample size or select a column that defines the sample size.

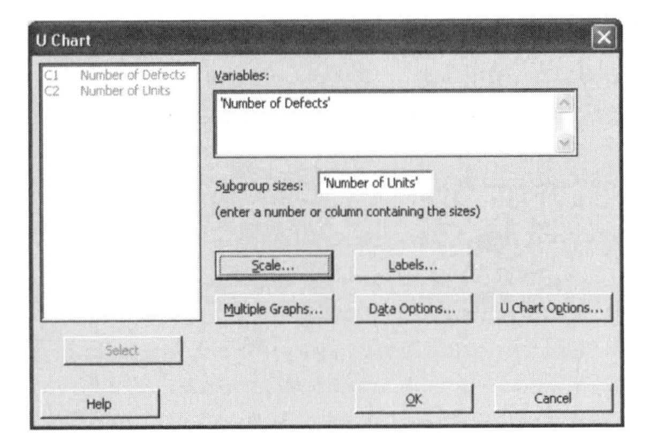

3. Click on the *u* chart options to select various options for the chart.

4. Once all options are defined, select OK to generate the chart.

SPC XL

1. Arrange the data per the table below. Note that the number of units inspected must be to the left of the number of defects column.

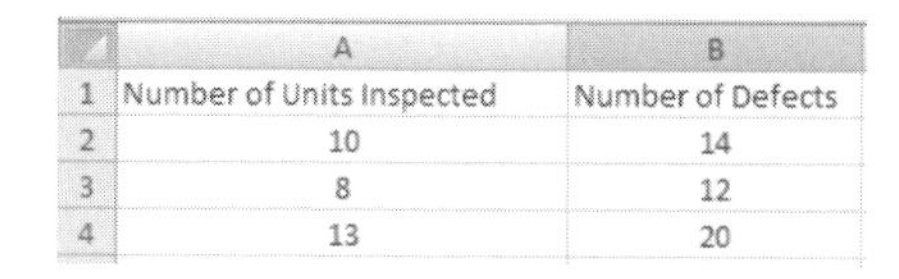

	A	B
1	Number of Units Inspected	Number of Defects
2	10	14
3	8	12
4	13	20

2. Select the control charts icon from the Sigma Zone Ribbon, then select *u* chart. You will be prompted to select a range for where your data resides. Then select next.

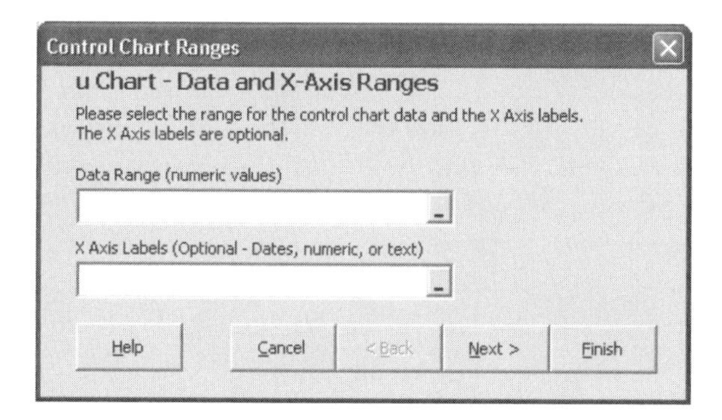

3. On the next screen, specify whether or not you would like the zones shown on the control chart. You can also request a histogram and process capability analysis.

4. Select Next to generate the control chart.

Important Terms and Concepts

Cause-and-effect diagram
Choice between attributes and variables data
Control chart for defects or nonconformities per unit or *u* chart
Control chart for fraction nonconforming or *p* chart
Control chart for nonconformities or *c* chart
Control chart for number nonconforming or *np* chart
Defect
Design of attributes control charts
Fraction defective
Fraction nonconforming
Nonconformity
Variable sample size for attributes control chart

chapter Six Exercises

6.1 The data in Table 6E.1 give the number of nonconforming bearing and seal assemblies in samples of size 100. Construct a fraction nonconforming control chart for these data. If any points plot out of control, assume that assignable causes can be found and determine the revised control limits.

Table 6E.1

Data for Exercise 6.1

Sample Number	Number of Nonconforming Assemblies	Sample Number	Number of Nonconforming Assemblies
1	7	11	6
2	4	12	15
3	1	13	0
4	3	14	9
5	6	15	5
6	8	16	1
7	10	17	4
8	5	18	5
9	2	19	7
10	7	20	12

6.2 The number of nonconforming switches in samples of size 150 is shown in Table 6E.2. Construct a fraction nonconforming control chart for these data.

Table 6E.2

Data for Exercise 6.2

Sample Number	Number of Nonconforming Switches	Sample Number	Number of Nonconforming Switches
1	8	11	6
2	1	12	0
3	3	13	4
4	0	14	0
5	2	15	3
6	4	16	1
7	0	17	15
8	1	18	2
9	10	19	3
10	6	20	0

Does the process appear to be in control? If not, assume that assignable causes can be found for all points outside the control limits, and calculate the revised control limits.

6.3 The data in Table 6E.3 represent the results of inspecting all units of a personal computer produced

Table 6E.3

Data for Exercise 6.3

Day	Inspected Units	Nonconforming Units	Fraction Nonconforming
1	80	4	0.050
2	110	7	0.064
3	90	5	0.056
4	75	8	0.107
5	130	6	0.046
6	120	6	0.050
7	70	4	0.057

(Continued)

Table 6E.3

Continued

Day	Inspected Units	Nonconforming Units	Fraction Nonconforming
8	125	5	0.040
9	105	8	0.076
10	95	7	0.074

for the past 10 days. Does the process appear to be in control?

6.4 A process that produces titanium forgings for automobile turbocharger wheels is to be controlled through the use of a fraction nonconforming chart. Initially, one sample of size 150 is taken each day for 20 days, and the results shown in Table 6E.4 are observed.

a. Establish a control chart to monitor future production.

b. What is the smallest sample size that could be used for this process and still give a positive lower control limit on the chart?

Table 6E.4

Data for Exercise 6.4

Day	Nonconforming Units	Day	Nonconforming Units
1	3	11	2
2	2	12	4
3	4	13	1
4	2	14	3
5	5	15	6
6	2	16	0
7	1	17	1
8	2	18	2
9	0	19	3
10	5	20	2

6.5 A process produces rubber belts in lots of size 2,500. Inspection records on the last 20 lots reveal the data in Table 6E.5.

a. Compute trial control limits for a fraction nonconforming control chart.

b. If you wanted to set up a control chart for controlling future production, how would you use these data to obtain the center line and control limits for the chart?

Table 6E.5

Data for Exercise 6.5

Lot Number	Number of Nonconforming Belts	Lot Number	Number of Nonconforming Belts
1	230	11	456
2	435	12	394
3	221	13	285
4	346	14	331
5	230	15	198
6	327	16	414
7	285	17	131
8	311	18	269
9	342	19	221
10	308	20	407

6.6 Based on the data in Table 6E.6 if an *np* chart is to be established, what would you recommend as the center line and control limits? Assume that $n = 500$.

Table 6E.6

Data for Exercise 6.6

Day	Number of Nonconforming Units
1	3
2	4

(*Continued*)

Table 6E.6

Continued

3	3
4	2
5	6
6	12
7	5
8	1
9	2
10	2

6.7 A company purchases a small metal bracket in containers of 5,000 each. Ten containers have arrived at the unloading facility, and 250 brackets are selected at random from each container. The fraction nonconforming in each sample are 0, 0, 0, 0.004, 0.008, 0.02, 0.004, 0, 0, and 0.008. Do the data from this shipment indicate statistical control?

6.8 A control chart for the fraction nonconforming is to be established using a center line of $p = 0.10$. What sample size is required if we wish to detect a shift in the process fraction nonconforming to 0.20 with probability 0.50?

6.9 A maintenance group improves the effectiveness of its repair work by monitoring the number of maintenance requests that require a second call to complete the repair. Twenty weeks of data are shown in Table 6E.7.

a. Find trial control limits for this process.
b. Design a control chart for controlling future production.

Table 6E.7

Data for Exercise 6.9

Sample Number	Sample Size	Number Nonconforming
1	100	10
2	100	15
3	100	31
4	100	18
5	100	24
6	100	12
7	100	23
8	100	15
9	100	8
10	100	8

6.10 Why is the *np* chart not appropriate with variable sample sizes?

6.11 A process that produces bearing housings is controlled with a fraction nonconforming control chart, using sample size $n = 100$ and a center line $\bar{p} = 0.02$.

a. Find the three-sigma limits for this chart.
b. Analyze the ten new samples ($n = 100$) shown in Table 6E.8 for statistical control. What conclusions can you draw about the process now?

Table 6E.8

Data for Exercise 6.11

Sample Number	Number Nonconforming	Sample Number	Number Nonconforming
1	5	6	1
2	2	7	2
3	3	8	6
4	8	9	3
5	4	10	4

6.12 Consider the fraction nonconforming control chart in Exercise 6.4. Find the equivalent *np* chart.

6.13 Consider the fraction nonconforming control chart in Exercise 6.5. Find the equivalent *np* chart.

6.14 Surface defects have been counted on 25 rectangular steel plates, and the data are shown in Table 6E.9. Set up a control chart for nonconformities using these data. Does the process producing the plates appear to be in statistical control?

Table 6E.9

Data for Exercise 6.14

Plate Number	Number of Nonconformities	Plate Number	Number of Nonconformities
1	1	14	0
2	0	15	2
3	4	16	1
4	3	17	3
5	1	18	5
6	2	19	4
7	5	20	6
8	0	21	3
9	2	22	1
10	1	23	0
11	1	24	2
12	0	25	4
13	8		

6.15 A paper mill uses a control chart to monitor the imperfection in finished rolls of paper. Production output is inspected for 20 days, and the resulting data are shown in Table 6E.10. Use these data to set up a control chart for nonconformities per roll of paper. Does the process appear to be in statistical control? What center line and control limits would you recommend for controlling current production?

Table 6E.10

Data for Exercise 6.15

Day	Number of Rolls Produced	Total Number of Imperfections	Day	Number of Rolls Produced	Total Number of Imperfections
1	18	12	11	18	18
2	18	14	12	18	14
3	24	20	13	18	9
4	22	18	14	20	10
5	22	15	15	20	14
6	22	12	16	20	13
7	20	11	17	24	16
8	20	15	18	24	18
9	20	12	19	22	20
10	20	10	20	21	17

6.16 Consider the papermaking process in Exercise 6.15. Set up a u chart based on an average sample size to control this process.

6.17 The number of nonconformities found on final inspection of a tape deck is shown in Table 6E.11. Can you conclude that the process is in statistical control?

Table 6E.11

Data for Exercise 6.17

Deck Number	Number of Nonconformities	Deck Number	Number of Nonconformities
2412	0	2421	1
2413	1	2422	0
2414	1	2423	3
2415	0	2424	2

(Continued)

Table 6E.11

Continued

Deck Number	Number of Nonconformities	Deck Number	Number of Nonconformities
2416	2	2425	5
2417	1	2426	1
2418	1	2427	2
2419	3	2428	1
2420	2	2429	1

What center line and control limits would you recommend for controlling future production?

6.18 The data in Table 6E.12 represent the number of nonconformities per 1,000 meters of telephone cable. From an analysis of these data, would you conclude that the process is in statistical control. What control limits would you recommend for future production?

Table 6E.12

Data for Exercise 6.18

Sample Number	Number of Nonconformities	Sample Number	Number of Nonconformities
1	1	12	6
2	1	13	9
3	3	14	11
4	7	15	15
5	8	16	8
6	10	17	3
7	5	18	6
8	13	19	7
9	0	20	4
10	19	21	9
11	24	22	20

6.19 Consider the data in Exercise 6.17. Suppose we wish to define a new inspection unit of four tape decks.

a. What are the center line and control limits for a control chart for monitoring future production based on the total number of defects in the new inspection unit?
b. What are the center line and control limits for a control chart for nonconformities per unit used to monitor future production?

6.20 Consider the data in Exercise 6.18. Suppose a new inspection unit is defined as 2,500 meters of wire.

a. What are the center line and control limits for a control chart for monitoring future production based on the total number of nonconformities in the new inspection unit?
b. What are the center line and control limits for a control chart for average nonconformities per unit used to monitor future production?

6.21 An automobile manufacturer wishes to control the number of nonconformities in a subassembly area producing manual transmissions. The inspection unit is defined as four transmissions, and data from 16 samples (each of size 4) are shown in Table 6E.13.

a. Set up a control chart for nonconformities per unit.

Table 6E.13

Data for Exercise 6.21

Sample Number	Number of Nonconformities	Sample Number	Number of Nonconformities
1	1	9	2
2	3	10	1
3	2	11	0
4	1	12	2
5	0	13	1
6	2	14	1
7	1	15	2
8	5	16	3

b. Do these data come from a controlled process? If not, assume that assignable causes can be found for all out-of-control points and calculate the revised control chart parameters.
c. Suppose the inspection unit is redefined as eight transmissions. Design an appropriate control chart for monitoring future production.

6.22 The number of workmanship nonconformities observed in the final inspection of disk-drive assemblies has been tabulated as shown in Table 6E.14. Does the process appear to be in control?

Table 6E.14

Data for Exercise 6.22

Day	Number of Assemblies Inspected	Total Number of Imperfections	Day	Number of Assemblies Inspected	Total Number of Imperfections
1	2	10	6	4	24
2	4	30	7	2	15
3	2	18	8	4	26
4	1	10	9	3	21
5	3	20	10	1	8

6.23 The manufacturer wishes to set up a control chart at the final inspection station for a gas water heater. Defects in workmanship and visual quality features are checked in this inspection. For the past 22 working days, 176 water heaters were inspected and a total of 924 nonconformities reported.

a. What type of control chart would you recommend here and how would you use it?
b. Using two water heaters as the inspection unit, calculate the center line and control limits that are consistent with the past 22 days of inspection data.

6.24 Assembled portable television sets are subjected to a final inspection for surface defects. A total procedure is established based on the requirement that the average number of nonconformities per unit be 4.0 units. What is the appropriate type of control chart?

6.25 A control chart for nonconformities is to be established in conjunction with the final inspection of a radio. The inspection unit is to be a group of ten radios. The average number of nonconformities per radio has, in the past, been 0.5. Find three-sigma control limits for a c chart based on this size of inspection unit.

6.26 A production line assembles electric clocks. The average number of nonconformities per clock is estimated to be 0.75. The quality engineer wishes to establish a c chart for this operation using an inspection unit of six clocks. Find the three-sigma limits for this chart.

6.27 Kittlitz (1999) presents data on homicides in Waco, Texas, for the years 1980–1989 (data taken from the *Waco Tribune-Herald,* December 29, 1989). There were 29 homicides in 1989. Table 6E.15 gives

Table 6E.15

Data for Exercise 6.27

Month	Date	Days Between	Month	Date	Days Between
Jan.	20		July	8	2
Feb.	23	34	July	9	1
Feb.	25	2	July	26	17
March	5	8	Sep.	9	45
March	10	5	Sep.	22	13
April	4	25	Sep.	24	2
May	7	33	Oct.	1	7
May	24	17	Oct.	4	3
May	28	4	Oct.	8	4
June	7	10	Oct.	19	11
June	16*	9.25	Nov.	2	14
June	16*	0.50	Nov.	25	23
June	22	5.25	Dec.	28	33
June	25	3	Dec.	29	1
July	6	11			

the dates of the 1989 homicides and the number of days between each homicide. The asterisks refer to the fact that two homicides occurred on June 16 and were determined to have occurred 12 hours apart.

a. Plot the days-between-homicides data on a normal probability plot. Does the assumption of a normal distribution seem reasonable for these data?
b. Transform the data using the 0.2777 root of the data. Plot the transformed data on a normal probability plot. Does this plot indicate that the transformation has been successful in making the new data more closely resemble data from a normal distribution?
c. Transform the data using the fourth root (0.25) of the data. Plot the transformed data on a normal probability plot. Does this plot indicate that the transformation has been successful in making the new data resemble more closely data from a normal distribution? Is the plot very different from the one in part (b)?
d. Construct an individual control chart using the transformed data from part (b).
e. Construct an individual control chart using the transformed data from part (c). How similar is it to the one you constructed in part (d)?
f. Is the process stable? Provide a practical interpretation of the control chart.

6.28 Suggest at least two nonmanufacturing scenarios in which attributes control charts could be useful for process monitoring.

6.29 What practical difficulties could be encountered in monitoring the days-between-events data?

6.30 A paper by R. N. Rodriguez ("Health Care Applications of Statistical Process Control: Examples Using the SAS® System," *SAS Users Group International: Proceedings of the 21st Annual Conference*, 1996) illustrated several informative applications of control charts to the health care environment. One of these showed how a control chart was employed to analyze the rate of CAT scans performed each month at a clinic. The data used in this example are shown in Table 6E.16.

NSCANB is the number of CAT scans performed each month and MMSB is the number of members enrolled in the health care plan each month, in units of member months. "Days" is the number of days in each month. The variable NYRSB converts MMSB to units of thousand members per year, and is computed as follows: NYRSB = MMSB(days/30)/12000. NYRSB represents the "area of opportunity."

Table 6E.16

Data for Exercise 6.30

Month	NSCANB	MMSB	Days	NYRSB
Jan. 94	50	26838	31	2.31105
Feb. 94	44	26903	28	2.09246
March 94	71	26895	31	2.31596
Apr. 94	53	26289	30	2.19075
May 94	53	26149	31	2.25172
Jun. 94	40	26185	30	2.18208
July 94	41	26142	31	2.25112
Aug. 94	57	26092	31	2.24681
Sept. 94	49	25958	30	2.16317
Oct. 94	63	25957	31	2.23519
Nov. 94	64	25920	30	2.16000
Dec. 94	62	25907	31	2.23088
Jan. 95	67	26754	31	2.30382
Feb. 95	58	26696	28	2.07636
March 95	89	26565	31	2.28754

Construct an appropriate control chart to monitor the rate at which CAT scans are performed at this clinic.

6.31 A paper by R. N. Rodriguez ("Health Care Applications of Statistical Process Control: Examples Using the SAS® System," *SAS Users Group International: Proceedings of the 21st Annual Conference*, 1996) illustrated several informative applications of control charts to the health care environment. One of these showed how a control chart was employed to analyze the number of office visits by health care plan members. The data for clinic E are shown in Table 6E.17.

Table 6E.17

Data for Exercise 6.31

Month	Phase	NVISITE	NYRSE	Days	MMSE
Jan. 94	1	1421	0.66099	31	7676
Feb. 94	1	1303	0.59718	28	7678
Mar. 94	1	1569	0.66219	31	7690
Apr. 94	1	1576	0.64608	30	7753
May 94	1	1567	0.66779	31	7755
Jun. 94	1	1450	0.65575	30	7869
July 94	1	1532	0.68105	31	7909
Aug. 94	1	1694	0.68820	31	7992
Sep. 94	2	1721	0.66717	30	8006
Oct. 94	2	1762	0.69612	31	8084
Nov. 94	2	1853	0.68233	30	8188
Dec. 94	2	1770	0.70809	31	8223
Jan. 95	2	2024	0.78215	31	9083
Feb. 95	2	1975	0.70684	28	9088
Mar. 95	2	2097	0.78947	31	9168

The variable NVISITE is the number of visits to clinic E each month, and MMSE is the number of members enrolled in the health care plan each month, in units of member months. "Days" is the number of days in each month. The variable NYRSE converts MMSE to units of thousand members per year and is computed as follows: NYRSE = MMSE(days/30)/12000. NYRSE represents the "area of opportunity." The variable Phase separates the data into two time periods.

a. Use the data from P1 to construct a control chart for monitoring the rate of office visits performed at clinic E. Does this chart exhibit control?

b. Plot the data from P2 on the chart constructed in part (a). Is there a difference in the two phases?

c. Consider only the P2 data. Do these data exhibit control?

6.32 The data in Table 6E.18 are the number of information errors found in customer records in a marketing company database. Five records were sampled each day. Set up a c chart for the total number of errors. Is the process in control?

Table 6E.18

Data for Exercise 6.32

Day	Record 1	Record 2	Record 3	Record 4	Record 5
1	8	7	1	11	17
2	11	1	11	2	9
3	1	1	8	2	5
4	3	2	5	1	4
5	3	2	13	6	5
6	6	3	3	3	1
7	8	8	2	1	5
8	4	10	2	6	4
9	1	6	1	3	2
10	15	1	3	2	8
11	1	7	13	5	1
12	6	7	9	3	1
13	7	6	3	3	1
14	2	9	3	8	7
15	6	14	7	1	8
16	2	9	4	2	1
17	11	1	1	3	2
18	5	5	19	1	3
19	6	15	5	6	6
20	2	7	9	2	8
21	7	5	6	14	10
22	4	3	8	1	2
23	4	1	4	20	5
24	15	2	7	10	17
25	2	15	3	11	2

Table 6E.19

Data for Exercise 6.33

Truck	No. of Orders	Truck	No. of Orders	Truck	No. of Orders	Truck	No. of Orders
1	22	9	5	17	8	25	6
2	58	10	26	18	35	26	13
3	7	11	12	19	6	27	9
4	39	12	26	20	23	28	21
5	7	13	10	21	10	29	8
6	33	14	30	22	17	30	12
7	8	15	5	23	7	31	4
8	23	16	24	24	10	32	18

6.33 Kaminski et al. (1992) present data on the number of orders per truck at a distribution center. Some of this data is shown in Table 6E.19. Set up a c chart for the number of orders per truck. Is the process in control?

chapter Seven Lot-by-Lot Acceptance-Sampling Procedures

Chapter Outline

Ensuring Quality Standards through Acceptance-Sampling

In September 2008, it was discovered that China-based dairy farmers and distributors had been adding melamine (a plastics manufacturing byproduct) to milk to falsely inflate protein readings. They would water down the milk, add melamine, and pass the protein tests.

Why were they able to pass the tests? While the tests ensured that protein levels were sufficient, they did *not* ensure that all ingredients were unadulterated. As such, the tainted milk quickly made its way into such products as candies, cakes, and baby formula.

Upon discovering this serious error, the Chinese government ordered a recall of all dairy products produced after September 2008. This action was taken because melamine causes kidney stones, particularly in young children, whose anatomy is smaller. The kidney stones have sharp edges that can cause severe damage when they pass. In fact, as a result of this particular episode, 300,000 children were sickened, 50,000 were admitted to the hospital, and six died. More than 20 million children were tested for kidney stones. The testing was urgent and choked hospitals as worried parents rushed to have their children examined.

To ensure top-quality milk going forward, the Chinese government took three bold actions: issued a list of banned food additives, overhauled the industry to move it away from local farmers and toward mass production, and increased testing on banned substances.

It is important to note, however, that acceptance-sampling can only confirm quality characteristics of those items that are tested. It *cannot* confirm overall quality. Therefore, companies utilizing this approach must put as much effort into selecting the testing attributes as they do into designing the acceptance plan.

Chapter Overview and Learning Objectives

This chapter presents lot-by-lot acceptance-sampling plans for both attributes and variables quality characteristics. Key topics include the design and operation of single-sampling plans, the use of the operating characteristic curve, and the concepts of rectifying inspection, average outgoing quality, and average total inspection. Similar concepts are briefly introduced for types of sampling plans where more than one sample may be taken to determine the disposition of a lot (double, multiple, and sequential-sampling). Three systems of standard sampling plans are also presented: the military standard plans known as MIL STD 105E and the Dodge–Romig plans for attributors and MIL STD 414 for variables. These plans are designed around different philosophies: MIL STD 105E and MIL STD 414 have an acceptable quality level focus, whereas the Dodge–Romig plans are oriented around either the lot tolerance percent defective or the average outgoing quality limit perspective.

After careful study of this chapter, you should be able to do the following:

1. Understand the role of acceptance-sampling in modern quality control systems
2. Understand the advantages and disadvantages of sampling
3. Understand the difference between attributes and variables sampling plans and the major types of acceptance-sampling procedures
4. Know how single-, double-, and sequential-sampling plans are used
5. Understand the importance of random-sampling
6. Know how to determine the OC curve for a single-sampling plan for attributes
7. Understand the effects of the sampling plan parameters on sampling plan performance
8. Know how to design single-, double-, and sequential-sampling plans for attributes
9. Know how rectifying inspection is used
10. Understand the structure and use of MIL STD 105E and its civilian counterpart plans
11. Understand the structure and use of the Dodge–Romig system of sampling plans
12. Understand the structure and use of MIL STD 414 and its civilian counterpart plans

7.1 The Acceptance-Sampling Problem

As we observed in Chapter 1, acceptance-sampling is concerned with inspection and decision making regarding products, one of the oldest aspects of quality assurance. In the 1930s and 1940s, acceptance-sampling was one of the major components of the field of statistical quality control and was used primarily for incoming or receiving inspection. In more recent years, it has become typical to work with suppliers to improve their process performance through the use of SPC and designed experiments and not to rely as much on acceptance-sampling as a primary quality assurance tool.

A typical application of acceptance-sampling is as follows: A company receives a shipment of product from a supplier. This product is often a component or raw material used in the company's manufacturing process. A sample is taken from the lot, and some quality characteristic of the units in the sample is inspected. On the basis of the information in this sample, a decision is made regarding **lot disposition.** Usually, this decision is either to accept or to reject the lot. Sometimes we refer to this decision as **lot sentencing.** Accepted lots are put into production; rejected lots may be returned to the supplier or may be subjected to some other **lot disposition action.**

The purpose of acceptance-sampling is to disposition or sentence lots

Although it is customary to think of acceptance-sampling as a receiving inspection activity, there are other uses of sampling methods. For example, frequently a manufacturer will sample and inspect its own product at various stages of production. Lots that are accepted are sent forward for further processing, and rejected lots may be reworked or scrapped.

Three aspects of sampling are important:

1. It is the purpose of acceptance-sampling to sentence lots, not to estimate the lot quality. Most acceptance-sampling plans are not designed for estimation purposes.
2. Acceptance-sampling plans do not provide any *direct* form of quality control. Acceptance-sampling simply accepts and rejects lots. Even if all lots are of the same quality, sampling will accept some lots and reject others, the accepted lots being no better than the rejected ones. Process controls are used to control and systematically improve quality, but acceptance-sampling does not.
3. The most effective use of acceptance-sampling is *not* to "inspect quality into the product," but rather as an audit tool to ensure that the output of a process conforms to requirements.

Acceptance-sampling is an indirect form of quality control

Generally, there are three approaches to lot sentencing: (1) accept with no inspection; (2) 100% inspection—that is, inspect every item in the lot, removing all defective units found (defectives may be returned to the supplier, reworked, replaced with known good items, or discarded); and (3) acceptance-sampling. The no-inspection alternative is useful in situations where either the supplier's process is so good that defective units are almost never encountered or where there is no economic justification to look for defective units. For example, if the supplier's process capability ratio is 3 or 4, acceptance-sampling is unlikely to discover any defective units. We generally use 100% inspection in situations where the component is extremely critical and

In previous chapters, the terms "nonconforming" and "nonconformity" were used instead of "defective" and "defect". This is because the popular meanings of "defective" and "defect" differ from their technical meanings and have caused considerable misunderstanding, particularly in product liability litigation. In the field of sampling inspection, however, "defective" and "defect" continue to be used in their technical sense—that is, nonconformance to requirements.

Acceptance-sampling is most likely to be useful in the following situations:

1. When testing is destructive.
2. When the cost of 100% inspection is extremely high.
3. When 100% inspection is not technologically feasible or would require so much calendar time that production scheduling would be seriously impacted.
4. When there are many items to be inspected and the inspection error rate is sufficiently high that 100% inspection might cause a higher percentage of defective units to be passed than would occur with the use of a sampling plan.
5. When the supplier has an excellent quality history and some reduction in inspection from 100% is desired but the supplier's process capability is sufficiently low as to make no inspection an unsatisfactory alternative.
6. When there are potentially serious product liability risks and although the supplier's process is satisfactory, a program for continuously monitoring the product is necessary.

passing any defectives would result in an unacceptably high failure cost at subsequent stages, or where the supplier's process capability is inadequate to meet specifications.

7.1.1 ADVANTAGES AND DISADVANTAGES OF SAMPLING

When acceptance-sampling is contrasted with 100% inspection, it has the following advantages:

1. It is usually less expensive, because there is less inspection.
2. There is less handling of the product, and thus reduced damage.
3. It is applicable to destructive testing.
4. Fewer personnel are involved in inspection activities.
5. It often greatly reduces the amount of inspection error.
6. The rejection of entire lots as opposed to the simple return of defectives often provides a stronger motivation to the supplier for quality improvements.

Acceptance-sampling also has several disadvantages, however. These include the following:

1. There are risks of accepting "bad" lots and rejecting "good" lots.
2. Less information is usually generated about the product or the process that manufactured the product.
3. Acceptance-sampling requires planning and documentation of the acceptance-sampling procedure, whereas 100% inspection does not.

Although this last point is often mentioned as a disadvantage of acceptance-sampling, proper design of an acceptance-sampling plan usually requires study of the actual level of quality required by the consumer. This resulting knowledge is often a useful input into the overall quality planning and engineering process. Thus, in many applications, it may not be a significant disadvantage.

We have pointed out that acceptance-sampling is a middle ground between the extremes of 100% inspection and no inspection. It often provides a methodology for moving between these extremes as sufficient information is obtained on the control of the manufacturing process that produces the product. Although there is no direct control of quality in the application of an acceptance-sampling plan to an isolated lot, when that plan is applied to a stream of lots from a supplier, it becomes a means of providing protection for both the producer of the lot and the consumer. It also provides for an accumulation of quality history regarding the process that produces the lot, and it may provide feedback that is useful in process control, such as determining when process controls at the supplier's plant are not adequate. Finally, it may place economic or psychological pressure on the supplier to improve the production process.

7.1.2 TYPES OF SAMPLING PLANS

There are a number of different ways to classify acceptance-sampling plans. One major classification is by variables and attributes. **Variables,** of course, are quality characteristics that are measured on a numerical scale. **Attributes** are quality characteristics that are expressed on a "go, no-go" basis.

A **single-sampling plan** is a lot-sentencing procedure in which one sample of n units is selected at random from the lot, and the disposition of the lot is determined based on the information contained in that sample (see Figure 7.1a). For example, a single-sampling plan for attributes would consist of a sample size n and an acceptance number c. The procedure would operate as follows: Select n items at random from the lot. If there are c or fewer defectives in the sample, accept the lot, and if there are more than c defective items in the sample, reject the lot. We investigate this type of sampling plan extensively in Section 7.2.

Double-sampling plans are somewhat more complicated. Following an initial sample, a decision based on the information in that sample is made either to (1) accept the lot, (2) reject the lot, or (3) take a second sample. If the second sample is taken, the information from both the first and second sample is combined in order to reach a decision whether to accept or reject the lot (see Figure 7.1b). Double-sampling plans are discussed in Section 7.3.

A **multiple-sampling plan** is an extension of the double-sampling concept, in that more than two samples may be required in order to reach a decision regarding the disposition of the lot (see Figure 7.1c). Sample sizes in multiple-sampling are usually smaller than they are in either single or double-sampling. The ultimate extension of multiple-sampling is **sequential-sampling,** in which units are selected from the lot one at a time, and following inspection of each unit, a decision is made either to accept the lot, reject the lot, or select another unit (see Figure 7.1d). Multiple- and sequential-sampling plans are also discussed in Section 7.3.

Single-, double-, multiple-, and sequential-sampling plans can be designed so that they produce equivalent results. That is, these procedures can be designed so that a lot of specified quality has exactly the same probability of acceptance under all four types of sampling plans. Consequently, when selecting the type of sampling procedure, one must consider factors such as the administrative efficiency, the type of information produced by the plan, the average amount of inspection required by the procedure, and the impact that a given procedure may have on the material flow in the manufacturing organization. These issues are discussed in more detail in Section 7.3.

7.1.3 LOT FORMATION

How the lot is formed can influence the effectiveness of the acceptance-sampling plan. There are a number of important considerations in forming lots for inspection. Some of these are as follows:

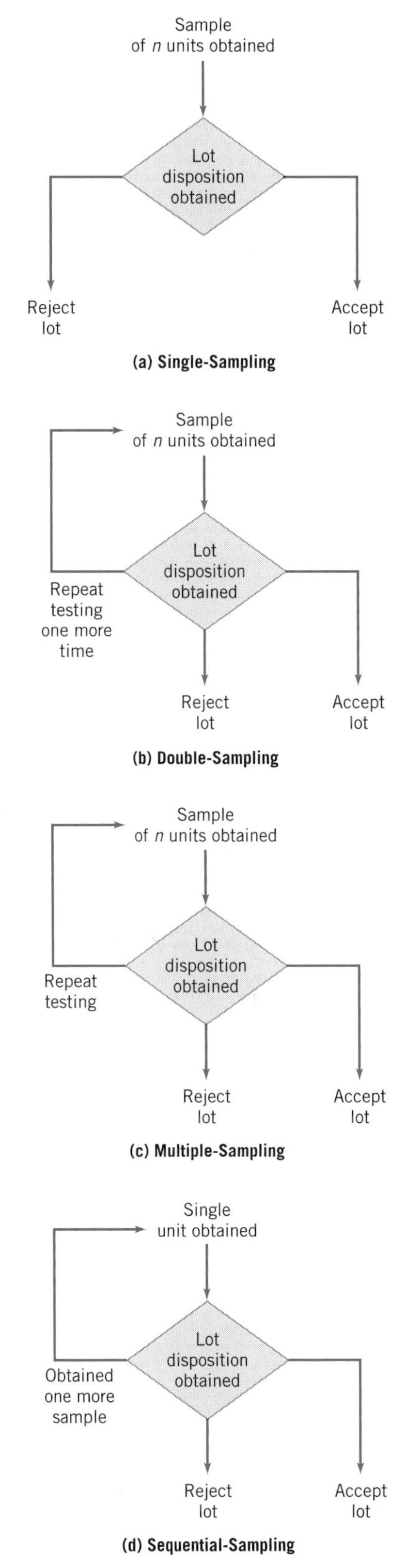

Figure 7.1 Types of Sampling Plans

1. **Lots should be homogeneous.**	The units in the lot should be produced by the same machines, the same operators, and from common raw materials at approximately the same time. When lots are nonhomogeneous, such as when the output of two different production lines is mixed, the acceptance-sampling scheme may not function as effectively as it could. Nonhomogeneous lots also make it more difficult to take corrective action to eliminate the source of defective products.
2. **Larger lots are preferred over smaller ones.**	It is usually more economically efficient to inspect large lots than small ones.
3. **Lots should be conformable to the materials-handling systems used in both the supplier and consumer facilities.**	In addition, the items in the lots should be packaged so as to minimize shipping and handling risks, and to make selection of the units in the sample relatively easy.

7.1.4 RANDOM-SAMPLING

The units selected for inspection from the lot should be chosen at random, and they should be representative of all the items in the lot. The random-sampling concept is extremely important in acceptance-sampling. Unless random samples are used, bias will be introduced. For example, the supplier may ensure that the units packaged at the top of the lot are of extremely good quality, knowing that the inspector will select the sample from the top layer. "Salting" a lot in this manner is not a common practice, but if it occurs and nonrandom-sampling methods are used, the effectiveness of the inspection process is destroyed.

The technique often suggested for drawing a random sample is to first assign a number to each item in the lot. Then n random numbers are drawn, where the range of these numbers is from 1 to the maximum number of units in the lot. This sequence of random numbers determines which units in the lot will constitute the sample. If products have serial or other code numbers, these numbers can be used to avoid the process of actually assigning numbers to each unit. Another possibility would be to use a three-digit random number to represent the length, width, and depth in a container.

In situations where we cannot assign a number to each unit, utilize serial or code numbers, or randomly determine the location of the sample unit, some other technique must be employed to ensure that the sample is random or representative. Sometimes the inspector may **stratify** the lot. This consists of dividing the lot into strata or layers and then subdividing each strata into cubes, as shown in Figure 7.2. Units are then selected from within each cube. Although this stratification of the lot is usually an imaginary activity performed by the inspector and does not necessarily ensure random samples, at least it ensures that units are selected from all locations in the lot.

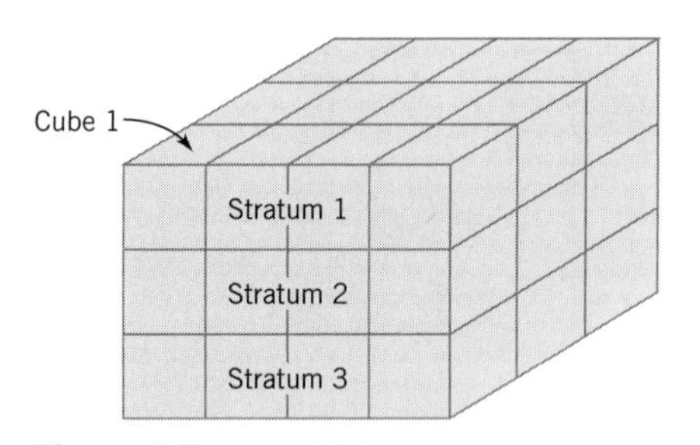

Figure 7.2 **Stratifying a Lot**

We cannot overemphasize the importance of random-sampling. If judgment methods are used to select the sample, the statistical basis of the acceptance-sampling procedure is lost.

7.1.5 GUIDELINES FOR USING ACCEPTANCE-SAMPLING

An acceptance-sampling plan is a statement of the sample size to be used and the associated acceptance or rejection criteria for sentencing individual lots. A sampling scheme is defined as a set of procedures consisting of acceptance-sampling plans in which lot sizes, sample sizes, and acceptance or rejection criteria along with the amount of 100% inspection and sampling are related. Finally, a sampling system is a unified collection of one or more acceptance-sampling schemes. In this chapter, we see examples of sampling plans, sampling schemes, and sampling systems.

The major types of acceptance-sampling procedures and their applications are shown in Table 7.1. In general, the selection of an acceptance-sampling procedure depends on both the objective of the sampling organization and the history of the organization whose product is sampled. Furthermore, the application of sampling methodology is not static; that is, there is a natural evolution from one level of sampling effort to another. For example, if we are dealing with a supplier who enjoys an excellent quality history, we might begin with an attributes sampling plan. As our experience with the supplier grows, and its good-quality reputation is proved by the results of our sampling activities, we might transition to a sampling procedure that requires much less inspection, such as skip-lot sampling. Finally, after extensive experience with the supplier, and if its process capability is extremely good, we might stop all acceptance-sampling activities on the product. In another situation, where we have little knowledge of or experience with the supplier's quality-assurance efforts, we might begin with attributes sampling using a plan that assures us that the quality of accepted lots is no worse than a specified target value. If this plan proves successful, and if the supplier's performance is satisfactory, we might transition from attributes to

Acceptance-Sampling Plan
Statement of the sample size to be used and the associated acceptance or rejection criteria for sentencing individual lots.

Sampling Scheme
Set of procedures consisting of acceptance-sampling plans in which lot sizes, sample sizes, and acceptance or rejection criteria along with the amount of 100% inspection and sampling are related.

Sampling System
A unified collection of one or more acceptance-sampling schemes.

Table 7.1

Acceptance-Sampling Procedures

Objective	Attributes Procedure	Variables Procedure
Assure quality levels for consumer/producer	Select plan for specific OC curve	Select plan for specific OC curve
Maintain quality at a target	AQL system; MIL STD 105E, ANSI/ASQC Z1.4	AQL system; MIL STD 414, ANSI/ASQC Z1.9
Assure average outgoing quality level	AOQL system; Dodge–Romig plans	AOQL system
Reduce inspection, with small sample sizes, good-quality history	Chain sampling	Narrow-limit gauging
Reduce inspection after good-quality history	Skip-lot sampling; double-sampling	Skip-lot sampling; double-sampling
Assure quality no worse than target	LTPD plan; Dodge–Romig plans	LTPD plan; hypothesis testing

variables inspection, particularly as we learn more about the nature of the supplier's process. Finally, we might use the information gathered in variables sampling plans in conjunction with efforts directly at the supplier's manufacturing facility to assist in the installation of process controls. A successful program of process controls at the supplier level might improve the supplier's process capability to the point where inspection could be discontinued.

These examples illustrate that there is a life cycle of application of acceptance-sampling techniques. Typically, we find that organizations with relatively new quality-assurance efforts place a great deal of reliance on acceptance-sampling. As their maturity grows and the quality organization develops, they begin to rely less on acceptance-sampling and more on statistical process control and experimental design.

Manufacturers try to improve the quality of their products by reducing the number of suppliers from whom they buy their components, and by working more closely with the ones they retain. Once again, the key tool in this effort to improve quality is statistical process control. Acceptance-sampling can be an important ingredient of any quality-assurance program; however, remember that it is an activity that you try to avoid doing. It is much more cost effective to use statistically based process monitoring at the appropriate stage of the manufacturing process. Sampling methods can in some cases be a tool that you employ along the road to that ultimate goal.

7.2 Single-Sampling Plans for Attributes

7.2.1 DEFINITION OF A SINGLE-SAMPLING PLAN

Key Variables in Sampling Plan

N = lot size

n = number inspected

d = number of defectives observed

c = max #observed defects

Suppose that a lot of size N has been submitted for inspection. A **single-sampling plan** is defined by the sample size n and the acceptance number c. Thus, if the lot size is $N = 10{,}000$, then the sampling plan

$$n = 89$$
$$c = 2$$

means that from a lot of size 10,000 a random sample of $n = 89$ units is inspected and the number of nonconforming or defective items d observed. If the number of observed defectives d is less than or equal to $c = 2$, the lot will be accepted. If the number of observed defectives d is greater than $c = 2$, the lot will be rejected. Since the quality characteristic inspected is an attribute, each unit in the sample is judged to be either conforming or nonconforming. One or several attributes can be inspected in the same sample; generally, a unit that is nonconforming to specifications on one or more attributes is said to be a defective unit. This procedure is called a single-sampling plan because the lot is sentenced based on the information contained in one sample of size n.

7.2.2 THE OC CURVE

An important measure of the performance of an acceptance-sampling plan is the **operating-characteristic (OC) curve.** This curve plots the probability of accepting the lot versus the lot fraction defective. Thus, the OC curve displays the discriminatory power of the sampling plan. That is, it shows the probability that a lot submitted with a certain fraction defective will be either accepted or rejected. The OC curve of the sampling plan $n = 89$, $c = 2$ is shown in Figure 7.3. It is easy to demonstrate how the points on this curve are obtained. Suppose that the lot size N is large (theoretically infinite). Under this condition, the distribution of the number of defectives d in a random sample of n items is binomial with parameters n and p, where p is the fraction of defective items in the lot. An equivalent way to conceptualize this is to draw lots of N items at random from a theoretically infinite process, and then to draw random samples of n from these lots. Sampling from the lot in this manner is the equivalent of sampling directly from the process. The probability of observing exactly d defectives is

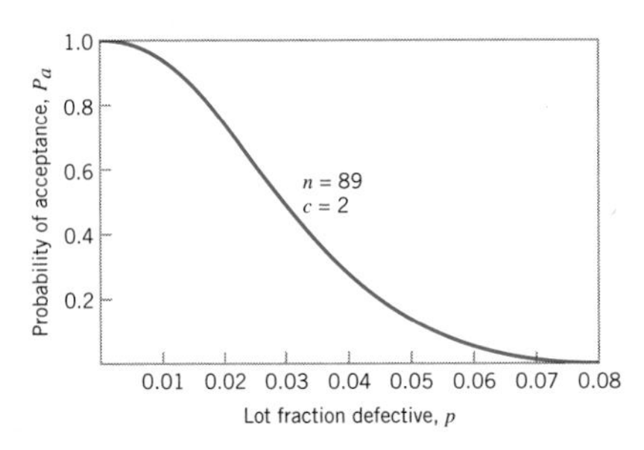

Figure 7.3 **OC Curve of the Single-Sampling Plan $n = 89$, $c = 2$**

$$P\{d \text{ defectives}\} = f(d) = \frac{n!}{d!(n-d)!}p^d(1-p)^{n-d} \tag{7.1}$$

The probability of acceptance is simply the probability that d is less than or equal to c, or

$$P_a = P\{d \leq c\} = \sum_{d=0}^{c}\frac{n!}{d!(n-d)!}p^d(1-p)^{n-d} \tag{7.2}$$

For example, if the lot fraction defective is $p = 0.01$, $n = 89$, and $c = 2$, then

$$P_a = P\{d \leq 2\} = \sum_{d=0}^{2}\frac{89!}{d!(89-d)!}(0.01)^d(0.99)^{89-d}$$

$$= \frac{89!}{0!89!}(0.01)^0(0.99)^{89} + \frac{89!}{1!88!}(0.01)^1(0.99)^{88} + \frac{89!}{2!(87)!}(0.01)^2(0.99)^{87}$$

$$= 0.9397$$

The OC curve is developed by evaluating Equation 7.2 for various values of p. Table 7.2 displays the calculated value of several points on the curve. Minitab and Excel can be used to calculate these binomial probabilities. See Appendix One.

The OC curve shows the **discriminatory power** of the sampling plan. For example, in the sampling plan $n = 89$, $c = 2$, if the lots are 2% defective, the probability of acceptance is approximately 0.74. This means that if 100 lots from a process that manufactures 2% defective product are submitted to this sampling plan, we will expect to accept 74 of the lots and reject 26 of them.

Table 7.2

Probabilities of Acceptance for the Single-Sampling Plan $n = 89$, $c = 2$

Fraction Defective, p	Probability of Acceptance, P_a
0.005	0.9897
0.010	0.9397
0.020	0.7366
0.030	0.4985
0.040	0.3042
0.050	0.1721
0.060	0.0919
0.070	0.0468
0.080	0.0230
0.090	0.0109

EFFECT OF n AND c ON OC CURVES. A sampling plan that discriminated perfectly between good and bad lots would have an OC curve that looks like Figure 7.4. The OC curve runs horizontally at a probability of acceptance $P_a = 1.00$ until a level of lot quality that is considered "bad" is reached, at which point the curve drops vertically to a probability of acceptance $P_a = 0.00$, and then the curve runs horizontally again for all lot fraction defectives greater than the undesirable level. If such a sampling plan could be employed, all lots of "bad" quality would be rejected, and all lots of "good" quality would be accepted.

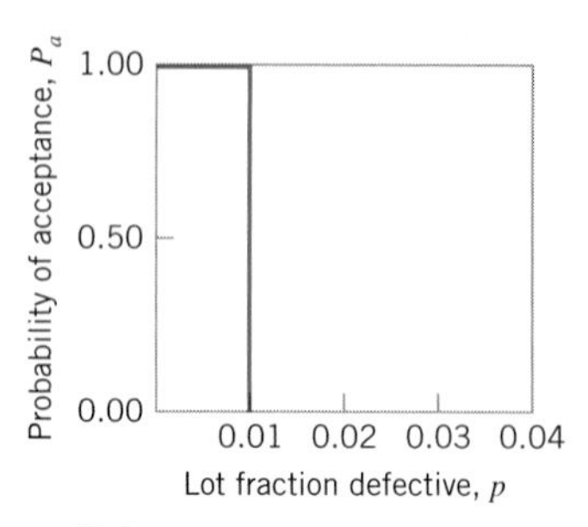

Figure 7.4 Ideal OC Curve

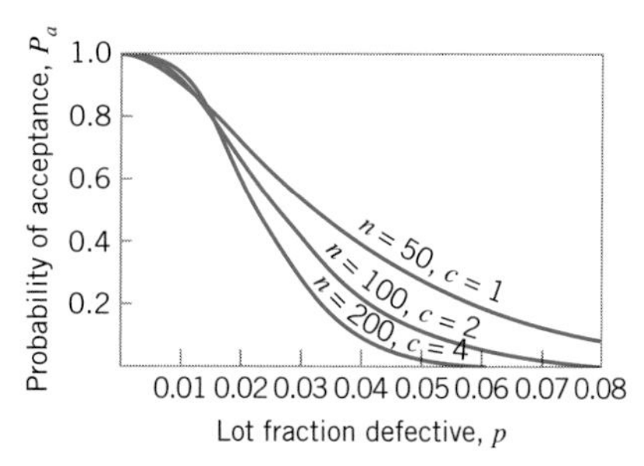

Figure 7.5 OC Curves for Different Sample Sizes

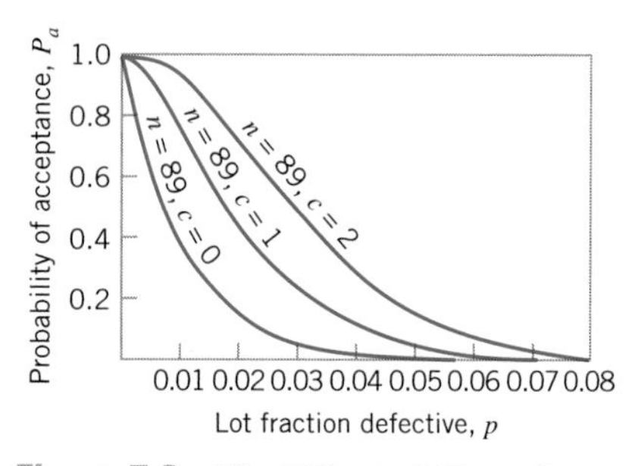

Figure 7.6 The Effect of Changing the Acceptance Number on the OC Curve

Unfortunately, the **ideal OC curve** in Figure 7.4 can almost never be obtained in practice. In theory, it could be realized by 100% inspection, if the inspection were error-free. The ideal OC curve shape can be approached, however, by increasing the sample size. Figure 7.5 shows that the OC curve becomes more like the idealized OC curve shape as the sample size increases. (Note that the acceptance number c is kept proportional to n.) Thus, the precision with which a sampling plan differentiates between good and bad lots increases with the size of the sample. The greater the slope of the OC curve, the greater the discriminatory power.

Figure 7.6 shows how the OC curve changes as the acceptance number changes. Generally, changing the acceptance number does not dramatically change the slope of the OC curve. As the acceptance number is decreased, the OC curve is shifted to the left. Plans with smaller values of c provide discrimination at lower levels of lot fraction defective than do plans with larger values of c.

SPECIFIC POINTS ON THE OC CURVE. Frequently, the quality engineer's interest focuses on certain points on the OC curve. The supplier is usually interested in knowing what level of lot or process quality would yield a high probability of acceptance. For example, the supplier might be interested in the 0.95 probability of acceptance point. This would indicate the level of process fallout that could be experienced and still have a 95% chance that the lots would be accepted. Conversely, the consumer might be interested in the other end of the OC curve. That is, what level of lot or process quality will yield a low probability of acceptance?

A consumer often establishes a sampling plan for a continuing supply of components or raw material with reference to an **acceptable quality level (AQL).** The AQL represents the poorest level of quality for the supplier's process that the consumer would consider to be acceptable as a process average. Note that the AQL is a property of the supplier's manufacturing process; it is not a property of the sampling plan. The consumer will often design the sampling procedure so that the OC curve gives a high probability of acceptance at the AQL. Furthermore, the AQL is not usually intended to be a specification on the product, nor is it a target value for the supplier's production process. It is simply a standard against which to judge the lots. It is hoped that the supplier's process will operate at a fallout level that is considerably better than the AQL.

Average Quality Level (AQL)
Poorest level of quality for the supplier's process that the consumer would consider to be acceptable as a process average.

The consumer will also be interested in the other end of the OC curve—that is, in the protection that is obtained for individual lots of poor quality. In such a situation, the consumer may establish a **lot tolerance percent defective (LTPD).** The LTPD is the poorest level of quality that the consumer is willing to accept in an individual lot. Note that the lot tolerance percent defective is not a characteristic of the sampling plan but is a level of lot quality specified by the consumer. Alternate names for the LTPD are the **rejectable quality level (RQL)** and the **limiting quality level (LQL).** It is possible to design acceptance-sampling plans that give specified probabilities of acceptance at the LTPD point. Subsequently, we will see how to design sampling plans that have specified performance at the AQL and LTPD points.

Lot Tolerance Percent Defective
Poorest level of quality that the consumer is willing to accept in an individual lot. Also called rejectable quality level (RQL) and limiting quality level (LQL)

TYPE-A AND TYPE-B OC CURVES. The OC curves that were constructed in the previous examples are called **type-B OC curves.** In the construction of the OC curve it was assumed that the samples came from a large lot or that we were sampling from a stream of lots selected at random from a process. In this situation, the **binomial distribution** is the exact probability distribution for calculating the probability of lot acceptance. Such an OC curve is referred to as a type-B OC curve.

The **type-A OC curve** is used to calculate probabilities of acceptance for an isolated lot of finite size. Suppose that the lot size is N, the sample size is n, and the acceptance number is c. The exact sampling distribution of the number of defective items in the sample is the **hypergeometric distribution.**

Figure 7.7 shows the type-A OC curve for a single-sampling plan with $n = 50$, $c = 1$, where the lot size is $N = 500$. The probabilities of acceptance defining the OC curve are calculated using the hypergeometric distribution. Also shown on this graph is the type-A OC curve for $N = 2{,}000$, $n = 50$, and $c = 1$. Note that the two OC curves are very similar. Generally, as the size of the lot increases, the lot size has a decreasing impact on the OC curve. In fact, if the lot size is at least ten times the sample size ($n/N \leq 0.10$), the type-A and type-B OC curves are virtually indistinguishable. As an illustration, the type-B OC curve for the sampling plan $n = 50$, $c = 1$ is also shown in Figure 7.6. Note that it is identical to the type-A OC curve based on a lot size of $N = 2{,}000$.

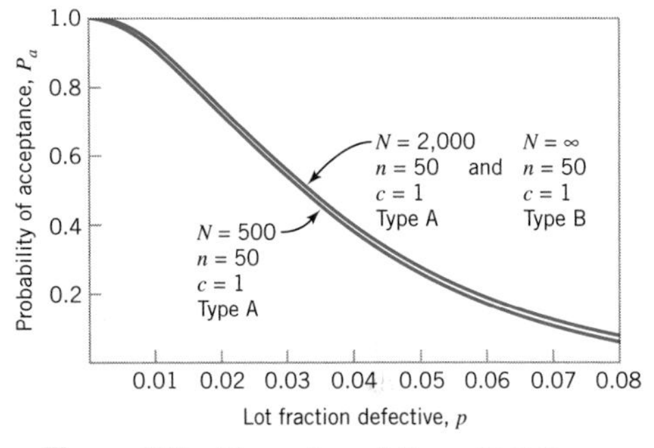

Figure 7.7 **Type-A and Type-B OC Curves**

The type-A OC curve will always lie below the type-B OC curve. That is, if a type-B OC curve is used as an approximation for a type-A curve, the probabilities of acceptance calculated for the type-B curve will always be higher than they would have been if the type-A curve had been used instead. However, this difference is only significant when the lot size is small relative to the sample size. Unless otherwise stated, all discussion of OC curves in this text is in terms of the type-B OC curve.

OTHER ASPECTS OF OC CURVE BEHAVIOR. Two approaches to designing sampling plans that are encountered in practice have certain implications for the behavior of the OC curve. Since not all of these implications are positive, it is worthwhile to briefly mention these two approaches to sampling plan design. These approaches are the use of sampling plans with zero acceptance numbers ($c = 0$) and the use of sample sizes that are a fixed percentage of the lot size.

Figure 7.8 shows several OC curves for acceptance-sampling plans with $c = 0$. By comparing Figure 7.8 with Figure 7.6, it is easy to see that plans with zero acceptance numbers have OC curves that have a very different shape than the OC curves of sampling plans for which $c > 0$. Generally, sampling plans with $c = 0$ have OC curves that are convex throughout their range. As a result of this shape, the probability of acceptance begins to drop very rapidly, even for small values of the lot fraction defective. This is extremely hard on the supplier, and in some circumstances, it may be extremely uneconomical for the consumer. For example, consider the sampling plans in Figure 7.6. Suppose the acceptable quality level is 1%. This implies that we would like to accept lots that are 1% defective or better. Note that if sampling

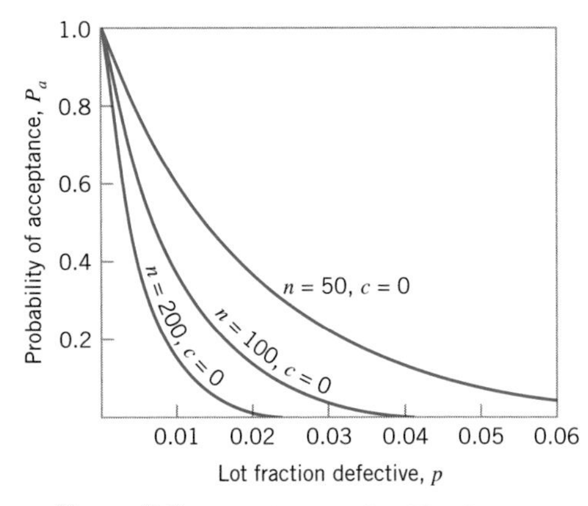

Figure 7.8 **OC Curves for Single-Sampling Plan with $c = 0$**

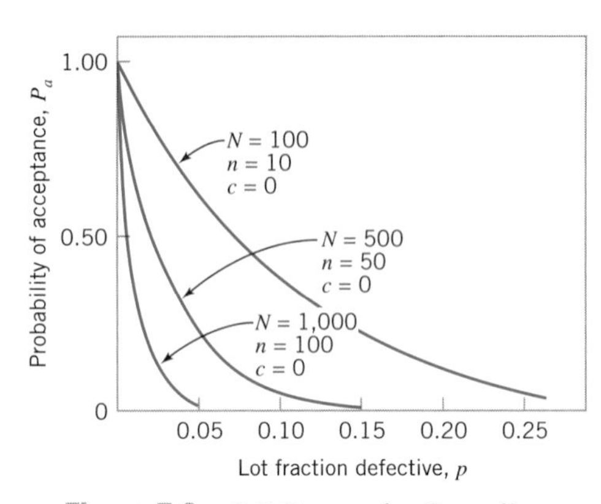

Figure 7.9 OC Curves for Sampling Plans where Sample Size n is 10% of the Lot Size

plan $n = 89$, $c = 1$ is used, the probability of lot acceptance at the AQL is about 0.78. On the other hand, if the plan $n = 89$, $c = 0$ is used, the probability of acceptance at the AQL is approximately 0.41. That is, nearly 60% of the lots of AQL quality will be rejected if we use an acceptance number of zero. If rejected lots are returned to the supplier, then a large number of lots will be unnecessarily returned, perhaps creating production delays at the consumer's manufacturing site. If the consumer screens or 100% inspects all rejected lots, a large number of lots that are of acceptable quality will be screened. This is, at best, an inefficient use of sampling resources. An alternative approach to using zero acceptance numbers is to use **chain-sampling plans.** Under certain circumstances, chain sampling works considerably better than acceptance-sampling plans with $c = 0$.

Figure 7.9 presents the OC curves for sampling plans in which the sample size is a fixed percentage of the lot size. The principal disadvantage of this approach is that the different sample sizes offer different levels of protection. It is illogical for the level of protection that the consumer enjoys for a critical part or component to vary as the size of the lot varies. Although sampling procedures such as this one were in wide use before the statistical principles of acceptance-sampling were generally known, their use has (unfortunately) not entirely disappeared.

7.2.3 DESIGNING A SINGLE-SAMPLING PLAN WITH A SPECIFIED OC CURVE

A common approach to the design of an acceptance-sampling plan is to require that the OC curve pass through two designated points. Note that one point is not enough to fully specify the sampling plan; however, two points are sufficient. In general, it does not matter which two points are specified.

Suppose that we wish to construct a sampling plan such that the probability of acceptance is $1 - \alpha$ for lots with fraction defective p_1, and the probability of acceptance is β for lots with fraction defective p_2. Assuming that binomial sampling (with type-B OC curves) is appropriate, we see that the sample size n and acceptance number c are the solution to

$$
\begin{aligned}
1 - \alpha &= \sum_{d=0}^{c} \frac{n!}{d!(n-d)!} p_1^d (1 - p_1)^{n-d} \\
\beta &= \sum_{d=0}^{c} \frac{n!}{d!(n-d)!} p_2^d (1 - p_2)^{n-d}
\end{aligned}
\qquad (7.3)
$$

Equation (7.3) was obtained by writing out the two points on the OC curve using the binomial distribution. The two simultaneous equations in Equation (7.3) are nonlinear, and there is no simple, direct solution.

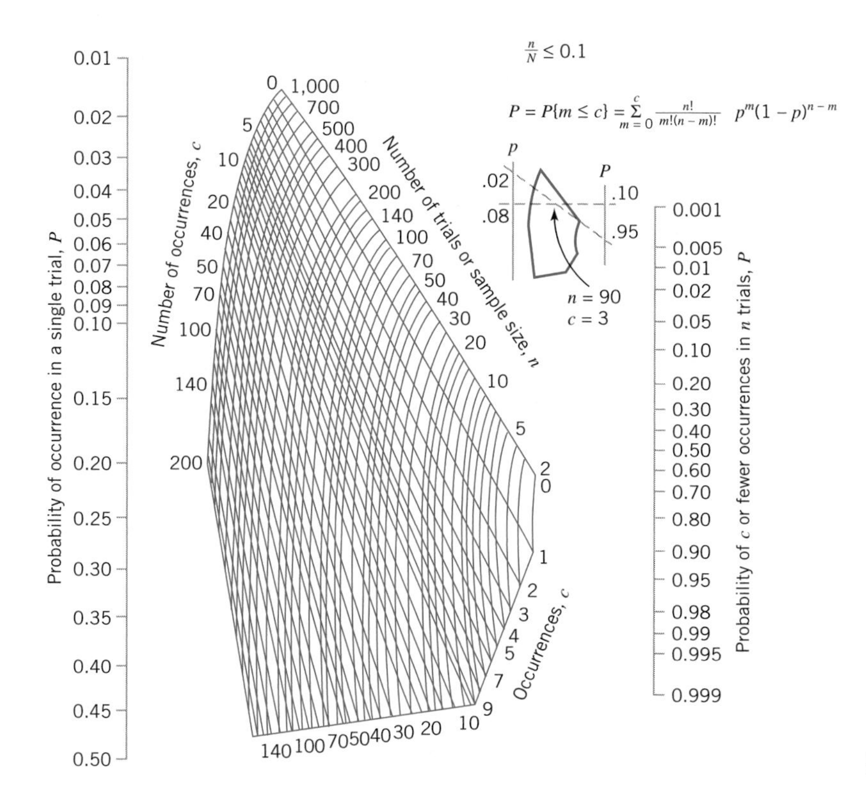

Figure 7.10 **Binomial Nomograph**

The nomograph in Figure 7.10 can be used for solving these equations. The procedure for using the nomograph is very simple. Two lines are drawn on the nomograph, one connecting p_1 and $1 - \alpha$, and the other connecting p_2 and β. The intersection of these two lines gives the region of the nomograph in which the desired sampling plan is located.

example 7.1 Using a Monograph

To illustrate the use of the nomograph, suppose we wish to construct a sampling plan for which $p_1 = 0.01$, $\alpha = 0.05$, $p_2 = 0.06$, and $\beta = 0.10$. Locating the intersection of the lines connecting ($p_1 = 0.01$, $1 - \alpha = 0.95$) and ($p_2 = 0.06$, $\beta = 0.10$) on the nomograph indicates that the plan $n = 89$, $c = 2$ is very close to passing through these two points on the OC curve. Obviously, since n and c must be integers, this procedure will actually produce several plans that have OC curves that pass close to the desired points. For instance, if the first line is followed either to the c line just above the intersection point or to the c line just below it, and the alternate sample sizes are read from the chart, this will produce two plans that pass almost exactly through the p_1, $1 - \alpha$ point but that may deviate somewhat from the p_2, β point. A similar procedure could be followed with the p_2, β line. The result of following both of these lines would be four plans that pass approximately through the two points specified on the OC curve.

In addition to the graphical procedure that we have described for designing sampling plans with specified OC curves, tabular procedures are also available for the same purpose. Duncan (1986) gives a good description of these techniques.

Although any two points on the OC curve could be used to define the sampling plan, it is customary in many industries to use the AQL and LTPD points for this purpose. When the levels of lot quality specified are p_1 = AQL and p_2 = LTPD, the corresponding points on the OC curve are usually referred to as the producer's risk point and the consumer's risk point, respectively. Thus, a would be called the producer's risk and β would be called the consumer's risk.

7.2.4 RECTIFYING INSPECTION

Acceptance-sampling programs usually require corrective action when lots are rejected. This generally takes the form of 100% inspection or **screening** of rejected lots, with all discovered defective items either removed for subsequent rework or return to the supplier, or replaced from a stock of known good items. Such sampling programs are called **rectifying inspection programs,** because the inspection activity affects the final quality of the outgoing product. This is illustrated in Figure 7.11. Suppose that incoming lots to the inspection activity have fraction defective p_0. Some of these lots will be accepted, and others will be rejected. The rejected lots will be screened, and their final fraction defective will be zero. However, accepted lots have fraction defective p_0. Consequently, the outgoing lots from the inspection activity are a mixture of lots with fraction defective p_0 and fraction defective zero, so the average fraction defective in the stream of outgoing lots is p_1, which is less than p_0. Thus, a rectifying inspection program serves to "correct" lot quality.

Rectifying inspection programs are used in situations where the manufacturer wishes to know the average level of quality that is likely to result at a given stage of the manufacturing operations. Thus, rectifying inspection programs are used either at receiving inspection, at in-process inspection of semifinished products, or at final inspection of finished goods. The objective of in-plant usage is to give assurance regarding the average quality of material used in the next stage of the manufacturing operations.

Rejected lots may be handled in a number of ways. The best approach is to return rejected lots to the supplier and require it to perform the screening and rework activities. This has the psychological effect of making the supplier responsible for poor quality and may exert pressure on the supplier to improve its manufacturing processes or to install better process controls. However, in many situations, because the components or raw materials are required in order to meet production schedules, screening and rework take place at the consumer level. This is not the most desirable situation.

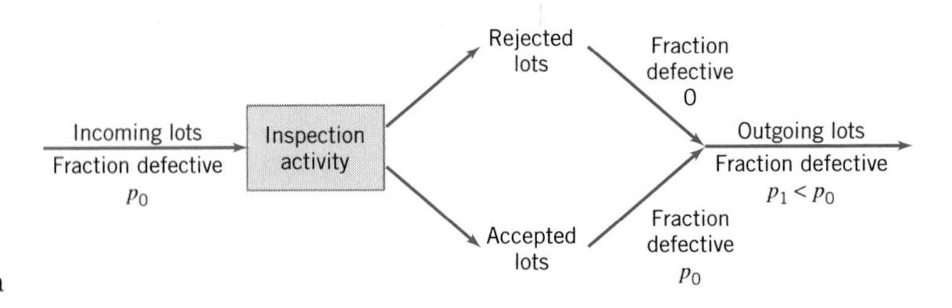

Figure 7.11 Rectifying Inspection

Average outgoing quality (AOQ) is widely used for the evaluation of a rectifying sampling plan. The average outgoing quality is the quality in the lot that results from the application of rectifying inspection. It is the average value of lot quality that would be obtained over a long sequence of lots from a process with fraction defective p. It is simple to develop a formula for AOQ. Assume that the lot size is N and that all discovered defectives are replaced with good units. Then in lots of size N, we have

1. n items in the sample that, after inspection, contain no defectives, because all discovered defectives are replaced
2. $N - n$ items that, if the lot is rejected, also contain no defectives
3. $N - n$ items that, if the lot is accepted, contain $p(N - n)$ defectives

Thus, lots in the outgoing stage of inspection have an expected number of defective units equal to $P_a p(N - n)$, which we may express as an *average fraction defective*, called the **average outgoing quality** or

$$\text{AOQ} = \frac{P_a p(N - n)}{N} \tag{7.4}$$

example 7.2 Calculating AOQ

To illustrate the use of Equation (7.4), suppose that $N = 10{,}000$, $n = 89$, and $c = 2$, and that the incoming lots are of quality $p = 0.01$. Now at $p = 0.01$, we have $P_a = 0.9397$, and the AOQ is

$$\begin{aligned} \text{AOQ} &= \frac{P_a p(N - n)}{N} \\ &= \frac{(0.9397)(0.01)(10{,}000 - 89)}{10{,}000} \\ &= 0.0093 \end{aligned}$$

That is, the average outgoing quality is 0.93% defective.

Note that as the lot size N becomes large relative to the sample size n, we may write Equation (7.4) as

$$\text{AOQ} = P_a p \tag{7.5}$$

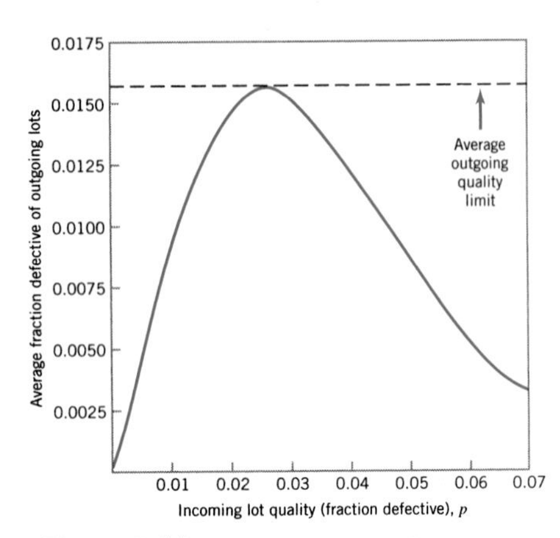

Figure 7.12 Average Outgoing Quality Curve for $n = 89$, $c = 2$

Average Outgoing Quality Limit

Remember, the AOQL is an average level of quality, across a large stream of lots.

Average outgoing quality will vary as the fraction defective of the incoming lots varies. The curve that plots average outgoing quality against incoming lot quality is called an *AOQ curve*. The AOQ curve for the sampling plan $n = 89$, $c = 2$ is shown in Figure 7.12. From examining this curve we note that when the incoming quality is very good, the average outgoing quality is also very good. In contrast, when the incoming lot quality is very bad, most of the lots are rejected and screened, which leads to a very good level of quality in the outgoing lots. In between these extremes, the AOQ curve rises, passes through a maximum, and descends. The maximum ordinate on the AOQ curve represents the worst possible average quality that would result from the rectifying inspection program, and this point is called the **average outgoing quality limit (AOQL).** From examining Figure 7.12, the AOQL is seen to be approximately 0.0155. That is, no matter how bad the fraction defective is in the incoming lots, the outgoing lots will never have a worse quality level on the average than 1.55% defective. Let us emphasize that this AOQL is an *average level of quality, across a large stream of lots*. It does not give assurance that an isolated lot will have quality no worse than 1.55% defective.

Another important measure relative to rectifying inspection is the total amount of inspection required by the sampling program. If the lots contain no defective items, no lots will be rejected, and the amount of inspection per lot will be the sample size n. If the items are all defective, every lot will be submitted to 100% inspection, and the amount of inspection per lot will be the lot size N. If the lot quality is $0 < p < 1$, the average amount of inspection per lot will vary between the sample size n and the lot size N. If the lot is of quality p and the probability of lot acceptance is P_a, then the **average total inspection (ATI)** per lot will be

$$\text{ATI} = n + (1 - P_a)(N - n) \tag{7.6}$$

example 7.3 Calculating ATI

To illustrate the use of Equation (7.6), consider Example 7.2 with $N = 10{,}000$, $n = 89$, $c = 2$, and $p = 0.01$. Then, since $P_a = 0.9397$, we have

$$\begin{aligned}\text{ATI} &= n + (1 - P_a)(N - n)\\ &= 89 + (1 - 0.9397)(10{,}000 - 89)\\ &= 687\end{aligned}$$

Remember that this is an average number of units inspected over *many* lots with fraction defective $p = 0.01$.

It is possible to draw a curve of average total inspection as a function of lot quality. Average total inspection curves for the sampling plan $n = 89$, $c = 2$ for lot sizes of 1,000, 5,000, and 10,000, are shown in Figure 7.13.

The AOQL of a rectifying inspection plan is a very important characteristic. It is possible to design rectifying inspection programs that have specified values of AOQL. However, specification of the AOQL is not sufficient to determine a unique sampling plan. Therefore, it is relatively common practice to choose the sampling plan that has a specified AOQL and, in addition, yields a minimum ATI at a particular level of lot quality. The level of lot quality usually chosen is the most likely level of incoming lot quality, which is generally called the *process average*. The procedure for generating these plans is relatively straightforward and is illustrated in Duncan (1986). Generally, it is unnecessary to go through this procedure, because tables of sampling plans that minimize ATI for a given AOQL and a specified process average p have been developed by Dodge and Romig. We describe the use of these tables in Section 7.5.

It is also possible to design a rectifying inspection program that gives a specified level of protection at the LTPD point and that minimizes the average total inspection for a specified process average p. The Dodge–Romig sampling inspection tables also provide these LTPD plans. Section 7.5 discusses the use of the Dodge–Romig tables to find plans that offer specified LTPD protection.

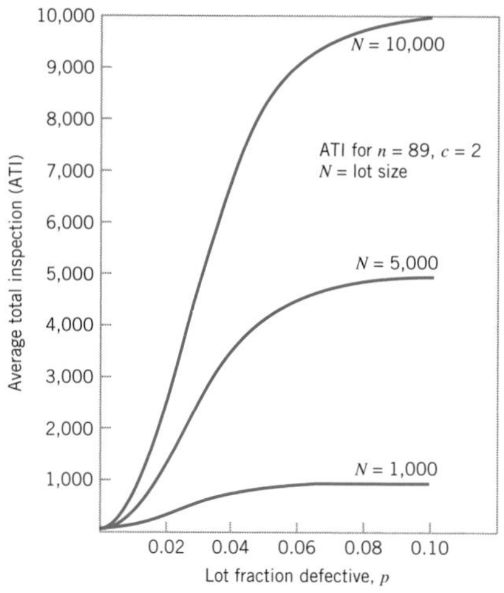

Figure 7.13 **Average Total Inspection (ATI) Curves for Sampling Plan $n = 89$, $c = 2$, for Lot Sizes of 1000, 5000, and 10,000**

7.3 Double-, Multiple-, and Sequential-Sampling

A number of extensions of single-sampling plans for attributes are useful. These include **double-sampling plans, multiple-sampling plans,** and **sequential-sampling plans.** This section discusses the design and application of these sampling plans.

Double-Sampling Plan Parameters

n_1 = sample size on the first sample
c_1 = acceptance number of the first sample
n_2 = sample size on second sample
c_2 = acceptance number for both samples

7.3.1 DOUBLE-SAMPLING PLANS

A double-sampling plan is a procedure in which, under certain circumstances, a second sample is required before the lot can be sentenced. A double-sampling plan is defined by four parameters.[2]

example 7.4 Illustration of a Double-Sampling Plan

Suppose $n_1 = 50$, $c_1 = 1$, $n_2 = 100$, and $c_2 = 3$. Thus, a random sample of $n_1 = 50$ items is selected from the lot, and the number of defectives in the sample, d_1, is observed.

If $d_1 \leq c_1 = 1$, the lot is accepted on the first sample. If $d_1 > c_2 = 3$, the lot is rejected on the first sample.

If $c_1 < d_1 \leq c_2$, a second random sample of size $n_2 = 100$ is drawn from the lot, and the number of defectives in this second sample, d_2, is observed.

Now the combined number of observed defectives from both the first and second sample, $d_1 + d_2$, is used to determine the lot sentence. If $d_1 + d_2 \leq c_2 = 3$, the

example 7.4 Continued

lot is accepted. However, if $d_1 + d_2 > c_2 = 3$, the lot is rejected. The operation of this double-sampling plan is illustrated graphically in Figure 7.14.

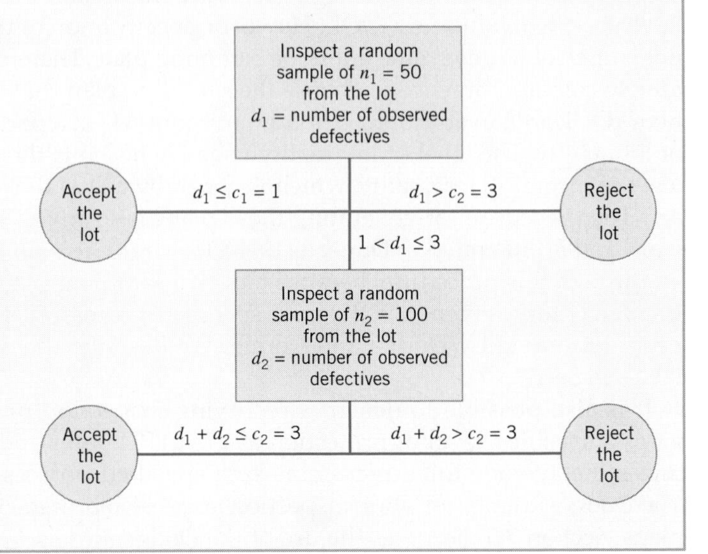

Figure 7.14 Operation of the Double-Sampling Plan, $n_1 = 50$, $c_1 = 1$, $n_2 = 100$, $c_2 = 3$

Some authors prefer the notation n_1, Ac_1, Re_1, n_2, Ac_2, $Re_2 = Ac_2 + 1$. Since the rejection number on the first sample Re_1 is not necessarily equal to Re_2, this gives some additional flexibility in designing double-sampling plans. MIL STD 105E and ANSI/ASQC Z1.4 currently use this notation. However, because assuming that $Re_1 = Re_2$ does not significantly affect the plans obtained, we have chosen to discuss this slightly simpler system.

The principal advantage of a double-sampling plan with respect to single-sampling is that it may reduce the total amount of required inspection. Suppose that the first sample taken under a double-sampling plan is smaller than the sample that would be required using a single-sampling plan that offers the consumer the same protection. In all cases, then, in which a lot is accepted or rejected on the first sample, the cost of inspection will be lower for double-sampling than it would be for single-sampling. It is also possible to reject a lot without complete inspection of the second sample. (This is called *curtailment* on the second sample.) Consequently, the use of double-sampling can often result in lower total inspection costs. Furthermore, in some situations, a double-sampling plan has the psychological advantage of giving a lot a second chance. This may have some appeal to the supplier. However, there is no real advantage to double-sampling in this regard, because single- and double-sampling plans can be chosen so that they have the same OC curves. Thus, both plans would offer the same risks of accepting or rejecting lots of specified quality.

Double-sampling has two potential disadvantages. First, unless curtailment is used on the second sample, under some circumstances double-sampling may require more total inspection than would be required in a single-sampling plan that offers the same protection. Thus, unless double-sampling is used carefully, its potential economic advantage may be lost. The second disadvantage of double-sampling is that it is administratively more complex, which may increase the opportunity for the occurrence of inspection errors. Furthermore, there may be problems in storing and handling raw materials or component parts for which one sample has been taken but that are awaiting a second sample before a final lot dispositioning decision can be made.

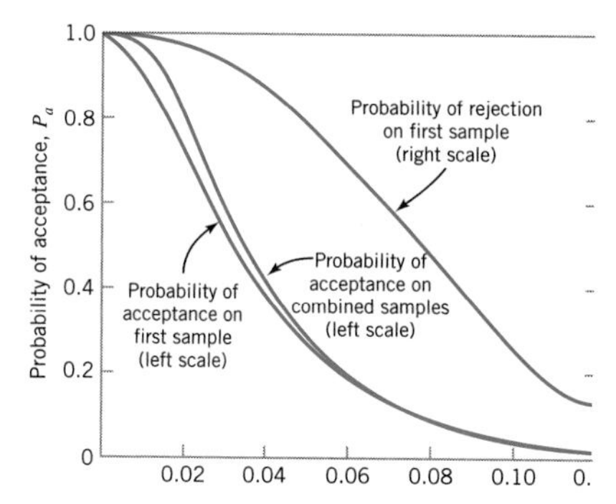

Figure 7.15 OC Curves for the Double-Sampling Plan, $n_1 = 50$, $c_1 = 1$, $n_2 = 100$, $c_2 = 3$

THE OC CURVE. The performance of a double-sampling plan can be conveniently summarized by means of its operating-characteristic (OC) curve. The OC curve for a double-sampling plan is somewhat more involved than the OC curve for single-sampling. In this section, we describe the construction of type-B OC curves for double-sampling. A double-sampling plan has a primary OC curve that gives the probability of acceptance as a function of lot or process quality. It also has supplementary OC curves that show the probability of lot acceptance and rejection on the first sample. The OC curve for the probability of rejection on the first sample is simply the OC curve for the single-sampling plan $n = n_1$ and $c = c_2$. Primary and supplementary OC curves for the plan $n_1 = 50$, $c_1 = 1$, $n_2 = 100$, $c_2 = 3$ are shown in Figure 7.15.

example 7.5 Computation of an OC curve for a Double-Sampling Plan

We now illustrate the computation of the OC curve for the plan $n_1 = 50$, $c_1 = 1$, $n_2 = 100$, $c_2 = 3$. If P_a denotes the probability of acceptance on the combined samples, and P_a^{I} and P_a^{II} denote the probability of acceptance on the first and second samples, respectively, then

$$P_a = P_a^{\mathrm{I}} + P_a^{\mathrm{II}}$$

P_a^{I} is just the probability that we will observe $d_1 \leq c_1 = 1$ defectives out of a random sample of $n_1 = 50$ items. Thus

$$P_a^{\mathrm{I}} = \sum_{d_1=0}^{1} \frac{50!}{d_1!(50 - d_1)!} p^{d_1}(1 - p)^{50-d_1}$$

If $p = 0.05$ is the fraction defective in the incoming lot, then

$$P_a^{\mathrm{I}} = \sum_{d_1=0}^{1} \frac{50!}{d_1!(50 - d_1)!}(0.05)^{d_1}(0.95)^{50-d_1} = 0.279$$

To obtain the probability of acceptance on the second sample, we must list the number of ways the second sample can be obtained. A second sample is drawn *only* if there are two or three defectives on the first sample—that is, if $c_1 < d_1 \leq c_2$.

1. $d_1 = 2$ *and* $d_2 = 0$ or 1; that is, we find two defectives on the first sample and one or less defectives on the second sample. The probability of this is

$$P\{d_1 = 2, d_2 \leq 1\} = P\{d_1 = 2\}P\{d_2 \leq 1\}$$

$$= \frac{50!}{2!\,48!}(0.05)^2(0.95)^{48} \sum_{d_2=0}^{1} \frac{100!}{d_2!(100 - d_2)!}(0.05)^{d_2}(0.95)^{100-d_2}$$

$$= (0.261)(0.037) = 0.0097$$

example 7.5 Continued

2. $d_1 = 3$ *and* $d_2 = 0$; that is, we find three defectives on the first sample and no defectives on the second sample. The probability of this is

$$P\{d_1 = 3, d_2 = 0\} = P\{d_1 = 3\} - P\{d_2 = 0\}$$
$$= \frac{50!}{3!\,(47)!}(0.05)^3(0.95)^{47}\frac{100!}{0!\,100!}(0.05)^0(0.95)^{100}$$
$$= (0.220)(0.0059) = 0.001$$

Thus, the probability of acceptance on the second sample is

$$P_a^{\mathrm{II}} = P\{d_1 = 2, d_2 \le 1\} + P\{d_1 = 3, d_2 = 0\}$$
$$= 0.0097 + 0.001 = 0.0107$$

The probability of acceptance of a lot that has fraction defective $p = 0.05$ is therefore

$$P_a = P_a^{\mathrm{I}} + P_a^{\mathrm{II}}$$
$$= 0.279 + 0.0107 = 0.2897$$

Other points on the OC curve are calculated similarly. Remember that these binomial probabilities can be calculated using Minitab on Excel.

THE AVERAGE SAMPLE NUMBER CURVE. The average sample number curve of a double-sampling plan is also usually of interest. In single-sampling, the size of the sample inspected from the lot is always constant, whereas in double-sampling, the size of the sample selected depends on whether or not the second sample is necessary. The probability of drawing a second sample varies with the fraction defective in the incoming lot. With complete inspection of the second sample, the average sample size in double-sampling is equal to the size of the first sample times the probability that there will only be one sample, plus the size of the combined samples times the probability that a second sample will be necessary. Therefore, a general formula for the average sample number in double-sampling, if we assume complete inspection of the second sample, is

$$\begin{aligned} \text{ASN} &= n_1 P_1 + (n_1 + n_2)(1 - P_1) \\ &= n_1 + n_2(1 - P_1) \end{aligned} \tag{7.7}$$

where P_1 is the probability of making a lot-dispositioning decision on the *first* sample. This is

$$P_1 = P\,\{\text{lot is accepted on the first sample}\} + P\,\{\text{lot is rejected on the first sample}\}$$

If Equation (7.7) is evaluated for various values of lot fraction defective p, the plot of ASN versus p is called an **average sample number curve.**

In practice, inspection of the second sample is usually terminated and the lot rejected as soon as the number of observed defective items in the combined sample exceeds the second acceptance number c_2. This is referred to as curtailment of the second sample. The use of curtailed inspection lowers the average sample number required in double-sampling. It is not recommended that curtailment be used in single sampling, or in the first sample of double-sampling, because it is usually desirable to have complete inspection of a fixed sample size in order to secure an unbiased estimate of the quality of the material supplied by the supplier. If curtailed inspection is used in single-sampling or on the first sample of double-sampling, the estimate of lot or process fallout obtained from these data is biased. For instance, suppose that the acceptance number is 1. If the first two items in the sample are defective, and the inspection process is curtailed, the estimate of lot or process fraction defective is 100%. Based on this information, even non-statistically trained managers or engineers will be very reluctant to believe that the lot is really 100% defective.

The ASN curve formula for a double-sampling plan with curtailment on the second sample is

$$\text{ASN} = n_1 + \sum_{j=c_1+1}^{c_1} P(n_1, j)\Big[n_2 P_L(n_2, c_2 - j) + \frac{c_2 - j + 1}{p} P_M(n_2 + 1, c_2 - j + 2)\Big] \tag{7.8}$$

In Equation (7.8), $P(n_1, j)$ is the probability of observing exactly j defectives in a sample of size n_1, $P_L(n_2, c_2 - j)$ is the probability of observing $c_2 - j$ or fewer defectives in a sample of size n_2, and $P_M(n_2 + 1, c_2 - j + 2)$ is the probability of observing $c_2 - j + 2$ defectives in a sample of size $n_2 + 1$.

Figure 7.15 compares the average sample number curves for complete and curtailed inspection for the double-sampling plan $n_1 = 60$, $c_1 = 2$, $n_2 = 120$, $c_3 = 3$ and the average sample number that would be used in single-sampling with $n = 89$, $c = 2$. Obviously, the sample size in the single-sampling plan is always constant. This double-sampling plan has been selected because it has an OC curve that is nearly identical to the OC curve for the single-sampling plan. That is, both plans offer equivalent protection to the producer and the consumer. Note from the inspection of Figure 7.16 that the ASN curve for double-sampling without curtailment on the second sample is not lower than the sample size used in single-sampling throughout the entire range of lot fraction defective. If lots are of very good quality, they will usually be accepted on the first sample, whereas if lots are of very bad quality, they will usually be rejected on the first sample. This gives an ASN for double-sampling that is smaller than the sample size used in single-sampling for lots that are

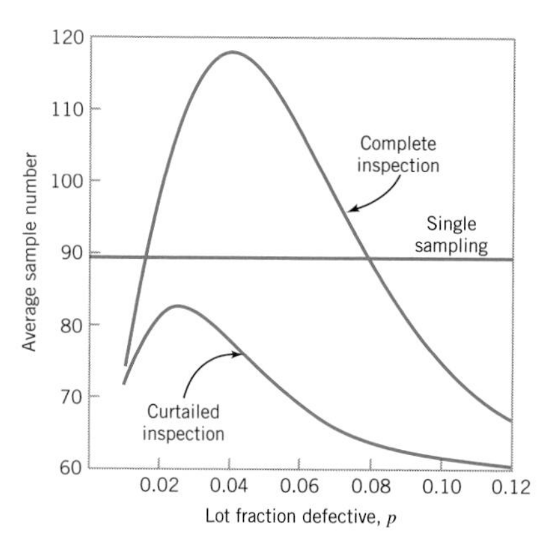

Figure 7.16 Average Sample Number Curves for Single and Double-Sampling

either very good or very bad. However, if lots are of intermediate quality, the second sample will be required in a large number of cases before a lot disposition decision can be made. In this range of lot quality, the ASN performance of double-sampling is worse than single-sampling.

This example points out that it is important to use double-sampling very carefully. Unless care is exercised to ensure that lot or process quality is in the range where double-sampling is most effective, then the economic advantages of double-sampling relative to single-sampling may be lost. It is a good idea to maintain a running estimate of the supplier's lot or process fallout so that if it shifts into a range where double-sampling is not economically effective, a change to single-sampling (or some other appropriate strategy) can be made. Another way to do this would be to record the proportion of times that the second sample is required in order to make a decision.

Figure 7.16 also shows the ASN curve using curtailment on the second sample. Note that if curtailment is used, the average sample number curve for double-sampling always lies below the sample size used in single-sampling.

DESIGNING DOUBLE-SAMPLING PLANS WITH SPECIFIED P_1, $1 - \alpha$, P_2, AND β. It is often necessary to be able to design a double-sampling plan that has a specified OC curve. Let $(p_1, 1 - \alpha)$ and (p_2, β) be the two points of interest on the OC curve. If, in addition, we impose another relationship on the parameters of the sampling plan, then a simple procedure can be used to obtain such plans. The most common constraint is to require that n_2 is a multiple of n_1. Refer to Duncan (1986) for a discussion of these techniques.

RECTIFYING INSPECTION. When rectifying inspection is performed with double-sampling, the AOQ curve is given by

$$\text{AOQ} = \frac{[P_a^{\text{I}}(N - n_1) + P_a^{\text{II}}(N - n_1 - n_2)]p}{N} \tag{7.9}$$

assuming that all defective items discovered, either in sampling or 100% inspection, are replaced with good ones. The average total inspection curve is given by

$$\text{ATI} = n_1 P_a^{\text{I}} + (n_1 + n_2)P_a^{\text{II}} + N(1 - P_a) \tag{7.10}$$

Remember that $P_a = P_a^{\text{I}} + P_a^{\text{II}}$ is the probability of final lot acceptance and that the acceptance probabilities depend on the level of lot or process quality p.

7.3.2 MULTIPLE-SAMPLING PLANS

A multiple-sampling plan is an extension of double-sampling in that more than two samples can be required to sentence a lot. An example of a multiple-sampling plan with five stages as shown in the adjacent display.

Cumulative Sample Size	Acceptance Number	Rejection Number
20	0	3
40	1	4
60	3	5
80	5	7
100	8	9

This plan will operate as follows: If, at the completion of any stage of sampling, the number of defective items is less than or equal to the acceptance number, the lot is accepted. If, during any stage, the number of defective items equals or exceeds the rejection number, the lot is rejected; otherwise the next sample is taken. The multiple-sampling procedure continues until the fifth sample is taken, at which time a lot disposition decision must be made. The first sample is usually inspected 100%, although subsequent samples are usually subject to curtailment.

The construction of OC curves for multiple-sampling is a straightforward extension of the approach used in double-sampling. Similarly, it is also possible to compute the average sample number curve of multiple-sampling plans. One may also design a multiple-sampling plan for specified values of p_1, $1 - \alpha$, p_2, and β. For an extensive discussion of these techniques, see Duncan (1986).

The principal advantage of multiple-sampling plans is that the samples required at each stage are usually smaller than those in single or double-sampling; thus, some economic efficiency is connected with the use of the procedure. However, multiple-sampling is much more complex to administer.

7.3.3 SEQUENTIAL-SAMPLING PLANS

Sequential-sampling is an extension of the double-sampling and multiple-sampling concept. In sequential-sampling, we take a sequence of samples from the lot and allow the number of samples to be determined entirely by the results of the sampling process. In practice, sequential sampling can theoretically continue indefinitely, until the lot is inspected 100%. In practice, sequential-sampling plans are usually truncated after the number inspected is equal to three times the number that would have been inspected using a corresponding single-sampling plan. If the sample size selected at each stage is greater than one, the process is usually called *group* sequential-sampling. If the sample size inspected at each stage is one, the procedure is usually called **item-by-item sequential-sampling.**

Item-by-item sequential-sampling is based on the sequential probability ratio test (SPRT), developed by Wald (1947). The operation of an item-by-item sequential-sampling plan is illustrated in Figure 7.17. The cumulative observed number of defectives is plotted on the chart. For each point, the abscissa is the total number of items selected up to that time, and the ordinate is the total number of observed defectives. If the plotted points stay within the boundaries of the acceptance and rejection lines, another sample must be drawn. As soon as a point falls on or above the upper line, the lot is rejected. When a sample point falls on or below the lower line, the lot is accepted. The equations for the two limit lines for specified values of p_1, $1 - \alpha$, p_2, and β are

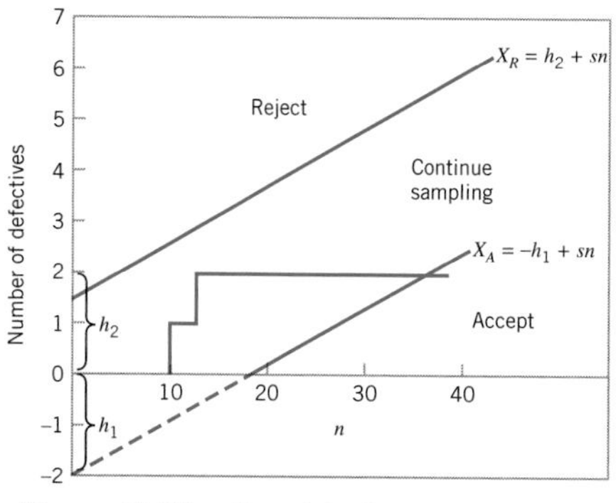

Figure 7.17 Graphical Performance of Sequential-Sampling

$$X_A = -h_1 + sn \quad \text{(acceptance line)} \tag{7.11a}$$

$$X_R = h_2 + sn \quad \text{(rejection line)} \tag{7.11b}$$

where

$$h_1 = \left(\log\frac{1-\alpha}{\beta}\right)/k \quad (7.12)$$

$$h_2 = \left(\log\frac{1-\beta}{\alpha}\right)/k \quad (7.13)$$

$$k = \log\frac{p_2(1-p_1)}{p_1(1-p_2)} \quad (7.14)$$

$$s = \log[(1-p_1)/(1-p_2)]/k \quad (7.15)$$

example 7.6 Developing a Sequential Sampling Plan

Suppose we wish to find a sequential-sampling plan for which $p_1 = 0.01$, $\alpha = 0.05$, $p_2 = 0.06$, and $\beta = 0.10$. Thus,

$$k = \log\frac{p_2(1-p_1)}{p_1(1-p_2)} = \log\frac{(0.06)(0.99)}{(0.01)(0.94)} = 0.80066$$

$$h_1 = \left(\log\frac{1-\alpha}{\beta}\right)/k = \left(\log\frac{0.95}{0.10}\right)/0.80066 = 1.22$$

$$h_2 = \left(\log\frac{1-\beta}{\alpha}\right)/k = \left(\log\frac{0.90}{0.05}\right)/0.80066 = 1.57$$

$$s = \log[(1-p_1)/(1-p_2)]/k = [\log(0.99/0.94)]/0.80066 = 0.028$$

Therefore, the limit lines are

$$X_A = -1.22 + 0.028n \quad \text{(accept)}$$

and

$$X_R = 1.57 + 0.028n \quad \text{(reject)}$$

Instead of using a graph to determine the lot disposition, the sequential-sampling plan can be displayed in a table such as Table 7.3. The entries in the table are found by substituting values of n into the equations for the acceptance and rejection lines and calculating acceptance and rejection numbers. For example, the calculations for $n = 45$ are

$$X_A = -1.22 + 0.028n = -1.22 + 0.028(45) = 0.04 \quad \text{(accept)}$$

$$X_R = 1.57 + 0.028n = 1.57 + 0.028(45) = 2.83 \quad \text{(reject)}$$

Table 7.3

Item-by-Item Sequential-Sampling Plan $p_1 = 0.01$, $\alpha = 0.05$, $p_2 = 0.06$, $\beta = 0.10$ (First 46 Units Only)

Number of Items Inspected, *n*	Acceptance Number	Rejection Number	Number of Items Inspected, *n*	Acceptance Number	Rejection Number
1	a	b	24	a	3
2	a	2	25	a	3
3	a	2	26	a	3
4	a	2	27	a	3
5	a	2	28	a	3
6	a	2	29	a	3
7	a	2	30	a	3
8	a	2	31	a	3
9	a	2	32	a	3
10	a	2	33	a	3
11	a	2	34	a	3
12	a	2	35	a	3
13	a	2	36	a	3
14	a	2	37	a	3
15	a	2	38	a	3
16	a	3	39	a	3
17	a	3	40	a	3
18	a	3	41	a	3
19	a	3	42	a	3
20	a	3	43	a	3
21	a	3	44	0	3
22	a	3	45	0	3
23	a	3	46	0	3

"a" means acceptance not possible.
"b" means rejection not possible.

Acceptance and rejection numbers must be integers, so the acceptance number is the next integer less than or equal to X_A, and the rejection number is the next integer greater than or equal to X_R. Thus, for $n = 45$, the acceptance number is 0 and the rejection number is 3. Note that the lot cannot be accepted until at least 44 units have been tested. Table 7.3

shows only the first 46 units. Usually, the plan would be truncated after the inspection of 267 units, which is three times the sample size required for an equivalent single-sampling plan.

THE OC CURVE AND ASN CURVE FOR SEQUENTIAL SAMPLING. The OC curve for sequential-sampling can be easily obtained. Two points on the curve are $(p_1, 1 - \alpha)$ and (p_2, β). A third point, near the middle of the curve, is $p = s$ and $P_a = h_2/(h_1 + h2)$.

The average sample number taken under sequential-sampling is

$$\text{ASN} = P_a\left(\frac{A}{C}\right) + (1 - P_a)\frac{B}{C} \tag{7.16}$$

where

$$A = \log\frac{B}{1 - \alpha}$$

$$B = \log\frac{1 - \beta}{\alpha}$$

and

$$C = p\log\left(\frac{p_2}{p_1}\right) + (1 - p)\log\left(\frac{1 - p_2}{1 - p_1}\right)$$

RECTIFYING INSPECTION. The average outgoing quality (AOQ) for sequential-sampling is given approximately by

$$\text{AOQ} = P_a p \tag{7.17}$$

The average total inspection is also easily obtained. Note that the amount of sampling is A/C when a lot is accepted and N when it is rejected. Therefore, the average total inspection is

$$\text{ATI} = P_a\left(\frac{A}{C}\right) + (1 - P_a)N \tag{7.18}$$

7.4 Military Standard 105E (ANSI/ASQC Z1.4, ISO 2859)

7.4.1 DESCRIPTION OF THE STANDARD

Standard sampling procedures for inspection by attributes were developed during World War II. MIL STD 105E is the most widely used acceptance-sampling system for attributes in the world today. The original version of the standard, MIL STD 105A, was issued in 1950. Since then, there have been four revisions; the latest version, MIL STD 105E, was issued in 1989.

The sampling plans discussed in previous sections of this chapter are individual sampling plans. A sampling scheme is an overall strategy specifying the way in which sampling plans are to be used. MIL STD 105E is a collection of sampling schemes; therefore, it is an acceptance-sampling system. Our discussion will focus primarily on MIL STD 105E; however, there is a derivative civilian standard, ANSI/ASQC Z1.4, which is quite similar to the military standard. The standard was also adopted by the International Organization for Standardization as ISO 2859.

The standard provides for three types of sampling: single-sampling, double-sampling, and multiple-sampling. For each type of sampling plan, a provision is made for either normal inspection, tightened inspection, or reduced inspection. Normal inspection is used at the start of the inspection activity. Tightened inspection is instituted when the supplier's recent quality history has deteriorated. Acceptance requirements for lots under tightened inspection are more stringent than under normal inspection. Reduced inspection is instituted when the supplier's recent quality history has been exceptionally good. The sample size generally used under reduced inspection is less than that under normal inspection.

The primary focal point of MIL STD 105E is the acceptable quality level (AQL). The standard is indexed with respect to a series of AQLs. When the standard is used for percent defective plans, the AQLs range from 0.10% to 10%. For defects per units plans, there are an additional ten AQLs running up to 1,000 defects per 100 units. It should be noted that for the smaller AQL levels, the same sampling plan can be used to control either a fraction defective or a number of defects per unit. The AQLs are arranged in a progression, each AQL being approximately 1.585 times the preceding one.

The AQL is generally specified in the contract or by the authority responsible for sampling. Different AQLs may be designated for different types of defects. For example, the standard differentiates critical defects, major defects, and minor defects. It is relatively common practice to choose an AQL of 1% for major defects and an AQL of 2.5% for minor defects. No critical defects would be acceptable.

The sample size used in MIL STD 105E is determined by the lot size and by the choice of inspection level. Three general levels of inspection are provided. Level II is designated as normal. Level I requires about half the amount of inspection that Level II does and may be used when less discrimination is needed. Level III requires about twice as much inspection as Level II and should be used when more discrimination is needed. There are also four special inspection levels, S-1, S-2, S-3, and S-4. The special inspection levels use very small samples and should be employed only when the small sample sizes are necessary and when greater sampling risks can or must be tolerated.

For a specified AQL and inspection level and a given lot size, MIL STD 105E provides a normal sampling plan that is to be used as long as the supplier is producing the product at AQL quality or better. It also provides a procedure for switching to tightened and reduced inspection whenever there is an indication that the supplier's quality has changed. The switching procedures between normal, tightened, and reduced inspection are illustrated graphically in Figure 7.18 and are described next.

1. **Normal to tightened.**	When normal inspection is in effect, tightened inspection is instituted when two out of five consecutive lots have been rejected on original submission.
2. **Tightened to normal.**	When tightened inspection is in effect, normal inspection is instituted when five consecutive lots or batches are accepted on original inspection.
3. **Normal to reduced.**	When normal inspection is in effect, reduced inspection is instituted provided all four of the following conditions are satisfied. a. The preceding ten lots have been on normal inspection, and none of the lots has been rejected on original inspection. b. The total number of defectives in the samples from the preceding ten lots is less than or equal to the applicable limit number specified in the standard. c. Production is at a steady rate; that is, no difficulty such as machine breakdowns, material shortages, or other problems have recently occurred. d. Reduced inspection is considered desirable by the authority responsible for sampling.
4. **Reduced to normal.**	When reduced inspection is in effect, normal inspection is instituted when any of the following four conditions occur. a. A lot or batch is rejected. b. When the sampling procedure terminates with neither acceptance nor rejection criteria having been met, the lot or batch will be accepted, but normal inspection is reinstituted starting with the next lot. c. Production is irregular or delayed. d. Other conditions warrant that normal inspection be instituted.
5. **Discontinuance of inspection.**	In the event that ten consecutive lots remain on tightened inspection, inspection under the provision of MIL STD 105E should be terminated, and action should be taken at the supplier level to improve the quality of submitted lots.

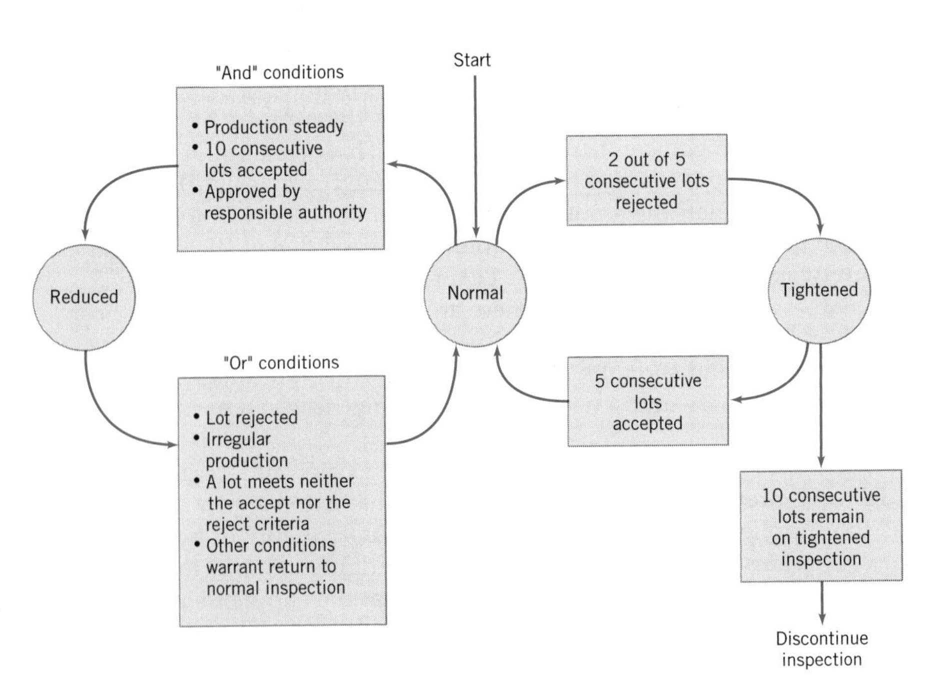

Figure 7.18 Switching Rules for Normal, Tightened, and Reduced Inspection, MIL STD 105E

7.4.2 PROCEDURE

A step-by-step procedure for using MIL STD 105E is as follows:

1. Choose the AQL.
2. Choose the inspection level.
3. Determine the lot size.
4. Find the appropriate sample size code letter from Table 7.4.
5. Determine the appropriate type of sampling plan to use (single, double, multiple).
6. Enter the appropriate table to find the type of plan to be used.
7. Determine the corresponding normal and reduced inspection plans to be used when required.

Table 7.4 presents the sample size code letters for MIL STD 105E. Tables 7.5, 7.6, and 7.7 present the single-sampling plans for normal inspection, tightened inspection, and reduced inspection, respectively. The standard also contains tables for double-sampling plans and multiple-sampling plans for normal, tightened, and reduced inspection.

example 7.7 Use of MIL STD 105C

Suppose that a product is submitted in lots of size $N = 2{,}000$. The acceptable quality level is 0.65%. We will use the standard to generate normal, tightened, and reduced single-sampling plans for this situation. For lots of size 2,000 under general inspection level II, Table 7.4 indicates that the appropriate sample size code letter is K. Therefore, from Table 7.5, for single-sampling plans under normal inspection, the normal inspection plan is $n = 125$, $c = 2$. Table 7.6 indicates that the corresponding tightened inspection plan is $n = 125$, $c = 1$. Note that in switching from normal to tightened inspection, the sample size remains the same, but the acceptance number is reduced by 1. This general strategy is used throughout MIL STD 105E for a transition to tightened inspection. If the normal inspection acceptance number is 1, 2, or 3, the acceptance number for the corresponding tightened inspection plan is reduced by 1. If the normal inspection acceptance number is 5, 7, 10, or 14, the reduction in acceptance number for tightened inspection is 2. For a normal acceptance number of 21, the reduction is 3. Table 7.7 indicates that under reduced inspection, the sample size for this example would be $n = 50$, the acceptance number would be $c = 1$, and the rejection number would be $r = 3$. Thus, if two defectives were encountered, the lot would be accepted, but the next lot would be inspected under normal inspection.

In examining the tables, note that if a vertical arrow is encountered, the first sampling plan above or below the arrow should be used. When this occurs, the sample size code letter and the sample size change. For example, if a single-sampling plan is indexed by an AQL of 1.5% and a sample size code letter of F, the code letter changes to G and the sample size changes from 20 to 32.

Table 7.4

Sample Size Code Letters (MIL STD 105E, Table 1)

	Special Inspection Levels				General Inspection Levels		
Lot or Batch Size	S-1	S-2	S-3	S-4	I	II	III
2 to 8	A	A	A	A	A	A	B
9 to 15	A	A	A	A	A	B	C
16 to 25	A	A	B	B	B	C	D
26 to 50	A	B	B	C	C	D	E
51 to 90	B	B	C	C	C	E	F
91 to 150	B	B	C	D	D	F	G
151 to 280	B	C	D	E	E	G	H
281 to 500	B	C	D	E	F	H	J
501 to 1200	C	C	E	F	G	J	K
1201 to 3200	C	D	E	G	H	K	L
3201 to 10,000	C	D	F	G	J	L	M
10,001 to 35,000	C	D	F	H	K	M	N
35,001 to 150,000	D	E	G	J	L	N	P
150,001 to 500,000	D	E	G	J	M	P	Q
500,001 and over	D	E	H	K	N	Q	R

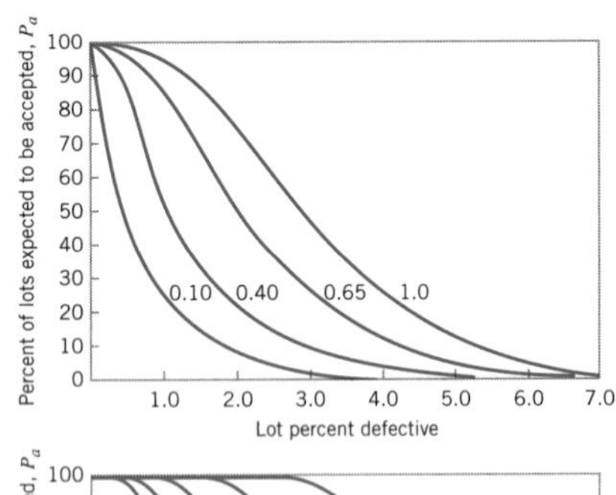

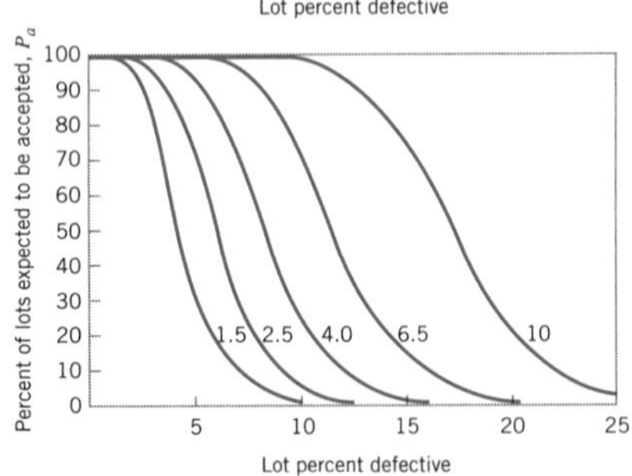

Figure 7.19 **OC Curves for Sample Size Code Letter K, MIL STD 105E**

7.4.3 DISCUSSION

MIL STD 105E presents the OC curves for single-sampling plans. These are all type-B OC curves. The OC curves for the matching double- and multiple-sampling plans are roughly comparable with those for the corresponding single-sampling plans. Figure 7.19 presents an example of these curves for code letter K. The OC curves presented in the standard are for the intitial-sampling plan only. They are not the OC curves for the overall inspection program, including shifts to and from tightened or reduced inspection.

Average sample number curves for double and multiple-sampling are given, assuming that no curtailment is used. These curves are useful in evaluating the average sample sizes that may be expected to occur under the various sampling plans for a given lot or process quality.

There are several points about MIL STD 105E that should be emphasized. First, MIL STD 105E is AQL-oriented. It focuses attention on the producer's risk end of the OC curve. The only control over the discriminatory power of the sampling plan (i.e., the steepness of the OC curve) is through the choice of inspection level.

Table 7.5

Master Table for Normal Inspection for Single-Sampling (U.S. Department of Defense MIL STD 105E, Table II-A)

Sample Size Code Letter	Sample Size	Acceptable Quality Levels (normal inspection)																									
		0.010	0.015	0.025	0.040	0.065	0.10	0.15	0.25	0.40	0.65	1.0	1.5	2.5	4.0	6.5	10	15	25	40	65	100	150	250	400	650	1000
		Ac Re	Ac Re	Ac Re	Ac Re	Ac Re	Ac Re	Ac Re	Ac Re	Ac Re	Ac Re	Ac Re	Ac Re	Ac Re	Ac Re	Ac Re	Ac Re	Ac Re	Ac Re	Ac Re	Ac Re	Ac Re	Ac Re	Ac Re	Ac Re	Ac Re	Ac Re
A	2	↓	↓	↓	↓	↓	↓	↓	↓	↓	↓	↓	↓	↓	↓	0 1	↓	↓	1 2	2 3	3 4	5 6	7 8	10 11	14 15	21 22	30 31
B	3	↓	↓	↓	↓	↓	↓	↓	↓	↓	↓	↓	↓	↓	0 1	↑	↓	1 2	2 3	3 4	5 6	7 8	10 11	14 15	21 22	30 31	44 45
C	5	↓	↓	↓	↓	↓	↓	↓	↓	↓	↓	↓	↓	0 1	↑	↓	1 2	2 3	3 4	5 6	7 8	10 11	14 15	21 22	30 31	44 45	
D	8	↓	↓	↓	↓	↓	↓	↓	↓	↓	↓	↓	0 1	↑	↓	1 2	2 3	3 4	5 6	7 8	10 11	14 15	21 22	30 31	44 45	↑	↑
E	13	↓	↓	↓	↓	↓	↓	↓	↓	↓	↓	0 1	↑	↓	1 2	2 3	3 4	5 6	7 8	10 11	14 15	21 22	30 31	44 45	↑	↑	↑
F	20	↓	↓	↓	↓	↓	↓	↓	↓	↓	0 1	↑	↓	1 2	2 3	3 4	5 6	7 8	10 11	14 15	21 22	↑	↑	↑	↑	↑	↑
G	32	↓	↓	↓	↓	↓	↓	↓	↓	0 1	↑	↓	1 2	2 3	3 4	5 6	7 8	10 11	14 15	21 22	↑	↑	↑	↑	↑	↑	↑
H	50	↓	↓	↓	↓	↓	↓	↓	0 1	↑	↓	1 2	2 3	3 4	5 6	7 8	10 11	14 15	21 22	↑	↑	↑	↑	↑	↑	↑	↑
J	80	↓	↓	↓	↓	↓	↓	0 1	↑	↓	1 2	2 3	3 4	5 6	7 8	10 11	14 15	21 22	↑	↑	↑	↑	↑	↑	↑	↑	↑
K	125	↓	↓	↓	↓	↓	0 1	↑	↓	1 2	2 3	3 4	5 6	7 8	10 11	14 15	21 22	↑	↑	↑	↑	↑	↑	↑	↑	↑	↑
L	200	↓	↓	↓	↓	0 1	↑	↓	1 2	2 3	3 4	5 6	7 8	10 11	14 15	21 22	↑	↑	↑	↑	↑	↑	↑	↑	↑	↑	↑
M	315	↓	↓	↓	0 1	↑	↓	1 2	2 3	3 4	5 6	7 8	10 11	14 15	21 22	↑	↑	↑	↑	↑	↑	↑	↑	↑	↑	↑	↑
N	500	↓	↓	0 1	↑	↓	1 2	2 3	3 4	5 6	7 8	10 11	14 15	21 22	↑	↑	↑	↑	↑	↑	↑	↑	↑	↑	↑	↑	↑
P	800	↓	0 1	↑	↓	1 2	2 3	3 4	5 6	7 8	10 11	14 15	21 22	↑	↑	↑	↑	↑	↑	↑	↑	↑	↑	↑	↑	↑	↑
Q	1250	0 1	↑	↓	1 2	2 3	3 4	5 6	7 8	10 11	14 15	21 22	↑	↑	↑	↑	↑	↑	↑	↑	↑	↑	↑	↑	↑	↑	↑
R	2000	↑	↑	1 2	2 3	3 4	5 6	7 8	10 11	14 15	21 22	↑	↑	↑	↑	↑	↑	↑	↑	↑	↑	↑	↑	↑	↑	↑	↑

↓ = Use first-sampling plan below arrow. If sample size equals, or exceeds, lot or batch size, do 100% inspection.

↑ = Use first-sampling plan above arrow.

Ac = Acceptance number.

Re = Rejection number.

Table 7.6

Master Table for Tightened Inspection—Single-Sampling (U.S. Department of Defense MIL STD 105E, Table II-B)

Sample Size Code Letter	Sample Size	Acceptable Quality Levels (tightened inspection)																									
		0.010	0.015	0.025	0.040	0.065	0.10	0.15	0.25	0.40	0.65	1.0	1.5	2.5	4.0	6.5	10	15	25	40	65	100	150	250	400	650	1000
		Ac Re	Ac Re	Ac Re	Ac Re	Ac Re	Ac Re	Ac Re	Ac Re	Ac Re	Ac Re	Ac Re	Ac Re	Ac Re	Ac Re	Ac Re	Ac Re	Ac Re	Ac Re	Ac Re	Ac Re	Ac Re	Ac Re	Ac Re	Ac Re	Ac Re	Ac Re
A	2	↓	↓	↓	↓	↓	↓	↓	↓	↓	↓	↓	↓	↓	↓	↓	↓	↓	↓	1 2	2 3	3 4	5 6	8 9	12 13	18 19	27 28
B	3	↓	↓	↓	↓	↓	↓	↓	↓	↓	↓	↓	↓	↓	↓	0 1	↓	↓	1 2	2 3	3 4	5 6	8 9	12 13	18 19	27 28	41 42
C	5	↓	↓	↓	↓	↓	↓	↓	↓	↓	↓	↓	↓	↓	0 1	↓	↓	1 2	2 3	3 4	5 6	8 9	12 13	18 19	27 28	41 42	↑
D	8	↓	↓	↓	↓	↓	↓	↓	↓	↓	↓	↓	↓	0 1	↓	↓	1 2	2 3	3 4	5 6	8 9	12 13	18 19	27 28	41 42	↑	↑
E	13	↓	↓	↓	↓	↓	↓	↓	↓	↓	↓	↓	0 1	↓	↓	1 2	2 3	3 4	5 6	8 9	12 13	18 19	27 28	41 42	↑	↑	↑
F	20	↓	↓	↓	↓	↓	↓	↓	↓	↓	↓	0 1	↓	↓	1 2	2 3	3 4	5 6	8 9	12 13	18 19	↑	↑	↑	↑	↑	↑
G	32	↓	↓	↓	↓	↓	↓	↓	↓	↓	0 1	↓	↓	1 2	2 3	3 4	5 6	8 9	12 13	18 19	↑	↑	↑	↑	↑	↑	↑
H	50	↓	↓	↓	↓	↓	↓	↓	↓	0 1	↓	↓	1 2	2 3	3 4	5 6	8 9	12 13	18 19	↑	↑	↑	↑	↑	↑	↑	↑
J	80	↓	↓	↓	↓	↓	↓	↓	0 1	↓	↓	1 2	2 3	3 4	5 6	8 9	12 13	18 19	↑	↑	↑	↑	↑	↑	↑	↑	↑
K	125	↓	↓	↓	↓	↓	↓	0 1	↓	↓	1 2	2 3	3 4	5 6	8 9	12 13	18 19	↑	↑	↑	↑	↑	↑	↑	↑	↑	↑
L	200	↓	↓	↓	↓	↓	0 1	↓	↓	1 2	2 3	3 4	5 6	8 9	12 13	18 19	↑	↑	↑	↑	↑	↑	↑	↑	↑	↑	↑
M	315	↓	↓	↓	↓	0 1	↓	↓	1 2	2 3	3 4	5 6	8 9	12 13	18 19	↑	↑	↑	↑	↑	↑	↑	↑	↑	↑	↑	↑
N	500	↓	↓	↓	0 1	↓	↓	1 2	2 3	3 4	5 6	8 9	12 11	18 19	↑	↑	↑	↑	↑	↑	↑	↑	↑	↑	↑	↑	↑
P	800	↓	↓	0 1	↓	↓	1 2	2 3	3 4	5 6	8 9	12 13	18 19	↑	↑	↑	↑	↑	↑	↑	↑	↑	↑	↑	↑	↑	↑
Q	1250	↓	0 1	↓	↓	1 2	2 3	3 4	5 6	8 9	12 13	18 19	↑	↑	↑	↑	↑	↑	↑	↑	↑	↑	↑	↑	↑	↑	↑
R	2000	0 1	↑	↓	1 2	2 3	3 4	5 6	8 9	12 13	18 19	↑	↑	↑	↑	↑	↑	↑	↑	↑	↑	↑	↑	↑	↑	↑	↑
S	3150			1 2																							

↓ = Use first-sampling plan below arrow. If sample size equals, or exceeds, lot or batch size, do 100% inspection.

↑ = Use first-sampling plan above arrow.

Ac = Acceptance number.

Re = Rejection number.

Table 7.7

Master Table for Reduced Inspection—Single-Sampling (U.S. Department of Defense MIL STD 105E, Table II-C)

Sample Size Code Letter	Sample Size	Acceptable Quality Levels (reduced inspection)†																									
		0.010	0.015	0.025	0.040	0.065	0.10	0.15	0.25	0.40	0.65	1.0	1.5	2.5	4.0	6.5	10	15	25	40	65	100	150	250	400	650	1000
		Ac Re	Ac Re	Ac Re	Ac Re	Ac Re	Ac Re	Ac Re	Ac Re	Ac Re	Ac Re	Ac Re	Ac Re	Ac Re	Ac Re	Ac Re	Ac Re	Ac Re	Ac Re	Ac Re	Ac Re	Ac Re	Ac Re	Ac Re	Ac Re	Ac Re	Ac Re
A	2	↓	↓	↓	↓	↓	↓	↓	↓	↓	↓	↓	↓	↓	↓	0 1	↓	↓	1 2	2 3	3 4	5 6	7 8	10 11	14 15	21 22	30 31
B	2	↓	↓	↓	↓	↓	↓	↓	↓	↓	↓	↓	↓	↓	0 1	↑	↓	0 2	1 3	2 4	3 5	5 6	7 8	10 11	14 15	21 22	30 31
C	2	↓	↓	↓	↓	↓	↓	↓	↓	↓	↓	↓	↓	0 1	↑	↓	0 2	1 3	1 4	2 5	3 6	5 8	7 10	10 13	14 17	21 24	↑
D	3	↓	↓	↓	↓	↓	↓	↓	↓	↓	↓	↓	0 1	↑	↓	0 2	1 3	1 4	2 5	3 6	5 8	7 10	10 13	14 17	21 24	↑	↑
E	5	↓	↓	↓	↓	↓	↓	↓	↓	↓	↓	0 1	↑	↓	0 2	1 3	1 4	2 5	3 6	5 8	7 10	10 13	14 17	21 24	↑	↑	↑
F	8	↓	↓	↓	↓	↓	↓	↓	↓	↓	0 1	↑	↓	0 2	1 3	1 4	2 5	3 6	5 8	7 10	10 13	↑	↑	↑	↑	↑	↑
G	13	↓	↓	↓	↓	↓	↓	↓	↓	0 1	↑	↓	0 2	1 3	1 4	2 5	3 6	5 8	7 10	10 13	↑	↑	↑	↑	↑	↑	↑
H	20	↓	↓	↓	↓	↓	↓	↓	0 1	↑	↓	0 2	1 3	1 4	2 5	3 6	5 8	7 10	10 13	↑	↑	↑	↑	↑	↑	↑	↑
J	32	↓	↓	↓	↓	↓	↓	0 1	↑	↓	0 2	1 3	1 4	2 5	3 6	5 8	7 10	10 13	↑	↑	↑	↑	↑	↑	↑	↑	↑
K	50	↓	↓	↓	↓	↓	0 1	↑	↓	0 2	1 3	1 4	2 5	3 6	5 8	7 10	10 13	↑	↑	↑	↑	↑	↑	↑	↑	↑	↑
L	80	↓	↓	↓	↓	0 1	↑	↓	0 2	1 3	1 4	2 5	3 6	5 8	7 10	10 13	↑	↑	↑	↑	↑	↑	↑	↑	↑	↑	↑
M	125	↓	↓	↓	0 1	↑	↓	0 2	1 3	1 4	2 5	3 6	5 8	7 10	10 13	↑	↑	↑	↑	↑	↑	↑	↑	↑	↑	↑	↑
N	200	↓	↓	0 1	↑	↓	0 2	1 3	1 4	2 5	3 6	5 8	7 10	10 13	↑	↑	↑	↑	↑	↑	↑	↑	↑	↑	↑	↑	↑
P	315	↓	0 1	↑	↓	0 2	1 3	1 4	2 5	3 6	5 8	7 10	10 13	↑	↑	↑	↑	↑	↑	↑	↑	↑	↑	↑	↑	↑	↑
Q	500	0 1	↑	↓	0 2	1 3	1 4	2 5	3 6	5 8	7 10	10 13	↑	↑	↑	↑	↑	↑	↑	↑	↑	↑	↑	↑	↑	↑	↑
R	800	↑	↑	0 2	1 3	1 4	2 5	3 6	5 8	7 10	10 13	↑	↑	↑	↑	↑	↑	↑	↑	↑	↑	↑	↑	↑	↑	↑	↑

↓ = Use first-sampling plan below arrow. If sample size equals, or exceeds, lot or batch size, do 100% inspection.

↑ = Use first-sampling plan above arrow.

Ac = Acceptance number.

Re = Rejection number.

† = If the acceptance number has been exceeded, but the rejection number has not been reached, accept the lot, but reinstate normal inspection.

ANSI/ASQC Z1.4 presents the scheme performance of the standard, giving scheme OC curves and the corresponding percentage points.

Second, the sample sizes selected for use in MIL STD 105E are 2, 3, 5, 8, 13, 20, 32, 50, 80, 125, 200, 315, 500, 800, 1250, and 2000. Thus, not all sample sizes are possible. Note that there are some rather significant gaps, such as between 125 and 200, and between 200 and 315.

Third, the sample sizes in MIL STD 105E are related to the lot sizes. To see the nature of this relationship, calculate the midpoint of each lot size range, and plot the logarithm of the sample size for that lot size range against the logarithm of the lot size range midpoint. Such a plot will follow roughly a straight line up to $n = 80$, and thereafter another straight line with a shallower slope. Thus, the sample size will increase as the lot size increases. However, the ratio of sample size to lot size will decrease rapidly. This gives significant economy in inspection costs per unit when the supplier submits large lots. For a given AQL, the effect of this increase in sample size as the lot size increases is to increase the probability of acceptance for submitted lots of AQL quality. The probability of acceptance at a given AQL will vary with increasing sample size from about 0.91 to about 0.99. This feature of the standard was and still is subject to some controversy. The argument in favor of the approach in MIL STD 105E is that rejection of a large lot has more serious consequences for the supplier than rejection of a small lot, and if the probability of acceptance at the AQL increases with sample size, this reduces the risk of false rejection of a large lot. Furthermore, the large sample also gives a more discriminating OC curve, which means that the protection that the consumer receives against accepting an isolated bad lot will also be increased.

Fourth, the switching rules from normal to tightened inspection and from tightened to normal inspection are also subject to some criticism. In particular, some engineers dislike the switching rules because there is often a considerable amount of misswitching from normal to tightened or normal to reduced inspection when the process is actually producing lots of AQL quality. Also, there is a significant probability that production would even be discontinued, even though there has been no actual quality deterioration.

Fifth, a **flagrant** and **common abuse** of MIL STD 105E is failure to use the switching rules at all. When this is done, it results in ineffective and deceptive inspection and a substantial increase in the consumer's risk. It is not recommended that MIL STD 105E be implemented without use of the switching rules from normal to tightened and normal to reduced inspection.

A civilian standard, ANSI/ASQC Z1.4 or ISO 2859, is the counterpart of MIL STD 105E. It seems appropriate to conclude our discussion of MIL STD 105E with a comparison of the military and civilian standards. ANSI/ASQC Z1.4 was adopted in 1981 and differs from MIL STD 105E in the following five ways:

1. The terminology "nonconformity," "nonconformance," and "percent nonconforming" is used.
2. The switching rules were changed slightly to provide an option for reduced inspection without the use of limit numbers.
3. Several tables that show measures of scheme performance (*including* the switching rules) were introduced. Some of these

performance measures include AOQL, limiting quality for which $P_a = 0.10$ and $P_a = 0.05$, ASN, and operating-characteristic curves.

4. A section was added describing proper use of individual sampling plans when extracted from the system.
5. A figure illustrating the switching rules was added.

These revisions modernize the terminology and emphasize the system concept of the civilian standard. All tables, numbers, and procedures used in MIL STD 105E are retained in ANSI/ASQC Z1.4 and ISO 2859.

7.5 The Dodge–Romig Sampling Plans

H. F. Dodge and H. G. Romig developed a set of sampling inspection tables for lot-by-lot inspection of product by attributes using two types of sampling plans: plans for lot tolerance percent defective (LTPD) protection and plans that provide a specified average outgoing quality limit (AOQL). For each of these approaches to sampling plan design, there are tables for single- and double-sampling.

Sampling plans that emphasize LTPD protection, such as the Dodge–Romig plans, are often preferred to AQL-oriented sampling plans, such as those in MIL STD 105E, particularly for critical components and parts. Many manufacturers believe that they have relied too much on AQLs in the past, and they are now emphasizing other measures of performance, such as defective parts per million (ppm). Consider the adjacent display. We see that even very small AQL imply large numbers of defective ppm. In complex products with many components, the effect of this can be devastating.

AQL	Defective Parts per Million
10%	100,000
1%	10,000
0.1%	1,000
0.01%	100
0.001%	10
0.0001%	1

The Dodge–Romig AOQL plans are designed so that the average total inspection for a given AOQL and a specified process average p will be minimized. Similarly, the LTPD plans are designed so that the average total inspection is a minimum. This makes the Dodge–Romig plans very useful for in-plant inspection of semifinished product.

Dodge-Roming Plans only apply to programs that submit rejected lots for 100% inspection

The Dodge–Romig plans apply only to programs that submit rejected lots to 100% inspection. Unless rectifying inspection is used, the AOQL concept is meaningless. Furthermore, to use the plans, we must know the process average—that is, the average fraction nonconforming of the incoming product. When a supplier is relatively new, we usually do not know its process fallout. Sometimes this may be estimated from a preliminary sample or from data provided by the supplier. Alternatively, the largest possible process average in the table can be used until enough information has been generated to provide a more accurate estimate of the supplier's process fallout. Obtaining a more accurate estimate of the incoming fraction nonconforming or process average will allow a more appropriate sampling plan to be adopted. It is not uncommon to find that sampling inspection begins with one plan, and after sufficient information is generated to reestimate the supplier's process fallout, a new plan is adopted. Control charts are usually employed to estimate the process average.

7.5.1 AOQL PLANS

The Dodge–Romig (1959) tables give AOQL sampling plans for AOQL values of 0.1%, 0.25%, 0.5%, 0.75%, 1%, 1.5%, 2%, 2.5%, 3%, 4%, 5%, 7%, and 10%. For each of these AOQL values, six classes of values for the process average are specified. Tables are provided for both single and double-sampling. These plans have been designed so that the average total inspection at the given AOQL and process average is approximately a minimum.

Tables 7.8 and 7.9 are adapted from H. F. Dodge and H. G. Romig, *Sampling Inspection Tables, Single and Double Sampling*, 2nd ed., John Wiley, New York, 1959, with the permission of the publisher.

An example of the Dodge–Romig sampling plans is shown in Table 7.8. To illustrate the use of the Dodge–Romig AOQL tables, suppose that we are inspecting LSI memory elements for a personal computer and that the elements are shipped in lots of size N = 5,000. The supplier's process average fallout is 1% nonconforming. We wish to find a single-sampling plan with an AOQL = 3%. From Table 7.8, we find that the plan is

$$n = 65 \quad c = 3$$

Table 7.8 also indicates that the LTPD for this sampling plan is 10.3%. This is the point on the OC curve for which $P_a = 0.10$. Therefore, the sampling plan $n = 65$, $c = 3$ gives an AOQL of 3% nonconforming and provides assurance that 90% of incoming lots that are as bad as 10.3% defective will be rejected. Assuming that incoming quality is equal to the process average and that the probability of lot acceptance at this level of quality is $P_a = 0.9957$, we find that the average total inspection for this plan is

$$\begin{aligned} \text{ATI} &= n \quad 1 - P_a \quad N - n \\ &= 65 \quad 1 - 0.9957 \quad 5{,}000 - 65 = 86.22 \end{aligned}$$

Thus, we will inspect approximately 86 units, on the average, in order to sentence a lot.

7.5.2 LTPD PLANS

The Dodge–Romig LTPD tables are designed so that the probability of lot acceptance at the LTPD is 0.1. Tables are provided for LTPD values of 0.5%, 1%, 2%, 3%, 4%, 5%, 7%, and 10%. Table 7.9 for an LTPD of 1% is representative of these Dodge–Romig tables.

To illustrate the use of these tables, suppose that LSI memory elements for a personal computer are shipped from the supplier in lots of size N = 5000. The supplier's process average fallout is 0.25% nonconforming, and we wish to use a single-sampling plan with an LTPD of 1%. From inspection of Table 7.9, the sampling plan that should be used is

$$n = 770 \quad c = 4$$

If we assume that rejected lots are screened 100% and that defective items are replaced with good ones, the AOQL for this plan is approximately 0.28%.

Table 7.8

Dodge–Romig Inspection Table—Single-Sampling Plans for AOQL = 3.0%

	Process Average																	
	0–0.06%			0.07–0.60%			0.61–1.20%			1.21–1.80%			1.81–2.40%			2.41–3.00%		
Lot Size	*n*	*c*	LTPD %	*n*	*c*	LTPD %	*n*	*c*	LTPD %	*n*	*c*	LTPD %	*n*	*c*	LTPD %	*n*	*c*	LTPD %
1–10	All	0	—	All	0	—	All	0	—	All	0	—	All	0	—	All	0	—
11–50	10	0	19.0	10	0	19.0	10	0	19.0	10	0	19.0	10	0	19.0	10	0	19.0
51–100	11	0	18.0	11	0	18.0	11	0	18.0	11	0	18.0	11	0	18.0	22	1	16.4
101–200	12	0	17.0	12	0	17.0	12	0	17.0	25	1	15.1	25	1	15.1	25	1	15.1
201–300	12	0	17.0	12	0	17.0	26	1	14.6	26	1	14.6	26	1	14.6	40	2	12.8
301–400	12	0	17.1	12	0	17.1	26	1	14.7	26	1	14.7	41	2	12.7	41	2	12.7
401–500	12	0	17.2	27	1	14.1	27	1	14.1	42	2	12.4	42	2	12.4	42	2	12.4
501–600	12	0	17.3	27	1	14.2	27	1	14.2	42	2	12.4	42	2	12.4	60	3	10.8
601–800	12	0	17.3	27	1	14.2	27	1	14.2	43	2	12.1	60	3	10.9	60	3	10.9
801–1,000	12	0	17.4	27	1	14.2	44	2	11.8	44	2	11.8	60	3	11.0	80	4	9.8
1,001–2,000	12	0	17.5	28	1	13.8	45	2	11.7	65	3	10.2	80	4	9.8	100	5	9.1
2,001–3,000	12	0	17.5	28	1	13.8	45	2	11.7	65	3	10.2	100	5	9.1	140	7	8.2
3,001–4,000	12	0	17.5	28	1	13.8	65	3	10.3	85	4	9.5	125	6	8.4	165	8	7.8
4,001–5,000	28	1	13.8	28	1	13.8	65	3	10.3	85	4	9.5	125	6	8.4	210	10	7.4
5,001–7,000	28	1	13.8	45	2	11.8	65	3	10.3	105	5	8.8	145	7	8.1	235	11	7.1
7,001–10,000	28	1	13.9	46	2	11.6	65	3	10.3	105	5	8.8	170	8	7.6	280	13	6.8
10,001–20,000	28	1	13.9	46	2	11.7	85	4	9.5	125	6	8.4	215	10	7.2	380	17	6.2
20,001–50,000	28	1	13.9	65	3	10.3	105	5	8.8	170	8	7.6	310	14	6.5	560	24	5.7
50,001–100,000	28	1	13.9	65	3	10.3	125	6	8.4	215	10	7.2	385	17	6.2	690	29	5.4

Table 7.9

Dodge–Romig Single-Sampling Table for Lot Tolerance Percent Defective (LTPD) = 1.0%

	Process Average																	
	0–0.01%			0.011%–0.10%			0.11–0.20%			0.21–0.30%			0.31–0.40%			0.41–0.50%		
Lot Size	n	c	AOQL %	n	c	AOQL %	n	c	AOQL %	n	c	AOQL %	n	c	AOQL %	n	c	AOQL %
1–120	All	0	0	All	0	0	All	0	0	All	0	0	All	0	0	All	0	0
121–150	120	0	0.06	120	0	0.06	120	0	0.06	120	0	0.06	120	0	0.06	120	0	0.06
151–200	140	0	0.08	140	0	0.08	140	0	0.08	140	0	0.08	140	0	0.08	140	0	0.08
201–300	165	0	0.10	165	0	0.10	165	0	0.10	165	0	0.10	165	0	0.10	165	0	0.10
301–400	175	0	0.12	175	0	0.12	175	0	0.12	175	0	0.12	175	0	0.12	175	0	0.12
401–500	180	0	0.13	180	0	0.13	180	0	0.13	180	0	0.13	180	0	0.13	180	0	0.13
501–600	190	0	0.13	190	0	0.13	190	0	0.13	190	0	0.13	190	0	0.13	305	1	0.14
601–800	200	0	0.14	200	0	0.14	200	0	0.14	330	1	0.15	330	1	0.15	330	1	0.15
801–1,000	205	0	0.14	205	0	0.14	205	0	0.14	335	1	0.17	335	1	0.17	335	1	0.17
1,001–2,000	220	0	0.15	220	0	0.15	360	1	0.19	490	2	0.21	490	2	0.21	610	3	0.22
2,001–3,000	220	0	0.15	375	1	0.20	505	2	0.23	630	3	0.24	745	4	0.26	870	5	0.26
3,001–4,000	225	0	0.15	380	1	0.20	510	2	0.23	645	3	0.25	880	5	0.28	1,000	6	0.29
4,001–5,000	225	0	0.16	380	1	0.20	520	2	0.24	770	4	0.28	895	5	0.29	1,120	7	0.31
5,001–7,000	230	0	0.16	385	1	0.21	655	3	0.27	780	4	0.29	1,020	6	0.32	1,260	8	0.34
7,001–10,000	230	0	0.16	520	2	0.25	660	3	0.28	910	5	0.32	1,150	7	0.34	1,500	10	0.37
10,001–20,000	390	1	0.21	525	2	0.26	785	4	0.31	1,040	6	0.35	1,400	9	0.39	1,980	14	0.43
20,001–50,000	390	1	0.21	530	2	0.26	920	5	0.34	1,300	8	0.39	1,890	13	0.44	2,570	19	0.48
50,001–100,000	390	1	0.21	670	3	0.29	1,040	6	0.36	1,420	9	0.41	2,120	15	0.47	3,150	23	0.50

Note from inspection of the Dodge–Romig LTPD tables that values of the process average cover the interval from zero to half the LTPD. Provision for larger process averages is unnecessary, since 100% inspection is more economically efficient than inspection sampling when the process average exceeds half the desired LTPD.

7.6 MIL STD 414 (ANSI/ASQC Z1.9)

7.6.1 GENERAL DESCRIPTION OF THE STANDARD

Many quality characteristics are expressed as a variable. The main advantage of variables-sampling plans is that the same operating characteristic curve can be obtained with a smaller sample size than would be required by attributes sampling. Thus, when this is a choice, sampling plans for variables should be given consideration.

MIL STD 414 is a lot-by-lot acceptance-sampling plan for variables. The standard was introduced in 1957. The focal point of this standard is the acceptable quality level (AQL), which ranges from 0.04% to 15%. There are five general levels of inspection, and level IV is designated as "normal." Inspection level V gives a steeper OC curve than level IV. When reduced sampling costs are necessary and when greater risks can or must be tolerated, lower inspection levels can be used. As with the attributes standard, MIL STD 105E, sample size code letters are used, but the same code letter does not imply the same sample size in both standards. In addition, the lot size classes are different in both standards. Sample sizes are a function of the lot size and the inspection level. Provision is made for normal, tightened, and reduced inspection. All the sampling plans and procedures in the standard assume that the quality characteristic of interest is normally distributed.

Figure 7.20 presents the organization of the standard. Note that acceptance-sampling plans can be designed for cases where the lot or process variability is either known or unknown, and where there are either single-specification limits or double-specification limits on the quality characteristic. The calculations can be organized in two ways:

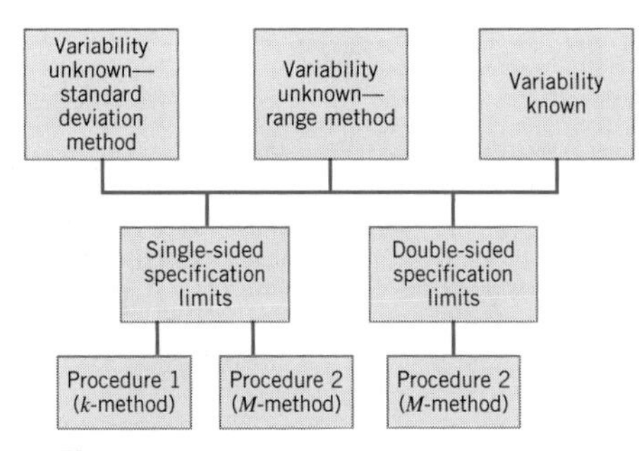

Figure 7.20 **Organization of MIL STD 414**

Procedure 1. Take a random sample of n items from the lot and compute the statistic

$$Z_{LSL} = \frac{\bar{x} - LSL}{\sigma} \tag{7.19}$$

Note that Z_{LSL} in Equation (7.19) simply expresses the distance between the sample average $\bar{x}$ and the lower specification limit in standard deviation units. The larger the value of Z_{LSL}, the farther the sample average $\bar{x}$ is from the lower specification limit, and consequently, the smaller is the lot fraction defective p. If there is a critical value of p of interest that should not be exceeded with stated probability, we can translate this value of p into a **critical distance**—say, k—for Z_{LSL}. Thus, if $Z_{LSL} \geq k$, we would accept the lot because the sample data imply that the lot mean is sufficiently far above the LSL to ensure that the lot fraction nonconforming is satisfactory. However, if $Z_{LSL} < k$, the mean is too close to the LSL, and the lot should be rejected.

Procedure 2. Take a random sample of n items from the lot and compute Z_{LSL} using Equation (7.19). Use Z_{LSL} to estimate the fraction defective of the lot or process as the area under the standard normal curve below Z_{LSL}. (Actually, using $Q_{LSL} = Z_{LSL}\sqrt{n/(n-1)}$ as a standard normal variable is slightly better, because it gives a better estimate of p.) Let $\hat{p}$ be the estimate of p so obtained. If the estimate $\hat{p}$ exceeds a specified maximum value M, reject the lot; otherwise, accept it.

In the case of single-specification limits, either Procedure 1 or Procedure 2 may be used. If there are double-specification limits, then Procedure 2 must be used. If the process or lot variability is known and stable, the variability known plans are the most economically efficient. When lot or process variability is unknown, either the standard deviation or the range of the sample may be used in operating the sampling plan. The range method requires a larger sample size, and we do not generally recommend its use.

MIL STD 414 is divided into four sections. Section A is a general description of the sampling plans, including definitions, sample size code letters, and OC curves for the various sampling plans. Section B of the standard gives variables sampling plans based on the sample standard deviation for the case in which the process or lot variability is unknown. Section C presents variables sampling plans based on the sample range method. Section D gives variables sampling plans for the case where the process standard deviation is known.

example 7.8 Using MIL STD 414

Consider a soft-drink bottler who is purchasing bottles from a supplier. The lower specification limit on bursting strength is 225 psi. Suppose that the AQL at this specification limit is 1%. Suppose that bottles are shipped in lots of size 100,000. Find a variables sampling plan that uses Procedure 1 from MIL STD 414. Assume that the lot standard deviation is unknown.

solution

From Table 7.10, if we use inspection level IV, the sample size code letter is O. From Table 7.11 we find that sample size code letter O implies a sample size of $n = 100$. For an acceptable quality level of 1%, on normal inspection, the value of k is 2.00. If tightened inspection is employed, the appropriate value of k is 2.14. Note that normal and tightened inspection use the same tables. The AQL values for normal inspection are indexed at the top of the table, and the AQL values for tightened inspection are indexed from the bottom of the table.

7.6.2 USE OF THE TABLES

Two typical tables from MIL STD 414 are reproduced in Tables 7.10 and 7.11. The following example illustrates the use of these tables.

MIL STD 414 contains a provision for a shift to tightened or reduced inspection when this is warranted. The process average is used as the basis for determining when such a shift is made. The process average is taken as the average of the sample estimates of percent defective computed for lots submitted on original inspection. Usually, the process average is computed using information from the preceding ten lots. Full details of the switching procedures are described in the standard and in a technical memorandum on MIL STD 414, published by the United States Department of the Navy, Bureau of Ordnance.

Table 7.10

Sample Size Code Letters (MIL STD 414, Table A.2)

	Inspection Levels				
Lot Size	I	II	III	IV	V
3 to 8	B	B	B	B	C
9 to 15	B	B	B	B	D
16 to 25	B	B	B	C	E
26 to 40	B	B	B	D	F
41 to 65	B	B	C	E	G
66 to 110	B	B	D	F	H
111 to 180	B	C	E	G	I
181 to 300	B	D	F	H	J
301 to 500	C	E	G	I	K
501 to 800	D	F	H	J	L
801 to 1,300	E	G	I	K	L
1,301 to 3,200	F	H	J	L	M
3,201 to 8,000	G	I	L	M	N
8,001 to 22,000	H	J	M	N	O
22,001 to 110,000	I	K	N	O	P
110,001 to 550,000	I	K	O	P	Q
550,001 and over	I	K	P	Q	Q

Table 7.11

Master Table for Normal and Tightened Inspection for Plans Based on Variability Unknown (Standard Deviation Method) (Single-Specification Limit—Form 1)(MIL STD 414, Table B.1)

Sample Size Code Letter	Sample Size	Acceptable Quality Levels (normal inspection)													
		0.04	0.065	0.10	0.15	0.25	0.40	0.65	1.00	1.50	2.50	4.00	6.50	10.00	15.00
		k	*k*	*k*	*k*	*k*	*k*	*k*	*k*	*k*	*k*	*k*	*k*	*k*	*k*
B	3	↓	↓	↓	↓	↓	↓	↓	▼	▼	1.12	0.958	0.765	0.566	0.341
C	4	↓	↓	↓	↓	↓	↓	↓	1.45	1.34	1.17	1.01	0.814	0.617	0.393
D	5	↓	↓	↓	↓	↓	↓	1.65	1.53	1.40	1.24	1.07	0.874	0.675	0.455
E	7	↓	↓	↓	↓	2.00	1.88	1.75	1.62	1.50	1.33	1.15	0.955	0.755	0.536
F	10	↓	↓	↓	2.24	2.11	1.98	1.84	1.72	1.58	1.41	1.23	1.03	0.828	0.611
G	15	2.64	2.53	2.42	2.32	2.20	2.06	1.91	1.79	1.65	1.47	1.30	1.09	0.886	0.664
H	20	2.69	2.58	2.47	2.36	2.24	2.11	1.96	1.82	1.69	1.51	1.33	1.12	0.917	0.695
I	25	2.72	2.61	2.50	2.40	2.26	2.14	1.98	1.85	1.72	1.53	1.35	1.14	0.936	0.712
J	30	2.73	2.61	2.51	2.41	2.28	2.15	2.00	1.86	1.73	1.55	1.36	1.15	0.946	0.723
K	35	2.77	2.65	2.54	2.45	2.31	2.18	2.03	1.89	1.76	1.57	1.39	1.18	0.969	0.745
L	40	2.77	2.66	2.55	2.44	2.31	2.18	2.03	1.89	1.76	1.58	1.39	1.18	0.971	0.746
M	50	2.83	2.71	2.60	2.50	2.35	2.22	2.08	1.93	1.80	1.61	1.42	1.21	1.00	0.774
N	75	2.90	2.77	2.66	2.55	2.41	2.27	2.12	1.98	1.84	1.65	1.46	1.24	1.03	0.804
O	100	2.92	2.80	2.69	2.58	2.43	2.29	2.14	2.00	1.86	1.67	1.48	1.26	1.05	0.819
P	150	2.96	2.84	2.73	2.61	2.47	2.33	2.18	2.03	1.89	1.70	1.51	1.29	1.07	0.841
Q	200	2.97	2.85	2.73	2.62	2.47	2.33	2.18	2.04	1.89	1.70	1.51	1.29	1.07	0.845
		0.065	0.10	0.15	0.25	0.40	0.65	1.00	1.50	2.50	4.00	6.50	10.00	15.00	
		Acceptable Quality Levels (tightened inspection)													

All AQL values are in percent defective.

↓ Use first sampling plan below arrow, that is, both sample size as well as k value. When sample size equals or exceeds lot size, every item in the lot must be inspected.

Estimation of the fraction defective is required in using Procedure 2 of MIL STD 414. It is also required in implementing the switching rules between normal, tightened, and reduced inspection. In the standard, three tables are provided for estimating the fraction defective.

When starting to use MIL STD 414, one can choose between the known standard deviation and unknown standard deviation procedures. When there is no basis for knowledge of σ, obviously the unknown standard deviation plan must be used. However, it is a good idea to maintain either an R or s chart on the results of each lot so that some information on the state of statistical control of the scatter in the manufacturing process can be collected. If this control chart indicates statistical control, it will be possible to switch to a known σ plan. Such a switch will reduce the required sample size. Even if the process were not perfectly controlled, the control chart could provide information leading to a conservative estimate of σ for use in a known σ plan. When a known s plan is used, it is also necessary to maintain a control chart on either R or s as a continuous check on the assumption of stable and known process variability.

MIL STD 414 contains a special procedure for application of mixed variables/attributes acceptance-sampling plans. If the lot does not meet the acceptability criterion of the variables plan, an attributes single-sampling plan, using tightened inspection and the same AQL, is obtained from MIL STD 105E. A lot can be accepted by either of the plans in sequence but must be rejected by both the variables and attributes plan.

7.6.3 DISCUSSION

In 1980, the American National Standards Institute and the American Society for Quality Control released an updated civilian version of MIL STD 414 known as ANSI/ASQC Z1.9. MIL STD 414 was originally structured to give protection essentially equivalent to that provided by MIL STD 105A (1950). When MIL STD 105D was adopted in 1963, this new standard contained substantially revised tables and procedures that led to differences in protection between it and MIL STD 414. Consequently, it is not possible to move directly from an attributes sampling plan in the current MIL STD 105E to a corresponding variables plan in MIL STD 414 if the assurance of continued protection is desired for certain lot sizes and AQLs.

The civilian counterpart of MIL STD 414, ANSI/ASQC Z1.9, restores this original match. That is, ANSI/ASQC Z1.9 is directly compatible with MIL STD 105E (and its equivalent civilian counterpart ANSI/ASQC Z1.4). This equivalence was obtained by incorporating the following revisions in ANSI/ASQC Z1.9:

1. Lot size ranges were adjusted to correspond to MIL STD 105D.
2. The code letters assigned to the various lot size ranges were arranged to make protection equal to that of MIL STD 105E.
3. AQLs of 0.04, 0.065, and 15 were deleted.

4. The original inspection levels I, II, III, IV, and V were relabeled S3, S4, I, II, III, respectively.

5. The original switching rules were replaced by those of MIL STD 105E, with slight revisions.

In addition, to modernizing terminology, the word *nonconformity* was substituted for defect, *nonconformance* was substituted for defective, and *percent nonconforming* was substituted for percent defective. The operating-characteristic curves were recomputed and replotted, and a number of editorial changes were made to the descriptive material of the standard to match MIL STD 105E as closely as possible. Finally, an appendix was included showing the match between ANSI/ASQC Z1.9, MIL STD 105E, and the corresponding civilian version ANSI Z1.4. This appendix also provided selected percentage points from the OC curves of these standards and their differences.

As of this writing, the Department of Defense has not officially adopted ANSI/ASQC Z1.9 and continues to use MIL STD 414. Both standards will probably be used in the immediate future. The principal advantage of the ANSI/ASQC Z1.9 standard is that it is possible to start inspection by using an attributes sampling scheme from MIL STD 105E or ANSI/ASQC Z1.4, collect sufficient information to use variables inspection, and then switch to the variables scheme, while maintaining the same AQL-code letter combination. It would then be possible to switch back to the attributes scheme if the assumption of the variables scheme appeared not to be satisfied. It is also possible to take advantage of the information gained in coordinated attributes and variables inspection to move in a logical manner from inspection sampling to statistical process control.

As in MIL STD 414, ANSI/ASQC Z1.9 assumes that the quality characteristic is normally distributed. This is an important assumption that we have commented on previously. We have suggested that a test for normality should be incorporated as part of the standard. One way this can be done is to plot a control chart for $\bar{x}$ and S (or $\bar{x}$ and R) from the variables data from each lot. After a sufficient number of observations have been obtained, a test for normality can be employed by plotting the individual measurements on normal probability paper or by conducting one of the specialized statistical tests for normality. It is recommended that a relatively large sample size be used in this statistical test. At least 100 observations should be collected before the test for normality is made, and it is our belief that the sample size should increase inversely with AQL. If the assumption of normality is badly violated, either a special variables sampling procedure must be developed or we must return to attributes inspection.

An additional advantage of applying a control chart to the result of each lot is that if the process variability has been in control for at least 30 samples, it will be possible to switch to a known standard deviation plan, thereby allowing a substantial reduction in sample size. Although this can be instituted in any combined program of attributes and variables inspection, it is easy to do so using the ANSI/ASQC standards because of the design equivalence between the attributes and variables procedures.

7.7 Chain Sampling

For situations in which testing is destructive or very expensive, sampling plans with small sample sizes are usually selected. These small sample size plans often have acceptance numbers of 0. Plans with zero acceptance numbers are often undesirable, however, in that their OC curves are convex throughout. This means that the probability of lot acceptance begins to drop very rapidly as the lot fraction defective becomes greater than 0. This is often unfair to the producer, and in situations where rectifying inspection is used, it can require the consumer to screen a large number of lots that are essentially of acceptable quality. Figures 7.6 and 7.8 present OC curves of sampling plans that have acceptance numbers of 0 and acceptance numbers that are greater than 0.

Dodge (1955) suggested an alternate procedure, known as **chain sampling,** that might be a substitute for ordinary single-sampling plans with zero acceptance numbers in certain circumstances. Chain-sampling plans make use of the cumulative results of several preceding lots. The general procedure is as follows:

1. For each lot, select the sample of size n and observe the number of defectives.
2. If the sample has zero defectives, accept the lot; if the sample has two or more defectives, reject the lot; and if the sample has one defective, accept the lot provided there have been no defectives in the previous i lots.

Thus, for a chain sampling plan given by $n = 5$, $i = 3$, a lot would be accepted if there were no defectives in the sample of five, or if there were one defective in the sample of five and no defectives had been observed in the samples from the previous three lots. This type of plan is known as a ChSP-1 plan.

The effect of chain sampling is to alter the shape of the OC curve near the origin so that it has a more desirable shape. That is, it is more difficult to reject lots with very small fraction defectives with a ChSP-1 plan than it is with ordinary single sampling. Figure 7.21 shows OC curves for ChSP-1 plans with $n = 5$, $c = 0$, and $i = 1, 2, 3$, and 5. The curve for $i = 1$ is dotted, and it is not a preferred choice. In practice, values of i usually vary between 3 and 5, since the OC curves of such plans approximate the single-sampling plan OC curve. The points on the OC curve of a ChSP-1 plan are given by the equation

$$P_a = P(0, n) + P(1, n)[P(0, n)]^I \tag{7.20}$$

where $P(0, n)$ and $P(1, n)$ are the probabilities of obtaining zero and one defectives, respectively, out of a random sample of size n. To illustrate the computations, consider the ChSP-1 plan with $n = 5$, $c = 0$, and $i = 3$. For $p = 0.10$, we have

$$P(0, n) = \frac{n!}{d!(n-d)!}p^d(1-p)^{n-d} = \frac{5!}{0!\,5!}(0.10)^0(0.90)^5 = 0.590$$

$$P(1, n) = \frac{n!}{d!(n-d)!}p^d(1-p)^{n-d} = \frac{5!}{1!\,(5-1)!}(0.10)^1(0.90)^4 = 0.328$$

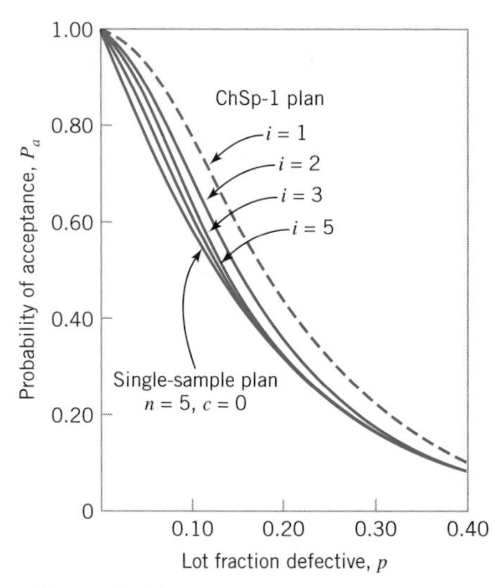

Figure 7.21 **OC Curves for ChSP-1 Plan with $n = 5$, $c = 0$, and $i = 1, 2, 3, 5$ (Reproduced with permission from H. F. Dodge, "Chain Sampling Inspection Plans," *Industrial Quality Control*, Vol. 11, No. 4, 1955)**

and

$$
\begin{aligned}
P_a &= P(0,n) + P(1,n)[P(0,n)]^i \\
&= 0.590 + (0.328)(0.590)^3 \\
&= 0.657
\end{aligned}
$$

The proper use of chain sampling requires that the following conditions be met:

1. The lot should be one of a series in a continuing stream of lots, from a process in which there is repetitive production under the same conditions, and in which the lots of products are offered for acceptance in substantially the order of production.
2. Lots should usually be expected to be of essentially the same quality.
3. The sampling agency should have no reason to believe that the current lot is of poorer quality than those immediately preceding.
4. There should be a good record of quality performance on the part of the supplier.
5. The sampling agency must have confidence that the supplier will not take advantage of its good record and occasionally send a bad lot when such a lot would have the best chance of acceptance.

7.8 Continuous-Sampling

All the sampling plans discussed previously are lot-by-lot plans. With these plans, there is an explicit assumption that the product is formed into lots, and the purpose of the sampling plan is to sentence the individual lots. However, many manufacturing operations, particularly complex assembly processes, do not result in the natural formation of lots. For example, the manufacture of many electronics products, such as personal computers, is performed on a conveyorized assembly line.

When production is continuous, two approaches may be used to form lots. The first procedure allows the accumulation of production at given points in the assembly process. This has the disadvantage of creating in-process inventory at various points, which requires additional space, may constitute a safety hazard, and is a generally inefficient approach to managing an assembly line. The second procedure arbitrarily marks off a given segment of production as a "lot." The disadvantage of this approach is that if a lot is ultimately rejected and 100% inspection of the lot is subsequently required, it may be necessary to recall products from manufacturing operations that are further downstream. This may require disassembly or at least partial destruction of semifinished items.

For these reasons, special sampling plans for continuous production have been developed. **Continuous-sampling plans** consist of alternating sequences of sampling inspection and screening (100% inspection). The plans usually begin with 100% inspection, and when a stated number of units is found to be free of defects (the number of units i is usually called the **clearance number**), sampling inspection is instituted. Sampling inspection continues until a specified number of defective units is found, at which time 100% inspection is resumed. Continuous-sampling plans are rectifying inspection plans, in that the quality of the product is improved by the partial screening.

7.8.1 CSP-1

Continuous-sampling plans were first proposed by Harold F. Dodge (1943). Dodge's initial plan is called CSP-1. At the start of the plan, all units are inspected 100%. As soon as the clearance number has been reached—that is, as soon as i consecutive units of product are found to be free of defects—100% inspection is discontinued, and only a fraction (f) of the units are inspected. These sample units are selected one at a time at random from the flow of production. If a sample unit is found to be defective, 100% inspection is resumed. All defective units found are either reworked or replaced with good ones. The procedure for CSP-1 is shown in Figure 7.22.

A CSP-1 plan has an overall AOQL. The value of the AOQL depends on the values of the clearance number i and the sampling fraction f. The same AOQL can be obtained by different combinations of i and f. Table 7.12 presents various values of i and f for CSP-1 that will lead to a stipulated AOQL. Note in the table that an AOQL of 0.79% could be obtained using a sampling plan with $i = 59$ and $f = \frac{1}{3}$, or with $i = 113$ and $f = \frac{1}{7}$.

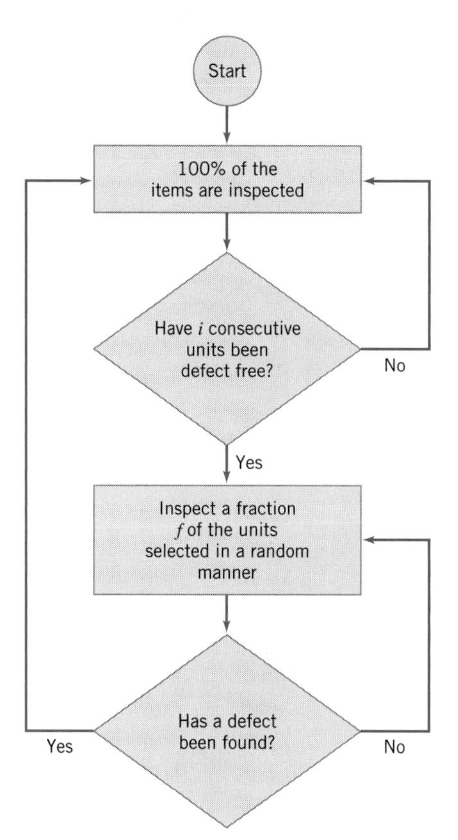

Figure 7.22 Procedure for CSP-1 Plans

Table 7.12
Values of i for CSP-1 Plans

	AOQL (%)															
f	0.018	0.033	0.046	0.074	0.113	0.143	0.198	0.33	0.53	0.79	1.22	1.90	2.90	4.94	7.12	11.46
1/2	1,540	840	600	375	245	194	140	84	53	36	23	15	10	6	5	3
1/3	2,550	1,390	1,000	620	405	321	232	140	87	59	38	25	16	10	7	5
1/4	3,340	1,820	1,310	810	530	420	303	182	113	76	49	32	21	13	9	6
1/5	3,960	2,160	1,550	965	630	498	360	217	135	91	58	38	25	15	11	7
1/7	4,950	2,700	1,940	1,205	790	623	450	270	168	113	73	47	31	18	13	8
1/10	6,050	3,300	2,370	1,470	965	762	550	335	207	138	89	57	38	22	16	10
1/15	7,390	4,030	2,890	1,800	1,180	930	672	410	255	170	108	70	46	27	19	12
1/25	9,110	4,970	3,570	2,215	1,450	1,147	828	500	315	210	134	86	57	33	23	14
1/50	11,730	6,400	4,590	2,855	1,870	1,477	1,067	640	400	270	175	110	72	42	29	18
1/100	14,320	7,810	5,600	3,485	2,305	1,820	1,302	790	500	330	215	135	89	52	36	22
1/200	17,420	9,500	6,810	4,235	2,760	2,178	1,583	950	590	400	255	165	106	62	43	26

The choice of i and f is usually based on practical considerations in the manufacturing process. For example, i and f may be influenced by the workload of the inspectors and operators in the system. It is a fairly common practice to use quality-assurance inspectors to do the sampling inspection, and place the burden of 100% inspection on manufacturing. As a general rule, however, it is not a good idea to choose values of f smaller than 1/200 because the protection against bad quality in a continuous run of production then becomes very poor.

The average number of units inspected in a 100% screening sequence following the occurrence of a defect is equal to

$$u = \frac{1 - q^i}{pq^i} \tag{7.21}$$

where $q = 1 - p$, and p is the fraction defective produced when the process is operating in control. The average number of units passed under the sampling inspection procedure before a defective unit is found is

$$v = \frac{1}{fp} \tag{7.22}$$

The average fraction of total manufactured units inspected in the long run is

$$\text{AFI} = \frac{u + fv}{u + v} \tag{7.23}$$

The average fraction of manufactured units passed under the sampling procedure is

$$P_a = \frac{v}{u + v} \tag{7.24}$$

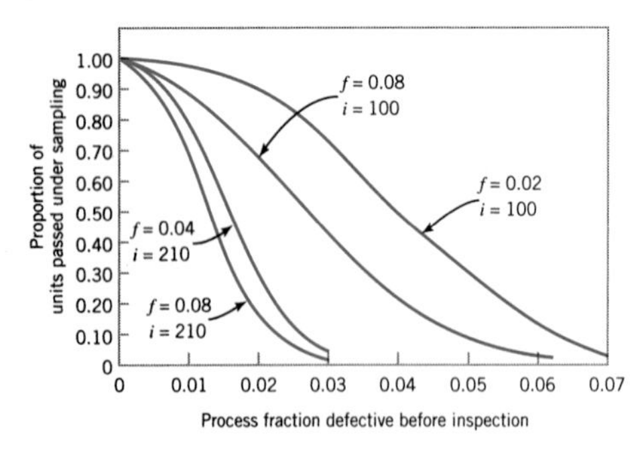

Figure 7.23 Operating-Characteristic Curves for Various Continuous-Sampling Plans, CSP-1 (Adapted with Permission from A. J. Duncan, *Quality Control and Industrial Statistics*, 5th ed., Irwin, Homewood, IL, 1986)

When P_a is plotted as a function of p, we obtain an operating-characteristic curve for a continuous-sampling plan. Note that whereas an OC curve for a lot-by-lot acceptance-sampling plan gives the percentage of lots that would be passed under sampling inspection, the OC curve for a continuous-sampling plan gives the percentage of units passed under sampling inspection. Graphs of operating-characteristic curves for several values of f and i for CSP-1 plans are shown in Figure 7.23. Note that for moderate-to-small values of f, i has much more effect on the shape of the curve than f.

7.8.2 OTHER CONTINUOUS-SAMPLING PLANS

There have been a number of variations of the original Dodge CSP-1 plan. One variation was designed to meet the objection that the occurrence of a single isolated defective unit sometimes does not warrant return to 100% inspection. This is particularly true when dealing with minor defects. To meet this objection, Dodge and Torrey (1951) proposed CSP-2 and CSP-3. Under CSP-2, 100% inspection will not be reinstated when production is under sampling inspection until two defective sample units have been found within a space of K sample units of each other. It is common practice to choose K equal to the clearance number i. CSP-2 plans are indexed by specific AOQLs that may be obtained by different combinations of i and f. CSP-3 is very similar to CSP-2, but is designed to give additional protection against spotty production. It requires that after a defective unit has been found in sampling inspection, the immediately following four units should be inspected. If any of these four units is defective, 100% inspection is immediately reinstituted. If no defectives are found, the plan continues as under CSP-2.

Another common objection to continuous-sampling plans is the abrupt transition between sampling inspection and 100% inspection. Lieberman and Solomon (1955) have designed multilevel continuous-sampling plans to overcome this objection. Multilevel continuous-sampling plans begin with 100% inspection, as does CSP-1, and then switch to inspecting a fraction f of the production as soon as the clearance number i has been reached. However, when, under sampling inspection at rate f, a run of i consecutive sample units is found free of defects, then sampling continues at the rate f^2. If a further run of i consecutive units is found to be free of defects, then sampling may continue at the rate f^3. This reduction in sampling frequency may be continued as far as the sampling agency wishes. If at any time sampling inspection reveals a defective unit, return is immediately made to the next lower level of sampling. This type of multilevel continuous-sampling plan greatly reduces the inspection effort when the manufacturing process is operating very well, and increases it during periods of poor production. This transition in inspection intensity is also accomplished without abrupt changes in the inspection load.

Much of the work on continuous-sampling plans has been incorporated into MIL STD 1235C. The standard provides for five different types of continuous-sampling plans. Tables to assist the analyst in designing sampling plans are presented in the standard. CSP-1 and CSP-2 are a part of MIL STD 1235C. In addition, there are two other single-level continuous-sampling procedures, CSP-F and CSP-V. The fifth plan in the standard is CSP-T, a multilevel continuous-sampling plan.

The sampling plans in MIL STD 1235C are indexed by sampling frequency code letter and AOQL. They are also indexed by the AQLs of MIL STD 105E. This aspect of MIL STD 1235C has sparked considerable controversy. CSP plans are not AQL plans and do not have AQLs naturally associated with them. MIL STD 105E, which does focus on the AQL, is designed for manufacturing situations in which lotting is a natural aspect of production, and provides a set of decision rules for sentencing lots so that certain AQL protection is obtained. CSP plans are designed for situations in which production is continuous and lotting is not a natural aspect of the manufacturing situations. In MIL STD 1235C, the sampling plan tables are footnoted and indicate that the AQLs have no meaning relative to the plan and are only an index.

7.9 Skip-Lot Sampling Plans

This section describes the development and evaluation of a system of lot-by-lot inspection plans in which a provision is made for inspecting only some fraction of the submitted lots. These plans are known as **skip-lot sampling plans.** Generally speaking, skip-lot sampling plans should be used only when the quality of the submitted product is good as demonstrated by the supplier's quality history.

Dodge (1956) initially presented skip-lot sampling plans as an extension of CSP-type continuous-sampling plans. In effect, a skip-lot sampling plan is the application of continuous-sampling to lots rather than to individual units of production on an assembly line. The version of skip-lot sampling initially proposed by Dodge required a single determination or analysis to ascertain the lot's acceptability or unacceptability. These plans are called SkSP-1. Skip-lot sampling plans designated SkSP-2 follow the next logical step; that is, each lot to be sentenced is sampled according to a particular attribute lot inspection plan. Perry (1973) gives a good discussion of these plans.

A skip-lot sampling plan of type SkSP-2 uses a specified lot inspection plan called the "reference-sampling plan," together with the following rules:

1. Begin with normal inspection, using the reference plan. At this stage of operation, every lot is inspected.
2. When i consecutive lots are accepted on normal inspection, switch to skipping inspection. In skipping inspection, a fraction f of the lots is inspected.
3. When a lot is rejected on skipping inspection, return to normal inspection.

The parameters f and i are the parameters of the skip-lot sampling plan SkSP-2. In general, the clearance number i is a positive integer, and the sampling fraction f lies in the interval $0 < f < 1$. When the sampling fraction $f = 1$, the skip-lot sampling plan reduces to the original reference-sampling plan. Let P denote the probability of acceptance of a lot from the reference-sampling plan. Then, $P_a(f, i)$ is the probability of acceptance for the skip-lot sampling plan SkSP-2, where

$$P_a(f, i) = \frac{fP + (1 - f)P^i}{f + (1 - f)P^i} \tag{7.25}$$

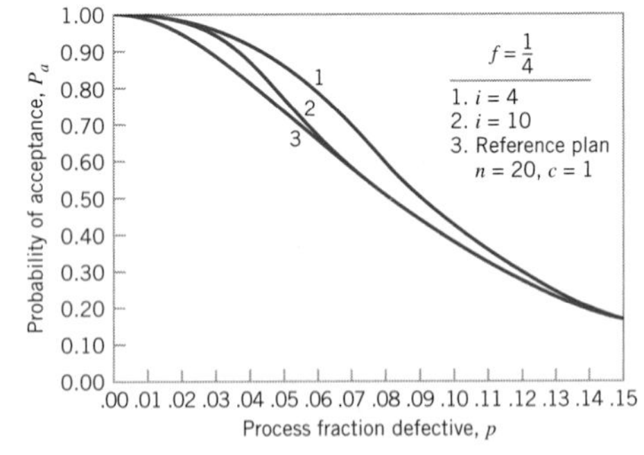

Figure 7.24 **OC Curves for SkSP-2 Skip-Lot Plans: Single-Sampling Reference Plan, Same f, Different i (from R. L. Perry, "Skip-Lot Sampling Plans," *Journal of Quality Technology*, Vol. 5, 1973, with permission of the American Society for Quality Control)**

It can be shown that for $f_2 < f_1$, a given value of the clearance number i, and a specified reference-sampling plan,

$$P_a(f_1, i) \leq P_a(f_2, i) \tag{7.26}$$

Furthermore, for integer clearance numbers $i < j$, a fixed value of f, and a given reference-sampling plan,

$$P_a(f, j) \leq P_a(f, i) \tag{7.27}$$

These properties of a skip-lot sampling plan are shown in Figures 7.24 and 7.25 for the reference-sampling plan $n = 20$, $c = 1$. The OC curve of the reference-sampling plan is also shown on these graphs.

A very important property of a skip-lot sampling plan is the average amount of inspection required. In general, skip-lot sampling plans are used where it is necessary to reduce the average amount of inspection required. The average sample number of a skip-lot sampling plan is

$$\text{ASN}(SkSP) = \text{ASN}(R)F \tag{7.28}$$

where F is the average fraction of submitted lots that are sampled and ASN(R) is the average sample number of the reference-sampling plan. It can be shown that

$$F = \frac{f}{(1-f)P^i + f} \tag{7.29}$$

Thus, since $0 < F < 1$, it follows that

$$\text{ASN}(SkSP) < \text{ASN}(R) \tag{7.30}$$

Therefore, skip-lot sampling yields a reduction in the **average sample number (ASN).** For situations in which the quality of incoming lots is very high, this reduction in inspection effort can be significant.

To illustrate the average sample number behavior of a skip-lot sampling plan, consider a reference-sampling plan of $n = 20$ and $c = 1$. Since the average sample number for a single-sampling plan is ASN $= n$, we have

$$\text{ASN}(SkSP) = n(F)$$

Figure 7.26 presents the ASN curve for the reference-sampling plan $n = 20$, $c = 1$ and the following skip-lot sampling plans:

1. $f = \frac{1}{3}, \quad i = 4$
2. $f = \frac{1}{3}, \quad i = 14$
3. $f = \frac{2}{3}, \quad i = 4$
4. $f = \frac{2}{3}, \quad i = 14$

From examining Figure 7.26, we note that for small values of incoming lot fraction defective, the reductions in average sample number are very substantial for the skip-lot sampling plans evaluated. If the incoming lot quality is very good, consistently close to zero fraction nonconforming, say, then a small value of f, perhaps $\frac{1}{4}$ or $\frac{1}{5}$, could be used. If incoming quality is slightly worse, then an appropriate value of f might be $\frac{1}{2}$.

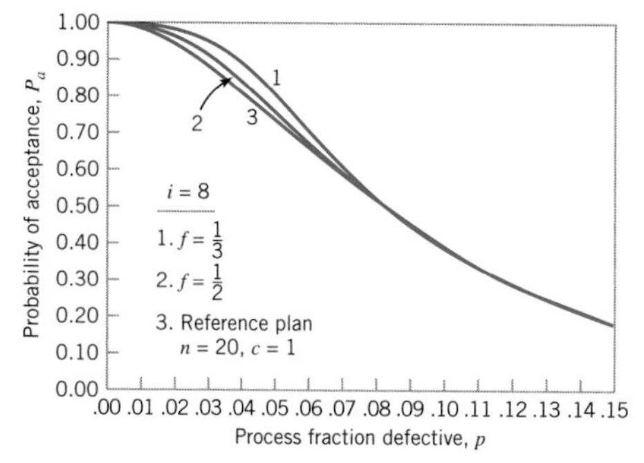

Figure 7.25 OC Curves for SkSP-2 Skip-Lot Plans: Single-Sampling Reference Plan, Same *i*, Different *f* (from R. L. Perry, "Skip-Lot Sampling Plans," *Journal of Quality Technology*, Vol. 5, 1973, with permission of the American Society for Quality Control)

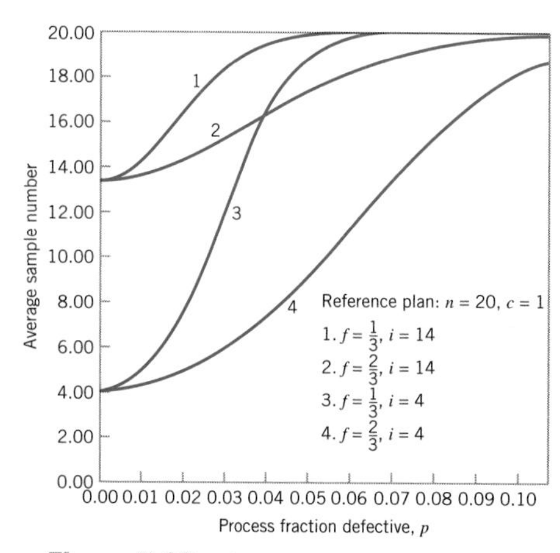

Figure 7.26 Average Sample Number (ASN) Curves for SkSP 2 Skip-Lot Plans with Single-Sampling Reference Plan (from R. L. Perry, "Skip-Lot Sampling Plans," *Journal of Quality Technology*, Vol. 5, 1973, with permission of the American Society for Quality Control)

Skip-lot sampling plans are an effective acceptance-sampling procedure and may be useful as a system of reduced inspection. Their effectiveness is partially good when the quality of submitted lots is very good. However, one should be careful to use skip-lot sampling plans only for situations in which there is a sufficient history of supplier quality to ensure that the quality of submitted lots is very good. Furthermore, if the supplier's process is highly erratic and there is a great deal of variability from lot to lot, skip-lot sampling plans are inappropriate. They seem to work best when the supplier's processes are in a state of statistical control and when the process capability is adequate to ensure virtually defect-free production.

Important Terms and Concepts

100% inspection
Acceptance-sampling plan
ISO 2859
ANSI/ASQ Z1.9
AOQL plans
Attributes data
Average outgoing quality
Average total inspection
Dodge–Romig sampling plans
Double-sampling plan
Ideal OC curve
Lot disposition actions
Lot sentencing
Lot tolerance percent defective (LTPD)
LTPD plans
MIL STD 105E
Multiple-sampling plan
Random-sampling
Rectifying inspection
Sequential-sampling plan
Single-sampling plan
Switching rules
Type-A and type-B OC curves

chapter Seven Exercises

7.1 Draw the type-B OC curve for the single-sampling plan $n = 50$, $c = 1$.

7.2 Draw the type-B OC curve for the single-sampling plan $n = 100$, $c = 2$.

7.3 Suppose that a product is shipped in lots of size $N = 5{,}000$. The receiving inspection procedure used is single-sampling with $n = 50$ and $c = 1$.

a. Draw the type-A OC curve for the plan.
b. Draw the type-B OC curve for this plan and compare it to the type-A OC curve found in part (a).
c. Which curve is appropriate for this situation?

7.4 Find a single-sampling plan for which $p_1 = 0.01$, $\alpha = 0.05$, $p_2 = 0.10$, and $\beta = 0.10$.

7.5 Find a single-sampling plan for which $p_1 = 0.05$, $\alpha = 0.05$, $p_2 = 0.15$, and $\beta = 0.10$.

7.6 Find a single-sampling plan for which $p_1 = 0.02$, $\alpha = 0.01$, $p_2 = 0.06$, and $\beta = 0.10$.

7.7 A company uses the following acceptance-sampling procedure. A sample equal to 10% of the lot is taken. If 2% or less of the items in the sample are defective, the lot is accepted; otherwise, it is rejected. If submitted lots vary in size from 5,000 to 10,000 units, what can you say about the protection by this plan? If 0.05 is the desired LTPD, does this scheme offer reasonable protection to the consumer?

7.8 A company uses a sample size equal to the square root of the lot size. If 1% or less of the items in the sample are defective, the lot is accepted; otherwise, it is rejected. Submitted lots vary in size from 1,000 to 5,000 units. Comment on the effectiveness of this procedure.

7.9 Consider the single-sampling plan found in Exercise 7.4. Suppose that lots of $N = 2{,}000$ are submitted. Draw the ATI curve for this plan. Draw the AOQ curve and find the AOQL.

7.10 Suppose that a single-sampling plan with $n = 150$ and $c = 2$ is being used for receiving inspection where the supplier ships the product in lots of size $N = 3{,}000$.

a. Draw the OC curve for this plan.
b. Draw the AOQ curve and find the AOQL.
c. Draw the ATI curve for this plan.

7.11 Suppose that a supplier ships components in lots of size 5,000. A single-sampling plan with $n = 50$ and $c = 2$ is being used for receiving inspection. Rejected lots are screened, and all defective items are reworked and returned to the lot.

a. Draw the OC curve for this plan.
b. Find the level of lot quality that will be rejected 90% of the time.
c. Management has objected to the use of the above sampling procedure and wants to use a plan with an acceptance number $c = 0$, arguing that this is more consistent with their Zero Defects program. What do you think of this?
d. Design a single-sampling plan with $c = 0$ that will give a 0.90 probability of rejection of lots having the quality level found in part (b). Note that the two plans are now matched at the LTPD point. Draw the OC curve for this plan and compare it to the one for $n = 50$, $c = 2$ in part (a).
e. Suppose that incoming lots are 0.5% nonconforming. What is the probability of rejecting these lots under both plans? Calculate the ATI at this point for both plans. Which plan do you prefer? Why?

7.12 Draw the primary and supplementary OC curves for a double-sampling plan with $n_1 = 50$, $c_1 = 2$, $n_2 = 100$, $c_2 = 6$. If the incoming lots have fraction nonconforming $p = 0.05$, what is the probability of acceptance on the first sample? What is the probability of final acceptance? Calculate the probability of rejection on the first sample.

7.13 a. Derive an item-by-item sequential-sampling plan for which $p_1 = 0.01$, $\alpha = 0.05$, $p_2 = 0.10$, and $\beta = 0.10$.

b. Draw the OC curve for this plan.

7.14 a. Derive an item-by-item sequential-sampling plan for which $p_1 = 0.02$, $\alpha = 0.05$, $p_2 = 0.15$, and $\beta = 0.10$.

b. Draw the OC curve for this plan.

7.15 Consider rectifying inspection for single-sampling. Develop an AOQ equation assuming that all defective items are removed but *not* replaced with good ones.

7.16 A supplier ships a component in lots of size $N = 3000$. The AQL has been established for this product at 1%. Find the normal, tightened, and reduced single-sampling plans for this situation from MIL STD 105E, assuming that general inspection level II is appropriate.

7.17 Repeat Exercise 7.16, using general inspection level I. Discuss the differences in the various sampling plans.

7.18 A product is supplied in lots of size $N = 10{,}000$. The AQL has been specified at 0.10%. Find the normal, tightened, and reduced single-sampling plans from MIL STD 105E, assuming general inspection level II.

7.19 MIL STD 105E is being used to inspect incoming lots of size $N = 5{,}000$. Single-sampling, general inspection level II, and an AQL of 0.65% are being used.

a. Find the normal, tightened, and reduced inspection plans.
b. Draw the OC curves of the normal, tightened, and reduced inspection plans on the same graph.

7.20 A product is shipped in lots of size $N = 2{,}000$. Find a Dodge–Romig single-sampling plan for which the LTPD = 1%, assuming that the process average is 0.25% defective. Draw the OC curve and the ATI curve for this plan. What is the AOQL for this sampling plan?

7.21 We wish to find a single-sampling plan for a situation where lots are shipped from a supplier. The supplier's process operates at a fallout level of 0.50% defective. We want the AOQL from the inspection activity to be 3%.

a. Find the appropriate Dodge–Romig plan.
b. Draw the OC curve and the ATI curve for this plan. How much inspection will be necessary, on the average, if the supplier's process operates close to the average fallout level?
c. What is the LTPD protection for this plan?

7.22 A supplier ships a product in lots of size $N = 8{,}000$. We wish to have an AOQL of 3%, and we are going to use single-sampling. We do not know the supplier's process fallout but suspect that it is at most 1% defective.

a. Find the appropriate Dodge–Romig plan.
b. Find the ATI for this plan, assuming that incoming lots are 1% defective.
c. Suppose that our estimate of the supplier's process average is incorrect and that it is really 0.25% defective. What sampling plan should we have used? What reduction in ATI would have been realized if we had used the correct plan?

7.23 An inspector wants to use a variables sampling plan with an AQL of 1.5%. Lots are of size 7,000 and the standard deviation is unknown. Find a sampling plan using Procedure 1 from MIL STD 414.

7.24 How does the sample size found in Exercise 7.23 compare with what would have been used under MIL STD 105E?

7.25 A lot of 500 items is submitted for inspection. Suppose that we wish to find a plan from MIL STD 414, using inspection level II. If the AQL is 4%, find the Procedure 1 sampling plan from the standard.

7.26 A soft-drink bottler purchases nonreturnable glass bottles from a supplier. The lower specification on the bursting strength of the bottles is 225 psi. The bottler wishes to use variables sampling to sentence the lots and has decided to use an AQL of 1%. Find an appropriate set of normal and tightened sampling plans from the standard. Suppose that a lot is submitted, and the sample results yield

$$\bar{x} = 255 \quad s = 10$$

Determine the disposition of the lot using Procedure 1. The lot size is $N = 100{,}000$.

7.27 A chemical ingredient is packed in metal containers. A large shipment of these containers has been delivered to a manufacturing facility. The mean bulk density of this ingredient should not be less than 0.15g/cm^3. Suppose that lots of this quality are to have a 0.95 probability of acceptance. If the mean bulk density is as low as 0.1450, the probability of acceptance of the lot should be 0.10. Suppose we know that the standard deviation of bulk density is approximately 0.005g/cm^3. Obtain a variables sampling plan that could be used to sentence the lots.

7.28 A standard of 0.3 ppm has been established for formaldehyde emission levels in wood products. Suppose that the standard deviation of emissions in an individual board is $\sigma = 0.10$ ppm. Any lot that contains 1% of its items above 0.3 ppm is considered acceptable. Any lot that has 8% or more of its items above 0.3 ppm is considered unacceptable. Good lots are to be accepted with probability 0.95, and bad lots are to be rejected with probability 0.90.

a. Using the 1% nonconformance level as an AQL, and assuming that lots consist of 5,000 panels, find an appropriate set of sampling plans from MIL STD 414, assuming σ is unknown.
b. Find an attributes-sampling plan that has the same OC curve as the variables sampling plan derived in part (a). Compare the sample sizes required for equivalent protection. Under what circumstances would variables sampling be more economically efficient?
c. Using the 1% nonconforming as an AQL, find an attributes sampling plan from MIL STD 105E. Compare the sample sizes and the protection obtained from this plan with the plans derived in parts (a) and (b).

7.29 Consider a single-sampling plan with $n = 25$, $c = 0$. Draw the OC curve for this plan. Now consider chain-sampling plans with $n = 25$, $c = 0$, and $i = 1, 2, 5, 7$. Sketch the OC curves for these chain-sampling plans on the same axis. Discuss the behavior of chain sampling in this situation compared to the conventional single-sampling plan with $c = 0$.

7.30 An electronics manufacturer buys memory devices in lots of 30,000 from a supplier. The supplier has a long record of good quality performance, with an average fraction defective of approximately 0.10%. The quality engineering department has suggested using a conventional acceptance-sampling plan with $n = 32$, $c = 0$.

a. Draw the OC curve of this sampling plan.
b. If lots are of a quality that is near the supplier's long-term process average, what is the average total inspection at that level of quality?
c. Consider a chain sampling plan with $n = 32$, $c = 0$, and $i = 3$. Contrast the performance of this plan with the conventional sampling plan $n = 32$, $c = 0$.
d. How would the performance of this chain sampling plan change if we substituted $i = 4$ in part (c)?

7.31 Suppose that a manufacturing process operates in continuous production, such that continuous sampling plans could be applied. Determine three different CSP-1 sampling plans that could be used for an AOQL of 0.198%.

7.32 For the sampling plans developed in Exercise 7.31, compare the plans' performance in terms of average fraction inspected, given that the process is in control at an average fallout level of 0.15%. Compare the plans in terms of their operating-characteristic curves.

7.33 Suppose that CSP-1 is used for a manufacturing process where it is desired to maintain an AOQL of 1.90%. Specify two CSP-1 plans that would meet this AOQL target.

7.34 Compare the plans developed in Exercise 7.33 in terms of average fraction inspected and their operating-characteristic curves. Which plan would you prefer if $p = 0.0375$?

chapter Eight Process Design and Improvement with Designed Experiments

Chapter Outline

Cork for wine—not always the best!

For more than three centuries, wine bottles around the world have been sealed with cork closures. But in the last 30 years, industry leaders have begun to use newer closures, such as screw caps and plastic stoppers.

Why the new approaches? Wineries strive to manage the amount of oxygen in bottles because oxidation impacts the flavor of the wine. Corks allow oxygen to enter the wine during closure (due to compression) and after closure (from air in the headspace). When bottles are sealed with screw caps, however, oxygen can only enter from air in the head space after closure. Which closure is best depends upon the wine. Fruity wines can benefit significantly from screw caps while wines that need more time to mature do well with corks.

Modesto, California-based G3 Enterprises, provides labeling, testing, and bottle closure services to the wine industry. They have been using the Design of Experiments (DOE) approach to both *determine* the best closure methods and *optimize* the resultant bottling processes.

As anyone in the industry can confirm, there are many variables in the bottling process, G3 utilizes screening designs to quickly and accurately identify such variables. Once the variables are identified, G3 runs additional experiments to understand the relationships between the variables.

Because the design of experiments methodology has been so successful, G3 now uses it for a wide variety of process modeling and product design projects, including for natural cork printing, aluminum cap liner design, and cork, capper, and capsule line packing line setups.

Chapter Overview and Learning Objectives

Quality improvement is most effective when it is an integral part of the product realization process. The formal introduction of experimental design methods at the earliest stage of the product development cycle is one of the keys to success. Designed experiments are widely used in six-sigma projects, most often during the improve step, and in design for six-sigma (DFSS).

Most experiments for process design and improvement involve several variables. Factorial experimental designs, and their variations, are used in such situations. This chapter gives an introduction to factorial designs, emphasizing their applications for process and quality improvement. We concentrate on experimental designs where all the factors have two levels and show how fractional versions of these designs can be used with great effectiveness in industrial experimentation. Important topics include the analysis of factorial experimental designs and the use of graphical methods in interpretation of the results. Both the interaction graph and a response surface plot are shown to be very useful in interpretation of results.

After careful study of this chapter you should be able to do the following:

1. Explain how designed experiments can be used to improve product design and improve process performance
2. Explain how designed experiments can be used to reduce the cycle time required to develop new products and processes
3. Understand how main effects and interactions of factors can be estimated
4. Understand the factorial design concept
5. Know how to use the analysis of variance (ANOVA) to analyze data from factorial designs
6. Know how residuals are used for model adequacy checking for factorial designs
7. Know how to use the 2^k system of factorial designs
8. Know how to construct and interpret contour plots and response surface plots
9. Know how to add center points to a 2^k factorial design to test for curvature and provide an estimate of pure experimental error
10. Understand how the blocking principal can be used in a factorial design to eliminate the effects of a nuisance factor
11. Know how to use the 2^{k-p} system of fractional factorial designs
12. Understand the basics of response surface methods
13. Understand the robust product and process design tools

8.1 What is Experimental Design?

As indicated in Chapter 1, a designed experiment is a test or series of tests in which purposeful changes are made to the input variables of a process so that we may observe and identify corresponding changes in the output response. The process, as shown in Figure 8.1, can be

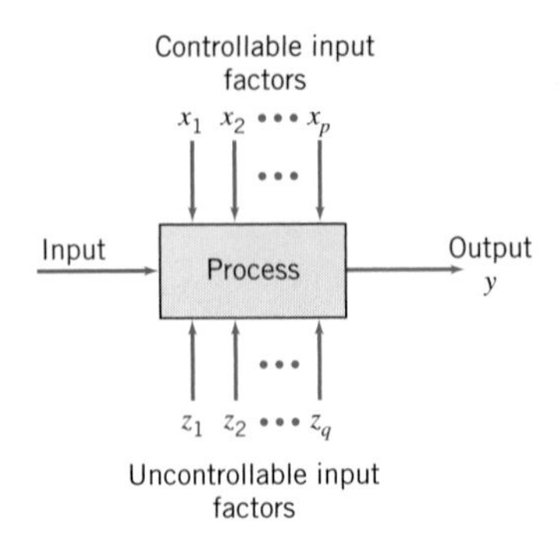

Figure 8.1 **General Model of a Process**

visualized as some combination of machines, methods, and people that transforms an input material into an output product. This output product has one or more observable quality characteristics or responses. Some of the process variables $x_1, x_2, \ldots, x_p$ are **controllable,** whereas others $z_1, z_2, \ldots, z_q$ are **uncontrollable** (although they may be controllable for purposes of the test). Sometimes these uncontrollable factors are called **noise** factors. The objectives of the experiment may include

1. Determining which variables are most influential on the response, y.
2. Determining where to set the influential x's so that y is near the nominal requirement.
3. Determining where to set the influential x's so that variability in y is small.
4. Determining where to set the influential x's so that the effects of the uncontrollable variables z are minimized.

Thus, experimental design methods may be used either in process development or process troubleshooting to improve process performance or to obtain a process that is **robust** or **insensitive** to external sources of variability.

Robust Designs
Design of experiments intended to provide a product that is robust to uncontrollable sources of variability. For example, cake mixes are designed to accommodate measurement error or baking temperature variances.

Statistical process-control methods and experimental design, two very powerful tools for the improvement and optimization of processes, are closely interrelated. For example, if a process is in statistical control but still has poor capability, then to improve process capability it will be necessary to reduce variability. Designed experiments may offer a more effective way to do this than SPC. Essentially, SPC is a **passive** statistical method: We watch the process and wait for some information that will lead to a useful change. However, if the process is in control, passive observation may not produce much useful information. On the other hand, experimental design is an **active** statistical method: We will actually perform a series of tests on the process or system, making changes in the inputs and observing the corresponding changes in the outputs, and this will produce information that can lead to process improvement.

Experimental design methods can also be very useful in establishing statistical control of a process. For example, suppose that a control chart indicates that the process is out of control, and the process has many controllable input variables. Unless we know *which* input variables are the important ones, it may be very difficult to bring the process under control. Experimental design methods can be used to identify these influential process variables.

Experimental design is a critically important engineering tool for improving a manufacturing process. It also has extensive application in the development of new processes. Application of these techniques early in process development can result in

1. Improved yield
2. Reduced variability and closer conformance to the nominal
3. Reduced development time
4. Reduced overall costs

Experimental design methods can also play a major role in **engineering design** activities, where new **products** are developed and existing ones improved. Designed experiments are widely used in design for six-sigma (DFSS) activities. Some applications of statistical experimental design in engineering design include

1. Evaluation and comparison of basic design configurations
2. Evaluation of material alternatives
3. Determination of key product design parameters that impact performance

Use of experimental design in these areas can result in improved manufacturability of the product, enhanced field performance and reliability, lower product cost, and shorter product development time.

8.2 Examples of Designed Experiments in Process and Product Improvement

In this section, we present several examples that illustrate the application of designed experiments in improving process and product quality. In subsequent sections, we will demonstrate the statistical methods used to analyze the data and draw conclusions from experiments such as these.

example 8.1 Characterizing a Process

An engineer has applied SPC to a process for soldering electronic components to printed circuit boards. Through the use of u charts and Pareto analysis he has established statistical control of the flow solder process and has reduced the average number of defective solder joints per board to around 1%. However, since the average board contains over 2,000 solder joints, even 1% defective presents far too many solder joints requiring rework. The engineer would like to reduce defect levels even further; however, since the process is in statistical control, it is not obvious what machine adjustments will be necessary.

The flow solder machine has several variables that can be controlled. They include

1. Solder temperature
2. Preheat temperature
3. Conveyor speed
4. Flux type
5. Flux specific gravity
6. Solder wave depth
7. Conveyor angle

example 8.1 Continued

In addition to these controllable factors, several others cannot be easily controlled during routine manufacturing, although they could be controlled for purposes of a test. They are

1. Thickness of the printed circuit board
2. Types of components used on the board
3. Layout of the components on the board
4. Operator
5. Production rate

In this situation, the engineer is interested in **characterizing** the flow solder machine; that is, he wants to determine which factors (both controllable and uncontrollable) affect the occurrence of defects on the printed circuit boards. To accomplish this task he can design an experiment that will enable him to estimate the magnitude and direction of the factor effects. That is, how much does the response variable (defects per unit) change when each factor is changed, and does changing the factors *together* produce different results than are obtained from individual factor adjustments? A factorial experiment will be required to do this. Sometimes we call this kind of factorial experiment a **screening experiment.**

The information from this screening or characterization experiment will be used to identify the critical process factors and to determine the direction of adjustment for these factors to further reduce the number of defects per unit. The experiment may also provide information about which factors should be more carefully controlled during routine manufacturing to prevent high defect levels and erratic process performance. Thus, one result of the experiment could be the application of control charts to one or more *process* variables (such as solder temperature) in addition to the u chart on process output. Over time, if the process is sufficiently improved, it may be possible to base most of the process-control plan on controlling process input variables instead of control-charting the output.

example 8.2 Optimizing a Process

In a characterization experiment, we are usually interested in determining which process variables affect the response. A logical next step is to **optimize**—that is, to determine the region in the important factors that lead to the best possible response. For example, if the response is yield, we will look for a region of maximum yield, and if the response is variability in a critical product dimension, we will look for a region of minimum variability.

Suppose we are interested in improving the yield of a chemical process. Let's say that we know from the results of a characterization experiment that the two most important process variables that influence yield are operating temperature and reaction time. The process currently runs at 155°F and 1.7 hours of reaction time, producing yields around 75%. Figure 8.2 shows a view of the time-temperature region from above. In this graph the lines of constant yield are connected to form **response contours,** and we have shown the contour lines for 60, 70, 80, 90, and 95% yield.

example 8.2 Continued

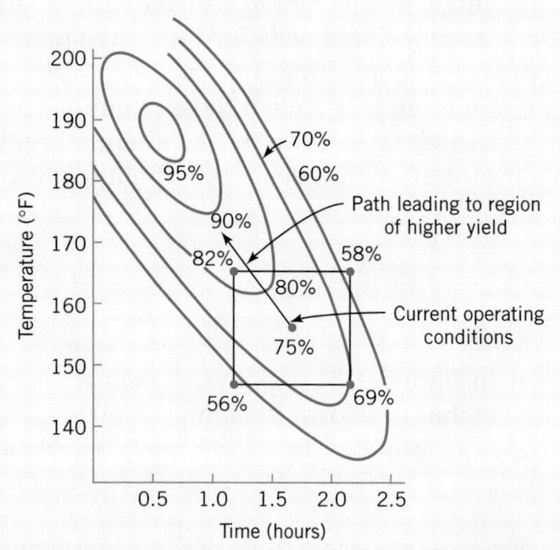

Figure 8.2 Contour Plot of Yield as a Function of Reaction Time and Reaction Temperature, Illustrating an Optimization Experiment

To locate the optimum, it is necessary to perform an experiment that varies time and temperature together. This type of experiment is called a **factorial experiment;** an example of a factorial experiment with both time and temperature run at two levels is shown in Figure 8.2. The responses observed at the four corners of the square indicate that we should move in the general direction of increased temperature and decreased reaction time to increase yield. A few additional runs could be performed in this direction, which would be sufficient to locate the region of maximum yield. Once we are in the region of the optimum, a more elaborate experiment could be performed to give a very precise estimate of the optimum operating condition. This type of experiment is called a **response surface experiment.**

example 8.3 A Product Design Example

Designed experiments can often be applied in the product design process. To illustrate, suppose that a group of engineers are designing a door hinge for an automobile. The quality characteristic of interest is the check effort, or the holding ability of the door latch that prevents the door from swinging closed when the vehicle is parked on a hill. The check mechanism consists of a spring and a roller. When the door is opened, the roller travels through an arc, causing the leaf spring to be compressed. To close the door, the spring must be forced aside, which creates the check effort. The engineering team believes the check effort is a function of the following factors:

1. Roller travel distance
2. Spring height pivot to base
3. Horizontal distance from pivot to spring
4. Free height of the reinforcement spring
5. Free height of the main spring

example 8.3 Continued

The engineers build a prototype hinge mechanism in which all these factors can be varied over certain ranges. Once appropriate levels for these five factors are identified, an experiment can be designed consisting of various combinations of the factor levels, and the prototype hinge can be tested at these combinations. This will produce information concerning which factors are most influential on latch check effort, and through use of this information, the design can be improved.

example 8.4 Determining System and Component Tolerances

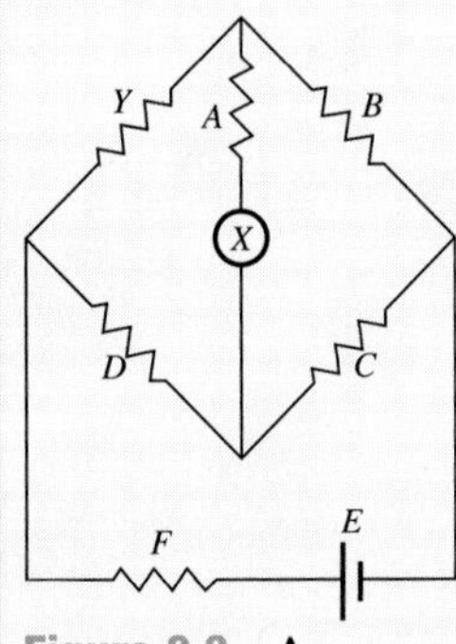

Figure 8.3 A Wheatstone Bridge

The Wheatstone bridge shown in Figure 8.3 is a device used for measuring an unknown resistance, Y. The adjustable resistor B is manipulated until a particular current flow is obtained through the ammeter (usually $X = 0$). Then the unknown resistance is calculated as

$$Y = \frac{BD}{C} - \frac{X^2}{C^2E}[A(D + C) + D(B + C)][B(C + D) + F(B + C)]$$

The engineer wants to design the circuit so that overall gauge capability is good; that is, he would like the standard deviation of measurement error to be small. He has decided that $A = 20\ \Omega$, $C = 2\ \Omega$, $D = 50\ \Omega$, $E = 1.5\ \Omega$, and $F = 2\ \Omega$ is the best choice of the design parameters as far as gauge capability is concerned, but the overall measurement error is still too high. This is likely due to the tolerances that have been specified on the circuit components. These tolerances are $\pm 1\%$ for each resistor A, B, C, D, and F, and $\pm 5\%$ for the power supply E. These tolerance bands can be used to define appropriate factor levels, and an experiment can be performed to determine which circuit components have the most critical tolerances and how much they must be tightened to produce adequate gauge capability. The information from this experiment will result in a design specification that tightens only the most critical tolerances the minimum amount possible consistent with desired measurement capability. Consequently, a lower-cost design that is easier to manufacture will be possible.

Notice that in this experiment it is unnecessary to build hardware, since the response from the circuit can be calculated. The actual response variable for the experiment should be the standard deviation of Y. However, an equation for the transmitted variation in Y from the circuit can be found using the methods of Section 8.9. Therefore, the entire experiment can be performed using a computer model of the Wheatstone bridge.

8.3 Guidelines for Designing Experiments

Designed experiments are a powerful approach to improving a process. To use this approach, it is necessary that everyone involved in the experiment have a clear idea in advance of the objective of the experiment, exactly what factors are to be studied, how the experiment is to be conducted, and at least a qualitative understanding of how the data will be analyzed. Montgomery (2009) gives an outline of the recommended procedure, reproduced in Figure 8.4. We now briefly amplify each point in this checklist.

1. **Recognition of and statement of the problem.**	In practice, it is often difficult to realize that a problem requiring formal designed experiments exists, so it may not be easy to develop a clear and generally accepted statement of the problem. However, it is absolutely essential to fully develop all ideas about the problem and about the specific objectives of the experiment. Usually, it is important to solicit input from all concerned parties—engineering, quality, marketing, the customer, management, and the operators (who usually have much insight that is all too often ignored). A clear statement of the problem and the objectives of the experiment often contributes substantially to better process understanding and eventual solution of the problem.
2. **Choice of factors and levels.**	The experimenter must choose the factors to be varied in the experiment, the ranges over which these factors will be varied, and the specific levels at which runs will be made. Process knowledge is required to do this. This process knowledge is usually a combination of practical experience and theoretical understanding. It is important to investigate all factors that may be of importance and to avoid being overly influenced by past experience, particularly when we are in the early stages of experimentation or when the process is not very mature. When the objective is factor screening or process characterization, it is usually best to keep the number of factor levels low. (Most often, two levels are used.) As noted in Figure 8.4, steps 2 and 3 are often carried out simultaneously, or step 3 may be done first in some applications.
3. **Selection of the response variable.**	In selecting the response variable, the experimenter should be certain that the variable really provides useful information about the process under study. Most often the average or standard deviation (or both) of the measured characteristic will be the response variable. Multiple responses are not unusual. Gauge capability is also an important factor. If gauge capability is poor, then only relatively large factor effects will be detected by the experiment, or additional replication will be required.
4. **Choice of experimental design.**	If the first three steps are done correctly, this step is relatively easy. Choice of design involves consideration of sample size (number of replicates), selection of a suitable run order for the experimental trials, and whether or not blocking or other randomization restrictions are involved. This chapter illustrates some of the more important types of experimental designs.
5. **Performing the experiment.**	When running the experiment, it is vital to carefully monitor the process to ensure that everything is being done according to plan. Errors in experimental procedure at this stage will usually destroy experimental validity. Up-front planning is crucial to success. It is easy to underestimate the logistical and planning aspects of running a designed experiment in a complex manufacturing environment.
6. **Data analysis.**	Statistical methods should be used to analyze the data so that results and conclusions are objective rather than judgmental. If the experiment has been designed correctly and if it has been performed according to the design, then the type of statistical method required is not elaborate. Many excellent software packages are available to assist in the data analysis, and simple graphical methods play an important role in data interpretation. Residual analysis and model validity checking are also important.
7. **Conclusions and recommendations.**	Once the data have been analyzed, the experiment must draw *practical* conclusions about the results and recommend a course of action. Graphical methods are often useful in this stage, particularly in presenting the results to others. Follow-up runs and confirmation testing should also be performed to validate the conclusions from the experiment.

Pre-experimental planning
1. Recognition of and statement of the problem
2. Choice of factors and levels
3. Selection of the response variable
(steps 2 and 3: often done simultaneously, or in reverse order)
4. Choice of experimental design
5. Performing the experiment
6. Data analysis
7. Conclusions and recommendations

Figure 8.4 **Procedure for Designing an Experiment**

Steps 1 to 3 are usually called **pre-experimental planning.** It is vital that these steps be performed as well as possible if the experiment is to be successful. Coleman and Montgomery (1993) discuss this in detail and offer more guidance in pre-experimental planning, including worksheets to assist the experimenter in obtaining and documenting the required information.

Throughout this entire process, it is important to keep in mind that experimentation is an important part of the learning process, where we tentatively formulate hypotheses about a system, perform experiments to investigate these hypotheses, and on the basis of the results formulate new hypotheses, and so on. This suggests that experimentation is **iterative**. It is usually a major mistake to design a single, large comprehensive experiment at the start of a study. A successful experiment requires knowledge of the important factors, the ranges over which these factors should be varied, the number of levels to use, and the proper units of measurement for the response. Generally, we do not know the answers to these questions at the start of an experiment, but we learn about them as we go along. Consequently, sequential experimentation is recommended, with no more than about 25% of available resources used for the first experiment.

8.4 The Analysis of Variance

ANOVA
Analysis of variance.

A powerful statistical technique called the **analysis of variance (ANOVA)** is used extensively in analyzing the data that results from a designed experiment. This chapter will illustrate the ANOVA for a variety of experimental design problems. This section presents the basic technique in the context of a simple example.

Designed experiments have had tremendous impact on manufacturing industries, including the design of new products and the improvement of existing ones, the development of new manufacturing processes, and process improvement. In the last 15 years, designed experiments have begun to be widely used outside of this traditional environment. These applications are in financial services, telecommunications, health care, e-commerce, legal services, marketing, logistics and transportation, and many of the nonmanufacturing components of manufacturing businesses. These types of businesses are sometimes referred to as the **real economy.** It has been estimated that manufacturing accounts for only about 20% of the total U.S. economy, so applications of experimental design in the real economy are of growing importance.

A web-based business depends on its World Wide Web page to attract customers. Consequently, the design of the Web site is extremely important. There are many components of the Web page that can be changed: the size and placement of the banner, different colors, different font sizes, different text messages, and so forth. Suppose that a company has designed three different Web pages and wants to determine if any one of them is most effective in attracting customers. The measure of success in attracting customers is click-through rate of the number of customers who, after visiting the Web site, click on one of the links to enter the Web site and explore the company's offerings. Each Web page design is going to be tested for five days, so the experiment will take 15 days to run. The different Web page designs are tested in random order. This will help to balance out the potential changes that may occur in the population of visitors over the test period. The data from this experiment is shown in Table 8.1.

This is an example of a **completely randomized** single-factor experiment. The single factor is the Web page design, and it has three levels. Factor levels are sometimes called **treatments,** a terminology that arose during the early agricultural uses of designed experiments in the 1920s. Each treatment has been **replicated** five times. Graphics are important in analyzing designed experiments. Figure 8.5 presents box plots of the click through data. Visual examination of these plots suggests that not all of the Web page designs perform identically with respect to click-through. The ANOVA can be used to investigate this analytically. We now give a brief general description of the ANOVA and illustrate how it can be applied to the Web page design experiment.

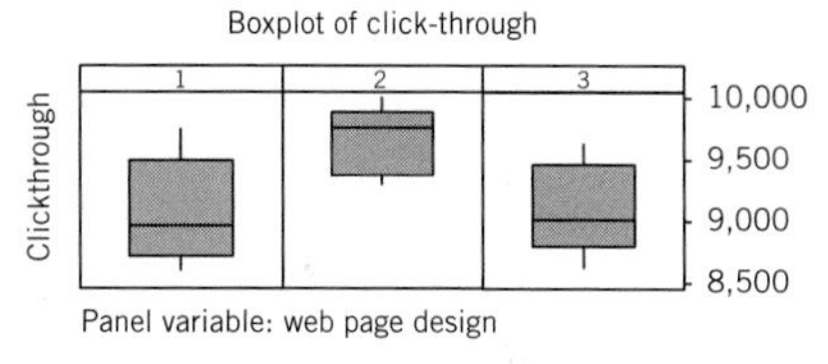

Figure 8.5 Box Plots of the Clickthrough Data

Suppose we have a different levels (treatments) of a single factor that we wish to compare. In the Web page desgin experiment, $a = 3$. The observed data would appear as shown in Table 8.2. An entry in Table 8.2—say, y_{ij}—represents the jth observation taken under treatment i. We initially consider the case in which there is an equal number of observations, n, on each treatment.

Table 8.1

The Web Page Design Experiment

Web Page Design	Number of Click-Throughs Each Day				
1	9,732	9,246	8,570	8,784	8,642
2	9,763	9,791	9,995	9,421	9,333
3	8,992	8,953	9,589	9,283	8,612

Table 8.2

Typical Data for a Single-Factor Experiment

Treatment	Observations				Totals	Averages
1	y_{11}	y_{12}	...	y_{1n}	$y_{1\cdot}$	$\bar{y}_{1\cdot}$
2	y_{21}	y_{22}	...	y_{2n}	$y_{2\cdot}$	$\bar{y}_{2\cdot}$
.	.	.	...	.	.	.
.	.	.	...	.	.	.
.	.	.	...	.	.	.
a	y_{a1}	y_{a2}	...	y_{an}	$y_{a\cdot}$	$\bar{y}_{a\cdot}$
					y	$\bar{y}_{\cdot\cdot}$

We may describe the observations in Table 8.2 by the **linear model**

$$y_{ij} = \mu + \tau_i + \varepsilon_{ij} \begin{cases} i = 1, 2, \ldots, a \\ j = 1, 2, \ldots, n \end{cases} \tag{8.1}$$

where y_{ij} is a random variable denoting the (ij)th observation, μ is a parameter common to all treatments called the **overall mean,** τ_i is a parameter associated with the ith treatment called the ith *treatment effect,* and ε_{ij} is a random error component. Note that the model could have been written as

$$y_{ij} = \mu_i + \varepsilon_{ij} \begin{cases} i = 1, 2, \ldots, a \\ j = 1, 2, \ldots, n \end{cases} \tag{8.2}$$

where $\mu_i = \mu + \tau_i$ is the mean of the ith treatment. In this form of the model, we see that each treatment defines a population that has mean μ_i, consisting of the overall mean μ plus an effect τ_i that is due to that particular treatment. We will assume that the errors ε_{ij} are normally and independently distributed with mean zero and variance σ^2. Therefore, each treatment can be thought of as a normal population with mean μ_i and variance σ^2.

Equation (8.1) is the underlying model for a single-factor experiment. Furthermore, since we require that the observations are taken in random order and that the environment (often called the experimental units) in which the treatments are used is as uniform as possible, this design is called a **completely randomized experimental design.**

We now present the analysis of variance for testing the equality of a population means. This is called a **fixed effects model analysis of variance (ANOVA)**. The treatment effects τ_i are usually defined as deviations from the overall mean μ, so that

$$\sum_{i=1}^{a} \tau_i = 0 \tag{8.3}$$

Let $\bar{y}_{i.}$ represent the total of the observations under the ith treatment and $\bar{y}_{i\cdot}$ represent the average of the observations under the ith treatment. Similarly, let $y_{..}$ represent the grand total of all observations and $\bar{y}_{..}$ represent the grand mean of all observations. Expressed mathematically,

$$y_{i.} = \sum_{j=1}^{n} y_{ij} \qquad \bar{y}_{i.} = y_{i.}/n \qquad i = 1, 2, \ldots, a$$

$$y_{..} = \sum_{i=1}^{a}\sum_{j=1}^{n} y_{ij} \qquad \bar{y}_{..} = y_{..}/n \tag{8.4}$$

where $N = an$ is the total number of observations. Thus, the "dot" subscript notation implies summation over the subscript that it replaces.

We are interested in testing the equality of the a treatment means $\mu_1, \mu_2, \ldots, \mu_a$. Using Equation (8.3), we find that this is equivalent to testing the hypotheses

$$\begin{aligned} H_0&: \ \tau_1 = \tau_2 = \ \ldots \ = \tau_a = 0 \\ H_1&: \ \tau_i \neq 0 \ \text{ for at least one } i \end{aligned} \tag{8.5}$$

Thus, if the null hypothesis is true, each observation consists of the overall mean μ plus a realization of the random error component ε_{ij}. This is equivalent to saying that all N observations are taken from a normal distribution with mean μ and variance σ^2. Therefore, if the null hypothesis is true, changing the levels of the factor has no effect on the mean response. Notice that the hypotheses here are similar to those in Chapter 4 where the t test was used, except in Chapter 4 we had $a = 2$ and we were testing the quality of two means.

The ANOVA partitions the total variability in the sample data into two component parts. Then, the test of the hypothesis in Equation (8.5) is based on a comparison of two independent estimates of the population variance. The total variability in the data is described by the **total sum of squares**

$$SS_T = \sum_{i=1}^{a}\sum_{j=1}^{n} (y_{ij} - \bar{y}_{..})^2$$

The basic ANOVA partition of the total sum of squares is given in the following definition:

Basic ANOVA Sum of Squares Identity

The **sum of squares identity** is

$$\sum_{i=1}^{a}\sum_{j=1}^{n} (y_{ij} - \bar{y}_{..})^2 = n\sum_{i=1}^{a} (\bar{y}_{i} - \bar{y}_{..})^2 + \sum_{i=1}^{a}\sum_{j=1}^{n} (y_{ij} - \bar{y}_{i\cdot})^2 \tag{8.6}$$

The identity in Equation (8.6) shows that the total variability in the data, measured by the total sum of squares, can be partitioned into a sum of squares of differences between treatment means and the grand mean and a sum of squares of differences of observations within a treatment from the treatment mean. Differences between observed treatment means and the grand mean measure the differences between treatments, whereas differences of observations within a treatment from the treatment mean can be due only to random error. Therefore, we write Equation (8.6) symbolically as

$$SS_T = SS_{\text{Treatments}} + SS_E \tag{8.7}$$

where

Definitions of Sums of Squares

$$SS_T = \sum_{i=1}^{a}\sum_{j=1}^{n}(y_{ij} - \bar{y}_{..})^2 = \text{total sum of squares}$$

$$SS_{\text{Treatments}} = n\sum_{i=1}^{a}(\bar{y}_{i.} - \bar{y}_{..})^2 = \text{treatment sum of squares}$$

and

$$SS_E = \sum_{i=1}^{a}\sum_{j=1}^{n}(y_{ij} - \bar{y}_{j\cdot})^2 = \text{error sum of squares}$$

Mean Squares are Sums of Squares Divided by Their Number of Degrees of Freedom

The quantities $MS_{\text{Treatments}} = SS_{\text{Treatments}}/(a - 1)$ and $MS_E = SS_E/[a(n - 1)]$ are called the mean squares for treatments and error respectively. If the null hypothesis of equal means is true, the error sum of squares always provides an estimate of the error variance σ^2. The treatment mean square estimates σ^2 if the null hypothesis H_0 is true but if H_1 is true, $MS_{\text{Treatments}}$ estimates σ^2 plus a positive term that incorporates variation due to the systematic difference in treatment means.

There is also a partition of the number of degrees of freedom that corresponds to the sum of squares identity in Equation (8.6). That is, there are $an = N$ observations; thus, SS_T has $an - 1$ degrees of freedom. There are a levels of the factor, so $SS_{\text{Treatments}}$ has $a - 1$ degrees of freedom. Finally, within any treatment there are n replicates providing $n - 1$ degrees of freedom with which to estimate the experimental error. Since there are a treatments, we have $a(n - 1)$ degrees of freedom for error. Therefore, the degrees of freedom partition is

$$an - 1 = a - 1 + a(n - 1)$$

Now assume that each of the a populations can be modeled as a normal distribution. Using this assumption we can show that if the null hypothesis H_0 is true, the ratio

$$F_0 = \frac{SS_{\text{Treatments}}/(a - 1)}{SS_E/[a(n - 1)]} = \frac{MS_{\text{Treatments}}}{MS_E} \tag{8.8}$$

Table 8.3

The Analysis of Variance for a Single-Factor Experiment

Source of Variation	Sum of Squares	Degrees of Freedom	Mean Square	F_0
Treatments	$SS_{\text{Treatments}}$	$a-1$	$MS_{\text{Treatments}}$	$\dfrac{MS_{\text{Treatments}}}{MS_E}$
Error	SS_E	$a(n-1)$	MS_E	
Total	SS_T	$an-1$		

has an F distribution with $a-1$ and $a(n-1)$ degrees of freedom. Furthermore, from our discussion of the mean squares, we know that MS_E is an unbiased estimator of σ^2. If the null hypothesis is false, then we expect $MS_{\text{Treatments}}$ to be greater than σ^2. Consequently, we should reject H_0 if the test statistic F_0 is large. Therefore, we would reject H_0 if $F_0 > F_{\alpha, a-1, a(n-1)}$ where F_0 is computed from Equation (8.8). A P value approach can also be used, with the P value equal to the probability above F_0 in the $F_{a-1, a(n-1)}$ distribution.

Efficient computational formulas for the sums of squares are shown in the adjacent display. The computations for this test procedure are usually summarized in tabular form as shown in Table 8.3. This is called an **analysis of variance (or ANOVA)** table.

The sums of squares computing formulas for the analysis of variance with equal sample sizes in each treatment are

$$SS_T = \sum_{i=1}^{a}\sum_{j=1}^{n} y_{ij}^2 - \frac{y_{..}^2}{N} \tag{8.9}$$

and

$$SS_{\text{Treatments}} = \sum_{i=1}^{a} \frac{y_{i.}^2}{n} - \frac{y_{..}^2}{N} \tag{8.10}$$

The error sum of squares is obtained by subtraction as

$$SS_E = SS_T - SS_{\text{Treatments}} \tag{8.11}$$

example 8.5 The Web Page Design Experiment

Consider the Web page design experiment described previously. Use the analysis of variance to test the hypothesis that different designs do not affect the mean click-through rate.

solution

The hypotheses are

H_0: $\tau_1 = \tau_2 = \tau_3 = 0$ or H_0: $\mu_1 = \mu_2 = \mu_3$

H_1: $\tau_i \neq 0$ for at least one i $\quad$ H_i: at least one mean is different

We will use $\alpha = 0.05$. The sums of squares for the ANOVA are computed from Equations (8.9), (8.10), and (8.11) as follows:

$$SS_T = \sum_{i=1}^{3}\sum_{j=1}^{5} y_{ij}^2 - \frac{y_{..}^2}{N}$$

$$= (9{,}732)^2 + (9{,}246)^2 + \cdots + (8{,}612)^2 - \frac{(139{,}106)^2}{15} = 2{,}713{,}943$$

example 8.5 Continued

$$SS_{\text{Treatments}} = \sum_{i=1}^{4} \frac{y_{i.}^2}{n} - \frac{y_{..}^2}{N}$$

$$= \frac{(45{,}374)^2 + (48{,}303)^2 + (45{,}429)^2}{5} - \frac{(139{,}106)^2}{15} = 1{,}122{,}796$$

$$SS_E = SS_T - SS_{\text{Treatments}}$$

$$= 2{,}713{,}9743 - 1{,}122{,}796 = 1{,}591{,}147$$

We usually do not perform these calculations by hand. The ANOVA from Minitab is presented in Table 8.4. Since $F_{0.05,3,12} = 3.89$, we reject H_0 and conclude that the web page design significantly affects the click-through rate. Note that the computer output reports a P value for the test statistic $F = 4.23$ in Table 8.4 of 0.041. Because the P value is smaller than $\alpha = 0.05$, we have sufficient evidence to conclude that H_0 is not true. Note that Minitab also provides some summary information about each click-through rate web page design, including the confidence interval on each mean. SPLC XL can also perform an ANOVA; see the chapter supplement.

Table 8.4

Minitab Analysis of Variance Output for the Web Page Design Experiment

One-way ANOVA: Click-through versus Web Page Design

```
Source           DF         SS        MS      F       P
Web Page Design   2    1122796    561398   4.23   0.041
Error            12    1591147    132596
Total            14    2713943

S = 364.1    R-Sq = 41.37%   R-Sq(adj) = 31.60%

                                  Individual 95% CIs for Mean Based on Pooled StDev
Level    N      Mean     StDev    -+------------+------------+------------+---------
1        5      9075     431      (-----------*-----------)
2        5      9661     276                     (-----------*-----------)
3        5      9086     368      (------------*-----------)
                                  -+------------+------------+------------+---------
                                  8750        9100         9450         9800

Pooled StDev = 364
```

(Continued)

Table 8.4

Continued

```
Fisher 95% Individual Confidence Intervals
All Pairwise Comparisons among Levels of Web Page Design

Simultaneous confidence level = 88.44%
Web Page Design = 1 subtracted from:
```

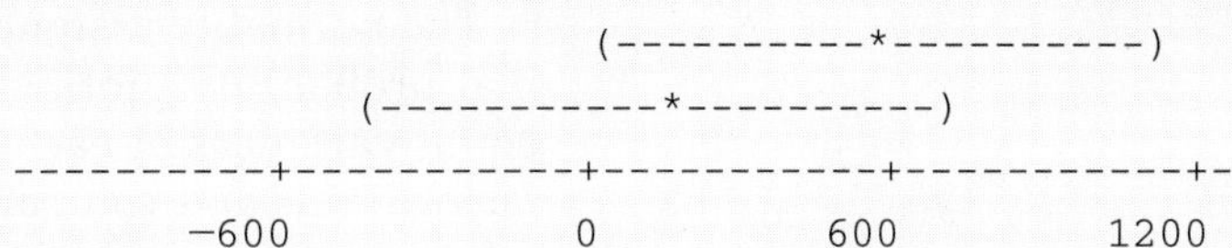

```
Web Page
Design    Lower   Center   Upper   ---------+---------+---------+---------+-
2          84.0    585.8  1087.6                (----------*----------)
3        -490.8     11.0   512.8          (-----------*----------)
                                   ---------+---------+---------+---------+-
                                         -600         0       600      1200
```

```
Web Page Design = 2 subtracted from:
```

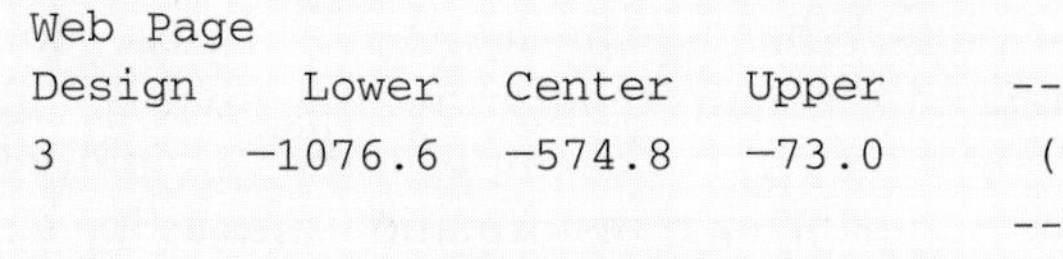

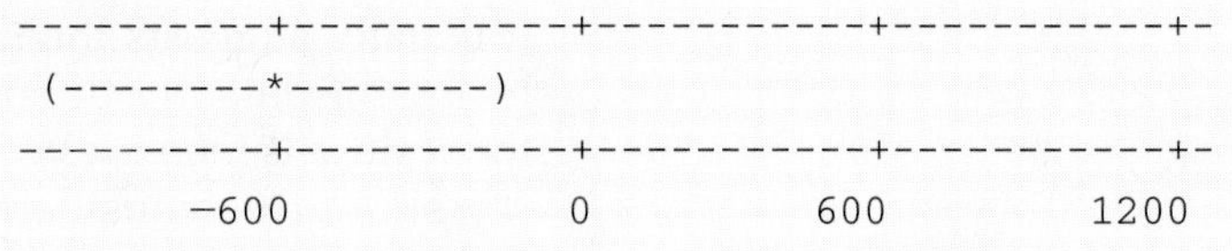

```
Web Page
Design    Lower   Center   Upper   ---------+---------+---------+---------+-
3       -1076.6   -574.8   -73.0   (--------*--------)
                                   ---------+---------+---------+---------+-
                                         -600         0       600      1200
```

Based on the ANOVA results, we would conclude that the three means μ_1, μ_2, and μ_3 are not all equal. But the ANOVA does not identify directly which means are different. This information is provided in the lower portion of Table 8.4, under the heading "Fisher 95% Individual Confidence Intervals." This is basically a set of confidence intervals (CI) on all possible differences in two of the three means. Notice that the CI for $\mu_3 - \mu_1$ overlaps zero, indicating that these two means are the same. However, the CI for $\mu_2 - \mu_1$, and $\mu_3 - \mu_2$ do not include zero. This leads to the conclusion at $\mu_2 \neq \mu_1$ and $\mu_2 \neq \mu_3$. Specifically, web page design 2 leads to a higher mean clickthrough rate than either design 1 or design 2.

Checking ANOVA Assumptions

The analysis of variance assumes that the model errors (and as a result, the observations) are normally and independently distributed with the same variance in each factor level. These assumptions can be checked by examining the residuals. The residual is the difference between the actual observation y_{ij} and the value $\bar{y}_{ij}$ that would be obtained from a least squares fit of the underlying analysis of variance model to the sample data. For the type of experimental design in this situation, the value $\bar{y}_{ij}$ is the factor-level mean $\bar{y}_{i.}$. Therefore, the residual is $e_{ij} = y_{ij} - \bar{y}_{i}$.—that is, the difference between an observation and the corresponding factor-level mean. The residuals for the web page design experiment are shown in Table 8.5.

Table 8.5

Residuals for the Web Page Design Experiment

Web Page Design	Residuals				
1	657.2	171.2	−104.8	−290.8	−432.8
2	102.4	130.4	334.4	−239.6	−327.6
3	−93.8	−132.8	503.2	197.2	−473.8

The normality assumption can be checked by constructing a normal probability plot of the residuals. To check the assumption of equal variances at each factor level, plot the residuals against the factor levels and compare the spread in the residuals. It is also useful to plot the residuals against $\bar{y}_{i.}$ (sometimes called the **fitted value**); the variability in the residuals should not depend in any way on the value of $\bar{y}_i$. Figure 8.6 shows these plots for the web page design experiment. There are no indications of problems with the assumptions from these plots. When a pattern appears in these plots, it usually suggests the need for data **transformation**—that is, analyzing the data in a different metric. For example, if the variability in the residuals increases with $\bar{y}_{i.}$, then a transformation such as log y or $\sqrt{y}$ should be considered. In some problems the dependency of residual scatter on $y_{i.}$ is very important information. It may be desirable to select the factor level that results in maximum mean response; however, this level may also cause more variation in response from run to run.

The independence assumption can be checked by plotting the residuals against the run order in which the experiment was performed. Figure 8.7 presents the plot for the web page design experiment. This plot exhibits essentially a random pattern. A pattern in this plot, such as sequences of positive and negative residuals, may indicate that the observations are not independent. This suggests that run order is important or that variables that change over time are important and have not been included in the experimental design.

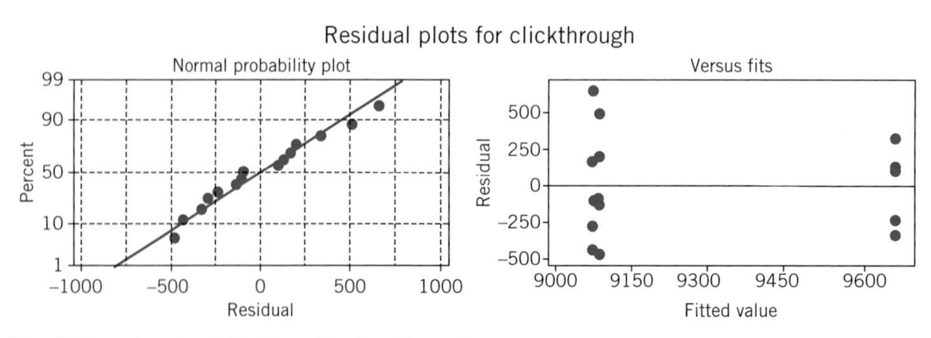

Figure 8.6 **Residual Plots for the Web Page Design Experiment**

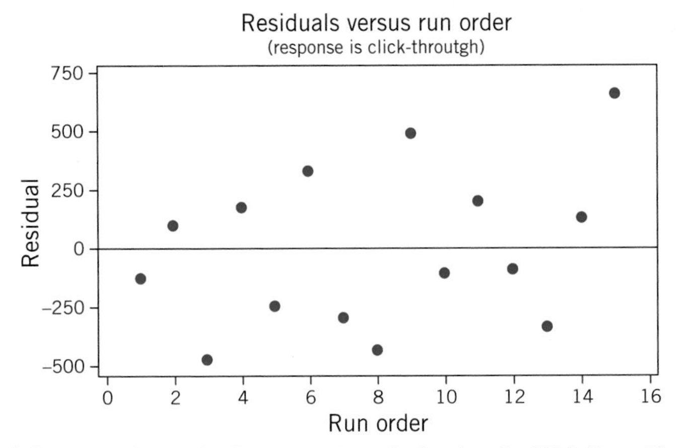

Figure 8.7 **Plot of Residuals versus Run Order for the Web Page Design Experiment**

8.5 Factorial Experiments

When there are several factors of interest in an experiment, a **factorial design** should be used. In such designs factors are varied together. Specifically, by a factorial experiment we mean that in each complete trial or replicate of the experiment, all possible combinations of the levels of the factors are investigated. Thus, if there are two factors A and B with a levels of factor A and b levels of factor B, then each replicate contains all ab possible combinations.

The effect of a factor is defined as the change in response produced by a change in the level of the factor. This is called a **main effect** because it refers to the primary factors in the study. For example, consider the data in Figure 8.8. In this factorial design, both the factors A and B have two levels, denoted by "−" and "+." These two levels are called "low" and "high," respectively. The main effect of factor A is the difference between the average response at the high level of A and the average response at the low level of A, or

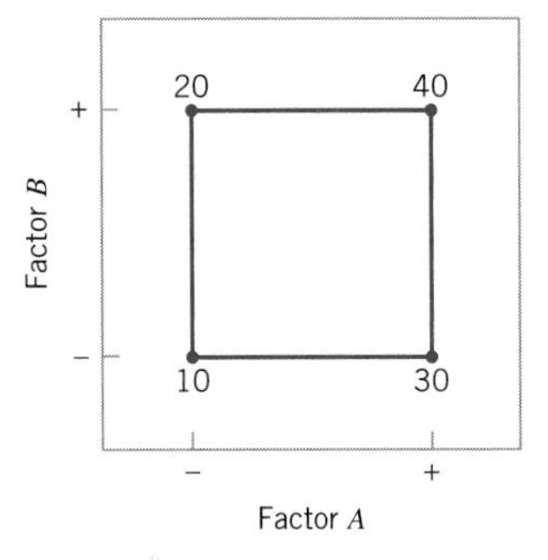

Figure 8.8 **A Factorial Experiment with Two Factors**

$$A = \bar{y}_{A^+} - \bar{y}_{A^-} = \frac{30 + 40}{2} - \frac{10 + 20}{2} = 20$$

That is, changing factor A from the low level (−) to the high level (+) causes an average response increase of 20 units. Similarly, the main effect of B is

$$B = \bar{y}_{B^+} - \bar{y}_{B^-} = \frac{20 + 40}{2} - \frac{10 + 30}{2} = 10$$

In some experiments, the difference in response between the levels of one factor is not the same at all levels of the other factors. When this occurs, there is an **interaction** between the factors. For example, consider the data in Figure 8.9. At the low level of factor B, the A effect is

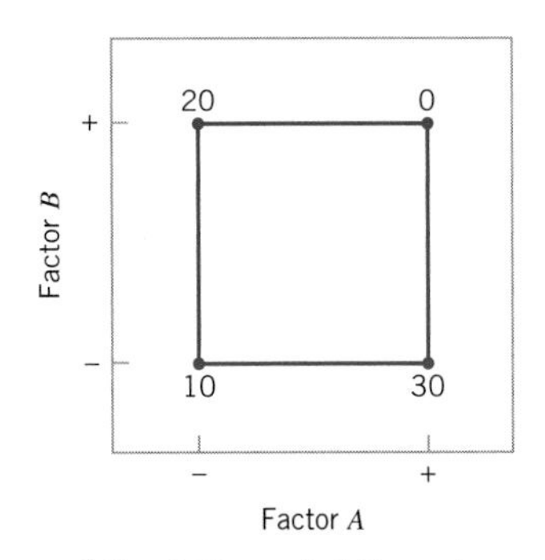

Figure 8.9 **A Factorial Experiment with Interaction**

$$A = 30 - 10 = 20$$

and at the high level of factor B, the A effect is

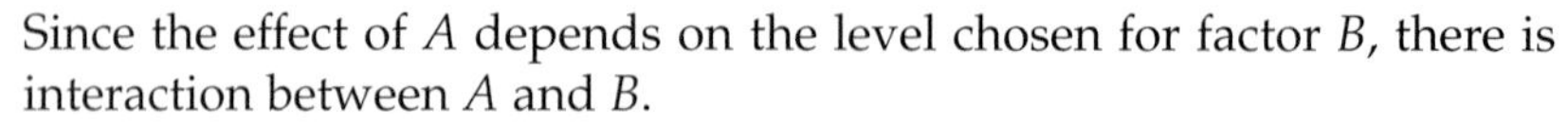

$$A = 0 - 20 = -20$$

Since the effect of A depends on the level chosen for factor B, there is interaction between A and B.

When an interaction is large, the corresponding main effects have little meaning. For example, by using the data in Figure 8.9, we find the main effect of A as

$$A = \frac{30 + 0}{2} - \frac{10 + 20}{2} = 0$$

and we would be tempted to conclude that there is no A effect. However, when we examine the main effect of A at *different levels of factor B*, we see that this is not the case. The effect of factor A depends on the levels of factor B. Thus, knowledge of the AB interaction is more useful than knowledge of the main effect. A significant interaction can mask the significance of main effects.

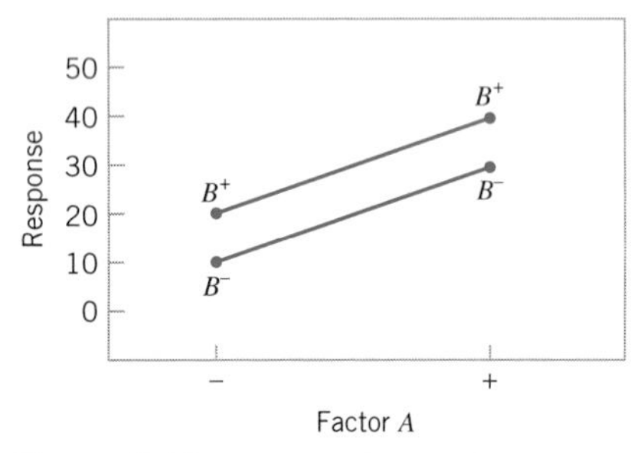

Figure 8.10 **Factorial Experiment No Interaction**

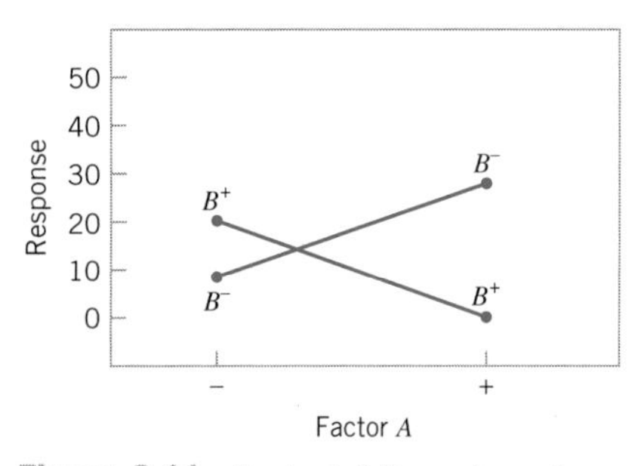

Figure 8.11 **Factorial Experiment with Interaction**

The concept of interaction can be illustrated graphically. Figure 8.10 plots the data in Figure 8.8 against the levels of A for both levels of B. Note that the B^- and B^+ lines are roughly parallel, indicating that factors A and B do not interact. Figure 8.11 plots the data in Figure 8.9. In Figure 8.11, the B^- and B^+ lines are not parallel, indicating the interaction between factors A and B. Such graphical displays are often useful in presenting the results of experiments.

An alternative to the factorial design that is (unfortunately) used in practice is to change the factors one at a time rather than to vary them simultaneously. The one-factor-at-a-time approach usually fails because it will not detect the interaction between factors, and interactions are fairly common. Factorial experiments are the only way to detect interactions. Furthermore, the one-factor-at-a-time method is inefficient; it will require more experimentation than a factorial, and there is no assurance that it will produce the correct results.

8.5.1 AN EXAMPLE

Aircraft primer paints are applied to aluminum surfaces by two methods—dipping and spraying. The purpose of the primer is to improve paint adhesion; some parts can be primed using either application method. A team using the DMAIC approach has identified three different primers that can be used with both application methods. Three specimens were painted with each primer using each application method, a finish paint was applied, and the adhesion force was measured. The 18 runs from this experiment were run in random order. The resulting data are shown in Table 8.6. The circled numbers in the cells are the cell totals. The objective of the experiment was to determine which combination of primer paint and application method produced the highest adhesion force. It would be desirable if at least one of the primers produced high adhesion force *regardless* of the application method, as this would add some flexibility to the manufacturing process.

Table 8.6

Adhesion Force Data

Primer Type	Application Method: Dipping	Spraying	$y_{i..}$
1	4.0, 4.5, 4.3 (12.8)	5.4, 4.9, 5.6 (15.9)	28.7
2	5.6, 4.9, 5.4 (15.9)	5.8, 6.1, 6.3 (18.2)	34.1
3	3.8, 3.7, 4.0 (11.5)	5.5, 5.0, 5.0 (15.5)	27.0
$y_{.j.}$	40.2	49.6	$89.8 = y_{...}$

8.5.2 STATISTICAL ANALYSIS

The analysis of variance (ANOVA) can be extended to handle the two-factor factorial experiment. Let the two factors be denoted A and B, with a levels of factor A and b levels of B. If the experiment is replicated n times, the data layout will look like Table 8.7. In general, the observation in the ijth cell in the kth replicate is y_{ijk}. In collecting the data, the abn observations would be run in *random* order. Thus, like the single-factor experiment, the two-factor factorial is a **completely randomized design.** Both factors are assumed to be fixed effects.

The observations from a two-factor factorial experiment may be described by the model

Model for a Two-Factor Factorial Experiment

$$y_{ijk} = \mu + \tau_i + \beta_j + (\tau\beta)_{ij} + \varepsilon_{ijk} \quad \begin{cases} i = 1, 2, \ldots, a \\ j = 1, 2, \ldots, b \\ k = 1, 2, \ldots, n \end{cases} \tag{8.12}$$

Table 8.7

Data for a Two-Factor Factorial Design

		Factor B: 1	2	...	b
Factor A	1	$y_{111}, y_{112}, \ldots, y_{11n}$	$y_{121}, y_{122}, \ldots, y_{12n}$	...	$y_{1b1}, y_{1b2}, \ldots, y_{1bn}$
	2	$y_{211}, y_{212}, \ldots, y_{21n}$	$y_{221}, y_{222}, \ldots, y_{22n}$	...	$y_{2b1}, y_{2b2}, \ldots, y_{2bn}$
	⋮	⋮	⋮	⋮	⋮
	a	$y_{a11}, y_{a12}, \ldots, y_{a1n}$	$y_{a21}, y_{a22}, \ldots, y_{a2n}$	...	$y_{ab1}, y_{ab2}, \ldots, y_{abn}$

where μ is the overall mean effect, τ_i is the effect of the ith level of factor A, β_j is the effect of the jth level of factor B, $(\tau\beta)_{ij}$ is the effect of the interaction between A and B, and ε_{ijk} is an NID(0, σ^2) random error component. We are interested in testing the hypotheses of no significant factor A effect, no significant factor B effect, and no significant AB interaction.

Let $y_{i..}$ denote the total of the observations at the ith level of factor A, $y_{.j.}$ denote the total of the observations at the jth level of factor B, $y_{ij.}$ denote the total of the observations in the ijth cell of Table 8.7, and $y_{...}$ denote the grand total of all the observations.

Define $\bar{y}_{i..}$, $\bar{y}_{.j.}$, $\bar{y}_{ij.}$, and $\bar{y}_{...}$ as the corresponding row, column, cell, and grand averages. That is,

$$
\begin{aligned}
y_{i..} &= \sum_{j=1}^{b}\sum_{k=1}^{n} y_{ijk} & \bar{y}_{i..} &= \frac{y_{i..}}{bn} & i &= 1,2,\ldots,a \\
y_{.j.} &= \sum_{i=1}^{a}\sum_{k=1}^{n} y_{ijk} & \bar{y}_{.j.} &= \frac{y_{.j.}}{an} & j &= 1,2,\ldots,b \\
y_{ij.} &= \sum_{k=1}^{n} y_{ijk} & \bar{y}_{ij.} &= \frac{y_{ij.}}{n} & i &= 1,2,\ldots,a \\
& & & & j &= 1,2,\ldots,b \\
y_{...} &= \sum_{i=1}^{a}\sum_{j=1}^{b}\sum_{k=1}^{n} y_{ijk} & \bar{y}_{...} &= \frac{y_{...}}{abn}
\end{aligned}
\tag{8.13}
$$

The analysis of variance decomposes the total corrected sum of squares

$$SS_T = \sum_{i=1}^{a}\sum_{j=1}^{b}\sum_{k=1}^{n}(y_{ijk} - \bar{y}_{...})^2$$

as follows:

ANOVA Sum of Squares Identity for a Two-Factor Factorial

$$
\begin{aligned}
\sum_{i=1}^{a}\sum_{j=1}^{b}\sum_{k=1}^{n}(y_{ijk} - \bar{y}_{...})^2 &= bn\sum_{i=1}^{a}(\bar{y}_{i..} - \bar{y}_{...})^2 + an\sum_{j=1}^{b}(\bar{y}_{.j.} - \bar{y}_{...})^2 \\
&\quad + n\sum_{i=1}^{a}\sum_{j=1}^{b}(\bar{y}_{ij.} - \bar{y}_{i..} - \bar{y}_{.j.} + \bar{y}_{...})^2 \\
&\quad + \sum_{i=1}^{a}\sum_{j=1}^{b}\sum_{k=1}^{n}(y_{ijk} - \bar{y}_{ij.})^2
\end{aligned}
$$

or symbolically,

$$SS_T = SS_A + SS_B + SS_{AB} + SS_E \tag{8.14}$$

The corresponding degree of freedom decomposition is

$$abn - 1 = (a - 1) + (b - 1) + (a - 1)(b - 1) + ab(n - 1) \tag{8.15}$$

Table 8.8

The ANOVA Table for a Two-Factor Factorial, Fixed Effects Model

Source of Variation	Sum of Squares	Degrees of Freedom	Mean Square	F_0
A	SS_A	$a - 1$	$MS_A = \dfrac{SS_A}{a-1}$	$F_0 = \dfrac{MS_A}{MS_E}$
B	SS_B	$b - 1$	$MS_B = \dfrac{SS_B}{b-1}$	$F_0 = \dfrac{MS_B}{MS_E}$
Interaction	SS_{AB}	$(a-1)(b-1)$	$MS_{AB} = \dfrac{SS_{AB}}{(a-1)(b-1)}$	$F_0 = \dfrac{MS_{AB}}{MS_E}$
Error	SS_E	$ab(n-1)$	$MS_E = \dfrac{SS_E}{ab(n-1)}$	
Total	SS_T	$abn - 1$		

This decomposition is usually summarized in an analysis of variance table such as the one shown in Table 8.8. Mean squares are sums of squares divided by their number of degrees of freedom.

To test for no row factor effects, no column factor effects, and no interaction effects, we would divide the corresponding mean square by mean square error. Each of these ratios will follow an F distribution, with numerator degrees of freedom equal to the number of degrees of freedom for the numerator mean square and $ab(n - 1)$ denominator degrees of freedom, when the null hypothesis of no factor effect is true. We would reject the corresponding hypothesis if the computed F exceeded the tabular value at an appropriate significance level, or alternatively, if the P value were smaller than the specified significance level.

The ANOVA is usually performed with computer software, although simple computing formulas for the sums of squares may be obtained easily. The computing formulas for these sums of squares follow.

$$SS_T = \sum_{i=1}^{a}\sum_{j=1}^{b}\sum_{k=1}^{n} y_{ijk}^2 - \frac{y_{...}^2}{abn} \tag{8.16}$$

Main effects

$$SS_A = \sum_{i=1}^{a}\frac{y_{i..}^2}{bn} - \frac{y_{...}^2}{abn} \tag{8.17}$$

$$SS_B = \sum_{j=1}^{b}\frac{y_{.j.}^2}{an} - \frac{y_{...}^2}{abn} \tag{8.18}$$

Interaction

$$SS_{AB} = \sum_{i=1}^{a}\sum_{j=1}^{b}\frac{y_{ij.}^2}{n} - \frac{y_{...}^2}{abn} - SS_A - SS_B \tag{8.19}$$

Error

$$SS_E = SS_T - SS_A - SS_B - SS_{AB} \tag{8.20}$$

example 8.6 The Aircraft Primer Paint Problem

Use the ANOVA described above to analyze the aircraft primer paint experiment described in Section 8.5.1.

solution

The sums of squares required are

$$SS_T = \sum_{i=1}^{a}\sum_{j=1}^{b}\sum_{k=1}^{n} y_{ijk}^2 - \frac{y_{...}^2}{abn}$$

$$= (4.0)^2 + (4.5)^2 + \cdots + (5.0)^2 - \frac{(89.8)^2}{18} = 10.72$$

$$SS_{\text{primers}} = \sum_{i=1}^{a}\frac{y_{i..}^2}{bn} - \frac{y_{...}^2}{abn}$$

$$= \frac{(28.7)^2 + (34.1)^2 + (27.0)^2}{6} - \frac{(89.8)^2}{18} = 4.58$$

$$SS_{\text{method}} = \sum_{j=1}^{b}\frac{y_{.j.}^2}{an} - \frac{y_{...}^2}{abn}$$

$$= \frac{(40.2)^2 + (49.6)^2}{9} - \frac{(89.8)^2}{18} = 4.91$$

$$SS_{\text{interaction}} = \sum_{i=1}^{a}\sum_{j=1}^{b}\frac{y_{ij.}^2}{n} - \frac{y_{...}^2}{abn} - SS_{\text{primers}} - SS_{\text{methods}}$$

$$= \frac{(12.8)^2 + (15.9)^2 + (11.5)^2 + (15.9)^2 + (18.2)^2}{3}$$

$$- \frac{(89.8)^2}{18} - 4.58 - 4.91 = 0.24$$

and

$$SS_E = SS_T - SS_{\text{primers}} - SS_{\text{methods}} - SS_{\text{interaction}}$$

$$= 10.72 - 4.58 - 4.91 - 0.24 = 0.99$$

example 8.6 Continued

Table 8.9

ANOVA for Example 8.6

Source of Variation	Sum of Squares	Degrees of Freedom	Mean Square	F_0	*P* value
Primer types	4.58	2	2.290	27.93	1.93×10^{-4}
Application methods	4.91	1	4.910	59.88	5.28×10^{-6}
Interaction	0.24	2	0.120	1.46	0.269
Error	0.99	12	0.082		
Total	10.72	17			

The ANOVA is summarized in Table 8.9. Note that the *P* values for both main effects are very small, indicating that the type of primer used and the application method significantly affect adhesion force. Since the *P* value for the interaction effect *F* ratio is relatively large, we would conclude that there is no interaction between primer type and application method. As an alternative to using the *P* values, we could compare the computed *F* ratios to a 5% (say) upper critical value of the *F* distribution. Since $F_{0.05,2,12} = 3.89$ and $F_{0.05,1,12} = 4.75$, we conclude that primer type and application method affect adhesion force. Furthermore, since $1.5 < F_{0.05,2,12}$, there is no indication of interaction between these factors.

In practice, ANOVA computations are performed on a computer using a statistics software package. Table 8.10 is the analysis of variance from Minitab. (SPC XL will also analyze factorial designs—see the chapter supplement). Note the similarity of this display to Table 8.9. Because the computer carries more decimal places than we did in the manual calculations, the *F* ratios in Tables 8.9 and 8.10 are slightly different. The *P* value for each *F* ratio in Table 8.11, is called "significance level" and when a *P*-value is less than 0.001, Minitab reports it as 0.000.

Table 8.10

ANOVA Output from Minitab, Example 8.6

```
Two-way ANOVA: Force versus Primer, Method Analysis
of Variance for Force

Source       DF       SS       MS       F      P
Primer        2   4.5811   2.2906   27.86  0.000
Method        1   4.9089   4.9089   59.70  0.000
Interaction   2   0.2411   0.1206    1.47  0.269
Error        12   0.9867   0.0822
Total        17  10.7178
```

example 8.6 Continued

Table 8.10
Continued

```
                         Individual 95% CI
Primer    Mean           -----+-----+-----+-----+-----
1         4.78                 (----*----)
2         5.68                                      (----*----)
3         4.50            (----*----)
                         -----+-----+-----+-----+-----
                            4.50       5.00       5.50     6.00

                         Individual 95% CI
Method     Mean          ------+------+------+------+--
Dip        4.467          (----*----)
Spray      5.511                          (----*----)
                         ------+------+------+------+--
                              4.550       4.900  5.250  5.600
```

Confidence intervals on each mean calculated using MS_E as an estimate of σ^2 and applying the standard confidence interval procedure for the mean of a normal distribution with unknown variance.

A graph of the adhesion force cell averages $\{\bar{y}_{ij.}\}$ versus the levels of primer type for each application method is shown in Figure 8.12. The absence of interaction is evident in the parallelism of the two lines. Furthermore, since a large response indicates greater adhesion force, we conclude that spraying is a superior application method and that primer type 2 is most effective. Therefore, if we wish to operate the process so as to attain maximum adhesion force, we should use primer type 2 and spray all parts.

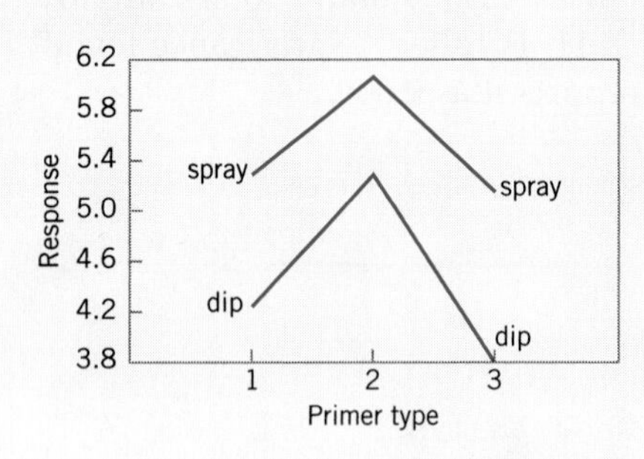

Figure 8.12 **Graph of Average Adhesion Force versus Primer Types for Example 8.6**

8.5.3 RESIDUAL ANALYSIS

Just as in the single-factor experiment, the **residuals** from a factorial experiment play an important role in assessing model adequacy. The residuals from a two-factor factorial are

$$
\begin{aligned}
e_{ijk} &= y_{ijk} - \hat{y}_{ijk} \\
&= y_{ijk} - \bar{y}_{ij.}
\end{aligned}
$$

Table 8.11

Residuals for the Aircraft Primer Paint Experiment

Primer Type	Application Method					
	Dipping			Spraying		
1	−0.26,	0.23,	0.03	0.10,	−0.40,	0.30
2	0.30,	−0.40,	0.10	−0.26,	0.03,	0.23
3	−0.03,	−0.13,	0.16	0.34,	−0.17,	−0.17

Figure 8.13 Normal Probability Plot of the Residuals from Example 8.6

That is, the residuals are simply the difference between the observations and the corresponding cell averages.

Table 8.11 presents the residuals for the aircraft primer paint data in Example 8.6. The normal probability plot of these residuals is shown in Figure 8.13. This plot has tails that do not fall exactly along a straight line passing through the center of the plot, indicating some small problems with the normality assumption, but the departure from normality is not serious. Figures 8.14 and 8.15 plot the residuals versus the levels of primer types and application methods, respectively. There is some indication that primer type 3 results in slightly lower variability in adhesion force than the other two primers. The graph of residuals versus fitted values in Figure 8.16 does not reveal any unusual or diagnostic pattern.

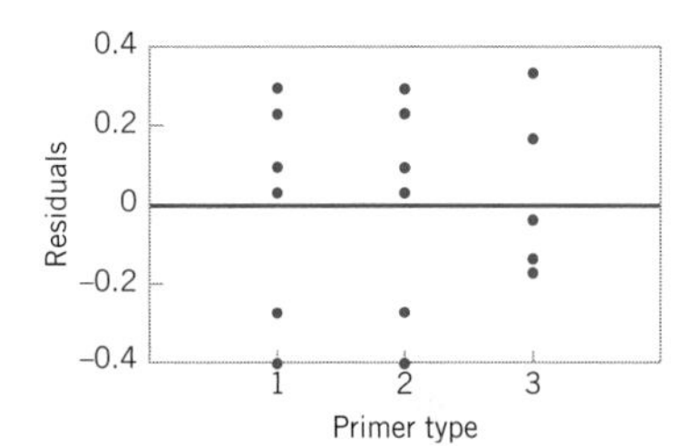

Figure 8.14 Plot of Residuals versus Primer Type

8.6 The 2^k Factorial Design

Certain special types of factorial designs are very useful in process development and improvement. One of these is a factorial design with k factors, each at two levels. Because each complete replicate of the design has 2^k runs, the arrangement is called a **2^k factorial design.** These designs have a greatly simplified analysis, and they also form the basis of many other useful designs.

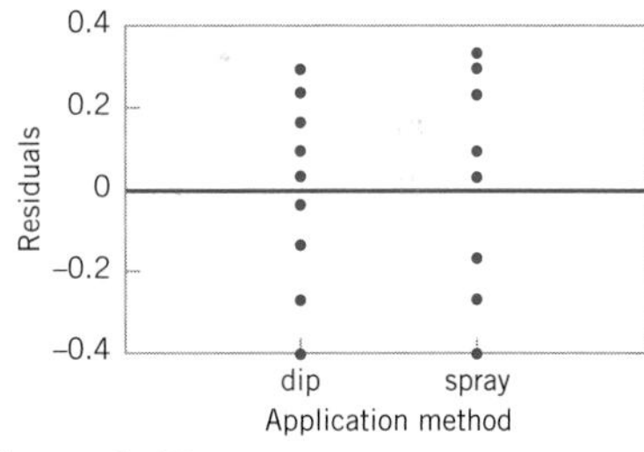

Figure 8.15 Plot of Residuals versus Application Method

8.6.1 THE 2^2 DESIGN

The simplest type of 2^k design is the 2^2—that is, two factors A and B, each at two levels. We usually think of these levels as the "low" or "−" and "high" or "+" levels of the factor. The geometry of the 2^2 design is shown in Figure 8.17*a*. Note that the design can be represented geometrically as a square with the $2^2 = 4$ runs forming the corners of the square. Figure 8.17*b* shows the four runs in a tabular format often called the **test matrix** or the **design matrix.** Each run of the test matrix is on the corners of the square and the − and + signs in each row show the settings for the variables A and B for that run.

Another notation is used to represent the runs. In general, a run is represented by a series of lowercase letters. If a letter is present, then the corresponding factor is set at the high level in that run; if it is absent, the

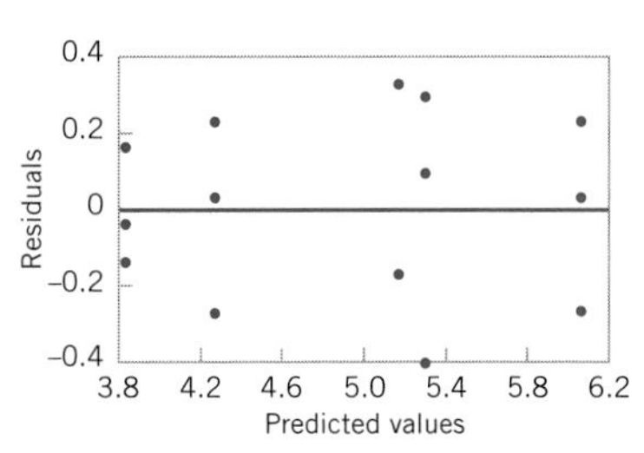

Figure 8.16 Plot of Residuals versus Predicted Values

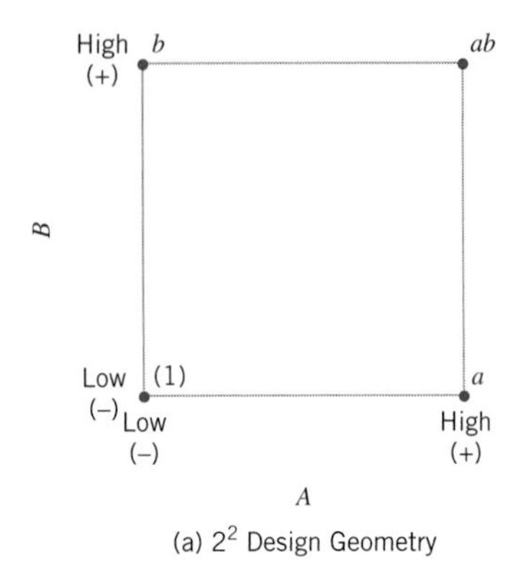

A	B
−	−
+	−
−	+
+	+

(b) 2^2 Test Matrix

Figure 8.17 **The 2^2 Factorial Design. (*a*) Design Geometry. (*b*) Test Matrix**

factor is run at its low level. For example, run a indicates that factor A is at the high level and factor B is at the low level. The run with both factors at the low level is represented by (1). This notation is used throughout the family of 2^k designs. For example, the run in a 2^4 with A and C at the high level and B and D at the low level is denoted by ac.

The effects of interest in the 2^2 design are the main effects A and B and the two-factor interaction AB. Let the letters (1), a, b, and ab also represent the totals of all n observations taken at these design points. It is easy to estimate the effects of these factors. To estimate the main effect of A, we would average the observations on the right side of the square when A is at the high level and subtract from this the average of the observations on the left side of the square where A is at the low level, or

$$\begin{aligned} A &= \bar{y}_{A^+} - \bar{y}_{A^-} \\ &= \frac{a + ab}{2n} - \frac{b + (1)}{2n} \\ &= \frac{1}{2n}[a + ab - b - (1)] \end{aligned} \tag{8.21}$$

Similarly, the main effect of B is found by averaging the observations on the top of the square where B is at the high level and subtracting the average of the observations on the bottom of the square where B is at the low level:

$$\begin{aligned} B &= \bar{y}_{B^+} - \bar{y}_{B^-} \\ &= \frac{b + ab}{2n} - \frac{a + (1)}{2n} \\ &= \frac{1}{2n}[b + ab - a - (1)] \end{aligned} \tag{8.22}$$

Finally, the AB interaction is estimated by taking the difference in the diagonal averages in Figure 8.17, or

$$\begin{aligned} AB &= \frac{ab + (1)}{2n} - \frac{a + b}{2n} \\ &= \frac{1}{2n}[ab + (1) - a - b] \end{aligned} \tag{8.23}$$

The quantities in brackets in Equations (8.21), (8.22), and (8.23) are called **contrasts.** For example, the A contrast is

$$\text{Contrast}_A = a + ab - b - (1)$$

In these equations, the contrast coefficients are always either +1 or −1. A table of plus and minus signs, such as Table 8.12, can be used to determine the sign on each run for a particular contrast. The

Table 8.12

Signs for Effects in the 2^2 Design

Run		Factorial Effect			
		I	A	B	AB
1	(1)	+	−	−	+
2	a	+	+	−	−
3	b	+	−	+	−
4	ab	+	+	+	+

column headings for the table are the main effects A and B, the AB interaction, and I, which represents the total. The row headings are the runs. Note that the signs in the AB column are the product of signs from columns A and B. To generate a contrast from this table, multiply the signs in the appropriate column of Table 8.12 by the runs listed in the rows and add.

To obtain the sums of squares for A, B, and AB, we use the following result:

$$SS = \frac{(\text{contrast})^2}{n\Sigma(\text{contrast coefficients})^2} \tag{8.24}$$

Therefore, the sums of squares for A, B, and AB are

$$\begin{aligned} SS_A &= \frac{[a + ab - b - (1)]^2}{4n} \\ SS_B &= \frac{[b + ab - a - (1)]^2}{4n} \\ SS_{AB} &= \frac{[ab + (1) - a - b]^2}{4n} \end{aligned} \tag{8.25}$$

The analysis of variance is completed by computing the total sum of squares SS_T (with $4n - 1$ degrees of freedom) as usual, and obtaining the error sum of squares SS_E [with $4(n - 1)$ degrees of freedom] by subtraction.

example 8.7 The Router Experiment

A router is used to cut registration notches in printed circuit boards. The average notch dimension is satisfactory, and the process is in statistical control (see the $\overline{X}$ and R control charts in Figure 8.18), but there is too much variability in the process.

example 8.7 Continued

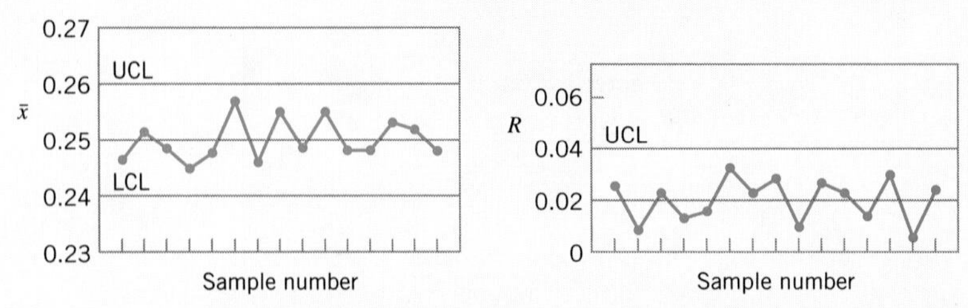

Figure 8.18 $\overline{X}$ **and** R **Control Charts on Notch Dimension, Example 8.7**

This excess variability leads to problems in board assembly. The components are inserted into the board using automatic equipment, and the variability in notch dimension causes improper board registration. As a result, the auto-insertion equipment does not work properly. How would you improve this process?

solution

Since the process is in statistical control, the quality improvement team assigned to this project decided to use a designed experiment to study the process. The team considered two factors: bit size (A) and speed (B). Two levels were chosen for each factor (bit size A at $\frac{1}{16}''$ and $\frac{1}{8}''$ and speed B at 40 rpm and 80 rpm), and a 2^2 design was set up. Since variation in notch dimension was difficult to measure directly, the team decided to measure it indirectly. Sixteen test boards were instrumented with accelerometers that allowed vibration on the (X, Y, Z) coordinate axes to be measured. The resultant vector of these three components was used as the response variable. Since vibration at the surface of the board when it is cut is directly related to variability in notch dimension, reducing vibration levels will also reduce the variability in notch dimension.

Four boards were tested at each of the four runs in the experiment, and the resulting data are shown in Table 8.13. Using Equations (8.21), (8.22), and (8.23), we can compute the factor effect estimates as follows:

$$A = \frac{1}{2n}[a + ab - b - (1)]$$

$$= \frac{1}{2(4)}[96.1 + 161.1 - 59.7 - 64.4] = \frac{133.1}{8} = 16.64$$

$$B = \frac{1}{2n}[b + ab - a - (1)]$$

$$= \frac{1}{2(4)}[59.7 + 161.1 - 96.1 - 64.4] = \frac{60.3}{8} = 7.54$$

$$AB = \frac{1}{2n}[ab + (1) - a - b]$$

$$= \frac{1}{2(4)}[161.1 + 64.4 - 96.1 - 59.7] = \frac{69.7}{8} = 8.71$$

example 8.7 Continued

Table 8.13

Data from the Router Experiment

Run		Factors A	B	Vibration				Total
1	(1)	–	–	18.2	18.9	12.9	14.4	64.4
2	a	+	–	27.2	24.0	22.4	22.5	96.1
3	b	–	+	15.9	14.5	15.1	14.2	59.7
4	ab	+	+	41.0	43.9	36.3	39.9	161.1

All the numerical effect estimates seem large. For example, when we change factor A from the low level to the high level (bit size from $\frac{1}{16}''$ to $\frac{1}{8}''$), the average vibration level increases by 16.64 cycles per second.

The magnitude of these effects may be confirmed with the analysis of variance, which is summarized in Table 8.14. The sums of squares in this table for main effects and interaction were computed Equation (8.25). The analysis of variance confirms our conclusions that were obtained by initially examining the magnitude and direction of the factor effects; both bit size and speed are important, and there is interaction between two variables. Table 8.14 was actually obtained using SPC XL. See the chapter appendix for details of how to do this.

Table 8.14

Analysis of Variance from the Router Experiment (Using SPC XL)

ANOVA TABLE

Response #1

Source	SS	df	MS	F	P	Contrib
A	1107.2	1	1107.2	185.252	0.000	64.76%
B	227.3	1	227.3	38.022	0.000	13.29%
AB	303.6	1	303.6	50.801	0.000	17.76%
Error	71.723	12	5.977			4.19%
Total	1709.834	15				

REGRESSION MODEL AND RESIDUAL ANALYSIS. It is easy to obtain the residuals from a 2^k design by fitting a regression model to the data. For the router experiment, the regression model is

$$y = \beta_0 + \beta_1 x_1 + \beta_2 x_2 + \beta_{12} x_1 x_2 + \varepsilon$$

where the factors A and B are represented by coded variables x_1 and x_2, and the AB interaction is represented by the cross-product term in the model, x_1x_2. The low and high levels of each factor are assigned the values $x_j = -1$ and $x_j = +1$, respectively. The coefficients β_0, β_1, β_2, and β_{12} are called **regression coefficients,** and ε is a random error term, similar to the error term in an analysis of variance model.

The fitted regression model is

$$\hat{y} = 23.83 + \left(\frac{16.64}{2}\right)x_1 + \left(\frac{7.54}{2}\right)x_2 + \left(\frac{8.71}{2}\right)x_1x_2$$

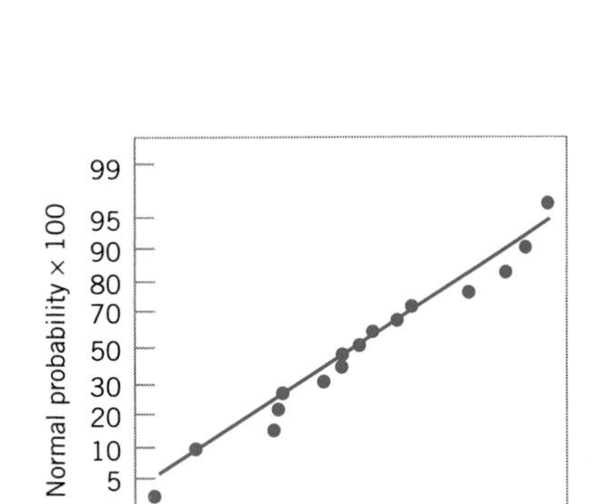

Figure 8.19 Normal Probability Plot, Example 8.7

where the estimate of the intercept $\hat{\beta}_0$ is the grand average of all 16 observations ($\bar{y}$) and the estimates of the other regression coefficients $\hat{\beta}_j$ are one-half the effect estimate for the corresponding factor. [Each regression coefficient estimate is one-half the effect estimate because regression coefficients measure the effect of a unit change in x_j on the mean of y, and the effect estimate is based on a two-unit change (from -1 to $+1$).]

This model can be used to obtain the predicted values of vibration level at any point in the region of experimentation, including the four points in the design. For example, consider the point with the small bit ($x_1 = -1$) and low speed ($x_2 = -1$). The predicted vibration level is

$$\hat{y} = 23.83 + \left(\frac{16.64}{2}\right)(-1) + \left(\frac{7.54}{2}\right)(-1) + \left(\frac{8.71}{2}\right)(-1)(-1) = 16.1$$

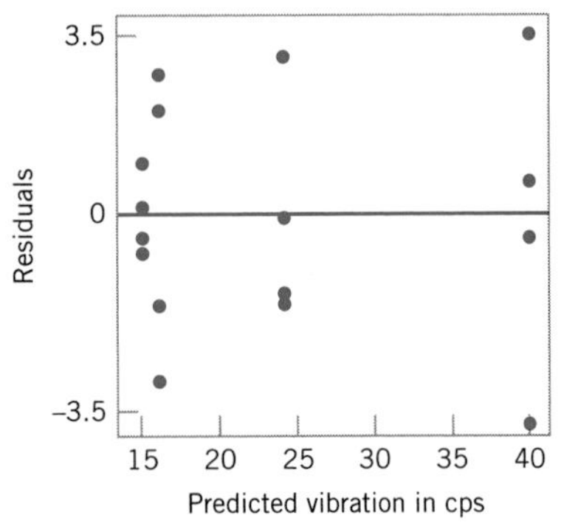

Figure 8.20 Plot of Residuals versus $\hat{y}$, Example 8.7

The four residuals corresponding to the observations at this design point are found by taking the difference between the actual observation and the predicted value as follows:

$$e_1 = 18.2 - 16.1 - 2.1 \quad e_3 = 12.9 - 16.1 = -3.2$$
$$e_1 = 18.9 - 16.1 = 2.8 \quad e_4 = 14.4 - 16.1 = -1.7$$

The residuals at the other three runs would be computed similarly.

Figures 8.19 and 8.20 present the normal probability plot and the plot of residuals versus the fitted values, respectively. The normal probability plot is satisfactory, as is the plot of residuals versus $\hat{y}$, although this latter plot does give some indication that there may be less variability in the data at the point of lowest predicted vibration level.

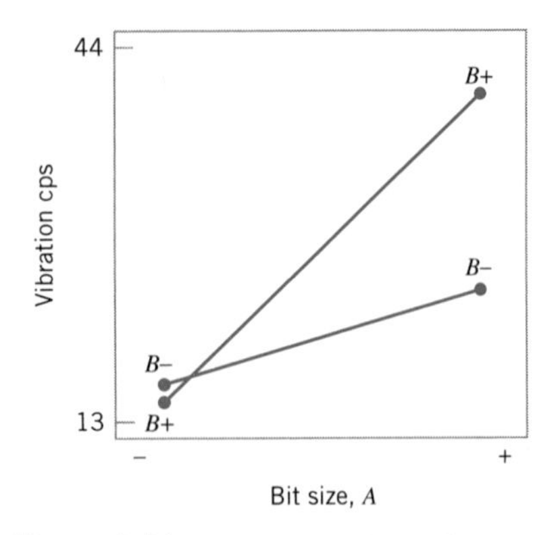

Figure 8.21 *AB* Interaction Plot

PRACTICAL INTERPRETATION OF EXAMPLE 8.7. Since both factors A (bit size) and B (speed) have large, positive effects, we could reduce vibration levels by running both factors at the low level. However, with both bit size and speed at low level, the production rate could be unacceptably low. The AB interaction provides a solution to this potential dilemma. Figure 8.21 presents the two-factor AB interaction plot. Note that the large positive effect of speed occurs primarily when bit size is at the high level. If we use the small bit, then either speed level will provide lower vibration levels. If we run with speed high and use the small bit, the production rate will be satisfactory.

When manufacturing implemented this set of operating conditions, the result was a dramatic reduction in variability in the registration notch dimension. The process remained in statistical control, as the control charts in Figure 8.22 imply, and the reduced variability dramatically improved the performance of the auto-insertion process.

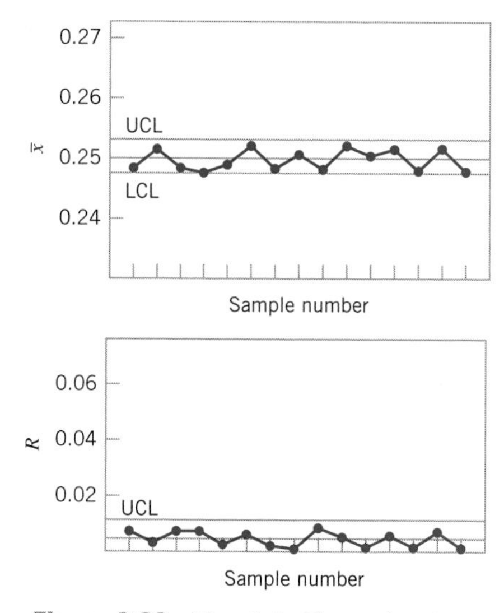

Figure 8.22 $\overline{X}$ **and** R **Charts for the Router Process after the Experiment**

ANALYSIS PROCEDURE FOR FACTORIAL EXPERIMENTS.

Table 8.15 summarizes the sequence of steps that is usually employed to analyze factorial experiments. These steps were followed in the analysis of the router experiment in Example 8.7. Recall that our first activity, after the experiment was run, was to estimate the effect of the factors bit size, speed, and the two-factor interaction. The preliminary model that we used in the analysis was the two-factor factorial model with interaction. Generally, in any factorial experiment with replication, we will almost always use the full factorial model as the preliminary model. We tested for significance of factor effects by using the analysis of variance. Since the residual analysis was satisfactory, and both main effects and the interaction term were significant, there was no need to refine the model. Therefore, we were able to interpret the results in terms of the original full factorial model, using the two-factor interaction graph in Figure 8.21. Sometimes refining the model includes deleting terms from the final model that are not significant, or taking other actions that may be indicated from the residual analysis.

Several statistics software packages include special routines for the analysis of two-level factorial designs. Many of these packages follow an analysis process similar to the one we have outlined. We will illustrate this analysis procedure again several times in this chapter.

Table 8.15

Analysis Procedure for Factorial Designs

1. Estimate the factor effects	4. Analyze residuals
2. Form preliminary model	5. Refine model, if necessary
3. Test for significance of factor effects	6. Interpret results

8.6.2 THE 2^k DESIGN FOR $k \geq 3$ FACTORS

The methods presented in the previous section for factorial designs with $k = 2$ factors each at two levels can be easily extended to more than two factors. For example, consider $k = 3$ factors, each at two levels. This design is a 2^3 factorial design, and it has eight factor-level combinations. Geometrically, the design is a cube as shown in Figure 8.23*a*, with the eight runs forming the corners of the cube. Figure 8.23*b* shows the **test design matrix.** This design allows three main effects to be estimated (A, B, and C) along with three two-factor

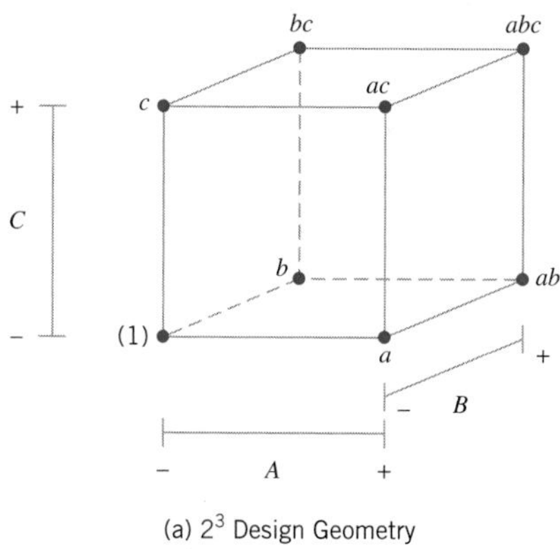

A	B	C
−	−	−
+	−	−
−	+	−
+	+	−
−	−	+
+	−	+
−	+	+
+	+	+

(b) 2^3 Test Matrix

Figure 8.23 **The 2^3 Factorial Design. (*a*) Design Geometry. (*b*) Test Matrix**

interactions (AB, AC, and BC) and a three-factor interaction (ABC). Thus, the full factorial model could be written symbolically as

$$y = \mu + A + B + C + AB + AC + BC + ABC + \varepsilon$$

where μ is an overall mean, ε is a random error term assumed to be NID$(0, \sigma^2)$, and the uppercase letters represent the main effects and interactions of the factors [note that we could have used Greek letters for the main effects and interactions, as in Equation (8.12)].

The main effects can be estimated easily. Remember that the lowercase letters (1), a, b, ab, c, ac, bc, and abc represent the total of all n replicates at each of the eight runs in the design. Referring to the cube in Figure 8.23, we would estimate the main effect of A by averaging the four runs on the right side of the cube where A is at the high level and subtracting from that quantity the average of the four runs on the left side of the cube where A is at the low level. This gives

Estimation of Main Effects

$$A = \bar{y}_{A^+} - \bar{y}_{A^-} = \frac{1}{4n}[a + ab + ac + abc - b - c - bc - (1)] \tag{8.26}$$

In a similar manner, the effect of B is the average difference of the four runs in the back face of the cube and the four in the front, or

$$B = \bar{y}_{B^+} - \bar{y}_{B^-} = \frac{1}{4n}[b + ab + bc + abc - a - c - ac - (1)] \tag{8.27}$$

and the effect of C is the average difference between the four runs in the top face of the cube and the four in the bottom, or

$$C = \bar{y}_{C^+} - \bar{y}_{C^-} = \frac{1}{4n}[c + ac + bc + abc - a - b - ab - (1)] \tag{8.28}$$

The top row of Figure 8.24 shows how the main effects of the three factors are computed.

Now consider the two-factor interaction AB. When C is at the low level, AB is simply the average difference in the A effect at the two levels of B, or

$$AB(C\text{ low}) = \frac{1}{2n}[ab - b] = \frac{1}{2n}[a - (1)]$$

Similarly, when C is at the high level, the AB interaction is

$$AB(C\text{ high}) = \frac{1}{2n}[abc - bc] - \frac{1}{2n}[ac - c]$$

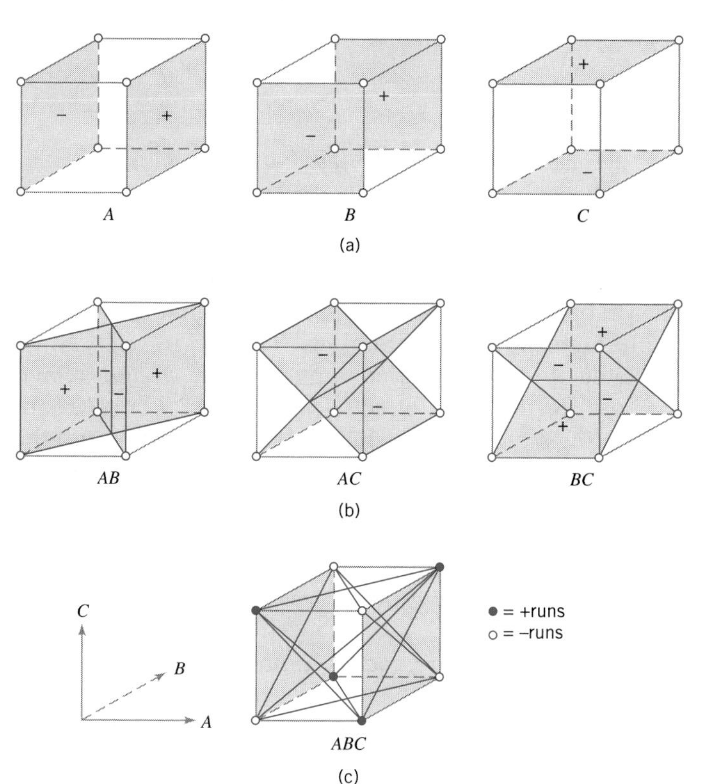

Figure 8.24 **Geometric Presentation of Contrasts Corresponding to the Main Effects and Interaction in the 2^3 Design. (*a*) Main Effects. (*b*) Two-Factor Interactions. (*c*) Three-Factor Interaction**

Interaction Effects

The AB interaction is the average of these two components, or

$$AB = \frac{1}{4n}[ab + (1) + abc + c - b - a - bc - ac] \tag{8.29}$$

Note that the AB interaction is simply the difference in averages on two diagonal planes in the cube (refer to the left-most cube in the middle row of Figure 8.24).

Using a similar approach, we see from the middle row of Figure 8.24 that the AC and BC interaction effect estimates are as follows:

$$AC = \frac{1}{4n}[ac + (1) + abc + b - a - c - ab - bc] \tag{8.30}$$

$$BC = \frac{1}{4n}[bc + (1) + abc + a - b - c - ab - ac] \tag{8.31}$$

The ABC interaction effect is the average difference between the AB interaction at the two levels of C. Thus

$$ABC = \frac{1}{4n}\{[abc - bc] - [ac - c] - [ab - b] + [a - (1)]\}$$

or

$$ABC = \frac{1}{4n}[abc - bc - ac + c - ab + b + a - (1)] \quad (8.32)$$

This effect estimate is illustrated in the bottom row of Figure 8.24.

The quantities in brackets in Equations (8.26) through (8.32) are contrasts in the eight factor-level combinations. These contrasts can be obtained from a table of plus and minus signs for the 2^3 design, shown in Table 8.16. Signs for the main effects (columns A, B, and C) are obtained by associating a plus with the high level and a minus with the low level. Once the signs for the main effects have been established, the signs for the remaining columns are found by multiplying the appropriate preceding columns, row by row. For example, the signs in column AB are the product of the signs in columns A and B.

Table 8.16 has several interesting properties:

1. Except for the identity column I, each column has an equal number of plus and minus signs.
2. The sum of products of signs in any two columns is 0; that is, the columns in the table are *orthogonal.*
3. Multiplying any column by column I leaves the column unchanged; that is, I is an *identity element.*
4. The product of any two columns yields a column in the table; for example, $A \times B = AB$, and $AB \times ABC = A^2B^2C = C$, since any column multiplied by itself is the identity column.

Table 8.16

Signs for Effects in the 2^3 Design

Treatment Combination	Factorial Effect							
	I	A	B	AB	C	AC	BC	ABC
(1)	+	–	–	+	–	+	+	–
a	+	+	–	–	–	–	+	+
b	+	–	+	–	–	+	–	+
ab	+	+	+	+	–	–	–	–
c	+	–	–	+	+	–	–	+
ac	+	+	–	–	+	+	–	–
bc	+	–	+	–	+	–	+	–
abc	+	+	+	+	+	+	+	+

The estimate of any main effect or interaction is determined by multiplying the factor-level combinations in the first column of the table by the signs in the corresponding main effect or interaction column, adding the result to produce a contrast, and then dividing the contrast by one-half the total number of runs in the experiment. Expressed mathematically,

$$\text{Effect} = \frac{\text{Contrast}}{n2^{k-1}} \tag{8.33}$$

The sum of squares for any effect is

$$SS = \frac{(\text{Contrast})^2}{n2^k} \tag{8.34}$$

example 8.8 A 2^3 Factorial Design

An experiment was performed to investigate the surface finish of a metal part. The experiment is a 2^3 factorial design in the factors feed rate (A), depth of cut (B), and tool angle (C), with $n = 2$ replicates. Table 8.17 presents the observed surface-finish data for this experiment, and the design is shown graphically in Figure 8.25. Analyze and interpret the data from this experiment.

Table 8.17

Surface-Finish Data for Example 8.8

Run		Design Factors			Surface Finish	Totals
		A	B	C		
1	(1)	−1	−1	−1	9, 7	16
2	a	1	−1	−1	10, 12	22
3	b	−1	1	−1	9, 11	20
4	ab	1	1	−1	12, 15	27
5	c	−1	−1	1	11, 10	21
6	ac	1	−1	1	10, 13	23
7	bc	−1	1	1	10, 8	18
8	abc	1	1	1	16, 14	30

example 8.8 Continued

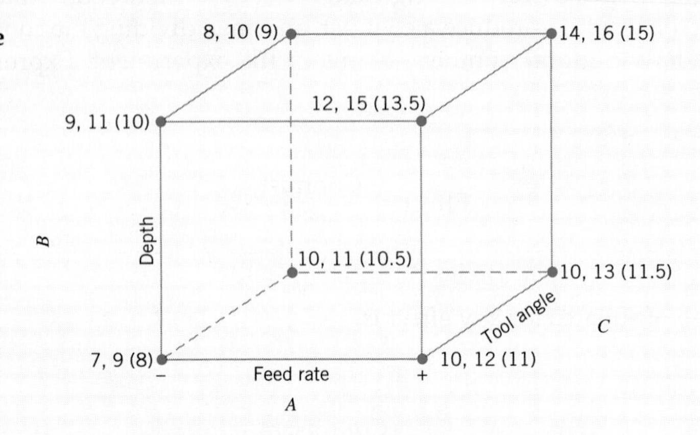

Figure 8.25 2^3 **Design for the Surface Finish Experiment in Example 8.38 (the Numbers in Parentheses are the Average Responses at Each Design Point)**

solution

The main effects may be estimated using Equations (8.26) through (8.32). The effect of A, for example, is

$$A = \frac{1}{4n}[a + ab + ac + abc - b - c - bc - (1)]$$

$$= \frac{1}{4(2)}[22 + 27 + 23 + 30 - 20 - 21 - 18 - 16]$$

$$= \frac{1}{8}[27] = 3.375$$

and the sum of squares for A is found using Equation (8.34):

$$SS_A = \frac{(\text{Contrast}_A)^2}{n2^k}$$

$$= \frac{(27)^2}{2(8)} = 45.5625$$

It is easy to verify that the other effect estimates and sums of squares are

$$\begin{aligned} B &= 1.625 & SS_B &= 10.5625 \\ C &= 0.875 & SS_C &= 3.0625 \\ AB &= 1.375 & SS_{AB} &= 7.5625 \\ AC &= 0.125 & SS_{AC} &= 0.0625 \\ BC &= -0.625 & SS_{BC} &= 1.5625 \\ ABC &= 1.125 & SS_{ABC} &= 5.0625 \end{aligned}$$

From examining the magnitude of the effects, feed rate (factor A) is clearly dominant, followed by depth of cut (B) and the AB interaction, although the interaction effect is relatively small. The analysis of variance for the full factorial model is summarized in Table 8.18. Based on the P values, it is clear that the feed rate (A) is highly significant.

Many computer programs analyze the 2^k factorial design. Table 8.19 is the output from Minitab. Although at first glance the two tables seem somewhat different, they

example 8.8 Continued

Table 8.18

Analysis of Variance for the Surface Finish Experiment

Source of Variation	Sum of Squares	Degrees of Freedom	Mean Square	F_0	*P* value
A	45.5625	1	45.5625	18.69	2.54×10^{-3}
B	10.5625	1	10.5625	4.33	0.07
C	3.0625	1	3.0625	1.26	0.29
AB	7.5625	1	7.5625	3.10	0.12
AC	0.0625	1	0.0625	0.03	0.88
BC	1.5625	1	1.5625	0.64	0.45
ABC	5.0625	1	5.0625	2.08	0.19
Error	19.5000	8	2.4375		
Total	92.9375	15			

Table 8.19

Analysis of Variance from Minitab for the Surface Finish Experiment

```
Factorial Design

Full Factorial Design

Factors:    3   Base Design:          3,8
Runs:      16   Replicates:             2
Blocks:  none   Center pts (total):     0

All terms are free from aliasing

Fractional Factorial Fit: Finish versus A, B, C

Estimated Effects and Coefficients for Finish (coded
units)

Term        Effect      Coef    SE Coef       T       P
Constant             11.0625     0.3903   28.34   0.000
A           3.3750    1.6875     0.3903    4.32   0.003
B           1.6250    0.8125     0.3903    2.08   0.071
C           0.8750    0.4375     0.3903    1.12   0.295
A*B         1.3750    0.6875     0.3903    1.76   0.116
```

(Continued)

example 8.8 Continued

Table 8.19

Continued

A*C	0.1250	0.0625	0.3903	0.16	0.877
B*C	−0.6250	−0.3125	0.3903	−0.80	0.446
A*B*C	1.1250	0.5625	0.3903	1.44	0.188

Analysis of Variance for Finish (coded units)

Source	DF	Seq SS	Adj SS	Adj MS	F	P
Main Effects	3	59.187	59.187	19.729	8.09	0.008
2-Way Interactions	3	9.187	9.187	3.062	1.26	0.352
3-Way Interactions	1	5.062	5.062	5.062	2.08	0.188
Residual Error	8	19.500	19.500	2.437		
Pure Error	8	19.500	19.500	2.438		
Total	15	92.937				

actually provide the same information. The analysis of variance displayed in the lower portion of Table 8.19 presents F ratios computed on important groups of model terms: main effects, two-way interactions, and the three-way interaction. The mean square for each group of model terms was obtained by combining the sums of squares for each model component and dividing by the number of degrees of freedom associated with that group of model terms.

A t test is used to test the significance of each individual term in the model. These t tests are shown in the upper portion of Table 8.19. Note that a "coefficient estimate" is given for each variable in the full factorial model. These are actually the estimates of the coefficients in the regression model that would be used to predict surface finish in terms of the variables in the full factorial model. Each t value is computed according to

$$t_0 = \frac{\hat{\beta}}{s.e.\,(\hat{\beta})}$$

where $\hat{\beta}$ is the coefficient estimate and $s.e.\,(\hat{\beta})$ is the estimated standard error of the coefficient. For a 2^k factorial design, the estimated standard error of the coefficient is

$$s.e.\,(\hat{\beta}) = \sqrt{\frac{\hat{\sigma}^2}{n2^k}}$$

We use the error or residual mean square from the analysis of variance as the estimate $\hat{\sigma}^2$. In our example,

$$s.e.\,(\hat{\beta}) = \sqrt{\frac{2.4375}{2(2^3)}} = 0.390312$$

example 8.8 Continued

as shown in Table 8.19. It is easy to verify that dividing any coefficient estimate by its estimated standard error produces the t value for testing whether the corresponding regression coefficient is zero.

The t tests in Table 8.19 are equivalent to the ANOVA F tests in Table 8.18. You may have suspected this already, since the P values in the two tables are identical to two decimal places. Furthermore, note that the square of any t value in Table 8.19 produces the corresponding F ratio value in Table 8.18. In general, the square of a t random variable with v degrees of freedom results in an F random variable with one numerator degree of freedom and v denominator degrees of freedom. This explains the equivalence of the two procedures used to conduct the analysis of variance for the surface finish experiment data.

Based on the ANOVA results, we conclude that the full factorial model in all these factors is unnecessary, and that a reduced model including fewer variables is more appropriate. The main effects of A and B both have relatively small P values (< 0.10), and this AB interaction is the next most important effect (P value $\cong 0.12$). The regression model that we would use to represent this process is

$$y = \beta_0 + \beta_1 x_1 + \beta_2 x_2 + \beta_{12} x_1 x_{12} + \varepsilon$$

where x_1 represents factor A, x_2 represents factor B, and $x_1 x_2$ represents the AB interaction. The regression coefficients $\hat{\beta}_1$, $\hat{\beta}_2$, and $\hat{\beta}_{12}$ are one-half the corresponding effect estimates, and $\hat{\beta}_0$ is the grand average. Thus

$$\begin{aligned} \hat{y} &= 11.0625 + \left(\frac{3.375}{2}\right)x_1 + \left(\frac{1.625}{2}\right)x_2 + \left(\frac{1.35}{2}\right)x_1 x \\ &= 11.0625 + 1.6875x_1 + 0.8125x_2 + 0.6875x_1x_2 \end{aligned}$$

Note that we can read the values of $\hat{\beta}_0$, $\hat{\beta}_1$, $\hat{\beta}_2$, and $\hat{\beta}_{12}$ directly from the "coefficient" column of Table 8.18.

This regression model can be used to predict surface finish at any point in the original experimental region. For example, consider the point where all three variables are at the low level. At this point, $x_1 = x_2 = -1$, and the predicted value is

$$\begin{aligned} \hat{y} &= 11.0625 + 1.6875(-1) + 0.8125(-1) + 0.6875(-1)(-1) \\ &= 9.25 \end{aligned}$$

Figure 8.26 shows the predicted values at each point in the original experimental design.

The residuals can be obtained as the difference between the observed and predicted values of surface finish at each design point. For the point where all three factors A, B, and C are at the low level, the observed values of surface finish are 9 and 7, so the residuals are $9 - 9.25 = -0.25$ and $7 - 9.25 = -2.25$.

A normal probability plot of the residuals is shown in Figure 8.27. Since the residuals lie approximately along a straight line, we do not suspect any severe nonnormality in the data. There are also no indications of outliers. It would also be helpful to plot the residuals versus the predicted values and against each of the factors A, B, and C. These plots do not indicate any potential model problems.

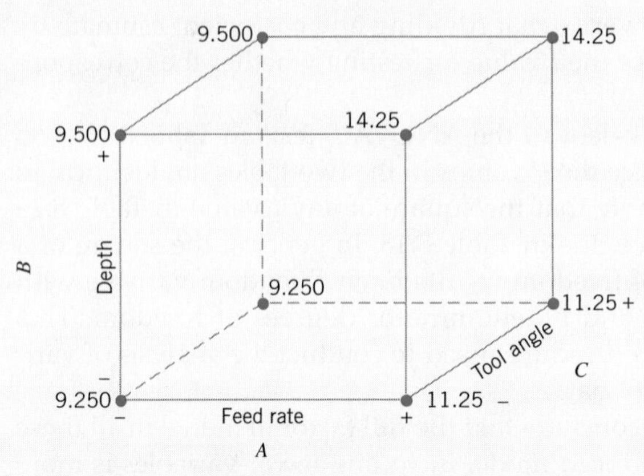

Figure 8.26 Predicted Values of Surface Finish at Each Point in the Original Design, Example 8.8

Finally, we can provide a practical interpretation of the results of our experiment. Both main effects A and B are positive, and since small values of the surface finish response are desirable, this would suggest that both A (feed rate) and B (depth of cut) should be run at the low level. However, the model has an interaction term, and the effect of this interaction should be taken into account when drawing conclusions. We could do this by examining an interaction plot, as in Example 8.7 (see Figure 8.21). Alternatively, the cube plot of predicted responses in Figure 8.26 can also be used for model interpretation. This figure indicates that the lowest values of predicted surface finish will be obtained when A and B are at the low level.

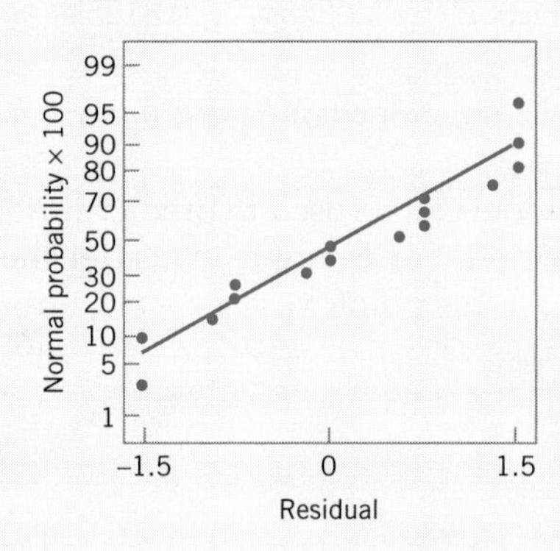

Figure 8.27 Normal Probability Plot of Residuals, Example 8.8

SOME COMMENTS ON THE REGRESSION MODEL. In the two previous examples, we used a regression model to summarize the results of the experiment. In general, a regression model is an equation of the form

$$y = \beta_0 + \beta_1 x_1 + \beta_2 x_2 + \cdots \beta_k x_k + \varepsilon \tag{8.35}$$

where y is the response variable, the x's are a set of regressor or predictor variables, the β's are the regression coefficients, and ε is an error term, assumed to be NID(0, σ^2). In our examples, we had $k = 2$ factors and the models had an interaction term, so the specific form of the regression model that we fit was

$$y = \beta_0 + \beta_1 x_1 + \beta_2 x_2 + \beta_{12} x_1 x_2 + \varepsilon$$

2^k Factorial Designs are Orthogonal Designs

In general, the regression coefficients in these models are estimated using the **method of least squares;** that is, the $\hat{\beta}$'s are chosen so as to minimize the sum of the squares of the errors (the ε's). However, in the special case of a 2^k design, it is extremely easy to find the least squares estimates of the β's. The least squares estimate of any regression coefficient β is simply one-half of the corresponding factor effect estimate. Recall that we have used this result to obtain the regression models in Examples 8.7 and 8.8. Also, please remember that this result only works for a 2^k factorial design, and it assumes that the x's are coded variables over the range $-1 \leq x \leq +1$ that represent the design factors.

It is very useful to express the results of a designed experiment in terms of a **model,** which will be a valuable aid in interpreting the experiment. Recall that we used the cube plot of predicted values from the model in Figure 8.25 to find appropriate settings for feed rate and depth of cut in Example 8.8. More general graphical displays can also be useful. For example, consider the model for surface finish in terms of feed rate (x_1) and depth of cut (x_2) without the interaction term

$$\hat{y} = 11.0625 + 1.6875x_1 + 0.8125x_2$$

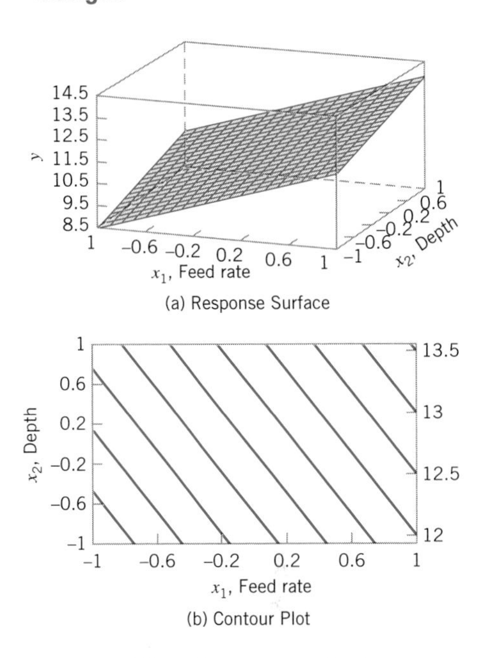

Figure 8.28 **(*a*) Response Surface for the Model $\hat{y} = 11.0625 + 1.6875x_1 + 0.8125x_2$. (*b*) The Contour Plot**

Note that the model was obtained simply by deleting the interaction term from the original model. This can only be done if the variables in the experimental design are **orthogonal,** as they are in a 2^k design. Figure 8.28 plots the predicted value of surface finish ($\hat{y}$) in terms of the two process variables x_1 and x_2. Figure 8.28*a* is a three-dimensional plot showing the plane of predicted response values generated by the regression model. This type of display is called a **response surface plot,** and the regression model used to generate the graph is often called a **first-order response surface model.** The graph in Figure 8.28*b* is a two-dimensional **contour plot** obtained by looking down on the three-dimensional response surface plot and connecting points of constant surface finish (response) in the x_1–x_2 plane. The lines of constant response are straight lines because the response surface is first-order; that is, it contains only the main effects x_1 and x_2.

In Example 8.8, we fit a first-order model with interaction:

$$\hat{y} = 11.0625 + 1.6875x_1 + 0.8125x_2 + 0.6875x_1x_2$$

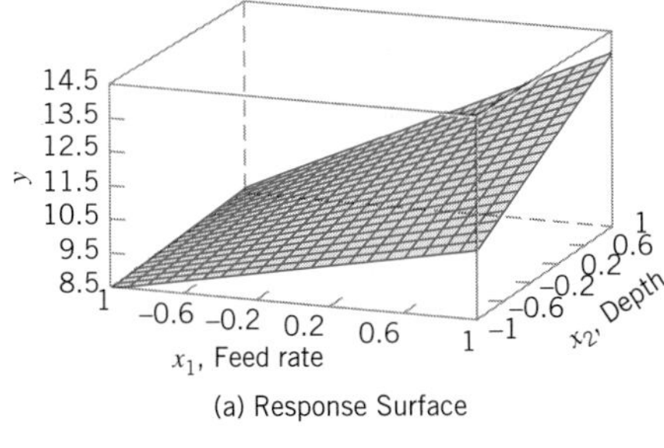

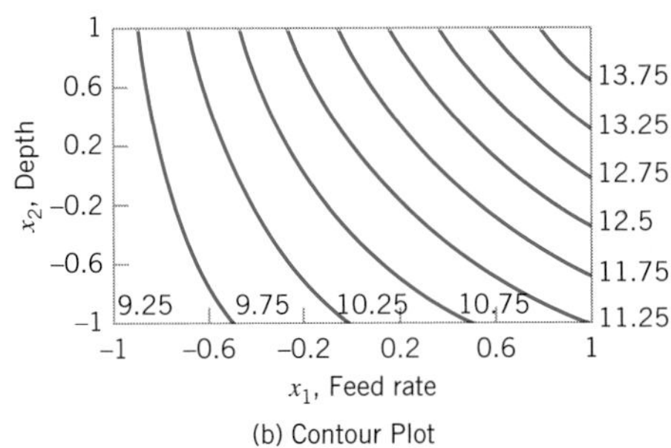

Figure 8.29 **(*a*) Response Surface for the Model $\hat{y} = 11.0625 + 1.6875x_1 + 0.8125x_2 + 0.6875x_1x_2$. (b) The Contour Plot**

Figure 8.29*a* is the three-dimensional response surface plot for this model and Figure 8.29*b* is the contour plot. Note that the effect of adding the interaction term to the model is to introduce **curvature** into the response surface; in effect, the plane is "twisted" by the interaction effect.

Inspection of a response surface makes interpretation of the results of an experiment very simple. For example, note from Figure 8.29 that if we wish to minimize the surface-finish response, we need to run x_1 and x_2 at (or near) their low levels. We reached the same conclusion by inspection of the cube plot in Figure 8.26. However, suppose we needed to obtain a particular value of surface finish, say 10.25 (the surface might need to be this rough so that a coating will adhere properly). Figure 8.29*b* indicates that there are many combinations of x_1 and x_2 that will allow the process to operate on the contour line $\hat{y} = 10.25$. The experimenter might select a set of operating conditions that maximized x_1 subject to x_1 and x_2 giving a predicted response on or near to the contour $\hat{y} = 10.25$, as this would satisfy the surface-finish objective while simultaneously making the feed rate as large as possible, which would maximize the production rate.

Response surface models have many uses. In Section 8.8 we will give an overview of some aspects of response surfaces and how they can be used for process improvement and optimization. However, note how useful the response surface was, even in this simple example. This is why we tell experimenters that **the objective of every designed experiment is a quantitative model of the process.**

PROJECTION OF 2^K DESIGNS. Any 2^k design will collapse or project into another two-level factorial design in fewer variables if one or more of the original factors are dropped. Usually this will provide additional insight into the remaining factors. For example, consider the surface finish experiment. Since factor *C* and all its interactions are negligible, we could eliminate factor *C* from the design. The result is to collapse the cube in Figure 8.25 into a square in the *A*–*B* plane; however, each of the four runs in the new design has four replicates. In general, if we delete *h* factors so that $r = k - h$ factors remain, the original 2^k design with *n* replicates will project into a 2^r design with $n2^h$ replicates.

8.6.3 A SINGLE REPLICATE OF THE 2^k DESIGN

Sparsity of Effects: Only a Few Factor Impact Performance

As the number of factors in a factorial experiment grows, the number of effects that can be estimated also grows. For example, a 2^4 experiment has four main effects, six two-factor interactions, four three-factor interactions, and one four-factor interaction, whereas a 2^6 experiment has six main effects, 15 two-factor interactions, 20 three-factor interactions, 15 four-factor interactions, six five-factor interactions, and one six-factor interaction. In most situations the **sparsity of effects principle** applies; that is, the system is usually dominated by the main effects and low-order interactions. Three-factor and higher interactions are usually negligible. Therefore, when the number of factors is moderately large—say, $k \geq 4$ or 5—a common practice is to run only a single replicate of the 2^k design and then pool or combine the higher-order interactions as an estimate of error.

example 8.9 Characterizing a Plasma Etching Process

An article in *Solid State Technology* ("Orthogonal Design for Process Optimization and Its Application in Plasma Etching," May 1987, pp. 127–132) describes the application of factorial designs in developing a nitride etch process on a single-wafer plasma etcher. The process uses C_2F_6 as the reactant gas. It is possible to vary the gas flow, the power applied to the cathode, the pressure in the reactor chamber, and

example 8.9 Continued

the spacing between the anode and the cathode (gap). Several response variables would usually be of interest in this process, but in this example we will concentrate on etch rate for silicon nitride. Perform an appropriate experiment to characterize the performance of this etching process with respect to the four process variables.

solution

The authors used a single replicate of a 2^4 design to investigate this process. Since it is unlikely that the three-factor and four-factor interactions are significant, we will tentatively plan to combine them as an estimate of error. The factor levels used in the design are shown here:

Design Factor Level	Gap A (cm)	Pressure B (m Torr)	C_2F_6 Flow C (SCCM)	Power D (W)
Low (−)	0.80	450	125	275
High (+)	1.20	550	200	325

Table 8.20 presents the data from the 16 runs of the 2^4 design. The design is shown geometrically in Figure 8.30. Table 8.21 is the table of plus and minus signs for the 2^4 design. The signs in the columns of this table can be used to estimate the factor effects. To illustrate, the estimate of the effect of gap on factor A is

$$\begin{aligned} A &= \frac{1}{8}[a + ab + ac + abc + ad + abd + acd + abcd - (1) - b \\ &\quad - c - d - bc - bd - cd - bcd] \\ &= \frac{1}{8}[669 + 650 + 642 + 635 + 749 + 868 + 860 + 729 - 550 \\ &\quad - 604 - 633 - 601 - 1037 - 1052 - 1075 - 1063] \\ &= -101.625 \end{aligned}$$

Thus, the effect of increasing the gap between the anode and the cathode from 0.80cm to 1.20cm is to decrease the etch rate by 101.625 angstroms per minute. It is easy to verify that the complete set of effect estimates is

$$\begin{aligned} A &= -101.625 & AD &= -153.625 \\ B &= -1.625 & BD &= -0.625 \\ AB &= -7.875 & ABD &= 4.125 \\ C &= 7.375 & CD &= -2.125 \\ AC &= -24.875 & ACD &= 5.625 \\ BC &= -43.875 & BCD &= -25.375 \\ ABC &= -15.625 & ABCD &= -40.125 \\ D &= 306.125 & & \end{aligned}$$

example 8.9 Continued

Table 8.20

The 2^4 Design for the Plasma Etch Experiment

Run	A (Gap)	B (Pressure)	C (C_2F_6 flow)	D (Power)	Etch Rate (Å/min)
1	−1	−1	−1	−1	550
2	1	−1	−1	−1	669
3	−1	1	−1	−1	604
4	1	1	−1	−1	650
5	−1	−1	1	−1	633
6	1	−1	1	−1	642
7	−1	1	1	−1	601
8	1	1	1	−1	635
9	−1	−1	−1	1	1,037
10	1	−1	−1	1	749
11	−1	1	−1	1	1,052
12	1	1	−1	1	868
13	−1	−1	1	1	1,075
14	1	−1	1	1	860
15	−1	1	1	1	1,063
16	1	1	1	1	729

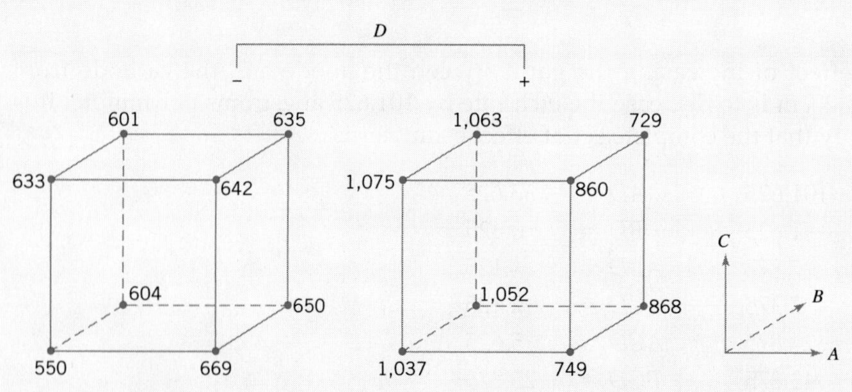

Figure 8.30 **The 2^4 Design for Example 8.9. The Etch Rate Response is Shown at the Corners of the Cubes**

example 8.9 Continued

Table 8.21

Contrast Constants for the 2^4 Design

Run		A	B	AB	C	AC	BC	ABC	D	AD	BD	ABD	CD	ACD	BCD	ABCD
1	(1)	−	−	+	−	+	+	−	−	+	+	−	+	−	−	+
2	*a*	+	−	−	−	−	+	+	−	−	+	+	+	+	−	−
3	*b*	−	+	−	−	+	−	+	−	+	−	+	+	−	+	−
4	*ab*	+	+	+	−	−	−	−	−	−	−	−	+	+	+	+
5	*c*	−	−	+	+	−	−	+	−	+	+	−	−	+	+	−
6	*ac*	+	−	−	+	+	−	−	−	−	+	+	−	−	+	+
7	*bc*	−	+	−	+	−	+	−	−	+	−	+	−	+	−	+
8	*abc*	+	+	+	+	+	+	+	−	−	−	−	−	−	−	−
9	*d*	−	−	+	−	+	+	−	+	−	−	+	−	+	+	−
10	*ad*	+	−	−	−	−	+	+	+	+	−	−	−	−	+	+
11	*bd*	−	+	−	−	+	−	+	+	−	+	−	−	+	−	+
12	*abd*	+	+	+	−	−	−	−	+	+	+	+	−	−	−	−
13	*cd*	−	−	+	+	−	−	+	+	−	−	+	+	−	−	+
14	*acd*	+	−	−	+	+	−	−	+	+	−	−	+	+	−	−
15	*bcd*	−	+	−	+	−	+	−	+	−	+	−	+	−	+	−
16	*abcd*	+	+	+	+	+	+	+	+	+	+	+	+	+	+	+

A very helpful method in judging the significance of factors in a 2^k experiment is to construct a **normal probability plot** of the effect estimates. If none of the effects is significant, then the estimates will behave like a random sample drawn from a normal distribution with zero mean, and the plotted effects will lie approximately along a straight line. Those effects that do not plot on the line are significant factors.

The normal probability plot of effect estimates from the plasma etch experiment is shown in Figure 8.31. Clearly, the main effects of *A* and *D* and the *AD* interaction are significant, as they fall far from the line passing through the other points. The analysis of variance summarized in Table 8.22 confirms these findings. Note that in the analysis of variance, we have pooled the three- and four-factor interactions to form the error mean square. If the normal probability plot had indicated that any of these interactions were important, they would not be included in the error term.

Table 8.22

Analysis of Variance for the Plasma Etch Experiment

Source of Variation	Sum of Squares	Degrees of Freedom	Mean Square	F_0
A	41,310.563	1	41,310.563	20.28
B	10.563	1	10.563	< 1
C	217.563	1	217.563	< 1
D	374,850.063	1	374,850.063	183.99
AB	248.063	1	248.063	< 1
AC	2,475.063	1	2,475.063	1.21
AD	94,402.563	1	99,402.563	46.34
BC	7,700.063	1	7,700.063	3.78
BD	1.563	1	1.563	< 1
CD	18.063	1	18.063	< 1
Error	10,186.815	5	2,037.363	
Total	531,420.938	15		

Since $A = -101.625$, the effect of increasing the gap between the cathode and anode is to decrease the etch rate. However, $D = 306.125$, so applying higher power levels will increase the etch rate. Figure 8.32 is a plot of the AD interaction. This plot indicates that the effect of changing the gap width at low power settings is small, but that increasing the gap at high power settings dramatically reduces the etch rate. High etch rates are obtained at high power settings and narrow gap widths.

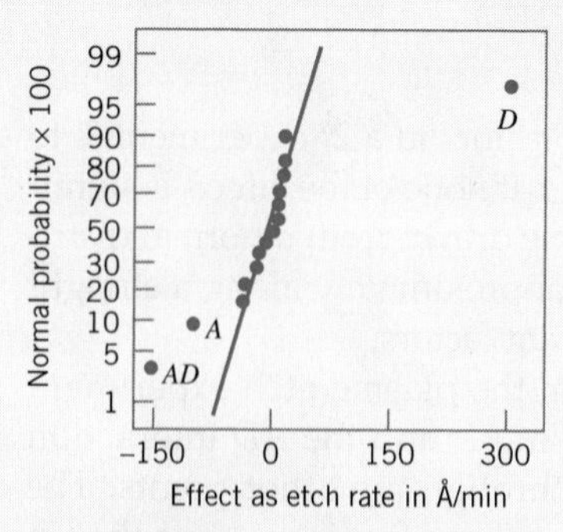

Figure 8.31 Normal Probability Plot of Effects, Example 8.9

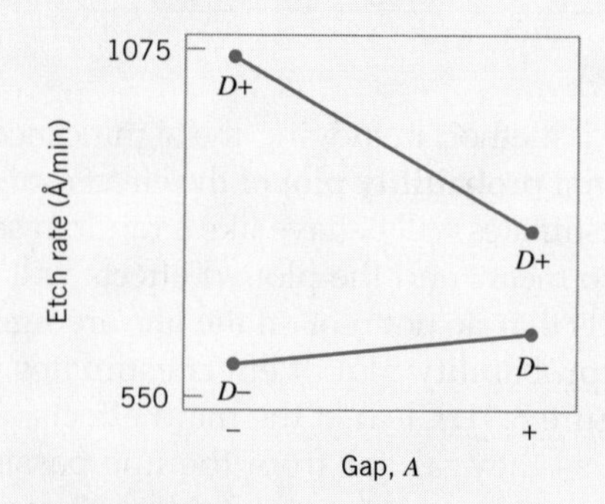

Figure 8.32 *AD* Interaction in the Plasma Etch Experiment

example 8.9 Continued

The regression model for this experiment is

$$\hat{y} = 776.0625 - \left(\frac{101.625}{2}\right)x_1 + \left(\frac{306.125}{2}\right)x_4 - \left(\frac{153.625}{2}\right)x_1x_4$$

For example, when both A and D are at the low level, the predicted value from this model is

$$\hat{y} = 776.0625 - \left(\frac{101.625}{2}\right)(-1) + \left(\frac{306.125}{2}\right)(-1)\left(\frac{153.625}{2}\right)(-1)(-1)$$
$$= 597$$

The four residuals at this run are

$$e_1 = 550 - 597 = -47$$
$$e_2 = 604 - 597 = 7$$
$$e_3 = 633 - 597 = 36$$
$$e_4 = 601 - 597 = 4$$

The residuals at the other three runs, (A high, D low), (A low, D high), and (A high, D high), are obtained similarly. A normal probability plot of the residuals is shown in Figure 8.33. The plot is satisfactory.

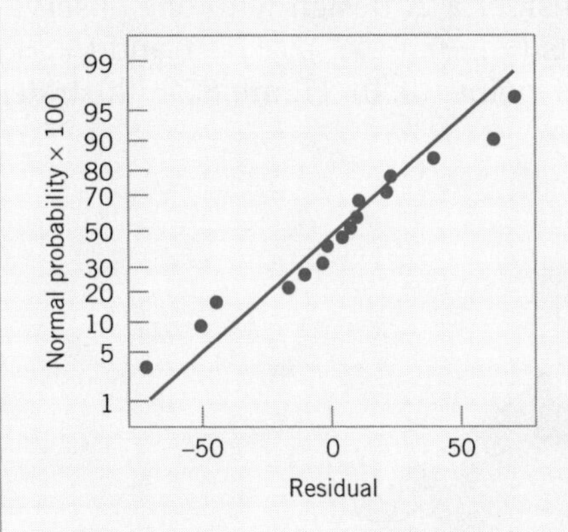

Figure 8.33 Normal Probability Plot of Residuals, Example 8.9

8.6.4 ADDITION OF CENTER POINTS TO THE 2^k DESIGN

A potential concern in the use of two-level factorial designs is the assumption of linearity in the factor effects. Of course, perfect linearity is unnecessary, and the 2^k system will work quite well even when the linearity assumption holds only approximately. In fact, we have already observed that when an interaction term is added to a main-effects model, curvature is introduced into the response surface. Since a 2^k design will support a main effects plus interactions model, some protection against curvature is already inherent in the design.

In some systems or processes, it will be necessary to incorporate **second-order effects** to obtain an adequate model. Consider the case of $k = 2$ factors. A model that includes second-order effects is

$$y = \beta_0 + \beta_1 x_1 + \beta_2 x_2 + \beta_{12} x_1 x_2 + \beta_{11} x_1^2 + \beta_{22} x_{22}^2 + \varepsilon \qquad (8.36)$$

where the coefficients β_{11} and β_{22} measure pure quadratic effects. Equation (8.36) is a **second-order response surface model.** This model cannot be fitted using a 2^2 design, because to fit a quadratic model, all factors must be run at at least three levels. It is important, however, to be able to determine whether the pure quadratic terms in Equation (8.36) are needed.

There is a method of adding one point to a 2^k factorial design that will provide some protection against pure quadratic effects (in the sense that one can test to determine if the quadratic terms are necessary). Furthermore, if this point is replicated, then an independent estimate of experimental error can be obtained. The method consists of adding **center points** to the 2^k design. These center points consist of n_C replicate runs at the point $x_i = 0$ $(i = 1, 2, \ldots, k)$. One important reason for adding the replicate runs at the design center is that center points do not impact the usual effect estimates in a 2^k design. We assume that the k factors are quantitative; otherwise, a "middle" or center level of the factor would not exist.

To illustrate the approach, consider a 2^2 design with one observation at each of the factorial points $(-, -)$, $(+, -)$, $(-, +)$, and $(+, +)$ and n_C observations at the center points (0, 0). Figure 8.34 illustrates

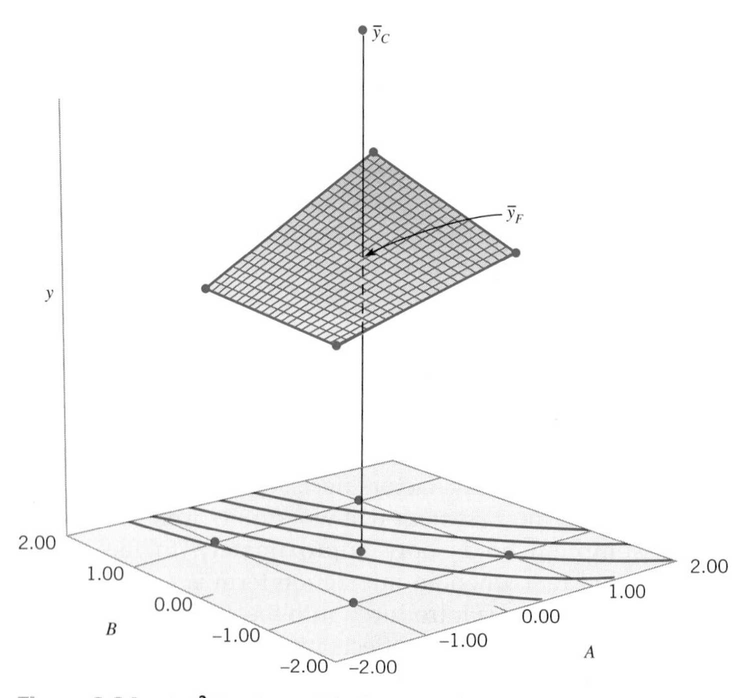

Figure 8.34 **A 2^2 Design with Center Points**

the situation. Let $\bar{y}_F$ be the average of the four runs at the four factorial points, and let $\bar{y}_C$ be the average of the n_C runs at the center point. If the difference $\bar{y}_F - \bar{y}_C$ is small, then the center points lie on or near the plane passing through the factorial points, and there is no curvature. On the other hand, if $\bar{y}_F - \bar{y}_C$ is large, then curvature is present. A single-degree-of-freedom sum of squares for pure quadratic curvature is given by

$$SS_{\text{pure qudratic}} = \frac{n_F n_C (\bar{y}_F - \bar{y}_C)^2}{n_F + n_C} \tag{8.35}$$

where, in general, n_F is the number of factorial design points. This quantity may be compared to the error mean square to test for curvature. More specifically, when points are added to the center of the 2^k design, then the model we may entertain is

Some computer programs use an equivalent *t* test.

$$y = \beta_0 + \sum_{j=1}^{k} \beta_j x_j + \sum_{i<j}\sum \beta_{ij} x_i x_j + \sum_{j=1}^{k} \beta_{jj} x_j^2 + \varepsilon$$

where β_{jj} are pure quadratic effects. The test for curvature actually tests the hypotheses

$$H_0\colon \sum_{j=1}^{k} \beta_{ij} = 0$$

$$H_1\colon \sum_{j=1}^{k} \beta_{ij} \neq 0$$

Furthermore, if the factorial points in the design are unreplicated, we may use the n_C center points to construct an estimate of error with $n_C - 1$ degrees of freedom.

example 8.10 Adding Center Points to a 2^k Design

Table 8.23 presents a modified version of the original unreplicated 2^4 design in Example 8.9 to which $n_C = 4$ center points have been added. Analyze the data and draw conclusions.

solution

The average of the center points is $\bar{y}_C = 752.75$ and the average of the 16 factorial points is $\bar{y}_F = 776.0625$. The curvature sum of squares is computed from Equation (8.35) as

$$SS_{\text{pure quadratic}} = \frac{n_F n_C (\bar{y}_F - \bar{y}_C)^2}{n_F + n_C} = \frac{16(4)(776.0625 - 752.75)^2}{16 + 4}$$
$$= 1{,}739.1$$

example 8.10 Continued

Table 8.23

The 2^4 Design for the Plasma Etch Experiment

Run	*A* (Gap)	*B* (Pressure)	*C* (C_2F_6 flow)	*D* (Power)	Etch Rate (Å/min)
1	−1	−1	−1	−1	550
2	1	−1	−1	−1	669
3	−1	1	−1	−1	604
4	1	1	−1	−1	650
5	−1	−1	1	−1	633
6	1	−1	1	−1	642
7	−1	1	1	−1	601
8	1	1	1	−1	635
9	−1	−1	−1	1	1,037
10	1	−1	−1	1	749
11	−1	1	−1	1	1,052
12	1	1	−1	1	868
13	−1	−1	1	1	1,075
14	1	−1	1	1	860
15	−1	1	1	1	1,063
16	1	1	1	1	729
17	0	0	0	0	706
18	0	0	0	0	764
19	0	0	0	0	780
20	0	0	0	0	761

Furthermore, an estimate of experimental error can be obtained by simply calculating the sample variance of the four center points as follows:

$$\hat{\sigma}^2 = \frac{\sum_{i=17}^{20}(y_i - 752.75)^2}{3} = \frac{3{,}122.7}{3} = 1040.92$$

This estimate of error has $n_C - 1 = 4 - 1 = 3$ degrees of freedom.

example 8.10 Continued

The pure quadratic sum of squares and the estimate of error may be incorporated into the analysis of variance for this experimental design. We would still use a normal probability plot of the effect estimates to preliminarily identify the important factors. The construction of this plot would not be affected by the addition of center points in the design, we would still identify *A* (Gap), *D* (Power), and the *AD* interaction as the most important effect.

Table 8.24 is the analysis of variance for this experiment obtained from Minitab. In the analysis, we included all four main effects and all six two-factor interactions in the model (just as we did in Example 8.9; see also Table 8.22). Note also that the pure quadratic sum of squares from equation 8.37 is called the "curvature" sum of squares, and the estimate of error calculated from the $n_C = 4$ center points is called the "pure error" sum of squares in Table 8.24. The "lack-of-fit" sum of squares in Table 8.24 is actually the total of the sums of squares for the three-factor and four-factor interactions. The *F*-test for lack of fit is computed as

$$F_0 = \frac{MS_{\text{lack of fit}}}{MS_{\text{pure error}}} = \frac{2037}{1041} = 1.96$$

Table 8.24

Analysis of Variance Output from Minitab for Example 8.10

Estimated Effects and Coefficients for Etch Rate (coded units)

Term	Effect	Coef	SE Coef	T	P
Constant		776.06	10.20	76.11	0.000
A	−101.62	−50.81	10.20	−4.98	0.001
B	−1.63	−0.81	10.20	−0.08	0.938
C	7.37	3.69	10.20	0.36	0.727
D	306.12	153.06	10.20	15.01	0.000
A*B	−7.88	−3.94	10.20	−0.39	0.709
A*C	−24.88	−12.44	10.20	−1.22	0.257
A*D	−153.63	−76.81	10.20	−7.53	0.000
B*C	−43.87	−21.94	10.20	−2.15	0.064
B*D	−0.63	−0.31	10.20	−0.03	0.976
C*D	−2.13	−1.06	10.20	−0.10	0.920
Ct Pt		−23.31	−2.80	−1.02	0.337

Analysis of Variance for Etch (coded units)

Source	DF	Seq SS	Adj SS	Adj MS	F	P
Main Effects	4	416389	416389	104097	62.57	0.000
2-Way Interactions	6	104845	104845	17474	10.50	0.002
Curvature	1	1739	1739	1739	1.05	0.337
Residual Error	8	13310	13310	1664		
Lack of Fit	5	10187	10187	2037	1.96	0.308
Pure Error	3	3123	3123	1041		
Total	19	536283				

example 8.10 Continued

and is not significant, indicating that none of the higher-order interaction terms is important. This computer program combines the pure error and lack-of-fit sum of squares to form a residual sum of squares with eight degrees of freedom. This residual sum of squares is used to test for pure quadratic curvature with

$$F_0 = \frac{MS_{\text{curvature}}}{MS_{\text{residual}}} = \frac{1739}{1664} = 1.05$$

The P-value in Table 8.23 associated with this F-ratio indicates that there is no evidence of pure quadratic curvature.

The upper portion of Table 8.23 shows the regression coefficient for each model effect, the corresponding t-value, and the P-value. Clearly the main effects of A and D and the AD interaction are the three largest effects.

8.6.5 BLOCKING AND CONFOUNDING IN THE 2^k DESIGN

Blocking
A technique often used when it is impossible to run all experiments under the some or homogeneous conditions. A **block** is a set of homogeneous conditions

Confounding
A design technique used to run factorial experiments in blocks where block size is smaller than a replicate. It causes the block effect to be indistinguishable (or confounded) with certain interactions.

It is often impossible to run all of the observations in a 2^k factorial design under constant or homogeneous conditions. For example, it might not be possible to conduct all the tests on one shift or use material from a single batch. When this problem occurs, **blocking** is an excellent technique for eliminating the unwanted variation that could be caused by the nonhomogeneous conditions. If the design is replicated, and if the block is of sufficient size, then one approach is to run each replicate in one **block** (set of homogeneous conditions). For example, consider a 2^3 design that is to be replicated twice. Suppose that it takes about 1 hour to complete each run. Then by running the eight runs from the first replicate on one day and the eight runs from the second replicate on another day, any time effect, or difference between how the process works on the two days, can be eliminated. Thus, the two days became the two blocks in the design. The average difference between the responses on the two days is the block effect.

Sometimes we cannot run a complete replicate of a factorial design under homogeneous experimental conditions. **Confounding** is a design technique for running a factorial experiment in blocks, where the block size is smaller than the number of runs in one complete replicate. The technique causes certain interactions to be indistinguishable from or **confounded with blocks.** We will illustrate confounding in the 2^k factorial design in 2^p blocks, where $p < k$.

Consider a 2^2 design. Suppose that each of the $2^2 = 4$ runs requires 4 h of laboratory analysis. Thus, two days are required to perform the experiment. If days are considered as blocks, then we must assign two of the four runs to each day.

Consider the design shown in Figure 8.35. Note that block 1 contains the runs (1) and ab and that block 2 contains a and b. The contrasts for estimating the main effects A and B are

$$\text{Contrast}_A = ab + a - b - (1)$$
$$\text{Contrast}_B = ab + b - a - (1)$$

These contrasts are unaffected by blocking since in each contrast there is one plus and one minus run from each block. That is, any difference between block 1 and block 2 will cancel out. The contrast for the AB interaction is

$$\text{Contrast}_{AB} = ab + (1) - a - b$$

Since the two runs with the plus sign, ab and (1), are in block 1 and the two with the minus sign, a and b, are in block 2, the block effect and AB interaction are identical. That is, AB is confounded with blocks.

The reason for this is apparent from the table of plus and minus signs for the 2^2 design. From this table, we see that all runs that have a plus on AB are assigned to block 1, and all runs that have a minus sign on AB are assigned to block 2.

This scheme can be used to confound any 2^k design in two blocks. As a second example, consider a 2^3 design, run in two blocks. Suppose we wish to confound the three-factor interaction ABC with blocks. From the table of plus and minus signs, shown in Table 8.16, we assign the runs that are minus on ABC to block 1 and those that are plus on ABC to block 2. The resulting design is shown in Figure 8.36.

For more information on confounding, refer to Montgomery (2009, Chapter 8). This book contains guidelines for selecting factors to confound with blocks so that main effects and low-order interactions are not confounded. In particular, the book contains a table of suggested confounding schemes for designs with up to seven factors and a range of block sizes, some as small as two runs.

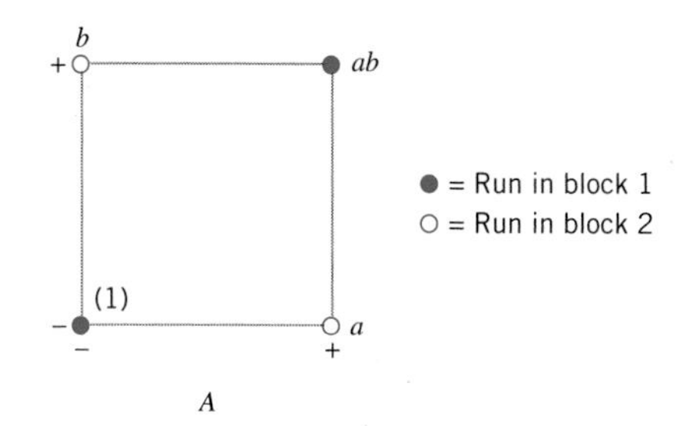

(a) Geometric view

Block 1	Block 2
(1)	a
ab	b

(b) Assignment of the four runs to two blocks

		Factorial Effect			
Run		*I*	*A*	*B*	*AB* = *Block*
1	(1)	1	2	2	1
2	a	1	1	2	2
3	b	1	2	1	2
4	ab	1	1	1	1

(c) Signs for effects of 2^2 Design in two blocks as AB confounded.

Figure 8.35 A 2^2 Design in Two Blocks.

8.7 Fractional Replication of the 2^k Design

As the number of factors in a 2^k design increases, the number of runs required increases rapidly. For example, a 2^5 requires 32 runs. In this design, only 5 degrees of freedom correspond to main effects and 10 degrees of freedom correspond to two-factor interactions. If we can assume that certain high-order interactions are negligible, then a fractional factorial design involving fewer than the complete set of 2^k runs can be used to obtain information on the main effects and low-order interactions. In this section, we will introduce fractional replication of the 2^k design. For a more complete discussion, see Montgomery (2009, Chapter 8).

8.7.1 THE ONE-HALF FRACTION OF THE 2^k DESIGN

A one-half fraction of the 2^k design contains 2^{k-1} runs and is often called a 2^{k-1} fractional factorial design. As an example, consider the 2^{3-1} design—that is, a one-half fraction of the 2^3. The table of plus and minus signs for the 2^3 design is shown in Table 8.25. Suppose we select the four runs a, b, c, and abc as our one-half fraction. These runs are

Table 8.25

Plus and Minus Signs for the 2^3 Factorial Design

	Factorial Effect							
Run	*I*	*A*	*B*	*C*	*AB*	*AC*	*BC*	*ABC*
a	+	+	−	−	−	−	+	+
b	+	−	+	−	−	+	−	+
c	+	−	−	+	+	−	−	+
abc	+	+	+	+	+	+	+	+
ab	+	+	+	−	+	−	−	−
ac	+	+	−	+	−	+	−	−
bc	+	−	+	+	−	−	+	−
(1)	+	−	−	−	+	+	+	−

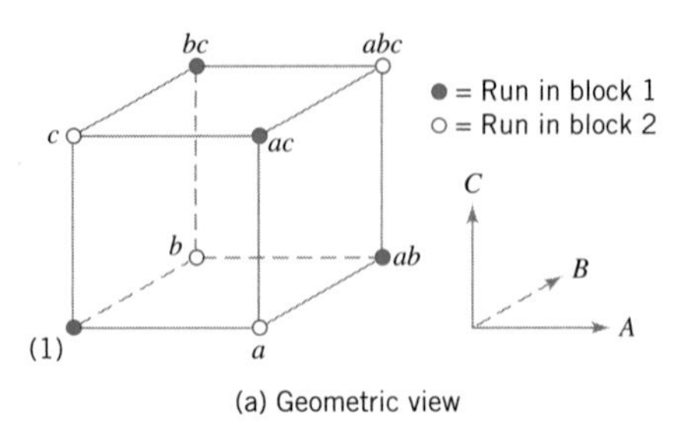

(a) Geometric view

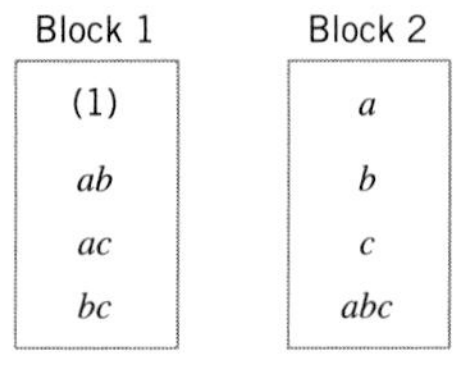

Block 1	Block 2
(1)	*a*
ab	*b*
ac	*c*
bc	*abc*

(b) Assignment of the eight runs to two blocks

Figure 8.36 **The 2^3 Design in Two Blocks with *ABC* Confounded.**

shown in the top half of Table 8.25. The design is shown geometrically in Figure 8.37*a*.

Note that the 2^{3-1} design is formed by selecting only those runs that yield a plus on the *ABC* effect. Thus, *ABC* is called the **generator** of this particular fraction. Furthermore, the identity element *I* is also plus for the four runs, so we call

$$I = ABC$$

the **defining relation** for the design.

The runs in the 2^{3-1} designs yield three degrees of freedom associated with the main effects. From Table 8.25, we obtain the estimates of the main effects as

$$A = \frac{1}{2}[a - b - c + abc]$$

$$B = \frac{1}{2}[-a + b - c + abc]$$

$$C = \frac{1}{2}[-a - b + c + abc]$$

It is also easy to verify that the estimates of the two-factor interactions are

$$BC = \frac{1}{2}[a - b - c + abc]$$

$$AC = \frac{1}{2}[-a + b - c + abc]$$

$$AB = \frac{1}{2}[-a - b + c + abc]$$

Thus, the linear combination of observations in column *A*—say, [*A*]—estimates $A + BC$. Similarly, [*B*] estimates $B + AC$, and [*C*] estimates $C + AB$. Two or more effects that have this property are called **aliases.** In our 2^{3-1} design, *A* and *BC* are aliases, *B* and *AC* are aliases, and *C* and *AB* are aliases. Aliasing is the direct result of fractional replication. In many practical situations, it will be possible to select the fraction so that the main effects and low-order interactions of interest will be aliased with high-order interactions (which are probably negligible).

The alias structure for this design is found by using the defining relation $I = ABC$. Multiplying any effect by the defining relation yields the aliases for that effect. In our example, the alias of *A* is

$$A = A \cdot ABC = A^2BC = BC$$

since $A \cdot I = A$ and $A^2 = I$. The aliases of *B* and *C* are

$$B = B \cdot ABC = AB^2C = AC$$

and

$$C = C \cdot ABC = ABC^2 = AB$$

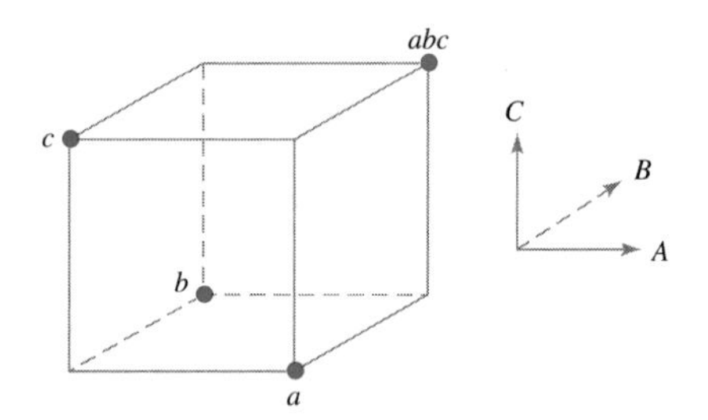

(a) The principal fraction, $I = +ABC$

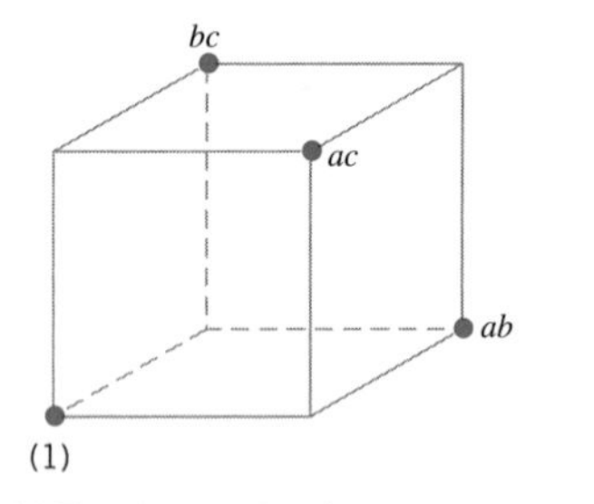

(b) The alternate fraction, $I = -ABC$

Figure 8.37 **The One-Half Fractions of the 2^3 Design**

Now suppose that we had chosen the other one-half fraction, that is, the runs in Table 8.25 associated with minus on ABC. This design is shown geometrically in Figure 8.37*b*. The defining relation for this design is $I = -ABC$. The aliases are $A = -BC$, $B = -AC$, and $C = -AB$. Thus, the effects A, B, and C with this particular fraction really estimate $A - BC$, $B - AC$, and $C - AB$. In practice, it usually does not matter which one-half fraction we select. The fraction with the plus sign in the defining relation is usually called the **principal fraction;** the other fraction is usually called the **alternate fraction.**

Aliases

Two of more effects that are calculated the same. These effects are "Aliases".

Principal Fraction

The fraction with the + sign.

Sometimes we use **sequences** of fractional factorial designs to estimate effects. For example, suppose we had run the principal fraction of the 2^{3-1} design. From this design we have the following effect estimates:

$$[A] = A + BC$$
$$[B] = B + AC$$
$$[C] = C + AB$$

Suppose we are willing to assume at this point that the two-factor interactions are negligible. If they are, then the 2^{3-1} design has produced estimates of the three main effects A, B, and C. However, if after running the principal fraction we are uncertain about the interactions, it is possible to estimate them by running the alternate fraction. The alternate fraction produces the following effect estimates:

$$[A]' = A - BC$$
$$[B]' = B - AC$$
$$[C]' = C - AC$$

If we combine the estimates from the two fractions, we obtain the following:

Effect, i	$\frac{1}{2}([i] - [i]')$	$\frac{1}{2}([i] - [i]')$
$i = A$	$\frac{1}{2}(A + BC + A - BC) = A$	$\frac{1}{2}[A + BC - (A - BC)] = BC$
$i = B$	$\frac{1}{2}(B + AC + B - AC) = B$	$\frac{1}{2}[B + AC - (B - AC)] = AC$
$i = C$	$\frac{1}{2}(C + AB + C - AB) = C$	$\frac{1}{2}[C + AB - (C - AB)] = AB$

Thus, by combining a sequence of two fractional factorial designs, we can isolate both the main effects and the two-factor interactions. This property makes the fractional factorial design highly useful in experimental problems because we can run sequences of small, efficient experiments, combine information across *several* experiments, and take advantage of learning about the process we are experimenting with as we go along.

A 2^{k-1} design may be constructed by writing down the treatment combinations for a full factorial in $k - 1$ factors and then adding the kth factor by identifying its plus and minus levels with the plus and minus signs of the highest-order interaction $\pm ABC \cdots (K-1)$. Therefore, a 2^{3-1} fractional factorial is obtained by writing down the full 2^2 factorial and then equating factor C to the $\pm AB$ interaction. Thus, to generate the principal fraction, we would use $C = +AB$ as follows:

Full 2^2		$2^{3-1}, I = ABC$		
A	B	A	B	$C = AB$
−	−	−	−	+
+	−	+	−	−
−	+	−	+	−
+	+	+	+	+

To generate the alternate fraction we would equate the last column to $C = -AB$.

example 8.11 A One-Half Fraction for the Plasma Etch Experiment

To illustrate the use of a one-half fraction, consider the plasma etch experiment described in Example 8.9. Suppose we decide to use a 2^{4-1} design with $I = ABCD$ to investigate the four factors gap (A), pressure (B), C_2F_6 flow rate (C), and power setting (D). Set up this design and analyze it using only the data from the full factorial that corresponds to the runs in the fraction.

solution

This design would be constructed by writing down a 2^3 in the factors A, B, and C and then setting $D = ABC$. The design and the resulting etch rates are shown in Table 8.26. The design is shown geometrically in Figure 8.38.

Table 8.26

The 2^{4-1} Design with Defining Relation $I = ABCD$

Run		A	B	C	$D = ABC$	Etch Rate
1	(1)	−	−	−	−	550
2	*ad*	+	−	−	+	749
3	*bd*	−	+	−	+	1052
4	*ab*	+	+	−	−	650
5	*cd*	−	−	+	+	1075
6	*ac*	+	−	+	−	642
7	*bc*	−	+	+	−	601
8	*abcd*	+	+	+	+	729

example 8.11 Continued

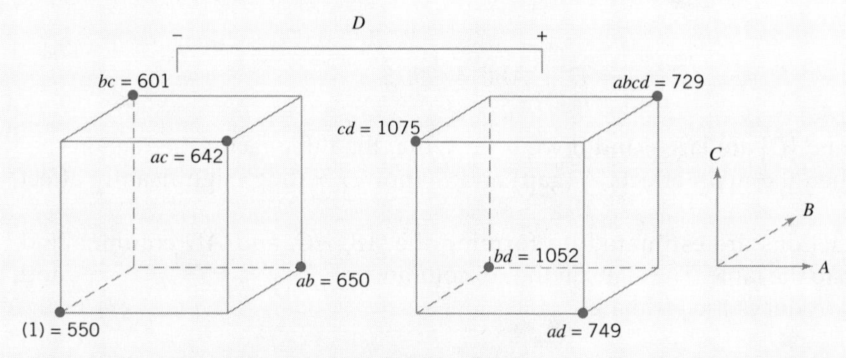

Figure 8.38 **The 2^{4-1} Design for Example 8.11**

In this design, the main effects are aliased with the three-factor interactions; note that the alias of A is:

$$A \cdot I = A \cdot ABCD$$
$$A = A^2BCD$$
$$A = BCD$$

Similarly,

$$B = ACD$$
$$C = ABD$$
$$D = ABC$$

The two-factor interactions are aliased with each other. For example, the alias of AB is CD:

$$AB \cdot I = AB \cdot ABCD$$
$$AB = A^2B^2CD$$
$$AB = CD$$

The other aliases are

$$AC = BD$$
$$AD = BC$$

The estimates of the main effects (and their aliases) are found using the four columns of signs in Table 8.24. For example, from column A we obtain:

$$[A] = A + BCD = \frac{1}{4}(-550 + 749 - 1052 + 650 - 1075 + 642 - 601 + 729)$$
$$= -127.00$$

The other columns produce:

$$[B] = B + ACD = 4.00$$
$$[C] = C + ABD = 11.50$$

example 8.11 Continued

and

$$[D] = D + ABC = 290.5$$

Clearly, $[A]$ and $[D]$ are large, and if we believe that the three-factor interactions are negligible, then the main effects A (gap) and D (power setting) significantly affect the etch rate.

The interactions are estimated by forming the AB, AC, and AD columns and adding them to the table. The signs in the AB column are $+, -, -, +, +, -, -, +$, and this column produces the estimate

$$[AB] = AB + CD = \frac{1}{4}(550 - 749 - 1052 + 650 + 1075 - 642 - 601 + 729) = -10.00$$

From the AC and AD columns we find

$$[AC] = AC + BD = -25.50$$
$$[AD] = AD + BD = -197.50$$

The $[AD]$ estimate is large; the most straightforward interpretation of the results is that this is the AD interaction. Thus, the results obtained from the 2^{4-1} design agree with the full factorial results in Example 8.9.

NORMAL PROBABILITY PLOTS AND RESIDUALS. The normal probability plot is very useful in assessing the significance of effects from a fractional factorial, especially when many effects are to be estimated. Residuals can be obtained from a fractional factorial by the regression model method shown previously. These residuals should be plotted against the predicted values, against the levels of the factors, and on normal probability paper as we have discussed before, both to assess the validity of the underlying model assumptions and to gain additional insight into the experimental situation.

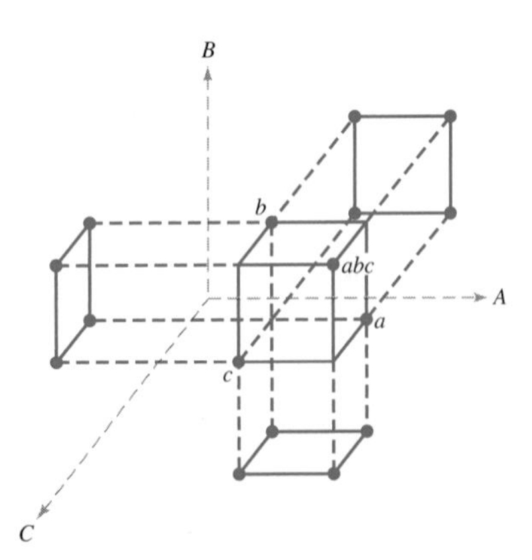

Figure 8.39 Projection of a 2^{3-1} Design into Three 2^2 Designs

PROJECTION OF THE 2^{k-1} DESIGN. If one or more factors from a one-half fraction of a 2^k can be dropped, the design will project into a full factorial design. For example, Figure 8.39 presents a 2^{3-1} design. Note that this design will project into a full factorial in any two of the three original factors. Thus, if we think that at most two of the three factors are important, the 2^{3-1} design is an excellent design for identifying the significant factors. Sometimes experiments that seek to identify a relatively few significant factors from a larger number of factors are called **screening experiments.** This projection property is highly useful in factor screening because it allows negligible factors to be eliminated, resulting in a stronger experiment in the active factors that remain.

In the 2^{4-1} design used in the plasma etch experiment in Example 8.11, we found that two of the four factors (B and C) could be dropped. If we eliminate these two factors, the remaining columns in Table 8.26 form a 2^2 design in the factors A and D, with two replicates. This design is shown in Figure 8.40.

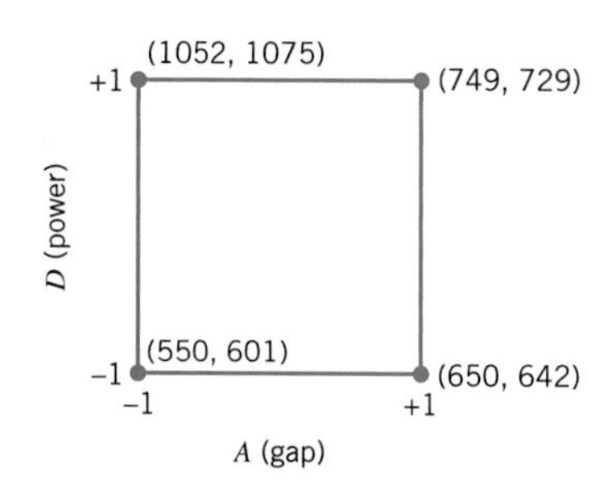

Figure 8.40 The 2^2 Design Obtained by Dropping Factors B and C from the Plasma Etch Experiment

DESIGN RESOLUTION. The concept of design resolution is a useful way to catalog fractional factorial designs according to the alias patterns they produce. Designs of resolution III, IV, and V are particularly important. The definitions of these terms and an example of each follow.

1. **Resolution III designs.**	In these designs, no main effects are aliased with any other main effect, but main effects are aliased with two-factor interactions and two-factor interactions may be aliased with each other. The 2^{3-1} design with $I = ABC$ is of resolution III. We usually employ a subscript Roman numeral to indicate design resolution; thus, this one-half fraction is a 2_{III}^{3-1} design.
2. **Resolution IV designs.**	In these designs, no main effect is aliased with any other main effect or two-factor interaction, but two-factor interactions are aliased with each other. The 2^{4-1} design with $I = ABCD$ used in Example 8.11 is of resolution IV (2_{IV}^{4-1}).
3. **Resolution V designs.**	In these designs, no main effect or two-factor interaction is aliased with any other main effect or two-factor interaction, but two-factor interactions are aliased with three-factor interactions. A 2^{5-1} design with $I = ABCDE$ is of resolution V (2_{V}^{5-1}).

Resolution III and IV designs are particularly useful in factor screening experiments. The resolution IV design provides very good information about main effects and will provide some information about two-factor interactions.

8.7.2 SMALLER FRACTIONS: THE 2^{k-p} FRACTIONAL FACTORIAL DESIGN

Although the 2^{k-1} design is valuable in reducing the number of runs required for an experiment, we frequently find that smaller fractions will provide almost as much useful information at even greater economy. In general, a 2^k design may be run in a $\frac{1}{2^p}$ fraction called a 2^{k-p} fractional factorial design. Thus, a $\frac{1}{4}$ fraction is called a 2^{k-2} fractional factorial design, a $\frac{1}{8}$ fraction is called a 2^{k-3} design, a $\frac{1}{16}$ fraction is called a 2^{k-4} design, and so on.

To illustrate a $\frac{1}{4}$ fraction, consider an experiment with six factors and suppose that the engineer is interested primarily in main effects but would also like to get some information about the two-factor interactions. A 2^{6-1} design would require 32 runs and would have 31 degrees of freedom for estimation of effects. Since there are only 6 main effects and 15 two-factor interactions, the one-half fraction is inefficient—it requires too many runs. Suppose we consider a $\frac{1}{4}$ fraction, or a 2^{6-2} design. This design contains 16 runs and with 15 degrees of freedom will allow estimation of all six main effects, with some capability for examination of the two-factor interactions. To generate this design we would write down a 2^4 design in the factors A, B, C, and D, and then add two columns for E and F. Refer to Table 8.26. To find the new columns, we would select the two design generators $I = ABCE$ and $I = BCDF$. Thus, column E would be found from $E = ABC$ and column F would be $F = BCD$. Thus, columns $ABCE$ and $BCDF$ are equal to the identity column. However, we know that the product of any two columns in the table of plus and minus signs for a 2^k is just another column in the table; therefore, the product of $ABCE$ and $BCDF$ or $ABCE\,(BCDF) = AB^2C^2DEF = ADEF$ is also an identity column. Consequently, the **complete defining relation** for the 2^{6-2} design is

$$I = ABCE = BCDF = ADEF$$

Table 8.27

Construction of the 2^{6-2} Design with Generators $I = ABCE$ and $I = BCDF$

Run	A	B	C	D	$E = ABC$	$F = BCD$
1	–	–	–	–	–	–
2	+	–	–	–	+	–
3	–	+	–	–	+	+
4	+	+	–	–	–	+
5	–	–	+	–	+	+
6	+	–	+	–	–	+
7	–	+	+	–	–	–
8	+	+	+	–	+	–
9	–	–	–	+	–	+
10	+	–	–	+	+	+
11	–	+	–	+	+	–
12	+	+	–	+	–	–
13	–	–	+	+	+	–
14	+	–	+	+	–	–
15	–	+	+	+	–	+
16	+	+	+	+	+	+

To find the alias of any effect, simply multiply the effect by each word in the above defining relation. The complete alias structure is shown here.

$$
\begin{array}{rll}
A = BCE = DEF = ABCDF & \quad & AB = CE = ACDF = BDEF \\
B = ACE = CDF = ABDEF & & AC = BE = ABDF = CDEF \\
C = ABE = BDF = ACDEF & & AD = EF = BCDE = ABCF \\
D = BCF = AEF = ABCDE & & AE = BC = DF = ABCDEF \\
E = ABC = ADF = BCDEF & & AF = DE = BCEF = ABCD \\
F = BCD = ADE = ABCEF & & BD = CF = ACDE = ABEF \\
ABD = CDE = ACF = BEF & & BF = CD = ACEF = ABDE \\
ACD = BDE = ABF = CEF & &
\end{array}
$$

Note that this is a resolution IV design; the main effects are aliased with three-factor and higher interactions, and two-factor interactions are aliased with each other. This design would provide very good information on the main effects and give some idea about the strength of the two-factor interactions.

SELECTION OF DESIGN GENERATORS. In the foregoing example, we selected $I = ABCD$ and $I = BCDF$ as the generators to construct the 2^{6-2} fractional factorial design. This choice is not arbitrary; some generators will produce designs with more attractive alias structures than other generators. For a given number of factors and number of runs we wish to make, we want to select the generators so that the design has the highest possible resolution. Montgomery (2009) presents a set of designs of maximum resolution for 2^{k-p} designs with $p \leq$ 10 factors. Many computer programs that support fractional factorials constant designs using these generators (or equivalent one).

8.8 Response Surface Methods

Response surface methodology (RSM) is a collection of mathematical and statistical techniques that are useful for modeling and analysis in applications where a response of interest is influenced by several variables and the objective is to **optimize this response.** The general RSM approach was developed in the early 1950s and was initially applied in the chemical industry with considerable success. Over the past 20 years RSM has found extensive application in a wide variety of industrial settings, far beyond its origins in chemical processes, including semiconductor and electronics manufacturing, machining, metal cutting, and joining processes, among many others. Many statistics software packages have included the experimental designs and optimization techniques that make up the basics of RSM as standard features. For a recent comprehensive presentation of RSM, see Myers and Montgomery (2002).

To illustrate the general idea of RSM, suppose that a chemical engineer wishes to find the levels of reaction temperature (x_1) and reaction time (x_2) that maximize the yield (y) of a process. The process yield is a function of the levels of temperature and time—say,

$$y = f(x_1, x_2) + \varepsilon$$

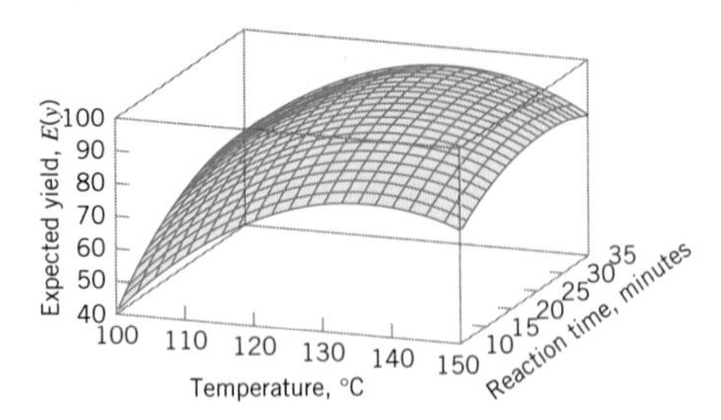

Figure 8.41 **A Three-Dimensional Response Surface Showing the Expected Yield as a Function of Reaction Temperature and Reaction Time**

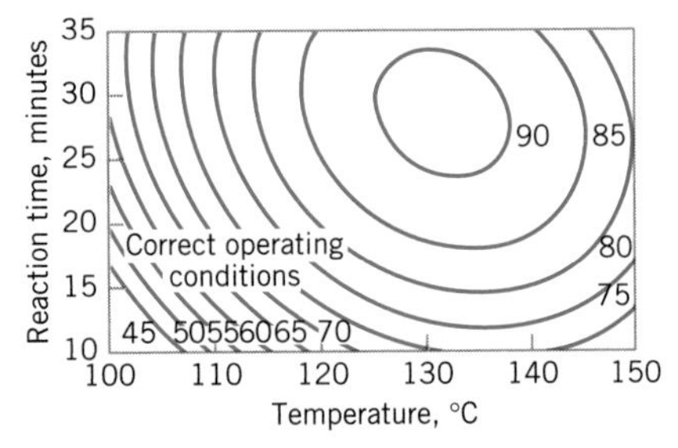

Figure 8.42 **A Contour plot of the Yield Response Surface in Figure 8.41**

where ε represents the noise or error observed in the response y. If we denote the expected value of the response by $E(y) = f(x_1, x_2)$, then the surface represented by

$$E(y) = f(x_1, x_2)$$

is called a **response surface.**

We may represent the response surface graphically as shown in Figure 8.41, where $E(y)$ is plotted versus the levels of x_1 and x_2. Note that the response is represented as a surface plot in three-dimensional space. To help visualize the shape of a response surface, we often plot the contours of the response surface as shown in Figure 8.42. In the **contour plot,** lines of constant response are drawn in the x_1, x_2 plane. Each contour corresponds to a particular height of the response surface. The contour plot is helpful in studying the levels of x_1, x_2 that result in changes in the shape or height of the response surface.

In most RSM problems, the form of the relationship between the response and the independent variables is unknown. Thus, the first step in RSM is to find a suitable approximation for the true relationship between y and the independent variables. Usually, a low-order polynomial in some region of the independent variables is employed. If the response is well modeled by a linear function of the independent variables, then the approximating function is the **first-order model**

$$y = \beta_0 + \beta_1 x_1 + \beta_2 x_2 + \cdots \beta_k x_k + \varepsilon \tag{8.37}$$

If there is curvature in the system, then a polynomial of higher degree must be used, such as the **second-order model**

$$y = \beta_0 + \sum_{i=1}^{k} \beta_i x_i + \sum_{i=1}^{k} \beta_{ii} x_i^2 + \sum_{i<j=2}\sum \beta_{ij} x_i x_j + \varepsilon \tag{8.38}$$

Many RSM problems utilize one or both of these approximating polynomials. Of course, it is unlikely that a polynomial model will be a reasonable approximation of the true functional relationship over the entire space of the independent variables, but for a relatively small region they usually work quite well.

The method of least squares is used to estimate the parameters in the approximating polynomials. That is, the estimates of the β's in equations (8.37) and (8.38) are those values of the parameters that minimize the sum of squares of the model errors. The response surface analysis is then done in terms of the fitted surface. If the fitted surface is an adequate approximation of the true response function, then analysis of the fitted surface will be approximately equivalent to analysis of the actual system.

Often in response surface applications, when we are fairly close to the optimum operating conditions, a second-order model is usually required to approximate the response because of curvature in the true response surface. The fitted second-order response surface model is

$$\hat{y} = \hat{\beta}_0 + \sum_{i=1}^{k} \hat{\beta}_{ii} x_i^2 + \sum_{i<j}^{k}\sum \hat{\beta}_{ij} x_i x_j$$

where $\hat{\beta}$ denotes the least squares estimate of β. In the next example, we illustrate how a fitted second-order model can be used to find the optimum operating conditions for a process, and how to describe the behavior of the response function.

example 8.12 Optimizing a Plasma Etcher

Recall that in Example 8.9 we used a factorial design in four factors to study etch rate in a plasma etcher we found that two factors, gap and power were important. Suppose that the engineers believe that a region near gap = 0.8 cm and power = 375 W will give etch rates near a desired target of between 1,100 and 1,150 Å/m. The experimenters decided to explore this region more closely by running an experiment that would support a second-order response surface model. Table 8.28 and Figure 8.43 show the experimental design, centered at gap = 0.8 cm and power = 375 W, which consists of a 2^2 factorial design with four center points and four runs located along the coordinate axes called axial runs. The resulting design is called a **central composite design,** and it is widely used in practice for fitting second-order response surfaces.

Table 8.28

Central Composite Design of Example 8.12

Observation	Gap (cm)	Power (W)	Coded x_1	Variables x_4	Etch Rate y_1(Å/m)	Uniformity y_2(Å/m)
1	0.600	350.0	−1.000	−1.000	1054.0	79.6
2	1.000	350.0	1.000	−1.000	936.0	81.3
3	0.600	400.0	−1.000	1.000	1179.0	78.5
4	1.000	400.0	1.000	1.000	1417.0	97.7
5	0.517	375.0	−1.414	0.000	1049.0	76.4
6	1.083	375.0	1.414	0.000	1287.0	88.1
7	0.800	339.6	0.000	−1.414	927.0	78.5
8	0.800	410.4	0.000	1.414	1345.0	92.3
9	0.800	375.0	0.000	0.000	1151.0	90.1
10	0.800	375.0	0.000	0.000	1150.0	88.3
11	0.800	375.0	0.000	0.000	1177.0	88.6
12	0.800	375.0	0.000	0.000	1196.0	90.1

example 8.12 Continued

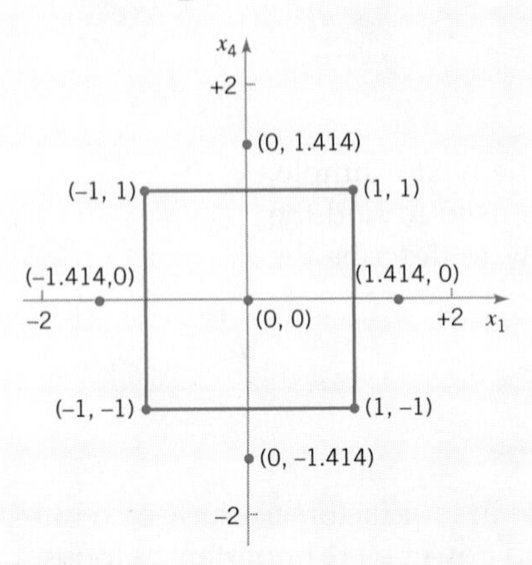

Figure 8.43 Central Composite Design in the Coded Variables for Example 8.12

Two response variables were measured during this phase of the study: etch rate (in Å/m) and etch uniformity (this is the standard deviation of the thickness of the layer of material applied to the surface of the wafer after it has been etched to a particular average (thickness). Determine optimal operating conditions for this process.

solution

Minitab can be used to analyze the data from this experiment. The Minitab output is in Table 8.29.

The second-order model fit to the etch rate response is

$$\hat{y}_1 = 1168.50 + 57.07x_1 + 149.64x_4 - 1.62x_1^2 - 17.63x_4^2 + 89.00x_1x_4$$

Table 8.29

Minitab Analysis of the Central Composite Design in Example 8.12

```
Response Surface Regression: Etch Rate versus A, B
The analysis was done using coded units.

Estimated Regression Coefficients for Etch Rate
Term        Coef   SE Coef       T       P
Constant  1168.50    17.59   66.417  0.000
A           57.07    12.44    4.588  0.004
B          149.64    12.44   12.029  0.000
A*A         -1.62    13.91   -0.117  0.911
B*B        -17.63    13.91   -1.267  0.252
A*B         89.00    17.59    5.059  0.002
S = 35.19     R - Sq = 97.0%     R - Sq(adj) = 94.5%
```

example 8.12 Continued

Table 8.29

Continued

Analysis of Variance for Etch Rate

Source	DF	Seq SS	Adj SS	Adj MS	F	P
Regression	5	238898	238898	47780	38.59	0.000
Linear	2	205202	205202	102601	82.87	0.000
Square	2	2012	2012	1006	0.81	0.487
Interaction	1	31684	31684	31684	25.59	0.002
Residual Error	6	7429	7429	1238		
Lack-of-Fit	3	5952	5952	1984	4.03	0.141
Pure Error	3	1477	1477	492		
Total	11	246327				

Response Surface Regression: Uniformity versus A, B

The analysis was done using coded units.

Estimated Regression Coefficients for Uniformity

Term	Coef	SE Coef	T	P
Constant	89.275	0.5688	156.963	0.000
A	4.681	0.4022	11.639	0.000
B	4.352	0.4022	10.821	0.000
A*A	−3.400	0.4496	−7.561	0.000
B*B	−1.825	0.4496	−4.059	0.007
A*B	4.375	0.5688	7.692	0.000

S = 1.138   R − Sq = 98.4%   R − Sq(adj) = 97.1%

Analysis of Variance for Uniformity

Source	DF	Seq SS	Adj SS	Adj MS	F	P
Regression	5	486.085	486.085	97.217	75.13	0.000
Linear	2	326.799	326.799	163.399	126.28	0.000
Square	2	82.724	82.724	41.362	31.97	0.001
Interaction	1	76.563	76.563	76.563	59.17	0.000
Residual Error	6	7.764	7.764	1.294		
Lack-of-Fit	3	4.996	4.996	1.665	1.81	0.320
Pure Error	3	2.768	2.768	0.923		
Total	11	493.849				

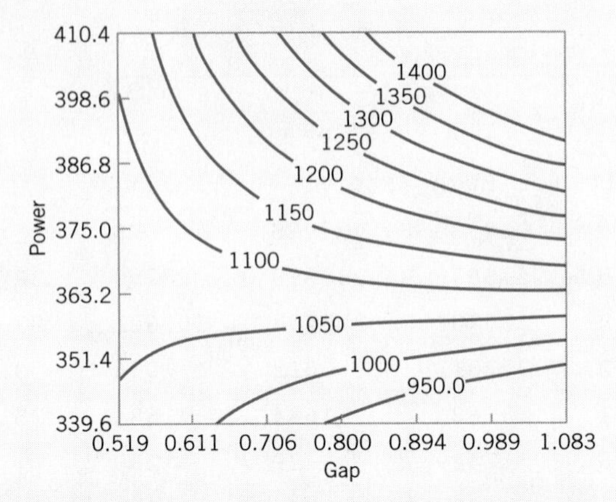

Figure 8.44 Contours of Constant Predicted Etch Rate, Example 8.12

However, we note from the t-test statistics in Table 8.29 that the quadratic terms x_1^2 and x_4^2 are not statistically significant. Therefore, the experimenters decided to model etch rate with a first-order model with interaction:

$$\hat{y}_1 = 1155.7 + 57.1x_1 + 149.7x_4 + 89x_1x_4$$

Figure 8.44 shows the contours of constant etch rate from this model. There are obviously many combinations of x_1 (gap) and x_4 (power) that will give an etch rate in the desired range of 1,100–1,150 Å/m.

The second-order model for uniformity is

$$\hat{y}_2 = 89.275 + 4.681x_1 + 4.352x_4 - 3.400x_1^2 - 1.825x_4^2 + 4.375x_1x_4$$

Table 8.29 gives the t-statistics for each model term. Since all terms are significant, the experimenters decided to use the quadratic model for uniformity. Figure 8.45 gives the contour plot and response surface for uniformity.

As in most response surface problems, the experimenter in this example had conflicting objectives regarding the two responses. One objective was to keep the etch rate

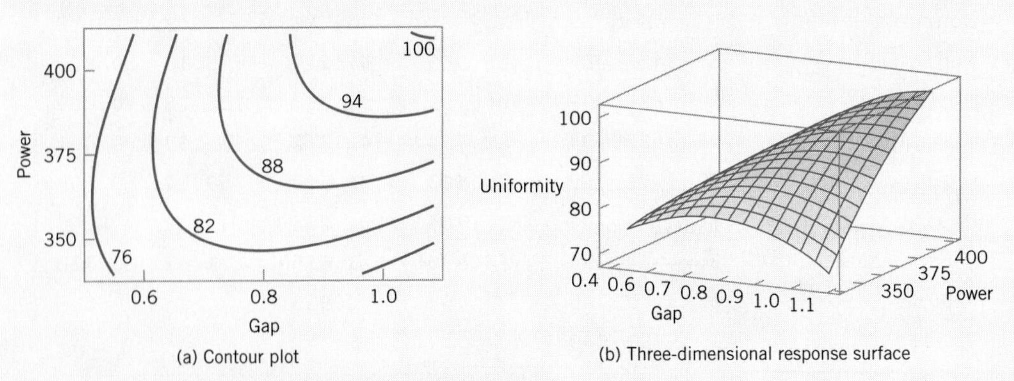

Figure 8.45 Plots of the Uniformity Response, Example 8.12

example 8.12 Continued

within the acceptable range of $1{,}100 \leq y_1 \leq 1{,}150$ but to simultaneously minimize the uniformity. Specifically, the uniformity must not exceed $y_2 = 80$, or many of the wafers will be defective in subsequent processing operations. When there are only a few independent variables, an easy way to solve this problem is to overlay the response surfaces to find the optimum. Figure 8.46 presents the overlay plot of both responses, with the contours of $\hat{y}_1 = 100$, $\hat{y}_2 = 150$, and $\hat{y}_2 = 80$ shown. The shaded areas on this plot identify infeasible combinations of gap and power. The graph indicates that several combinations of gap and power should result in acceptable process performance.

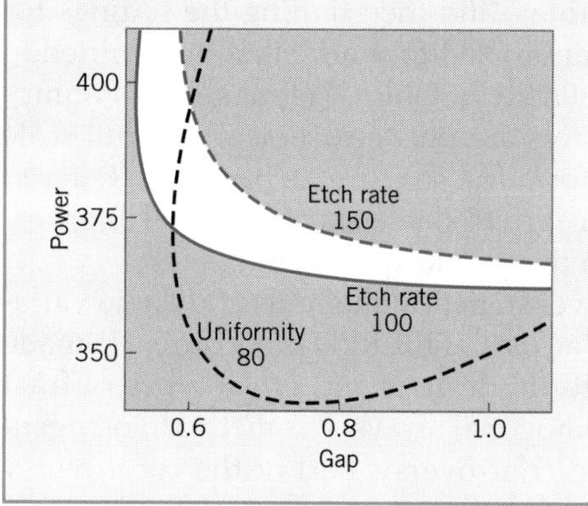

Figure 8.46 Overlay of the Etch Rate and Uniformity Response Surfaces in Example 8.12 Showing the Region of the Optimum (Unshaded Region)

Example 8.12 illustrates the use of a **central composite design (CCD)** for fitting a second-order response surface model. These designs are widely used in practice because they are relatively efficient with respect to the number of runs required. In general, a CCD in k factors requires 2^k factorial runs, $2k$ axial runs, and at least one center point (3 to 5 center points are typically used). Designs for $k = 2$ and $k = 3$ factors are shown in Figure 8.47.

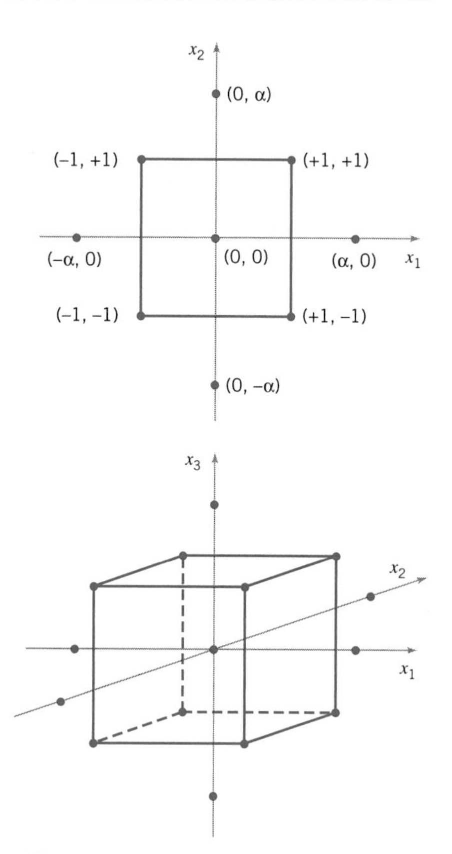

Figure 8.47 Central Composite Designs for $k = 2$ and $k = 3$

8.9 Robust Product and Process Design

We have emphasized the importance of using statistically designed experiments for process design, development, and improvement. Over the past 30 years, engineers and scientists have become increasingly aware of the benefits of using designed experiments, and as a consequence, there have been many new application areas. One of the most important of these is in **process robustness studies,** where the focus is on the following:

1. Designing processes so that the manufactured product will be as close as possible to the desired target specifications even though some process variables (such as temperature), environmental factors (such as relative humidity), or raw material characteristics are impossible to control precisely.

2. Determining the operating conditions for a process so that critical product characteristics are as close as possible to the desired target value and the variability around this target is minimized.

Examples of this type of problem occur frequently. For instance, in semiconductor manufacturing we would like the oxide thickness on a wafer to be as close as possible to the target mean thickness, and we would also like the variability in thickness across the wafer (a measure of uniformity) to be as small as possible.

In the early 1980s, a Japanese engineer, Genichi Taguchi, introduced an approach to solving these types of problems, which he referred to as the **robust parameter design (RPD)** problem [see Taguchi and Wu (1980), Taguchi (1986)]. His approach was based on classifying the variables in a process as either **control** (or **controllable**) **variables** and **noise** (or **uncontrollable**) **variables,** and then finding the settings for the controllable variables that minimized the variability transmitted to the response from the uncontrollable variables. We make the assumption that although the noise factors are uncontrollable in the full-scale system, they can be controlled for purposes of an experiment. Refer to Figure 8.1 for a graphical view of controllable and uncontrollable variables in the general context of a designed experiment.

Taguchi introduced some novel statistical methods and some variations on established techniques as part of his RPD procedure. He made use of highly fractionated factorial designs and other types of fractional designs obtained from orthogonal arrays. His methodology generated considerable debate and controversy. Part of the controversy arose because Taguchi's methodology was advocated in the West initially (and primarily) by consultants, and the underlying statistical science had not been adequately peer-reviewed. By the late 1980s, the results of a very thorough and comprehensive peer review indicated that although Taguchi's engineering concepts and the overall objective of RPD were well founded, there were substantial problems with his experimental strategy and methods of data analysis. For specific details of these issues, see Box, Bisgaard, and Fung (1988); Hunter (1985, 1987); Montgomery (1999); Myers, Montgomery and Anderson-Cook (2009); and Pignatiello and Ramberg (1991). Many of these concerns are also summarized in the extensive panel discussion in the May 1992 issue of *Technometrics* [see Nair et al. (1992)].

Taguchi's methodology for the RPD problem revolves around the use of an orthogonal design for the controllable factors that is "crossed" with a separate orthogonal design for the noise factors. Table 8.30 presents an example from Byrne and Taguchi (1987) that involved the development of a method to assemble an elastometric connector to a nylon tube that would deliver the required pull-off force. There are four controllable factors, each at three levels (A = interference, B = connector wall thickness, C = insertion depth, D = percent adhesive), and three noise or uncontrollable factors, each at two levels (E = conditioning time, F = conditioning temperature, G = conditioning relative humidity). Panel (a) of Table 8.30 contains the design for the controllable factors. Note that the design is a three-level fractional factorial; specifically, it is a 3^{4-2} design. Taguchi calls this the **inner array design.** Panel (b) of Table 8.30 contains a 2^3 design for the noise factors, which Taguchi calls the **outer array design.** Each run in the inner array is performed for all treatment combinations in the outer array, producing the 72 observations on pull-off force shown in the table. This type of design is called a **crossed array design.**

Table 8.30

Taguchi Parameter Design with Both Inner and Outer Arrays [Byrne and Taguchi (1987)]

		(a) Inner Array			(b) Outer Array							
				E	1	1	1	1	2	2	2	2
				F	1	1	2	2	1	1	2	2
				G	1	2	1	2	1	2	1	2
Run	*A*	*B*	*C*	*D*								
1	1	1	1	1	15.6	9.5	16.9	19.9	19.6	19.6	20.0	19.1
2	1	2	2	2	15.0	16.2	19.4	19.2	19.7	19.8	24.2	21.9
3	1	3	3	3	16.3	16.7	19.1	15.6	22.6	18.2	23.3	20.4
4	2	1	2	3	18.3	17.4	18.9	18.6	21.0	18.9	23.2	24.7
5	2	2	3	1	19.7	18.6	19.4	25.1	25.6	21.4	27.5	25.3
6	2	3	1	2	16.2	16.3	20.0	19.8	14.7	19.6	22.5	24.7
7	3	1	3	2	16.4	19.1	18.4	23.6	16.8	18.6	24.3	21.6
8	3	2	1	3	14.2	15.6	15.1	16.8	17.8	19.6	23.2	24.2
9	3	3	2	1	16.1	19.9	19.3	17.3	23.1	22.7	22.6	28.6

Taguchi suggested that we summarize the data from a crossed array experiment with two statistics: the average of each observation in the inner array across all runs in the outer array, and a summary statistic that attempted to combine information about the mean and variance, called the **signal-to-noise ratio.** These signal-to-noise ratios are purportedly defined so that a maximum value of the ratio minimizes variability transmitted from the noise variables. Then an analysis is performed to determine which settings of the controllable factors result in (1) the mean as close as possible to the desired target and (2) a maximum value of the signal-to-noise ratio.

Examination of Table 8.30 reveals a major problem with the Taguchi design strategy; namely, the crossed array approach will lead to a very large experiment. In our example, there are only seven factors, yet the design has 72 runs. Furthermore, the inner array design is a 3^{4-2} resolution III design, so in spite of the large number of runs, we cannot obtain *any* information about interactions among the controllable variables. Indeed, even information about the main effects is potentially tainted, because the main effects are heavily aliased with the two-factor interactions. It also turns out that the Taguchi signal-to-noise ratios are problematic; maximizing the ratio does not necessarily minimize variability.

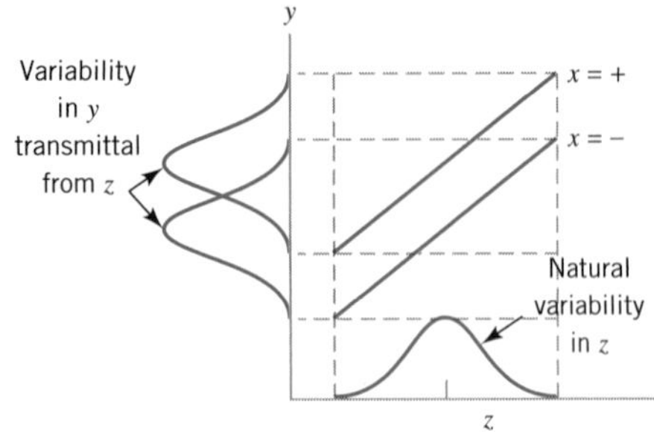

(a) No control × noise interaction

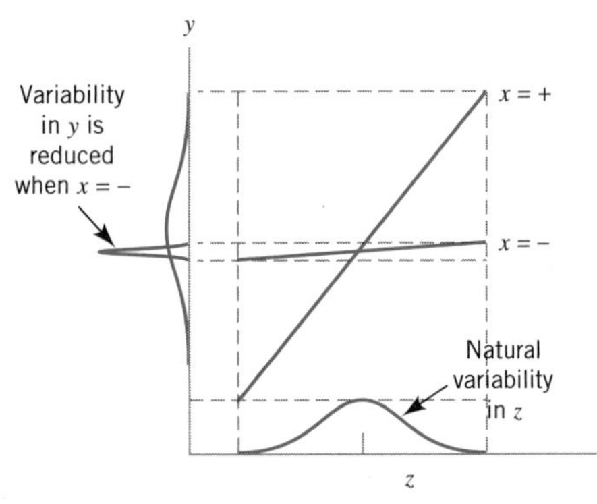

(b) Significant control × noise interaction

FIGURE 8.48 The Role of the Control × Noise Interaction in Robust Design

An important point about the crossed array design is that it *does* provide information about controllable factor × noise factor interactions. These interactions are crucial to the solution of an RPD problem. For example, consider the two-factor interaction graphs in Figure 8.48, where x is the controllable factor and z is the noise factor. In Figure 8.48*a*, there is no $x \times z$ interaction; therefore, there is no setting for the controllable variable x that will affect the variability transmitted to the response by the variability in z. However, in Figure 8.48*b* there is a strong $x \times z$ interaction. Note that when x is set to its low level there is much less variability in the response variable than when x is at the high level. Thus, unless there is at least one controllable factor × noise factor interaction, there is no robust design problem. Focusing on identifying and modeling these interactions is one of the keys to a more efficient and effective approach to investigating process robustness.

It is logical to utilize a **model** for the response that includes both controllable and noise factors and their interactions. To illustrate, suppose that we have two controllable factors x_1 and x_2 and a single noise factor z_1. We assume that both control and noise factors are expressed as the usual coded variables (that is, they are centered at zero and have lower and upper limits at ± 1). If we wish to consider a first-order model involving the controllable variables, then a logical model is

$$y = \beta_0 + \beta_1 x_1 + \beta_2 x_2 + \beta_{12} x_1 x_2 + \gamma_1 z_1 + \delta_{11} x_1 z_1 + \delta_{11} x_1 z_1 + \delta_{21} x_2 z + \varepsilon \qquad (8.39)$$

Note that this model has the main effects of both controllable factors, the main effect of the noise variable, and both interactions between the controllable and noise variables. This type of model incorporating both controllable and noise variables is often called a **response model.** Unless at least one of the regression coefficients δ_{11} and δ_{21} is nonzero, there will be no robust design problem.

An important advantage of the response model approach is that both the controllable factors and the noise factors can be placed in a single experimental design; that is, the inner and outer array structure of the Taguchi approach can be avoided. We usually call the design containing both controllable and noise factors a **combined array design.**

As mentioned previously, we assume that noise variables are random variables, although they are controllable for purposes of an experiment. Specifically, we assume that the noise variables are expressed in coded units, that they have expected value 0, variance σ_z^2, and that if there are several noise variables, they have zero covariances. Under these assumptions, it is easy to find a model for the mean response just by taking the expected value of y in equation (8.39). This yields

$$E_z(y) = \beta_0 + \beta_1 x_1 + \beta_2 x_2 + \beta_{12} x_1 x_2$$

where the z subscript on the expectation operator is a reminder to take the expected value with respect to *both* random variables in equation (8.39), z_1 and ε. To find a model for the variance of the response y, first rewrite equation (8.39) as follows:

$$y = \beta_0 + \beta_1 x_1 + \beta_2 x_2 + \beta_{12} x_1 x_2 + (\gamma_1 + \delta_{11} x_1 + \delta_{22} x_2) z_1 + \varepsilon$$

Now the variance of y can be obtained by applying the variance operator across this last expression. The resulting variance model is

$$V_z(y) = \sigma_z^2(\gamma_1 + \delta_{11}x_1 + \delta_{22}x_2)^2 + \sigma^2$$

Once again, we have used the z subscript on the variance operator as a reminder that *both* z_1 and ε are random variables.

We have derived simple models for the mean and variance of the response variable of interest. Note the following:

1. The mean and variance models involve **only the controllable variables.** This means that we can potentially set the controllable variables to achieve a target value of the mean and minimize the variability transmitted by the noise variable.
2. Although the variance model involves only the controllable variables, it also involves the *interaction regression coefficients* between the controllable and noise variables. This is how the noise variable influences the response.
3. The variance model is a **quadratic function** of the controllable variables.
4. The variance model (apart from σ^2) is simply the square of the slope of the fitted response model in the direction of the noise variable.

To use these models operationally, we would

1. Perform an experiment and fit an appropriate response model such as equation (8.39).
2. Replace the unknown regression coefficients in the mean and variance models with their least squares estimates from the response model and replace σ^2 in the variance model by the residual mean square found when fitting the response model.
3. Simultaneously optimize the mean and variance models. Often this can be done graphically. For more discussion of other optimization methods, refer to Myers, Montgomery and Anderson-Cook (2009).

It is very easy to generalize these results. Suppose that there are k controllable variables $\mathbf{x}' = [x_1, x_2, \ldots, x_k]$, and r noise variables $\mathbf{z}' = [z_1, z_2, \ldots, z_r]$. We will write the general response model involving these variables as

$$y(\mathbf{x}, \mathbf{z}) = f(\mathbf{x}) + h(\mathbf{x}, \mathbf{z}) + \varepsilon \qquad (8.40)$$

where $f(\mathbf{x})$ is the portion of the model that involves only the controllable variables and $h(\mathbf{x}, \mathbf{z})$ are the terms involving the main effects of the noise factors and the interactions between the controllable and noise factors. Typically, the structure for $h(\mathbf{x}, \mathbf{z})$ is

$$h(\mathbf{x}, \mathbf{z}) = \sum_{i=1}^{r} \gamma_i z_i + \sum_{i=1}^{k}\sum_{j=1}^{r} \delta_{ij} x_i z_j$$

The structure for $f(\mathbf{x})$ will depend on what type of model for the controllable variables the experimenter thinks is appropriate. The logical choices are the first-order model with interaction and the second-order model. If we assume that the noise variables have mean 0, variance σ_z^2, and have zero covariances, and that the noise variables and the random errors ε have zero covariances, then the mean model for the response is simply

$$E_z[y(\mathbf{x},\mathbf{z})] = f(\mathbf{x}) \tag{8.41}$$

To find the variance model, we will use a transmission of error approach. This involves first expanding equation (8.40) around $\mathbf{z} = \mathbf{0}$ in a first-order Taylor series:

$$y(\mathbf{x},\mathbf{z}) \cong f(\mathbf{x}) + \sum_{i=1}^{r} \frac{\partial h(\mathbf{x},\mathbf{z})}{\partial z_i}(z_i - 0) + R + \varepsilon$$

where R is the remainder. If we ignore the remainder and apply the variance operator to this last expression, the variance model for the response is

$$V_t[y(\mathbf{x},\mathbf{z})] = \sigma_t^2 \sum_{i=1}^{t} \left(\frac{\partial h(x,z)}{\partial z_i} \right)^2 + \sigma^2 \tag{8.42}$$

Myers and Montgomery (2002) give a slightly more general form for equation.

example 8.13 Robust Design

To illustrate a process robustness study, consider an experiment [described in detail in Montgomery (2009)] in which four factors were studied in a 2^4 factorial design to investigate their effect on the filtration rate of a chemical product. We will assume that factor A, temperature, is hard to control in the full-scale process but that it can be controlled during the experiment (which was performed in a pilot plant). The other three factors, pressure (B), concentration (C), and stirring rate (D), are easy to control. Thus the noise factor z_1 is temperature and the controllable variables x_1, x_2, and x_3 are pressure, concentration, and stirring rate, respectively. The experimenters conducted the (unreplicated) 2^4 design shown in Table 8.31. Since both the controllable factors and the noise factor are in the same design, the 2^4 factorial design used in this experiment is an example of a **combined array design.** We want to determine operating conditions that maximize the filtration rate and minimize the variability transmitted from the noise variable temperature.

example 8.13 Continued

Table 8.31

Pilot Plant Filtration Rate Experiment

Run Number	Factor *A*	*B*	*C*	*D*	Run Label	Filtration Rate (gal/h)
1	−	−	−	−	(1)	45
2	+	−	−	−	*a*	71
3	−	+	−	−	*b*	48
4	+	+	−	−	*ab*	65
5	−	−	+	−	*c*	68
6	+	−	+	−	*ac*	60
7	−	+	+	−	*bc*	80
8	+	+	+	−	*abc*	65
9	−	−	−	+	*d*	43
10	+	−	−	+	*ad*	100
11	−	+	−	+	*bd*	45
12	+	+	−	+	*abd*	104
13	−	−	+	+	*cd*	75
14	+	−	+	+	*acd*	86
15	−	+	+	+	*bcd*	70
16	+	+	+	+	*abcd*	96

solution

Using the methods for analysing a 2^k factorial design, the response model is

$$\begin{aligned}\hat{y}(x, z_1) &= 70.06 + \left(\frac{21.625}{2}\right)z_1 + \left(\frac{9.875}{2}\right)x_2 + \left(\frac{14.625}{2}\right)x_3 \\ &\quad - \left(\frac{18.125}{2}\right)x_2 z_1 + \left(\frac{16.625}{2}\right)x_3 z_1 \\ &= 70.06 + 10.81z_1 + 4.94x_2 + 7.31x_3 - 9.06x_2 z_1 + 8.31x_3 z_1\end{aligned}$$

Using equations (8.41) and (8.42), we can find the mean and variance models as

$$E_z[y(x, z_1)] = 70.06 + 4.94x_2 + 7.31x_3$$

example 8.13 Continued

and

$$V_z[y(x, z_1)] = \sigma_z^2(10.81 - 9.06x_2 + 8.31x_3)^2 + \sigma^2$$
$$= \sigma_z^2(116.91 + 82.08x_2^2 + 69.06x_3^2 - 195.88x_2$$
$$+ 179.66x_3 - 150.58x_2x_3) + \sigma^2$$

respectively. Now assume that the low and high levels of the noise variable temperature have been run at one standard deviation either side of its typical or average value, so that $\sigma_z^2 = 1$, and use $\hat{\sigma} = 19.51$ (this is the residual mean square obtained by fitting the response model). Therefore, the variance model becomes

$$V_z[y(x, z_1)] = 136.42 - 195.88x_2 + 179.66x_3 - 150.58x_2x_3$$
$$+ 82.08x_2^2 + 69.06x_3^2$$

Figure 8.49 presents a contour plot of the response contours from the mean model. To construct this plot, we held the noise factor (temperature) at zero and the nonsignificant controllable factor (pressure) at zero. Note that mean filtration rate increases as both concentration and stirring rate increase. The square root of $V_z[y(\mathbf{x}, \mathbf{z})]$ is plotted in Figure 8.50. Note that both a contour plot and a three-dimensional response surface

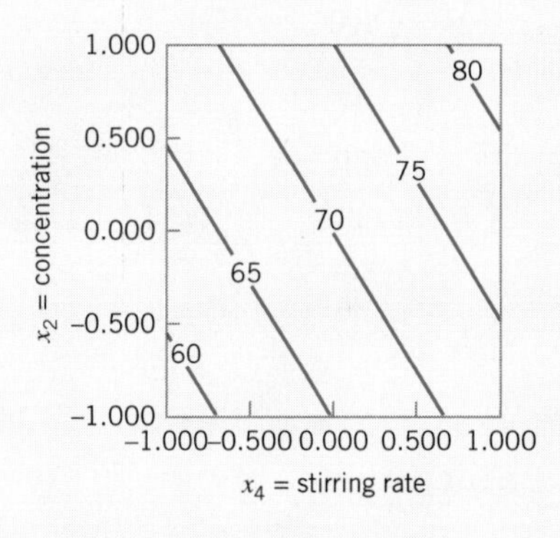

FIGURE 8.49 Contours of Constant Mean Filtration Rate, Example 8.13, with z_1 = Temperature = 0

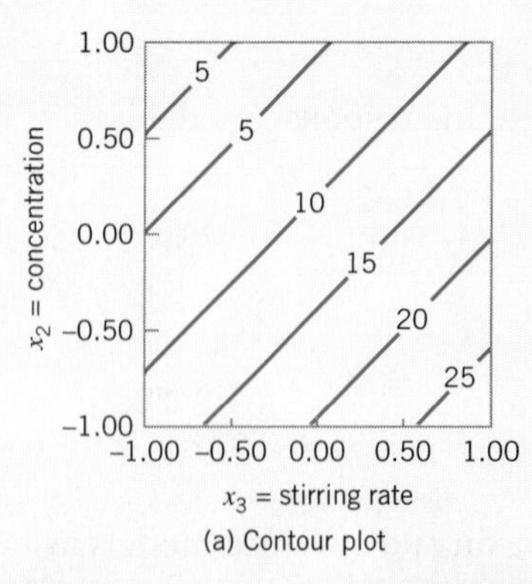

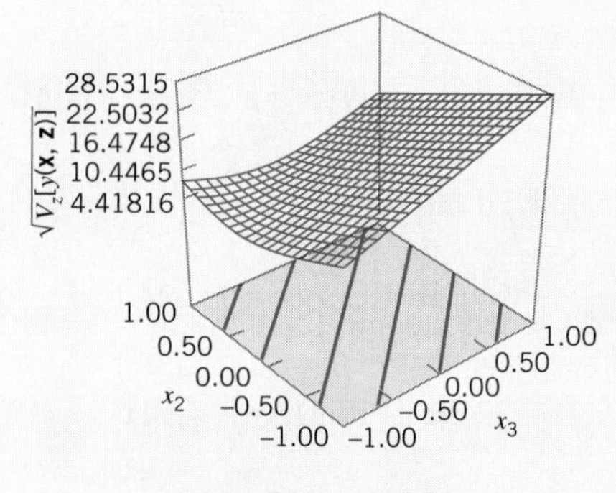

FIGURE 8.50 Contour Plot and Response Surface of $\sqrt{V_z[y(\mathbf{x}, \mathbf{z})]}$ for Example 8.13, with z_1 = Temperature = 0

example 8.13 Continued

plot are given. This plot was also constructed by setting the noise factor temperature and the nonsignificant controllable factor to 0.

Suppose that the experimenter wants to maintain mean filtration rate above 75 and minimize the variability around this value. Figure 8.51 shows an overlay plot of the contours of mean filtration rate and the square root of $V_z[y(\mathbf{x}, \mathbf{z})]$ as a function of concentration and stirring rate, the significant controllable variables. To achieve the desired objectives, it will be necessary to hold concentration at the high level and stirring rate very near the middle level.

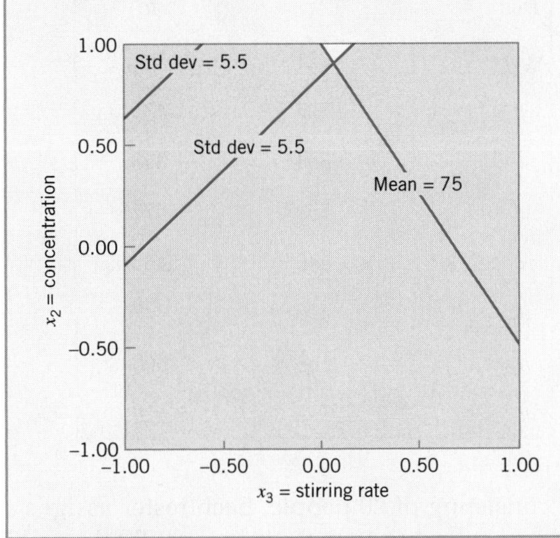

FIGURE 8.51 **Overlay Plot of Mean and Standard Deviation for Filtration Rate, Example 8.13, with z_1 = Temperature = 0**

Important Terms and Concepts

2^k factorial designs
2^{k-p} fractional factorial designs
Aliasing
Analysis of variance (ANOVA)
Analysis procedure for factorial designs
Blocking
Center points in 2^k and 2^{k-p} factorial designs
Central composite design
Combined array design
Completely randomized design
Confounding
Contour plot
Controllable process variables
Crossed array design
Curvature in the response function
Defining relation for a fractional factorial design
Factorial design
Fractional factorial design
Generators for a fractional factorial design
Guidelines for planning experiments
Interaction
Main effect of a factor
Noise variable
Normal probability plot of effects
Orthogonal design
Pre-experimental planning
Projection of fractional factorial designs
Regression model representation of experimental results
Residual analysis
Residuals
Resolution of a fractional factorial design
Response surface
Robust design
Screening experiments
Sequential experimentation
Sparsity of effects principle
Two-factor interaction

chapter Eight Exercises

8.1. An article in *Industrial Quality Control* (1956, pp. 5–8) describes an experiment to investigate the effect of glass type and phosphor type on the brightness of a television tube. The response measured is the current necessary (in microamps) to obtain a specified brightness level. The data are shown in Table 8E.1. Analyze the data and draw conclusions.

8.2. A process engineer is trying to improve the life of a cutting tool. He has run a 2^3 experiment using cutting speed (A), metal hardness (B), and cutting angle (C) as the factors. The data from two replicates are shown in Table 8E.2.

a. Do any of the three factors affect tool life?
b. What combination of factor levels produces the longest tool life?
c. Is there a combination of cutting speed and cutting angle that always gives good results regardless of metal hardness?

8.3. Find the residuals from the tool life experiment in Exercise 8.2. Construct a normal probability plot of the residuals. Plot the residuals versus the predicted values. Comment on the plots.

8.4. Four factors are thought to possibly influence the taste of a soft-drink beverage: type of sweetener (A), ratio of syrup to water (B), carbonation level (C), and temperature (D). Each factor can be run at two levels, producing a 2^4 design. At each run in the design, samples of the beverage are given to a test panel consisting of 20 people. Each tester assigns a point score from 1 to 10 to the beverage. Total score is the response variable, and the objective is to find a formulation that maximizes total score. Two replicates of this design are run, and the results are shown in Table 8E.3. Analyze the data and draw conclusions.

8.5. Consider the experiment in Exercise 8.4. Plot the residuals against the levels of factors A, B, C, and

Table 8E.1

Data for Exercise 8.1

	Phosphor Type		
Glass Type	1	2	3
1	280	300	290
	290	310	285
	285	295	290
2	230	260	220
	235	240	225
	240	235	230

Table 8E.2

Data for the Experiment in Exercise 8.2

	Replicate	
Run	I	II
(1)	221	311
a	325	435
b	354	348
ab	552	472
c	440	453
ac	406	377
bc	605	500
abc	392	419

Table 8E.3

Taste-Text Experiment for Exercise 8.4

	Replicate	
Treatment Combination	I	II
(1)	188	195
a	172	180
b	179	187
ab	185	178

(*Continued*)

Table 8E.3

Continued

Treatment Combination	Replicate I	Replicate II
c	175	180
ac	183	178
bc	190	180
abc	175	168
d	200	193
ad	170	178
bd	189	181
abd	183	188
cd	201	188
acd	181	173
bcd	189	182
abcd	178	182

D. Also construct a normal probability plot of the residuals. Comment on these plots.

8.6. Find the standard error of the effects for the experiment in Exercise 8.4. Using the standard errors as a guide, what factors appear significant?

8.7. Suppose that only the data from replicate I in Exercise 8.4 were available. Analyze the data and draw appropriate conclusions.

8.8. Suppose that only one replicate of the 2^4 design in Exercise 8.4 could be run, and that we could only conduct eight tests each day. Set up a design that would block out the day effect. Show specifically which runs would be made on each day.

8.9. Show how a 2^5 experiment could be set up in two blocks of 16 runs each. Specifically, which runs would be made in each block?

8.10. R. D. Snee ("Experimenting with a Large Number of Variables," in *Experiments in Industry: Design, Analysis and Interpretation of Results*, by R. D. Snee, L. B. Hare, and J. B. Trout, editors, ASQC, 1985) describes an experiment in which a 2^{5-1} design with $I = ABCDE$ was used to investigate the effects of five factors on the color of a chemical product. The factors were A = solvent/reactant, B = catalyst/reactant, C = temperature, D = reactant purity, and E = reactant pH. The results obtained are as follows:

$$\begin{aligned} e &= -0.63 & d &= 6.79 \\ a &= 2.51 & ade &= 6.47 \\ b &= 2.68 & bde &= 3.45 \\ abe &= 1.66 & abd &= 5.68 \\ c &= 2.06 & cde &= 5.22 \\ ace &= 1.22 & acd &= 4.38 \\ bce &= -2.09 & bcd &= 4.30 \\ abc &= 1.93 & abcde &= 4.05 \end{aligned}$$

a. Prepare a normal probability plot of the effects. Which effects seem active? Fit a model using these effects.

b. Calculate the residuals for the model you fit in part (a). Construct a normal probability plot of the residuals and plot the residuals versus the fitted values. Comment on the plots.

c. If any factors are negligible, collapse the 2^{5-1} design into a full factorial in the active factors. Comment on the resulting design and interpret the results.

8.11. An article in *Industrial and Engineering Chemistry* ("More on Planning Experiments to Increase Research Efficiency," 1970, pp. 60–65) uses a 2^{5-2} design to investigate the effect of A = condensation temperature, B = amount of material 1, C = solvent volume, D = condensation time, and E = amount of material 2, on yield. The results obtained are as follows:

$$\begin{aligned} e &= 23.2 & cd &= 23.8 \\ ab &= 15.5 & ace &= 23.4 \\ ad &= 16.9 & bde &= 16.8 \\ bc &= 16.2 & abcde &= 18.1 \end{aligned}$$

a. Verify that the design generators used were $I = ACE$ and $I = BDE$.

b. Write down the complete defining relation and the aliases from this design.

c. Estimate the main effects.

8.12. A 2^4 factorial design has been run in a pilot plant to investigate the effect of four factors on the molecular weight of a polymer. The data from this experiment are as follows (values are coded by dividing by 10).

$$\begin{aligned} (1) &= 88 & d &= 86 \\ a &= 80 & ad &= 81 \end{aligned}$$

$b = 89$	$bd = 85$
$ab = 87$	$abd = 86$
$c = 86$	$cd = 85$
$ac = 81$	$acd = 79$
$bc = 82$	$bcd = 84$
$abc = 80$	$abcd = 81$

a. Cnstruct a normal probability plot of the effects. Which effects are active?

b. Cnstruct an appropriate model. Fit this model and test for significant effects.

c. Analyze the residuals from this model by constructing a normal probability plot of the residuals and plotting the residuals versus the predicted values of y.

8.13. Reconsider the data in Exercise 8.12. Suppose that four center points were added to this experiment. The molecular weights at the center point are 90, 87, 86, and 93.

a. Analyze the data as you did in Exercise 8.12, but include a test for curvature.

b. If curvature is significant in an experiment such as this one, describe what strategy you would pursue next to improve your model of the process.

8.14. An engineer has performed an experiment to study the effect of four factors on the surface roughness of a machined part. The factors (and their levels) are A = tool angle (12, 15), B = cutting fluid viscosity (300, 400), C = feed rate (10, 15 in/min), and D = cutting fluid cooler used (no, yes). The data from this experiment (with the factors coded to the usual $+1$, -1 levels) are shown in Table 8E.4.

a. Estimate the factor effects. Plot the effect estimates on a normal probability plot and select a tentative model.

b. Fit the model identified in part (a) and analyze the residuals. Is there any indication of model inadequacy?

c. Repeat the analysis from parts (a) and (b) using $1/y$ as the response variable. Is there an indication that the transformation has been useful?

d. Fit the model in terms of the coded variables that you think can be used to provide the best predictions of the surface roughness. Convert this prediction equation into a model in the natural variables.

Table 8E.4

Surface Roughness Experiment for Exercise 8.14

Run	A	B	C	D	Surface Roughness
1	−	−	−	−	0.00340
2	+	−	−	−	0.00362
3	−	+	−	−	0.00301
4	+	+	−	−	0.00182
5	−	−	+	−	0.00280
6	+	−	+	−	0.00290
7	−	+	+	−	0.00252
8	+	+	+	−	0.00160
9	−	−	−	+	0.00336
10	+	−	−	+	0.00344
11	−	+	−	+	0.00308
12	+	+	−	+	0.00184
13	−	−	+	+	0.00269
14	+	−	+	+	0.00284
15	−	+	+	+	0.00253
16	+	+	+	+	0.00163

8.15. An experiment was run to study the effect of two factors, time and temperature, on the inorganic impurity levels in paper pulp. The results of this experiment are shown in Table 8E.5:

a. What type of experimental design has been used in this study?

b. Fit a quadratic model to the response, using the method of least squares.

c. Construct the fitted impurity response surface. What values of x_1 and x_2 would you recommend if you wanted to minimize the impurity level?

d. Suppose that

$$x_1 = \frac{\text{temp} - 750}{50} \qquad x_2 = \frac{\text{time} - 30}{15}$$

Table 8E.5

The Experiment for Exercise 8.15

x_1	x_2	y
−1	−1	210
1	−1	95
−1	1	218
1	1	100
−1.5	0	225
1.5	0	50
0	−1.5	175
0	1.5	180
0	0	145
0	0	175
0	0	158
0	0	166

where temperature is in °C and time is in hours. Find the optimum operating conditions in terms of the natural variables temperature and time.

8.16. An article in *Rubber Chemistry and Technology* (Vol. 47, 1974, pp. 825–836) describes an experiment that studies the relationship of the Mooney viscosity of rubber to several variables, including silica filler (parts per hundred) and oil filler (parts per hundred). Some of the data from this experiment are shown in Table 8E.6, where

$$x_1 = \frac{\text{silica} - 60}{15} \qquad x_2 = \frac{\text{oil} - 21}{1.5}$$

a. What type of experimental design has been used? Is it rotatable?

b. Fit a quadratic model to these data. What values of x_1 and x_2 will maximize the Mooney viscosity?

8.17. In their book *Empirical Model Building and Response Surfaces* (John Wiley, 1987), G. E. P. Box and N. R. Draper describe an experiment with three factors. The data shown in Table 8E.7 are a variation of

Table 8E.6

The Viscosity Experiment for Exercise 8.16

Coded Levels		
x_1	x_2	y
−1	−1	13.71
1	−1	14.15
−1	1	12.87
1	1	13.53
−1.4	0	12.99
1.4	0	13.89
0	−1.4	14.16
0	1.4	12.90
0	0	13.75
0	0	13.66
0	0	13.86
0	0	13.63
0	0	13.74

Table 8E.7

The Experiment for Exercise 8E.17

x_1	x_2	x_3	y_1	y_2
−1	−1	−1	24.00	12.49
0	−1	−1	120.33	8.39
1	−1	−1	213.67	42.83
−1	0	−1	86.00	3.46
0	0	−1	136.63	80.41
1	0	−1	340.67	16.17
−1	1	−1	112.33	27.57
0	1	−1	256.33	4.62
1	1	−1	271.67	23.63

(Continued)

Table 8E.7

Continued

x_1	x_2	x_3	y_1	y_2
−1	−1	0	81.00	0.00
0	−1	0	101.67	17.67
1	−1	0	357.00	32.91
−1	0	0	171.33	15.01
0	0	0	372.00	0.00
1	0	0	501.67	92.50
−1	1	0	264.00	63.50
0	1	0	427.00	88.61
1	1	0	730.67	21.08
−1	−1	1	220.67	133.82
0	−1	1	239.67	23.46
1	−1	1	422.00	18.52
−1	0	1	199.00	29.44
0	0	1	485.33	44.67
1	0	1	673.67	158.21
−1	1	1	176.67	55.51
0	1	1	501.00	138.94
1	1	1	1010.00	142.45

the original experiment on p. 247 of their book. Suppose that these data were collected in a semiconductor manufacturing process.

a. The response y_1 is the average of three readings on resistivity for a single wafer. Fit a quadratic model to this response.

b. The response y_2 is the standard deviation of the three resistivity measurements. Fit a first-order model to this response.

c. Where would you recommend that we set x_1, x_2, and x_3 if the objective is to hold mean resistivity at 500 and minimize the standard deviation?

8.18. The data shown in the Table 8E.8 were collected in an experiment to optimize crystal growth as

Table 8E.8

Crystal Growth Experiment for Exercise 8.18

x_1	x_2	x_3	y
−1	−1	−1	66
−1	−1	1	70
−1	1	−1	78
−1	1	1	60
1	−1	−1	80
1	−1	1	70
1	1	−1	100
1	1	1	75
−1.682	0	0	65
1.682	0	0	82
0	−1.682	0	68
0	1.682	0	63
0	0	−1.682	100
0	0	1.682	80
0	0	0	83
0	0	0	90
0	0	0	87
0	0	0	88
0	0	0	91
0	0	0	85

a function of three variables x_1, x_2, and x_3. Large values of y (yield in grams) are desirable. Fit a second-order model and analyze the fitted surface. Under what set of conditions is maximum growth achieved?

8.19. Reconsider the crystal growth experiment from Exercise 8.18. Suppose that $x_3 = z$ is now a noise variable, and that the modified experimental design shown in Table 8E.9 has been conducted. The experimenters want the growth rate to be as large as possible, but they also want the variability transmitted from z to be small. Under what set of conditions is growth greater than 90 with minimum variability achieved?

Table 8E.9

Crystal Growth Experiment for Exercise 8.19

x_1	x_2	z	y
−1	−1	−1	66
−1	−1	1	70
−1	1	−1	78
−1	1	1	60
1	−1	−1	80
1	−1	1	70
1	1	−1	100

Table 8E.9

Continued

x_1	x_2	z	y
1	1	1	75
−1.682	0	0	65
1.682	0	0	82
0	−1.682	0	68
0	1.682	0	63
0	0	0	83
0	0	0	90
0	0	0	87
0	0	0	88
0	0	0	91
0	0	0	85

supplement One One-Way ANOVA Analysis

MINITAB

1. Minitab analyzes one-way ANOVA data in two different formats; one requiring that all the data be stacked in a single column along with another column that identifies different factor levels and another that requires that the data from each factor level appear in its own column. The stacked format is preferred because it is consistent with how other more general experimental designs are handled.
2. Select Stat –> ANOVA –> One-Way and choose either stacked or unstacked.
3. Specify the response in the **Response OK:** input box.
4. Specify the treatment identification column in the **Factor:** input box.
5. Use the **Graphs:** menu to select the graphical diagnostics you want.
6. Enter the treatment variable in the **Residuals versus the variables:** input box.
7. You might want to turn on **Store residuals** so that they will be saved for follow-on analysis.

SPC XL

1. Arrange data in rows (data for example 8.5 is shown below).

	A	B	C	D	E	F
1	Design 1	9732	9246	8570	8784	8642
2	Design 2	9763	9791	9995	9421	9333
3	Design 3	8992	8953	9589	9283	8612

2. From the SPC XL ribbon, select Analysis Tools –> 1 Way ANOVA. When prompted select the data for the analysis (this is the region between B1–F3 above). Select OK.
3. When prompted, select that the data are grouped in rows.
4. SPC XL will return an ANOVA table and group statistics.

MINITAB

1. Two-factor experiments can be analyzed using three different platforms in Minitab: Stat –> ANOVA –> Two-way, Stat –> ANOVA –> Balanced ANOVA, and Stat –> ANOVA –> General Linear Model. The last two platforms can analyze experiments with more than two factors.
2. In the worksheet, specify the response in one column and the levels of the row and column variables associated with each observation on the response in the additional columns
3. From Stat –> ANOVA –> Two-way identify the columns containing the response and the two design factors as Rows and Columns. By default Minitab will include the interaction, but it can be omitted by choosing the **Fit additive Model option**.
4. Turn on the diagnostic plots by clicking on **Graphs**. Selecting all of the graphs is a good idea.

SPC XL

1. On a blank Excel worksheet, select the Create Design icon from the DOE Pro menu. Then select Computer Aided Design from the menu options. Select a z-level design.
2. Select the number of factors needed for your design, then select the next button.
3. Enter the factor names (and high and low values if necessary). Then select the next button.
4. Enter the number of responses and replications for your design. Then select the next button.
5. Enter the response names and select finish. DOE PRO will then generate the design for you. Enter the design responses in the yellow highlighted region.
   a. If a single replication is chosen and you would like to pool the higher order interactions, select Modify Design then interactions. Specify the desired interactions for the analysis. (Note: You can also add replications and change other design characteristics in this area.)
6. Once all of the data has been entered, select Analyze Design –> ANOVA. DOE PRO will then prepare an ANOVA Table.
7. To generate an interaction and marginal means plot, select Analyze Design –> Multiple Plots.
8. To generate residual plots, select Residual Analysis –> Generate Residual Sheet. DOE Pro will then generate the residuals for the design.
9. To create the residual plots, select Residual Analysis –> Residual Plots. Common Plots are shown in the table below. Note: DOE PRO does not have the capability to prepare a normal probability plot. These must be generated manually through the instructions within Chapter 4.

X	Y
Run #	Residual
Predicted	Residual
Factor	Residual

MINITAB

1. You can use Minitab to create both factorial and fractional factorial designs. Select Stat –> DOE –> Factorial. This will take you to a dialog box where you can select either full two-level or fractional two-level factorials, Plackett-Burman designs, or general factorial design.
2. After the design has been created and the response data entered, the analysis can be done automatically by selecting Stat –> DOE –> Factorial –> Factorial –> Analyze Factorial Design.

SPC XL

1. On a blank Excel worksheet, select the Create Design icon from the DOE PRO XL menu. Then select Fractional Factorial.
2. Select the design you would like based upon the number of factors, number of runs, and design resolution.
3. Reviewing the recommended aliasing pattern and indicate whether or not the pattern is acceptable by indicating yes or nor.
4. Follow steps 3 and beyond from Supplement One to complete the analysis.

MINITAB

1. Select Stat –> DOE –> Response Surface –> Create Response Surface Design
2. Select the central composite design you wish to run
3. Enter the data from the experiment
4. To analyze the data select Stat –> DOE –> Response Surface –> Analyze Response Surface Design

SPC XL

1. On a blank Excel worksheet, select the Create Design icon from the DOE PRO XL menu. Then select Fractional Factorial.
2. Select the design you would like based upon the # of factors, # of runs, and design resolution.
3. Reviewing the recommended aliasing pattern and indicate whether or not the pattern is acceptable by indicating yes or no.
4. Follow steps 3 and beyond from Supplement One to complete the analysis.

chapter Nine Reliability

Chapter Outline

What keeps CIOs up at night? System unavailability!

Ask a CIO what factors are important for business success and you are sure to hear "system reliability" as a key factor in achieving business goals and profits. However, achieving 99% uptime is a great challenge when dealing with complex information systems that must operate 24/7 with various applications and technologies.

In addition, 99% uptime doesn't mean much if the 1% downtime occurs during a mission-critical period or during a peak operating time. Jobs are on the line every time a system goes down. And in the case of hospitals and other life-supporting systems, lives are on the line as well.

Given the importance of system uptime, CIOs have defined metrics to help better measure reliability and system performance indicators. These are often uptime, availability, help desk performance, CPU utilization rates, network uptime, server uptime, bandwidth fluctuations, support center IT performance, and security. Security is a critical component. If companies are not keeping ahead of viruses and other malicious cracking attempts, they will not have reliable information technology systems.

Given the rapid evolution of technology, a significant focus is also placed upon system adaptability and manageability. More and more CIOs and IT departments are refusing to incorporate system components into existing networks unless vendors keep up with software updates and are compatible with other standardized systems.

After all, there is nothing a CIO wants more than a well-running, secure, stable, and efficient system. Why? Because when systems are running poorly, many CIOs find themselves involved in heavily tactical and time-consuming bug-fixing and back-office problem-solving. On the other hand, when systems are running as intended, CIOs get to contribute as intended: by being involved with and making strategic and board-level technology decisions. And, of course, they get a much better night's sleep!

Chapter Overview and Learning Objectives

The reliability of a product or system is one of the eight dimensions of quality. Reliability is the probability that a product or system performs its intended function in a satisfactory manner for a defined period of time under a stated set of operating conditions. Because reliability is a probability, we use probability distributions to describe reliability. This chapter describes and illustrates how two widely used distributions, the exponential and the Weibull, are used in reliability modeling and determination. We also show how the failure rate of a unit or component can be calculated from time-to-failure data. Determining the reliability of a system of connected subsystems or components from the component reliability data is illustrated. Various aspects of life testing are also discussed. The concepts of maintainability and availability for repairable systems are introduced. We also describe failure mode and effects analysis, a tool often useful in improving the reliability of a system.

After careful study of this chapter you should be able to do the following:

1. Know the definition of reliability
2. Explain how reliability is one of the eight dimensions of quality
3. Understand the probability distributions associated with reliability analysis
4. Understand the concept of failure rate
5. Perform reliability calculations for systems based on knowledge of the reliability of components
6. Understand the concepts of maintainability and availability
7. Understand failure mode and effects analysis

9.1 Basic Concepts of Reliability

We define **reliability** as the probability that a product or system will satisfactorily perform its intended function for a defined period of time under a stated set of operating conditions. This definition has several important components. First, note that reliability is expressed as a **probability**. That means that probability distributions and probability models are used in determining reliability. We will see that there are several important probability distributions that arise naturally in reliability analysis. Second, the **intended function** of the product or system is essential to the definition. Products are designed for specific purposes and the customer expects them to perform these functions in a satisfactory fashion. For example, an elevator is a system that is designed to move people and other loads from one level of a structure to another. Third, the concept of a defined period of **time** is important. This is a measure of how long the product or system is designed to last. For example, the life of automotive engine oil might be six months or 5,000 miles. Either a time period, or some measure of usage (or both) could be employed to define the anticipated life of a product. Finally, the stated **operating conditions** for the product or system are an important component of the definition. These conditions might be

Reliability
The probability that a product or system will satisfactorily perform its intended function for a defined period of time under a stated set of operating conditions.

environmental factors, such as temperature or relative humidity. There could also be storage, handling, or transportation conditions. They could also be conditions of use; for example, a standard two-wheel drive sedan is not intended for extensive off-road use.

There is a lot of emphasis today on achieving high levels of product reliability. High technology products, such as new generations of commercial aircraft, complex electromechanical devices, automobiles, computers and electronic products, and modern telecommunications systems require high levels of reliability for their safe and effective use by consumers. Military organizations and NASA have long been focused or reliability as essential to the systems that they operate. Most consumers consider reliability to be one of the eight dimensions of quality discussed in Chapter 1. Generally, **reliability** can be thought of as **quality over time**.

9.2 Life Distributions

Because reliability is expressed as a probability, probability distributions are employed in reliability analysis and calculations. We will let $t > 0$ be the time to failure of the unit of product or system. We assume that t is a continuous random variable, and $f(t)$ is the probability distribution of t. Sometimes $f(t)$ is called the life distribution, the survival distribution, or the time-to-failure distribution. The **reliability function** $R(t)$ is just the reliability of the product or system at time t; that is, it is the probability that the product has not failed at time t. This can be expressed as

$$R(t) = \int_t^{\infty} f(x)dx \tag{9.1}$$

Since the **cumulative distribution** is defined as the area to the left of t, that is,

$$F(t) = \int_0^t f(x)dx$$

The reliability function is just the reverse cumulative or the area to the right of t.

One of the most important life distributions is the **exponential distribution**, introduced briefly in Chapter 4:

$$f(t) = \lambda e^{-\lambda t}, \quad t > 0 \tag{9.2}$$

where $\lambda > 0$ is the only parameter that defines the distribution. The mean of the exponential distribution is $1/\lambda$ and the variance of the distribution is $1/\lambda^2$. The exponential distribution is shown in Figure 9.1.

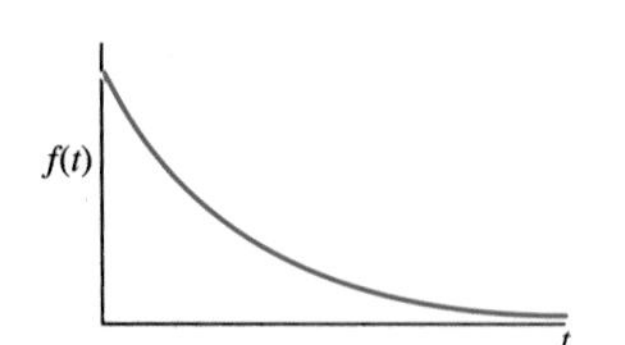

Figure 9.1 The Exponential Distribution

Sometimes the mean of the time-to-failure distribution is called the **mean time to failure**, abbreviated **MTTF**. If a unit is **repairable** and can be returned to a good-as-new condition following failure, the mean of the time to failure distribution is often called the **mean time between failures**, abbreviated **MTBF**.

If the time to failure is exponential, the reliability function is simple:

$$R(t) = e^{-\lambda t} \tag{9.3}$$

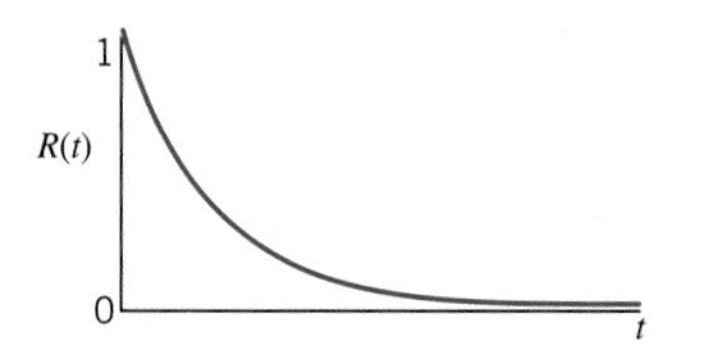

Figure 9.2 Reliability Function for the Exponential Distribution

The reliability function for the exponential distribution is shown in Figure 9.2. To illustrate the calculations, suppose that the time to failure in hours of a medical laser has an exponential distribution with $\lambda = 0.00005$. This is an exponential distribution with MTTF equal to $1/\lambda = 1/0.00005 = 20{,}000$ hours. Therefore, the reliability function is

$$R(t) = e^{-0.00005t}$$

The reliability of this laser at 10,000 hours is

$$R(10{,}000) = e^{-0.00005(10{,}000)} = e^{-0.5} = 0.6065.$$

That is, the reliability of the laser at 10,000 hours or, equivalently, the probability of the laser lasting at least 10,000 hours is 0.6065. In this example, the mean of the exponential distribution is 20,000 hours. This can be thought of as the mean life, or as noted above, the mean time to failure (MTTF).

It is interesting to find the reliability of the laser at 20,000 hours. Or, in other words, what is the probability that the laser will live at least as long as the mean life? We find the answer by computing the reliability at 20,000 hours to be

$$R(20{,}000) = e^{-0.00005(20{,}000)} = e^{-1} = 0.3679.$$

That is, the probability that the laser will fail **before** the mean life is $1 - 0.3679 = 0.6321$. At first glance, this may seem a little unusual—particularly if you were expecting the answer to be close to 0.5. However, the mean of a distribution is not in general equal to the median (for which the probability of occurrence either above or below the median is exactly 0.5), and in the case of the exponential distribution the mean is considerably to the right of the median because the distribution is highly skewed; that is, it has a long tail to the right.

Another important distribution used to model survival time in the reliability field is the **Weibull distribution**

$$f(t) = \frac{\beta}{\delta}\left(\frac{t}{\delta}\right)^{\beta-1} \exp\left[-\left(\frac{t}{\delta}\right)^{\beta}\right], t > 0 \tag{9.4}$$

where $\delta > 0$ is the **scale** parameter and $\beta > 0$ is the **shape** parameter. The mean and variance of the Weibull distribution are

$$\mu = \delta\Gamma\left(1 + \frac{1}{\beta}\right) \tag{9.5}$$

and

$$\sigma^2 = \delta^2\Gamma\left(1 + \frac{2}{\beta}\right) - \delta^2\left[\Gamma\left(1 + \frac{1}{\beta}\right)\right]^2 \tag{9.6}$$

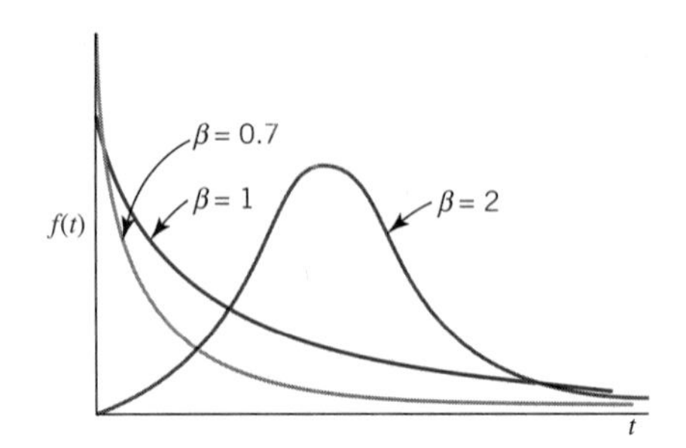

Figure 9.3 Several Weibull Distributions

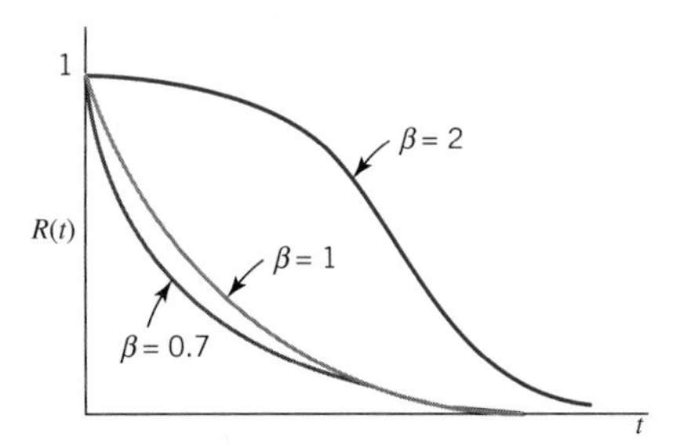

Figure 9.4 The Reliability Function for the Weibull Distribution

respectively. In equations (9.5) and (9.6) $\Gamma(r)$ is the **gamma function,** defined as

$$\Gamma(r) = \int_0^{\infty} x^{r-1}e^{-x}dx$$

If the argument r is a positive integer, then $\Gamma(r) = (r-1)! = (r-1) \times (r-2) \times \cdots \times 1$. For example, if $r = 4$, then $\Gamma(4) = 3! = 3 \times 2 \times 1 = 6$.

Several Weibull distributions are shown in Figure 9.3. When the shape parameter $\beta < 1$ the Weibull distribution is **hyperexponential** in shape; that is, the curve is steeper that the exponential distribution. If the shape parameter, $\beta = 1$, the Weibull reduces to the exponential distribution. When $\beta > 1$ the Weibull distribution has a single peak (it is **unimodal**) with a long tail to the right. At $\beta = 3.4$ the Weibull distribution can be well-approximated by the normal distribution.

The reliability function for the Weibull distribution is

$$R(t) = e^{-(t/\delta)^{\beta}} \tag{9.7}$$

The reliability function for the Weibull distribution is shown in Figure 9.4. Notice that the behavior of the reliability function depends on the value of the shape parameter.

To illustrate the use of the Weibull distribution, suppose that the life in hours of a cell phone component has a Weibull distribution with $\delta = 10{,}000$ and $\beta = 1/2$. For this component, we can calculate the mean (or MTTF) and variance of life as follows:

$$\begin{aligned}\mu &= \delta\Gamma\left(1 + \frac{1}{\beta}\right) = 10{,}000\Gamma\left(1 + \frac{1}{1/2}\right) = 10{,}000\Gamma(3)\\ &= 10{,}000 \times 2! = 20{,}000 \text{ hours}\end{aligned}$$

and

$$\begin{aligned}\sigma^2 &= \delta^2\Gamma\left(1 + \frac{2}{\beta}\right) - \delta^2\left[\Gamma\left(1 + \frac{1}{\beta}\right)\right]^2\\ &= 10{,}000^2\Gamma\left(1 + \frac{2}{1/2}\right) - 10{,}000^2\left[\Gamma\left(1 + \frac{1}{1/2}\right)\right]^2\\ &= 10{,}000^2\Gamma(5) - 10{,}000^2\Gamma(3)^2\\ &= 10{,}000^2[4! - (2!)^2]\\ &= 2{,}000{,}000{,}000 \text{ hours}^2\end{aligned}$$

The standard deviation of the lifetime is $\sigma = 44{,}721.36$ hours. Suppose that we are interested in the reliability of this component at 30,000 hours of use. This is computed from equation (9.7) as follows:

$$R(30{,}000) = e^{-(30{,}000/10{,}000)^{1/2}} = e^{-3^{1/2}} = e^{-1.7321} = 0.1769.$$

In other words, only 17.63% of all of these components will be expected to survive at least 30,000 hours.

There are other probability distributions that are used in reliability analysis, including the normal, the lognormal, and the gamma distributions. The uses of these distributions are more limited than the exponential and the Weibull and will not be illustrated in this section.

How do we know which distribution to use as a model for reliability? If a sample of data is available we can use **probability plotting** to determine the appropriate choice of distribution. This is a more general case of normal probability plotting, discussed in earlier chapters. Sample failure time data is plotted on a probability plot designed for a specific distribution (either exponential or Weibull), and if the data can be well approximated by a straight line on the plot, then that distribution is an appropriate model for the data.

For example, consider the failure data for sample 1 in Table 9.1. Figure 9.5 presents exponential and Weibull probability plots for these data, obtained from Minitab. The data plots approximately as a straight line on both plots. Therefore, the exponential is a reasonable choice of a probability distribution (recall that the exponential is a special case of the Weibull) for these data. Minitab also provides estimates of the distribution parameters. Notice that the estimate of the Weibull shape parameter is almost unity. This is what we would expect to find if the distribution really is exponential.

Figure 9.6 presents exponential and Weibull probability plots of the sample 2 failure data from Table 9.1. The data is not well approximated by the straight line in the exponential probability plot. However, the Weibull probability plot displays a reasonable straight line approximation for the sample 2 data. Furthermore, the estimate of the Weibull shape parameter is approximately 2, another strong indication that the exponential is not a good choice of probability model for the sample 2 data.

Table 9.1

Two Samples of Failure Data

Observation	Sample 1	Sample 2
1	13.833	126.527
2	14.690	148.849
3	36.802	59.799
4	14.394	78.171
5	80.100	51.375
6	124.211	104.816
7	5.091	87.696
8	1.499	181.002
9	65.779	312.005
10	7.809	193.807
11	2.844	149.153
12	15.942	38.914
13	63.776	152.639
14	35.460	112.504
15	27.885	135.564
16	53.075	131.475
17	21.769	102.436
18	34.248	64.772
19	7.109	75.665
20	3.017	80.306

9.3 Instantaneous Failure Rate

The **failure rate** can be a useful way to describe the life performance of a product over time. The failure rate over an interval of time $[t_1, t_2]$ is given by

$$FR(t_1, t_2) = \left[\frac{R(t_1) - R(t_2)}{R(t_1)}\right]\frac{1}{t_2 - t_1} \tag{9.8}$$

The term in brackets is the probability of a failure during the interval $\Delta t = t_2 - t_1$ given that the item didn't fail before time t_1, and the second term accounts for the dimensional characteristic. It is often useful to consider the **instantaneous failure rate**, which is the limiting form of equation (9.8) as $\Delta t = t_2 - t_1$ goes to 0. It can be shown that this instantaneous failure rate is

$$h(t) = \frac{f(t)}{R(t)} \tag{9.9}$$

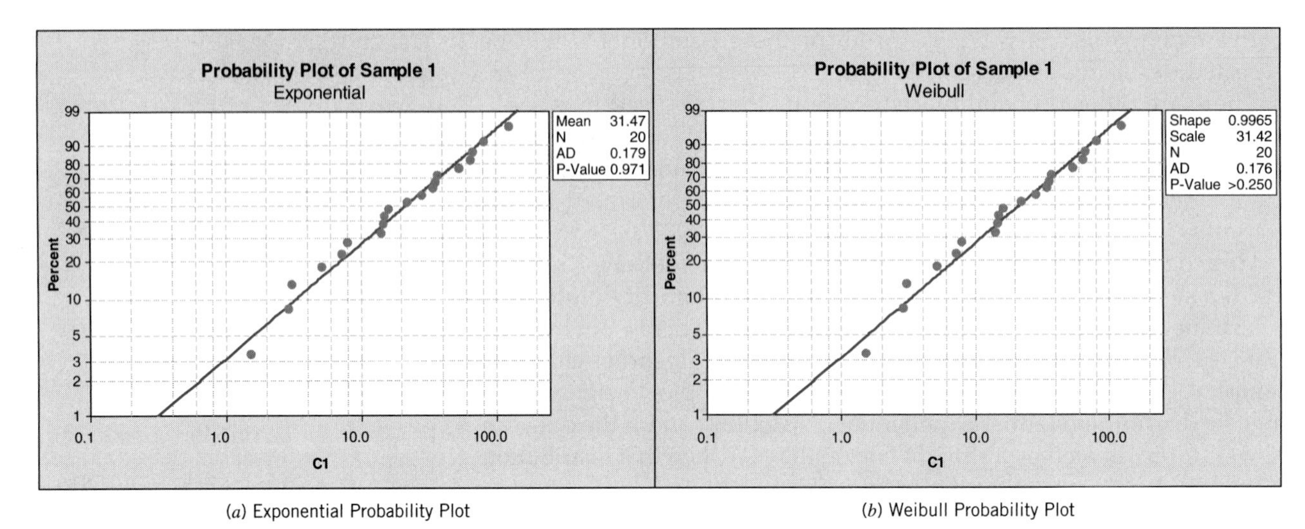

(*a*) Exponential Probability Plot (*b*) Weibull Probability Plot

Figure 9.5 Probability Plots of Sample 1 from Table 9.1. (a) Exponential Probability Plot (b) Weibull Probability Plot

The instantaneous failure rate is also usually called the **hazard function.** The hazard function can be thought of as the instantaneous probability of failure at time t, given that the unit has survived to time t.

If the time to failure is exponential, then the hazard function is

$$h(t) = \frac{f(t)}{R(t)} = \frac{\lambda e^{-\lambda t}}{e^{-\lambda t}} = \lambda \tag{9.10}$$

That is, the hazard function is constant. This means that the probability of failure is exactly the same at any time t during the life of the item. So if the life distribution for the processor in your personal computer is exponential, then the probability of failure is the same after 1,000 hours of use as it was when the processor was brand new. In other words, the failure mechanism is entirely random and not affected by wear or other stresses encountered in use.

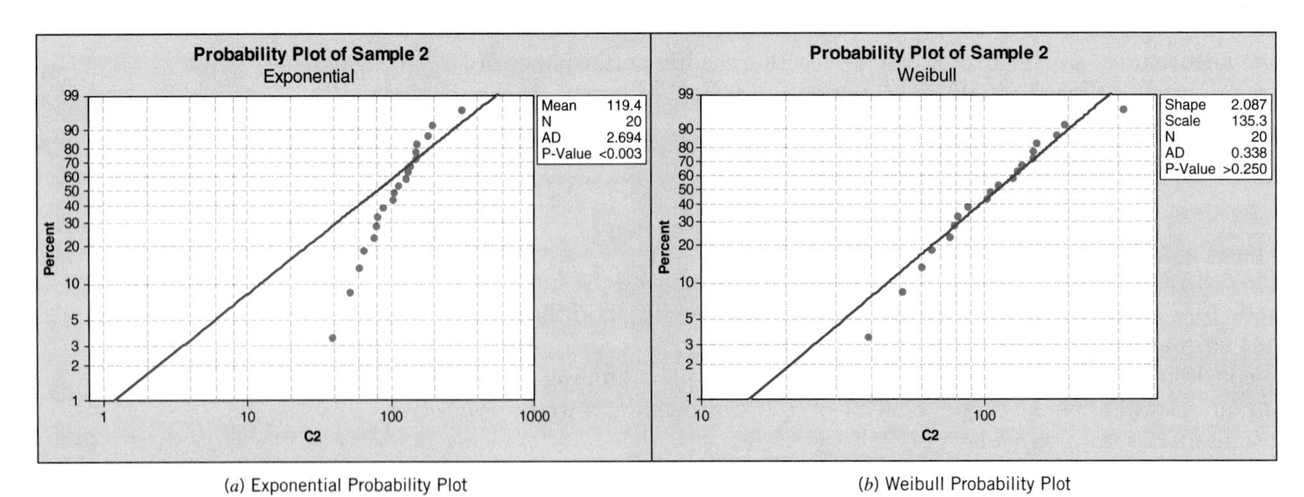

(*a*) Exponential Probability Plot (*b*) Weibull Probability Plot

Figure 9.6 Probability Plots of Sample 2 from Table 9.1. (a) Exponential Probability Plot (b) Weibull Probability Plot

The Weibull hazard function is

$$h(t) = \frac{\beta}{\delta}\left(\frac{t}{\delta}\right)^{\beta-1} \tag{9.11}$$

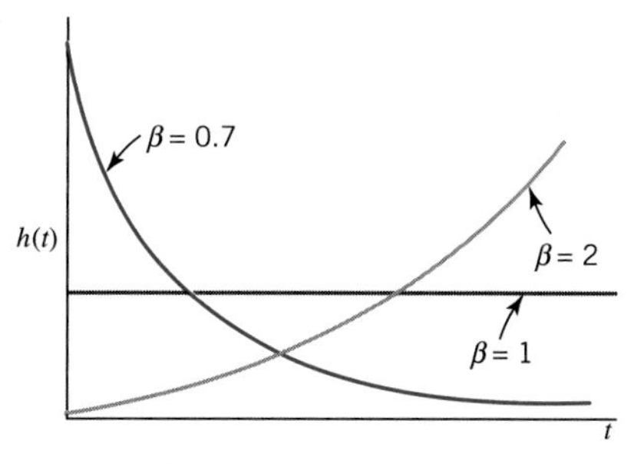

Figure 9.7 The Hazard Function for the Weibull Distribution

The shape of the Weibull hazard function depends on the shape parameter δ. Figure 9.7 shows the Weibull hazard function for several values of δ. If $\delta < 1$, the hazard function is decreasing. This case of a decreasing failure rate might be appropriate for the early life of a product, where reliability is being improved by burn-in or operational testing of units before they are released to customers, stress-test screening (such as exposure of units to higher than expected levels of vibration or elevated temperatures) to remove items that will fail early on due to manufacturing defects, and debugging of software. Sometimes this is called the **infant mortality** phase of product life. When $\delta = 1$ the Weibull hazard function is constant, because the Weibull distribution reduces to the exponential distribution in this case. When the shape parameter $\delta > 1$, the Weibull hazard function increases. An increasing failure rate is appropriate when units are failing because of accumulated stress, wear, or other effects of aging.

The failure rate of a unit can be estimated from reliability test data. Suppose that n units are placed on test and that the ith item fails at time t_i. Let r be the number of observed failures when the test is terminated at elapsed time T and that no units that failed during the test are replaced with new ones. Under these conditions, where the test is **time-terminated without replacement**, the failure rate is estimated as

$$h(t) = \frac{r}{\sum_{i=1}^{r} t_i + (n - r)T} = \frac{r}{Q} \tag{9.12}$$

where $Q = \sum_{i=1}^{r} t_i + (n - r)T$ is the total amount of test time. That is, the failure rate is the ratio of the total number of test failures to the total amount of test time.

example 9.1 Testing without Replacement

Suppose that $n = 10$ units are placed on test and that five of them fail at 25, 28, 35, 60, and 85 hours, respectively. The test is terminated at 100 hours. Find the failure rate.

Solution

The sum of the failure times for the five units that failed is $\sum_{i=1}^{5} t_i = 25 + 28 + 35 + 60 + 85 = 233$. Now, applying equation 9.12, we can calculate the failure rate as

$$h(t) = \frac{r}{Q} = \frac{r}{\sum_{i=1}^{r} t_i + (n - r)T} = \frac{5}{233 + (10 - 5)100} = \frac{5}{733} = 0.00682$$

That is, the instantaneous probability of failure of one of these units is 0.00682.

In some reliability testing situations, the test is time-terminated, but units that fail during the test are placed with new ones. For a **time-terminated with replacement** test, the failure rate is estimated by

$$h(t) = \frac{r}{\sum_{i=1}^{r} t_i} \tag{9.13}$$

This modified version of the original equation takes into account that the total test time Q is just $Q = \sum_{i=1}^{r} t_i$, the sum of the failure times for the r units that fail.

example 9.2 A Time-Terminated Test with Replacement

Suppose that 25 units are tested for 100 cycles of use. At the end of the 100 cycle test period, 4 of the units have failed. When a unit failed, it was replaced with a new one and the test was continued. Estimate the failure rate.

Solution

Because $r = 4$, the failure rate can be calculated from equation (9.13) as follows:

$$h(t) = \frac{r}{Q} = \frac{r}{\sum_{i=1}^{r} t_i} = \frac{4}{25(100)} = \frac{4}{2500} = 0.0016$$

For these units, the instantaneous probability of failure is 0.0016.

A third type of reliability test is a **failure-terminated** test; units are placed on test and the test is terminated when a certain number of them have failed. Equation (9.13) can be used to estimate the failure rate for these types of tests.

example 9.3 A Failure-Terminated Test

100 units are placed on test, and when five of them have failed, the test is terminated. The failure times in hours for the five units that failed are 1071, 1235, 1440, 1518, and 1620. Estimate the failure rate for these units.

example 9.3 Continued

Solution

We can apply equation (9.13) to these data as follows:

$$h(t) = \frac{r}{\sum_{i=1}^{r} t_i} = \frac{5}{1071 + 1235 + 1440 + 1518 + 1620} = \frac{5}{6884}$$
$$= 0.0001453$$

For these units, the instantaneous probability of failure is 0.0001453.

9.4 Life Cycle Reliability

The failure rate of a product is usually not constant over its lifetime. There are usually three distinct periods of reliability performance. Figure 9.8 presents the instantaneous failure rate (hazard function) for a typical complex product over the lifetime of a large number of units. This is sometimes called the "bathtub curve." During the infant mortality phase, failures are occurring due to early-life problems with the product, including manufacturing defects, and also due to random failures as well. After early-life failures are removed from the population, the product enters a random failure phase. The assumption of a constant failure rate is usually reasonable in this phase. The smaller the failure rate in this phase, the better the product. As the product approaches end of life, the failure rate increases due to wear-out, accumulated stress, or other accumulated effects of aging.

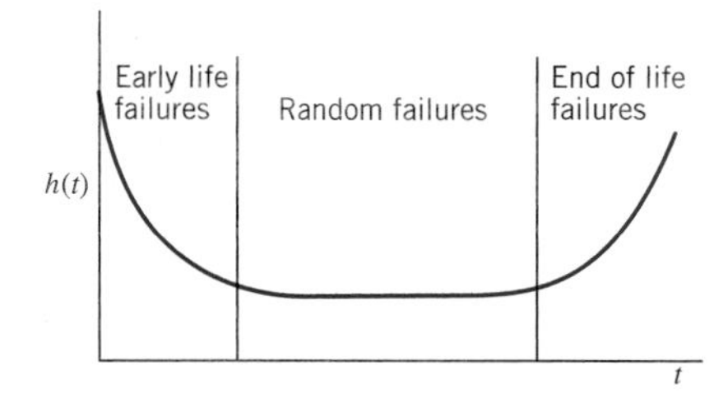

Figure 9.8 Failure Rate Over the Life of a Product (the Bathtub Curve)

The bathtub curve demonstrates why the exponential and Weibull distributions are so important in reliability analysis. The Weibull can have both a decreasing and an increasing hazard function, so it can be used as a model for reliability during the infant mortality and end of life periods. Both the exponential and the Weibull are good models for random failures. There are also other distributions that can be used to model reliability during the different phases of product life. For example, the normal distribution has an increasing hazard function, and is sometime suggested as a model for the end of life period. However, the great flexibility of the Weibull distribution allows its use in these situations almost regardless of the actual distribution of product life.

9.5 Determining System Reliability from Component Reliabilities

Many products are a collection of subsystems and components that must function in specific ways for the product to function satisfactorily. Very often, the system performance depends on the configuration of the

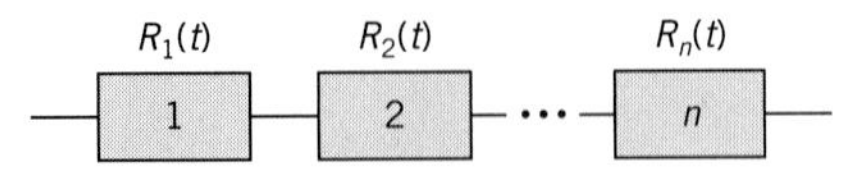

Figure 9.9 **A Series System**

subsystems and components. In this section we show how the reliability of a product is impacted by the configuration of the components.

In some products the components are arranged in **series,** and all components must function for the system to function. An example of a simple series system is shown in Figure 9.9. We assume that in such a system, the n components all function **independently**. Let $R_i(t)$, $i = 1, 2, \ldots, n$ be the reliability of components i at time t. Then the reliability of the system at time t, which we will denote by $R_s(t)$, is just the product of the individual component reliabilities:

$$R_S(t) = R_1(t) \cdot R_2(t) \cdot \cdots \cdot R_n(t) \tag{9.14}$$

Note that as components are added to the system, the system reliability will decrease. Also, the reliability of the system will always be less than the reliability of its least reliable component.

example 9.4 A Simple Series System

A series system is made up of three components, as shown in Figure 9.10. The reliabilities of each component at some pre-specified time t are shown in the figure. All components operate independently. What is the system reliability?

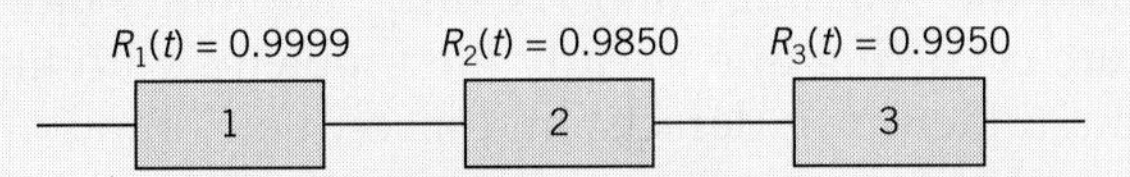

Figure 9.10 **The Series System for Example 9.4**

Solution

The system reliability is found from equation (9.14) as follows:

$$\begin{aligned} R_s(t) &= R_1(t) \cdot R_2(t) \cdot R_3(t) \\ &= 0.9999 \times 0.9850 \times 0.9950 \\ &= 0.979977 \end{aligned}$$

Notice that the reliability of the system (0.979977) is less than the reliability of the least reliable components ($R_2(t) = 0.9850$).

A special case involving the series system occurs when all components have an exponential failure distribution. Recall that for the exponential, the reliability of the ith component will be $R_i(t) = e^{-\lambda_i t}$ for components $i = 1, 2, \ldots, n$. Therefore the system reliability is

$$\begin{aligned} R_s(t) &= e^{-\lambda_1 t} \cdot e^{-\lambda_2 t} \cdot \cdots \cdot e^{-\lambda_n t} \\ &= e^{-(\lambda_1 + \lambda_2 + \cdots \lambda_n)t} \\ &= e^{-\lambda_s t} \end{aligned} \tag{9.15}$$

where $\lambda_s = \lambda_1 + \lambda_2 + \cdots + \lambda_n$ is the **system failure rate.**

example 9.5 Exponential Components in Series

Suppose that an electronic system is composed of two logic devices, four resistors, and two diodes. All components are in series and operate independently. The component reliabilities are all exponential and have the following reliability functions:

$$\text{Logic devices:} \quad R_L(t) = e^{-1.2\times 10^{-8}t}$$
$$\text{Resistors:} \quad R_R(t) = e^{-1.5\times 10^{-7}t}$$
$$\text{Diodes:} \quad R_D(t) = e^{-1.9\times 10^{-9}t}$$

Find the equation for system reliability and the system reliability at 10,000 hours.

Solution

Because all life distributions are exponential, we can use the result in equation (9.15). The system reliability is

$$\begin{aligned} \lambda_s &= 2(1.2 \times 10^{-8}) + 4(1.5 \times 10^{-7}) + 2(1.9 \times 10^{-9}) \\ &= 6.278 \times 10^{-7} \end{aligned}$$

The system reliability equation is

$$R_S(t) = e^{-\lambda_s t} = e^{-6.278\times 10^{-7}t}$$

The system reliability at 10,000 hours is

$$\begin{aligned} R_s(t) &= e^{-6.278\times 10^{-7}t} \\ &= e^{-6.278\times 10^{-7}(1\times 10^{4})} \\ &= e^{-6.278\times 10^{-3}} \\ &= 0.99374 \end{aligned}$$

The probability of a failure before 10,000 hours of operation for this system is relatively small (about 0.006).

Another common configuration of components is a **parallel system.** A system with the n components arranged in parallel is illustrated in Figure 9.11. There are a number of ways that parallel systems can be designed to operate. In the simplest configuration, all components begin operation at time zero and the system operates as long as at least one of the components operates. This is sometimes called **active redundancy**. The reliability of such a system, assuming that all components operate independently, is

$$R_S(t) = 1 - \prod_{i=1}^{n}[1 - R_i(t)] \tag{9.16}$$

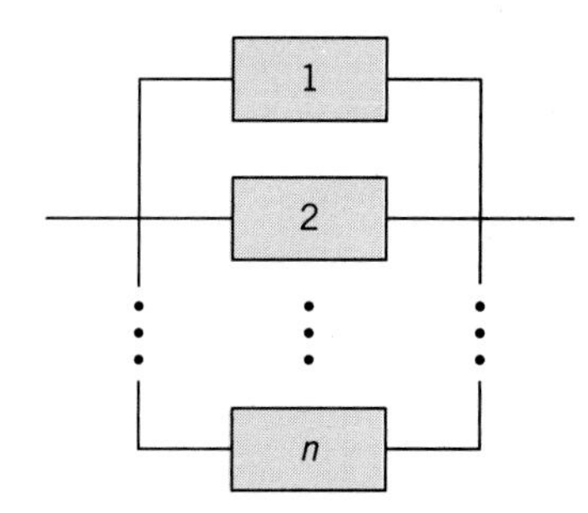

Figure 9.11 **A Parallel System**

The quantity $1 - R_i(t)$ in equation (9.16) is the probability that the ith component fails by time t. Therefore $\prod_{i=1}^{n}[1 - R_i(t)]$ is the probability that the system fails by time t and the reliability at time t is just 1 minus this probability.

As the number of parallel components in a system increases, the overall system reliability increases. The reliability for a parallel system is always greater than the reliabilities of the individual components. Parallel systems occur in many types of products. For example, aircraft typically have multiple parallel hydraulic systems for the flight controls. Automobile braking systems have some parallel components; each wheel has its own independent brake caliper and the braking system will work unless all four components fail.

example 9.6 A Simple Active Redundant System

Figure 9.12 illustrates a parallel system with four components. The reliabilities of the components are shown in the figure. The components operate independently and the system will operate as long as one of the four components operates. Find the system reliability.

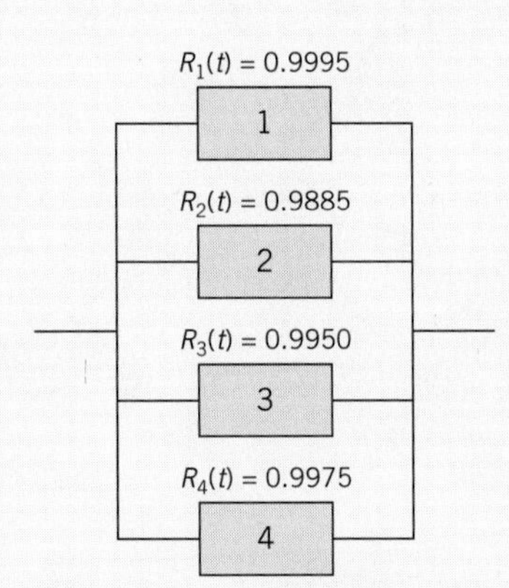

Figure 9.12 The Parallel System for Example 9.6

Solution

The solution is found by applying equation (9.16) as follows:

$$
\begin{aligned}
R_s(t) &= 1 - \prod_{i=1}^{4}[1 - R_i(t)] \\
&= 1 - [1 - 0.9995][1 - 0.9885][1 - 0.9950][1 - 0.9975] \\
&= 1 - 0.000000000072 \\
&= 0.999999999928
\end{aligned}
$$

Notice that arranging the components is parallel has greatly increased the reliability of the system in comparison to the reliabilities of the individual components.

Another variation of a redundant system occurs when the system operates if k out of the n parallel components operate, where $k < n$. Suppose that the components operate independently and assume that all components have the same reliability function; that is, $R_i(t) = R(t)$, $i = 1, 2, \ldots, n$. This is usually a reasonable assumption. Sometimes these types of systems are called ***k*-out-of-*n* systems**. In this case the system reliability is

$$\begin{aligned} R_s(t) &= \sum_{x=k}^{n} \binom{n}{x} [R(t)]^x [1 - R(t)]^{n-x} \\ &= 1 - \sum_{x=0}^{k-1} \binom{n}{x} [R(t)]^x [1 - R(t)]^{n-x} \end{aligned} \quad (9.17)$$

example 9.7 A *k*-out-of-*n* System

Suppose that we have a k-out-of-n system where there are three components ($n = 3$) and two of them must function for the system to function ($k = 2$). Find an equation for the system reliability.

Solution

From equation (9.17),

$$\begin{aligned} R_s(t) &= \sum_{x=2}^{3} \binom{3}{x} [R(t)]^x [1 - R(t)]^{n-x} \\ &= \binom{3}{2} [R(t)]^2 [1 - R(t)]^{3-2} + \binom{3}{3} [R(t)]^3 [1 - R(t)]^{3-3} \\ &= 3[R(t)]^2 [1 - R(t)] + [R(t)]^3 \\ &= [R(t)]^2 [3 - 2R(t)] \end{aligned}$$

If the reliability of an individual component is $R(t) = 0.99$, the reliability of the system is

$$\begin{aligned} R_s(t) &= [R(t)]^2 [3 - 2R(t)] \\ &= (0.99)^2 [3 - 2(0.99)] \\ &= 0.999702 \end{aligned}$$

Notice that as in any parallel system, the reliability of the system is greater than the reliability of any individual component.

A third variation of the parallel system is **stand-by redundancy.** In a standby redundant system there are several components arranged in parallel, but only one of the components operates until it fails. At that time a decision switch activates one of the other parallel or stand-by components. These remaining components are not subject to failure

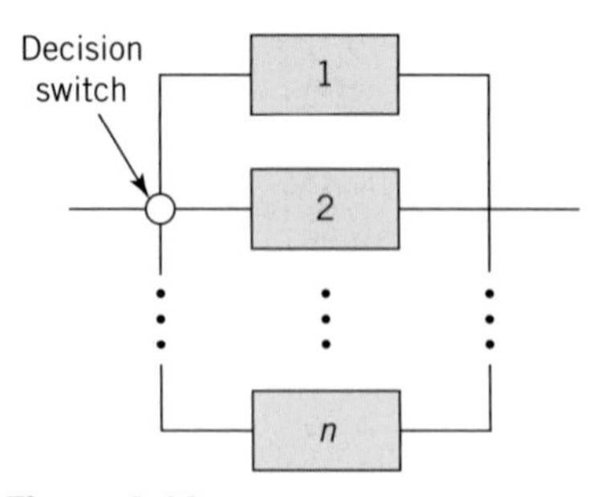

Figure 9.13 **A Standby Redundant System**

until they are activated. Figure 9.13 shows a standby redundant system. The most common situation is a decision switch that is assumed never to fail and all components are identical. There are stand-by systems that deviate from these assumptions, but their analysis is beyond the scope of this discussion. Refer to Tobias and Trindade (1995) for more details.

A very common situation is n identical components with each having an exponential time-to-failure distribution and a perfect switch. In this case the time-to-failure distribution for the system is a **gamma distribution**

$$f(t) = \frac{\lambda}{(n-1)!}(\lambda t)^{n-1}e^{-\lambda t}, \quad t > 0 \tag{9.18}$$

where λ is the failure rate of the individual components. The system reliability function for this case is

$$R_s(t) = e^{-\lambda t}\sum_{i=0}^{n-1}\frac{(\lambda t)^i}{i!} \tag{9.19}$$

The mean time to failure of the system is

$$\mu = \frac{n}{\lambda}$$

and the variance of the time to failure is

$$\sigma^2 = \frac{n}{\lambda^2}$$

example 9.8 A Stand-by Redundant System with Two Components

Suppose that two components with exponential time to failure distributions are arranged in a stand-by redundant system with a perfect decision switch. Suppose that the mean of the exponential distribution for each component is 1,000 hours. Find the reliability function, the mean and variance of the system time to failure, and the system reliability at 3,000 hours.

Solution

Because the component time-to-failure distributions with failure rate $\lambda = 1/1{,}000$ the reliability function is

$$\begin{aligned} R_S(t) &= e^{-\lambda t}\sum_{i=0}^{n-1}\frac{(\lambda t)^i}{i!} \\ &= e^{-t/1000}\sum_{i=0}^{1}\frac{(t/1000)^i}{i!} \\ &= e^{-t/1000}\left[1 + \frac{t}{1000}\right] \end{aligned}$$

example 9.8 Continued

The mean and variance of the system time-to-failure are

$$\mu = \frac{n}{\lambda} = \frac{2}{1/1000} = 2000 \text{ hr}$$

and

$$\sigma^2 = \frac{n}{\lambda^2} = \frac{2}{(1/1000)^2} = 2{,}000{,}000 \text{ hr}^2$$

The system reliability at 3,000 hours is

$$R_S(3000) = e^{-3000/1000}\left[1 + \frac{3000}{1000}\right] = e^{-3}(4) = 0.1991.$$

Many systems are a **combination** of series and parallel subsystems. In some cases the system reliability can be determined by determining the reliabilities of the serial and parallel subsystems and them combining them. For example, consider the system in Figure 9.14*a*. The first subsystem consists of two parallel active redundant units. The reliability of this group of components is

$$\begin{aligned} R(t) &= 1 - \prod_{i=1}^{2}[1 - R_i(t)] \\ &= 1 - (1 - 0.99)(1 - 0.99) \\ &= 1 - 0.0001 \\ &= 0.9999 \end{aligned}$$

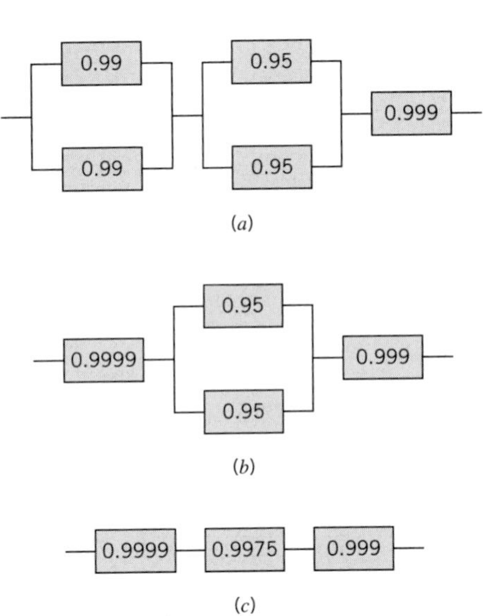

Figure 9.14 Reliability Calculations for a Complex System

In effect, we could replace the first two parallel components with a single component having reliability 0.9999. This leads to the situation shown in Figure 9.14*b*. The second subsystem also contains two parallel units. The reliability of this subsystem is

$$\begin{aligned} R(t) &= 1 - \prod_{i=1}^{2}[1 - R_i(t)] \\ &= 1 - (1 - 0.95)(1 - 0.95) \\ &= 1 - 0.0025 \\ &= 0.9975 \end{aligned}$$

The second subsystem of two parallel components could be replaced with a single component having reliability 0.9975. The system can now be seen to effectively consist of three series components as shown in Figure 9.14*c*, so the system reliability is

$$\begin{aligned} R_S(t) &= R_1(t) \cdot R_2(t) \cdot R_3(t) \\ &= 0.9999 \times 0.9975 \times 0.9990 \\ &= 0.996403 \end{aligned}$$

9.6 Life Testing and Reliability Estimation

Life testing is an integral part of the reliability field. Usually these tests consist of putting a certain number of units on test and aging them until some or perhaps many of the units have failed. Sometimes the purpose of the test is to obtain data that will enable the analyst to determine the appropriate life distribution for the data. Another objective is to estimate the reliability of the units.

A very simple way to do this is the following. Put n units on test. Suppose that our objective is to estimate the reliability of these units at a particular time, say 2,000 hours. Run the test for 2,000 hours and observe the number of units that fail at or before this time. We assume that failed units are not replaced. Suppose that the number of failed units is r. Then an estimate of the reliability at 2,000 hours is

$$\hat{R}(2{,}000) = 1 - \frac{r}{n} \tag{9.20}$$

When n is large, it is also possible to find a $100(1 - \alpha)\%$ lower confidence bound on the reliability from

$$LCB = \hat{R}(2{,}000) - Z_{\alpha}\sqrt{\frac{\hat{R}(2{,}000)[1 - \hat{R}(2{,}000)]}{n}} \tag{9.21}$$

For example, suppose that $n = 100$ and at the end of 2,000 hours there have been 10 failures. The estimate of the reliability at 2,000 hours is

$$\begin{aligned}\hat{R}(2{,}000) &= 1 - \frac{r}{n}\\ &= 1 - \frac{10}{100}\\ &= 0.90\end{aligned}$$

The 95% lower confidence bound on $R(2{,}000)$ is

$$\begin{aligned}LCB &= \hat{R}(2{,}000) - Z_{\alpha}\sqrt{\frac{\hat{R}(2{,}000)[1 - \hat{R}(2{,}000)]}{n}}\\ &= 0.90 - 1.645\sqrt{\frac{0.90(0.10)}{100}}\\ &= 0.90 - 0.0494\\ &= 0.8507\end{aligned}$$

This method relies on the normal approximation to the binomial distribution. When that assumption is not appropriate, the confidence bound can be found from a tabulation of the binomial distribution.

When an assumption can be made about the form of the life distribution some other methods can be used to estimate reliability. The most common situation is for the exponential distribution. In the exponential distribution with failure rate λ, the mean time to failure (or the MTTF) is

$$\theta = \frac{1}{\lambda}$$

Suppose that $\hat{\theta}$ is the estimate of the MTTF θ. Then the estimate of the reliability at time t is

$$\hat{R}(t) = e^{-t/\hat{\theta}} \tag{9.22}$$

It is also possible to find a $100(1 - \alpha)\%$ two-sided confidence interval on the reliability from

$$[e^{-t/\theta_L}, e^{-t/\theta_U}] \tag{9.23}$$

where θ_L and θ_U are the $100(1 - \alpha)\%$ confidence bounds on the MTTF θ. It turns out that the appropriate estimate for θ is the reciprocal of equation (9.12); that is

$$\hat{\theta} = \frac{\sum_{i=1}^{r} t_i + (n - r)T}{r} = \frac{Q}{r} \tag{9.24}$$

If the test is terminated after a fixed number of failures are observed, the $100(1 - \alpha)\%$ lower and upper confidence bounds on the MTTF θ are

$$\left[\frac{2Q}{\chi^2_{\alpha/2, 2r}}, \frac{2Q}{\chi^2_{1-\alpha/2, 2r}}\right] \tag{9.25}$$

and if the test is terminated after a fixed time, the $100(1 - \alpha)\%$ confidence bounds on the MTTF θ are

$$\left[\frac{2Q}{\chi^2_{\alpha/2, 2r+2}}, \frac{2Q}{\chi^2_{1-\alpha/2, 2r+2}}\right] \tag{9.26}$$

The attractive thing about knowing the form of the failure distribution is that it is not necessary to run the test for the length of time for which a reliability estimate is required. For example, you may want to obtain an estimate of reliability at 2,000 hours but if you can safely assume that the life distribution is exponential it may only be necessary to run the test for, say, 200 hours, a much shorter time.

example 9.9 Reliability Estimation with the Exponential Distribution

Suppose that $n = 100$ units are placed on test and that five of them fail at 25, 28, 35, 60, and 85 hours, respectively. The test is terminated at 100 hours, and no failed units are replaced during the test. Estimate the MTTF and the reliability of these units at 1,000 hours.

example 9.9 Continued

Solution

The sum of the failure times for the five units that failed is $Q = \sum_{i=1}^{5} t_i = 25 + 28 + 35 + 60 + 85 = 233$. Now from equation (9.13) we can calculate the MTTF as

$$\hat{\theta} = \frac{\sum_{i=1}^{r} t_i + (n-r)T}{r} = \frac{233 + (100-5)100}{5} = \frac{9733}{5}$$
$$= 1{,}946.60 \text{ hours}$$

The reliability at 1,000 hours is found from equation (9.22):

$$\begin{aligned}\hat{R}(1{,}000) &= e^{-t/\hat{\theta}} \\ &= e^{-1000/1946.60} \\ &= e^{-0.5137} \\ &= 0.598\end{aligned}$$

The 95% confidence interval on the MTTF is

$$\left[\frac{2Q}{\chi^2_{\alpha/2,2r+2}}, \frac{2Q}{\chi^2_{1-\alpha/2,2r+2}}\right]$$

$$\left[\frac{2(9733)}{\chi^2_{0.025,2(5)+2}}, \frac{2(9733)}{\chi^2_{0.975,2(5)+2}}\right]$$

$$\left[\frac{19{,}466}{\chi^2_{0.025,12}}, \frac{19{,}466}{\chi^2_{0.975,12}}\right]$$

$$\left[\frac{19{,}466}{23.34}, \frac{19{,}466}{4.40}\right]$$

$$[834.02, 4{,}424.09]$$

Therefore the lower and upper confidence bounds on the MTTF θ are $\theta_L = 834.02$ and $\theta_U = 4{,}424.09$, respectively. Consequently, the 95% confidence interval on the reliability at 1,000 hours is

$$[e^{-t/\theta_L}, e^{-t/\theta_U}]$$

$$[e^{-1000/834.02}, e^{-1000/4424.09}]$$

$$[e^{-1.20}, e^{-0.23}]$$

$$[0.30, 0.80]$$

There are many other types of reliability tests. A very widely used method for testing is **accelerated life testing**, where units are tested at higher levels of temperature, humidity vibration, or other stresses than the units would encounter in normal use. The accelerated stress levels are designed to make failures occur much sooner that they would under conditions of normal use. Then an acceleration model is used to extrapolate information about the failure at accelerated stress levels to failure under normal use conditions. Accelerated life testing is very useful for highly reliable long-life units. There are also standard demonstration testing and acceptance sampling plans for reliability. Widely used test plans based on the exponential distribution are in the Department of Defense Handbook H108 and Military Standard 781C.

9.7 Availability and Maintainability

When units or systems are repairable, there are other measures of reliability that are important. We assume that a failed system is subject to repair (or replacement) immediately upon failure. The **maintainability** of a system is the probability that the failed system is returned to its normal operating condition within a specified time period. The time to repair a failed system is usually a random variable. The mean of this distribution is usually called the mean time to repair, abbreviated MTTR.

Mean Time to Repair – MTTR

The failure of a unit or system can be affected by maintenance actions and policies. **Corrective maintenance** generally does not engage in any repairs, replacements, or other interventions in the operation of the unit until a failure occurs. Corrective maintenance can result in catastrophic failures and may not be economically the best strategy in the long run. **Preventive maintenance** requires maintaining the unit on a predetermined schedule, even when no problems are apparent. A simple example of preventive maintenance is changing the engine oil in an automobile at regular intervals of operation. Preventive maintenance usually reduces catastrophic repair costs, but it can result in excessive lost production time if the maintenance actions are extensive and take a great deal of time. They can also result in the creation of problems that can lead to failure if maintenance actions are not properly performed. Generally, preventive maintenance should be considered when the cost to maintain the system is less than the cost to repair the system after it has failed. Preventive maintenance is usually a good idea when the system has an increasing failure rate with time, particularly when the maintenance actions can return the system to a "good-as-new" condition. **Predictive maintenance** is a strategy that tries to predict the time to failure of the unit or system, based on identification of the failure modes and the development of techniques to monitor leading indicators of when these failure modes will occur. Monitoring a unit for corrosion, excessive vibration, lubricant condition, the appearance of cracks, temperature or an acoustic signature are approaches that might be used. When the system reaches an **alarm value** on a monitored characteristic the maintenance action is scheduled. If the monitored characteristic reaches a **breakdown value,** the system should be shut down for maintenance. Part of a predictive maintenance strategy involves determining appropriate thresholds for the alarm and breakdown values.

The **availability** of a repairable system at time t is the probability that the system is properly operating at that time. The steady-state availability at time t is computed as

$$A(t) = \frac{MTBF}{MTBF + MTTR} \tag{9.27}$$

example 9.10 Calculating Availability

The mean time to failure for a mechanical component is 200 hours. The mean time to repair the component is 5 hours. Determine the availability.

Solution

From equation (9.27), the availability is

$$A(t) = \frac{MTBF}{MTBF + MTTR} = \frac{200}{200 + 5} = 0.9756$$

9.8 Failure Mode and Effects Analysis

Failure mode and effects analysis (FMEA) is a problem-solving technique that is intended to help users or designers identify and either reduce or eliminate the impact of potential failures before they occur. It may be applied to systems, subsystems or components, in product or process design, or to services. FMEA can also be implemented in health care; for example, see The Joint Commission website (http://www.jointcommission.org) or the Institute for Health Care Improvement (http://www.ihi.org). FMEA is team-based and can be used as part of an overall quality management system or as a stand-alone tool. It is often used in design for six sigma and in the improve step of DMAIC. There is a military standard (MIL STD 1629) that addresses FMEA, but it is limited to design and process applications. Another useful reference is the FMEA manual published by the Automotive Industry Action Group (2001).

The FMEA team should represent a broad range of experience with the system. Typically team members are drawn from areas such as product design, operations, manufacturing or production engineering, quality and reliability engineering, marketing, distribution, sales, purchasing, and logistics and materials management. Customers should also be included on FMEA teams. Failure to include customers can led to an incomplete list of failure modes and underestimation of the severity of failures. One person on the team should be identified as the leader and should have the responsibility of making decisions and allocating the necessary resources to complete the FMEA. The team has the responsibilities of defining the scope of the FMEA, establishing expectations and deliverables, and setting milestones, due dates, and deadlines.

The input requirements to an FMEA are diverse and come from many sources. They may include

1. A process flow chart or diagram, or a block diagram identifying all major subsystems
2. Specifications
3. Customer requirements
4. Test data
5. Warranty data and related information
6. Information on field failures (if available)
7. Rework history (if available)
8. Design change histories
9. Prior related FMEAs
10. Process capability information
11. Information from related activities such as designed experiments or control charts
12. Information on related products, designs, or technologies

The outputs from a FMEA may include documentation for each system, subsystem, process, or service and recommendations for design changes or corrective actions.

The basic steps in a FMEA are

1. Identify potential failure modes, including
    a. Who is impacted by the failure?
    b. When does the failure occur?
    c. What happens in the event of a failure?
    d. Where does the failure occur?
    e. Why does the failure occur?
    f. How does the failure occur?
2. Quantify the risk that is associated with each potential failure. Risk is quantified by using a **risk priority number (RPN).** The risk is based on the severity, occurrence, and detection of a potential failure.
3. Develop a corrective action plan for the most serious risks.
4. Repeat the analysis until all potential failures are at an acceptable risk level. The acceptable level of risk must be defined by the agent that has authorized the FMEA.
5. Document and report the results.

Risk Priority Number – RPN

The risk priority numbers defined above have three components, severity, occurrence, and detection. Each of these components is

Table 9.2

Calculation of Risk Priority Numbers

Potential failure	Severity	Occurrence	Detection	RPN
1	4	1	1	4
2	5	5	5	125
3	10	5	5	250
4	2	10	1	20

Table 9.3

Calculation of Risk Priority Numbers

Potential failure	Severity	Occurrence	Detection	RPN
1	2	10	10	200
2	10	10	2	200
3	8	5	5	200
4	5	10	4	200

measured on a scale from 1 to 10, with 10 identifying the highest level. The RPN is obtained by multiplying the scores for each component; therefore, an RPN varies between 1 (lowest) and 1,000 (highest). Corrective actions are usually prioritized based on the values of the RPNs.

For example, consider the RPNs calculated in Table 9.2. Based on the RPNs, the order of priority in developing corrective actions for the potential failures would be 3, 2, 4, and 1. However, in some cases, it is not clear what the priority should be. This can occur when there are several potential failures that have identical RPNs, as in Table 9.3. Generally the approach is to try to eliminate the occurrence, reduce the severity, reduce the occurrence, and improve detection. For the situation in Table 9.3, if we could eliminate the occurrences, then all potential failure now have different RPNs, so a priority can be determined. Then we can focus on reducing the severity of the potential failures. If occurrences cannot be eliminated or reduced, then we should focus on reducing the severity of the failures and, finally, on improving detection.

Important Terms and Concepts

Availability
Exponential distribution
Failure mode and effects analysis (FMEA)
Failure rate
Failure-terminated tests
Hazard function
k-out-of-n system
Life cycle reliability
Life distribution
Maintainability
Mean time between failures (MTBF)
Mean time to failure (MTTF)
Mean time to repair (MTTR)
Parallel system
Reliability
Reliability function
Risk priority number (RPN)
Series system
Stand-by redundant system
Survival distribution
Testing with replacement
Testing without replacement
Time-terminated tests
Time-to-failure distribution
Weibull distribution

9.1 An electronic component in a dental x-ray system has an exponential time to failure distribution with $\lambda = 0.00004$. What are the mean and variance of the time to failure? What is the reliability at 30,000 hours?

9.2 Reconsider the component described in Exercise 9.1. What is the reliability of the component at 25,000 hours?

9.3 A component in an automobile door latch has an exponential time to failure distribution with $\lambda = 0.00125$ cycles. What are the mean and variance of the number of cycles to failure? What is the reliability at 8,000 cycles?

9.4 You are designing a system that must have reliability at least 0.95 at 10,000 hours of operation. If you can reasonably assume that the time to failure is exponential, what MTTF must you achieve in order to satisfy the reliability requirement?

9.5 A synthetic fiber is stressed by repeatedly applying a particular load. Suppose that the number of cycles to failure has an exponential distribution with mean 3,000 cycles. What is the probability that the fiber will break at 1,500 cycles? What is the probability that the fiber will break at 2,500 cycles?

9.6 An electronic component has an exponential time to failure distribution with $\lambda = 0.0002$ hours. What are the mean and variance of the time to failure? What is the reliability at 7,000 hours?

9.7 The life in hours of a mechanic assembly has a Weibull distribution with $\delta = 5{,}000$ and $\beta = 1/2$. Find the mean and variance of the time to failure. What is the reliability of the assembly at 4,000 hours? What is its reliability at 7,500 hours?

9.8 The life in hours of subsystem in an appliance has a Weibull distribution with $\delta = 7{,}000$ cycles and $\beta = 1.5$. What is the reliability of the appliance subsystem at 10,000 hours?

9.9 Suppose that the lifetime of a component has a Weibull distribution with shape parameter $\beta = 2$. If the system should have reliability 0.99 at 7,500 hours of use, what value of the scale parameter is required?

9.10 A component has a Weibull time-to-failure distribution. The value of the scale parameter is $\delta = 200$. Calculate the reliability at 250 hours for the following values of the shape parameter: $\beta = 0.5, 1, 1.5, 2,$ and 2.5. For the fixed value of the scale parameter, what impact does changing the shape parameter have on the reliability?

9.11 A component has a Weibull time-to-failure distribution. The value of the scale parameter is $\delta = 500$. Suppose that the shape parameter is $\beta = 3.4$. Find the mean and variance of the time to failure. What is the reliability at 600 hours?

9.12 **Continuation of Exercise 9.11.** If a Weibull distribution has a shape parameter of $\beta = 3.4$, it can be reasonably well approximated by a normal distribution with the same mean and variance. For the situation of Exercise 9.11, calculate the reliability at 600 hours using the normal distribution. How close is this to the reliability value calculated from the Weibull distribution in Exercise 9.11?

9.13 Suppose that a unit has a Weibull time-to-failure distribution. The value of the scale parameter is $\delta = 250$. Graph the hazard function for the following values of the shape parameter: $\beta = 0.5, 1, 1.5, 2,$ and 2.5. For the fixed value of the scale parameter, what impact does changing the shape parameter have on the hazard function?

9.14 Suppose that you have 15 observations on the number of hours to failure of a unit. The observations are 442, 381, 960, 571, 1861, 162, 334, 825, 2562, 324, 312, 368, 367, 968, and 15. Is the exponential distribution a reasonable choice for the time to failure distribution? Estimate the MTTF.

9.15 Suppose that you have 20 observations on the time to failure of a unit. The observations are 259, 53, 536, 1320, 341, 667, 538, 1713, 428, 152, 29, 445, 677, 637, 696, 540, 1392, 192, 1871, and 2469. Is the exponential distribution a reasonable choice for the time to failure distribution? Estimate the MTTF.

9.16 Twenty observations on the time to failure of a system are as follows: 1054, 320, 682, 1440, 1085, 938, 871, 471, 1053, 1103, 780, 665, 1218, 659, 393, 913, 566, 439, 533, and 813. Is the Weibull distribution a reasonable choice for the time to failure distribution? Estimate the scale and shape parameters.

9.17 Consider the following 10 observations on time to failure: 50, 191, 63, 174, 71, 62, 119, 89, 123, and 175. Is either the exponential or the Weibull a reasonable choice of the time to failure distribution?

9.18 Fifteen observations on the time to failure of a unit are as follows: 173, 235, 379, 439, 462, 455, 617, 41, 454, 1083, 371, 359, 588, 121, and 1066. Is the Weibull distribution a reasonable choice for the time to failure distribution? Estimate the scale and shape parameters.

9.19 Consider the following 20 observations on time to failure: 702, 507, 664, 491, 514, 323, 350, 681, 281, 599, 495, 254, 185, 608, 626, 622, 790, 248, 610, and 537. Is either the exponential or the Weibull a reasonable choice of the time to failure distribution?

9.20 Consider the time to failure data in Exercise 9.19. Is the normal distribution a reasonable model for these data? Why, or why not?

9.21 A simple series system is shown in the accompanying figure. The reliability of each component is shown in the figure. Assuming that the components operate independently, calculate the system reliability.

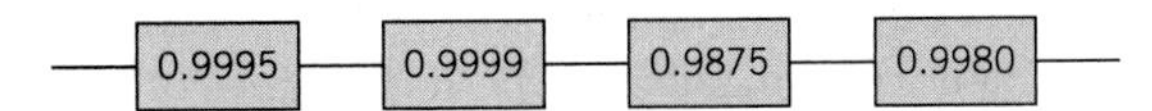

9.22 A series system is shown in the accompanying figure. The reliability of each component is shown in the figure. Assuming that the components operate independently, calculate the system reliability.

9.23 A simple series system is shown in the accompanying figure. The lifetime of each component is exponentially distributed and the λ for each component is shown in the figure. Assuming that the components operate independently, calculate the system reliability at 10,000 hours.

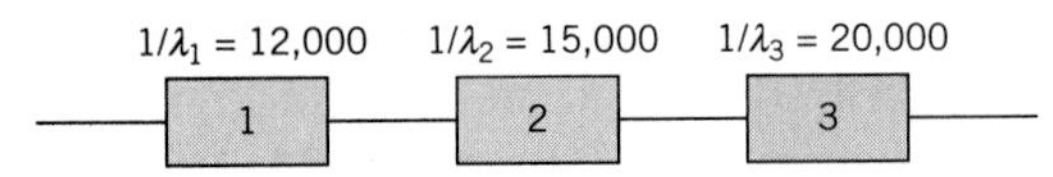

9.24 A series system has four independent identical components. The reliability of each component is exponentially distributed. If the reliability of the system at 8,000 hours must be at least 0.999, what MTTF must be specified for each component in the system?

9.25 Consider the parallel system shown in the accompanying figure. The reliability of each component is provided in the figure. Assuming that the components operate independently, calculate the system reliability.

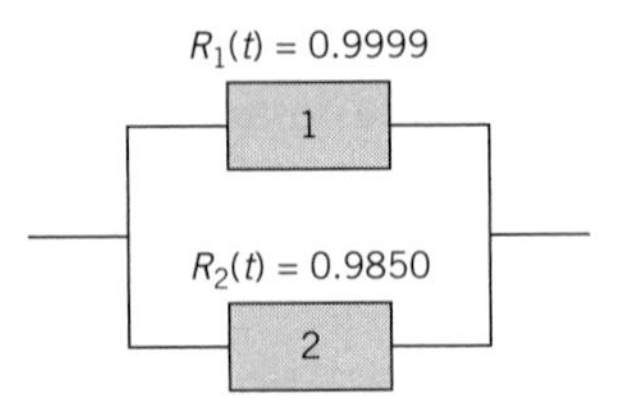

9.26 Consider the parallel system shown in the accompanying figure. The reliability of each component is shown in the figure. Assuming that the components operate independently, calculate the system reliability.

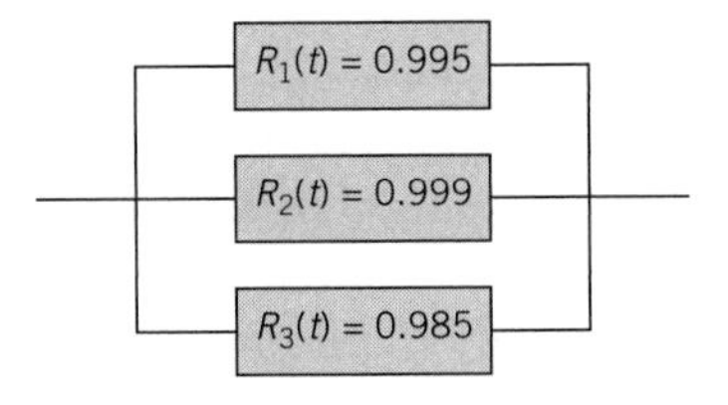

9.27 Consider the stand-by system in the accompanying figure. The components have an exponential lifetime distribution, and the decision switch operates perfectly. The MTTF of each component is 1,000 hours. Assuming that the components operate independently, calculate the system reliability at 2,000 hours.

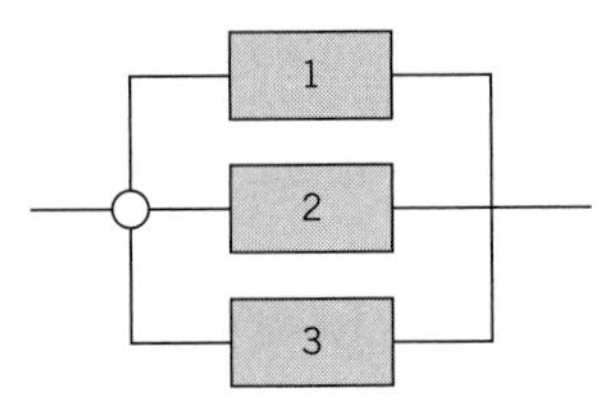

9.28 Consider the stand-by system shown in the accompanying figure. The components have an exponential lifetime distribution, and the decision switch operates perfectly. The MTTF of each component is 500 hours. Assuming that the components operate independently, calculate the system reliability at 1,500 hours.

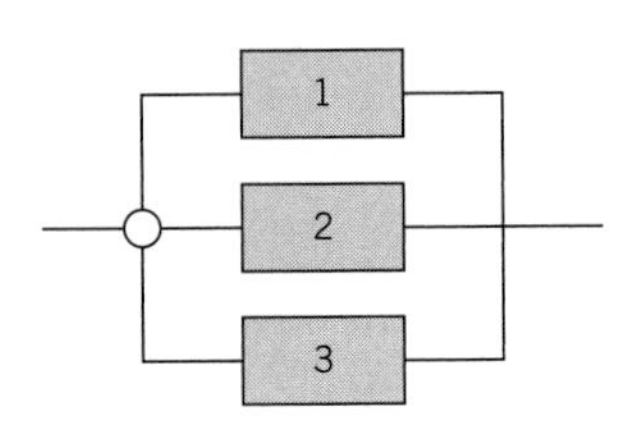

9.29 Consider the system shown in the accompanying figure. The reliability of each component is shown in the figure. Assuming that the components operate independently, calculate the system reliability.

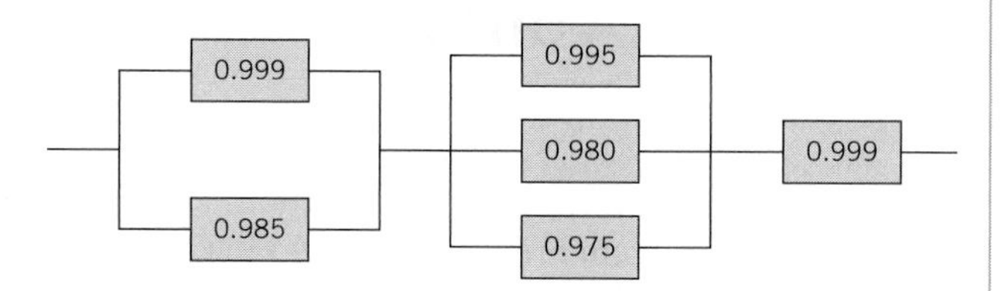

9.30 Consider the system shown in the accompanying figure. The reliability of each component is provided in the figure. Assuming that the components operate independently, calculate the system reliability.

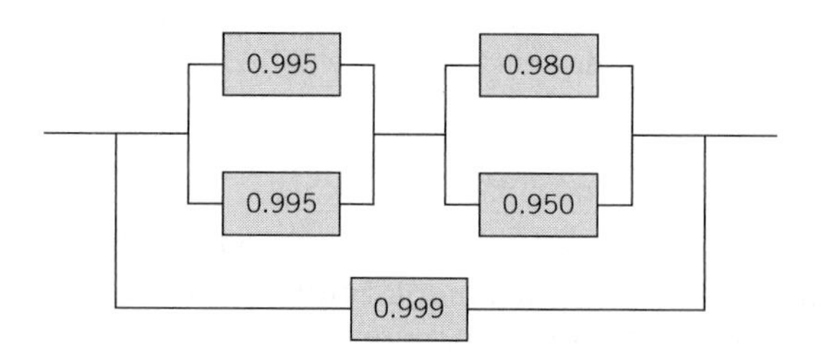

9.31 Consider the system shown in the accompanying figure. The reliability of each component is provided in the figure. Assuming that the components operate independently, calculate the system reliability.

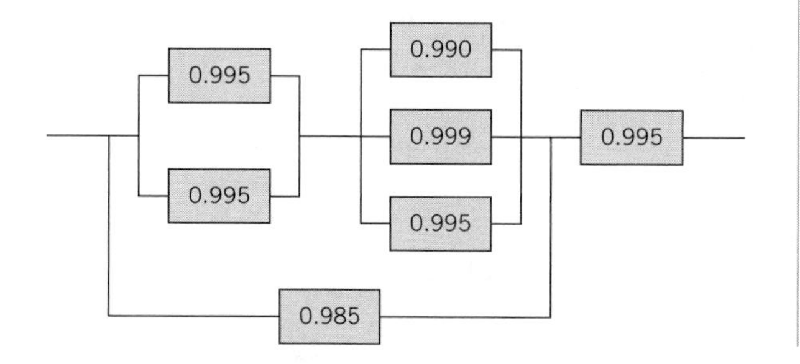

9.32 Suppose that $n = 20$ units are placed on test and that five of them fail at 20, 25, 40, 75, and 100 hours, respectively. The test is terminated at 100 hours without replacing any of the failed units. Find the failure rate.

9.33 **Continuation of Exercise 9.32.** For the data in Exercise 9.32, find a 95% confidence interval on the MTTF and the reliability at 200 hours, assuming that the lifetime has an exponential distribution.

9.34 Suppose that 50 units are tested for 1,000 cycles of use. At the end of the 1,000-cycle test period, 5 of the units have failed. When a unit failed, it was replaced with a new one and the test was continued. Estimate the failure rate.

9.35 Suppose that 50 units are placed on test, and when 3 of them have failed, the test is terminated. The failure times in hours for the 3 units that failed are 950, 1050, and 1525. Estimate the failure rate for these units.

9.36 Suppose that $n = 25$ units are placed on test and that two of them fail at 200 and 400 hours, respectively. The test is terminated at 1,000 hours without replacing any of the failed units. Determine the failure rate.

9.37 Suppose that $n = 25$ units are placed on test and that three of them fail at 150, 300, and 500 hours, respectively. The test is terminated at 2,000 hours, and no failed units are replaced during the test. Estimate the MTTF and the reliability of these units at 5,000 hours. Assuming that the lifetime has an exponential distribution, find 95% confidence intervals on these quantities.

Appendix One Summary of Common Probability Distributions Often Used in Quality Control and Improvement

Appendix Outline

Several discrete probability distributions arise frequently in statistical quality control. These distributions were briefly introduced in Chapter 4 but are described in greater detail within this supplement.

A.1 Important Discrete Distributions

Several discrete probability distributions arise frequently in statistical quality control. In this section, we discuss the hypergeometric distribution, the binomial distribution, the Poisson distribution, and the Pascal, or negative binomial, distribution.

A.1.1 THE HYPERGEOMETRIC DISTRIBUTION

The hypergeometric distribution is often used when we want to sample from a population without replacement. Suppose that there is a finite population consisting of N items. Some number, D, of these items fall into a class of interest.

A random sample of n items is selected from the population without replacement, and the number of items in the sample that fall into the class of interest—say, x—is observed. Then x is a hypergeometric random variable with the probability distribution defined as follows:

The **hypergeometric probability distribution** is

$$P(x) = \frac{\binom{D}{x}\binom{N-D}{n-x}}{\binom{N}{n}} \qquad x = 0, 1, 2, \ldots, \min(n, D)$$

The mean and variance of the distribution are

$$\mu = \frac{nD}{N}$$

and

$$\sigma^2 = \frac{nD}{N}\left(1 - \frac{D}{N}\right)\left(\frac{N-n}{N-1}\right)$$

In the definition, the quantity

$$\binom{a}{b} = \frac{a!}{b!(a-b)!}$$

is the number of ways a items can taken b at a time.

example A.1

Suppose a lot contains 100 items, 5 of which do not conform to requirements. If 10 items are selected at random without replacement, then the probability of finding 1 or fewer nonconforming items in the sample is

Probability of finding 1 or fewer = Probability of finding 0 + Probability of finding 1

$$= \frac{\binom{5}{0}\binom{95}{10}}{\binom{100}{10}} + \frac{\binom{5}{1}\binom{95}{9}}{\binom{100}{10}} = 0.92314$$

Some computer programs can perform these calculations. The display below is the output from SPC XL for calculating hypergeometric probabilities with $N = 100$, $D = 5$, $n=10$.

Hypergeometric Distribution

User−defined parameters	
Population size (N)	100
Size of sub−population of interest (D)	5
Sample size (n)	10
# of elements from sub−population of interest (x)	1
Probabilities	
P(X = 1)	0.33939

Note that the shape of a hypergeometric distribution depend upon the parameters N, n, and D. Some examples are shown in Figure A1-1.

A.1.2 THE BINOMIAL DISTRIBUTION

The binomial distribution is often used to describe the number of times an event will occur in a series of events (i.e., if you flip a coin five times, how many heads will you receive?).

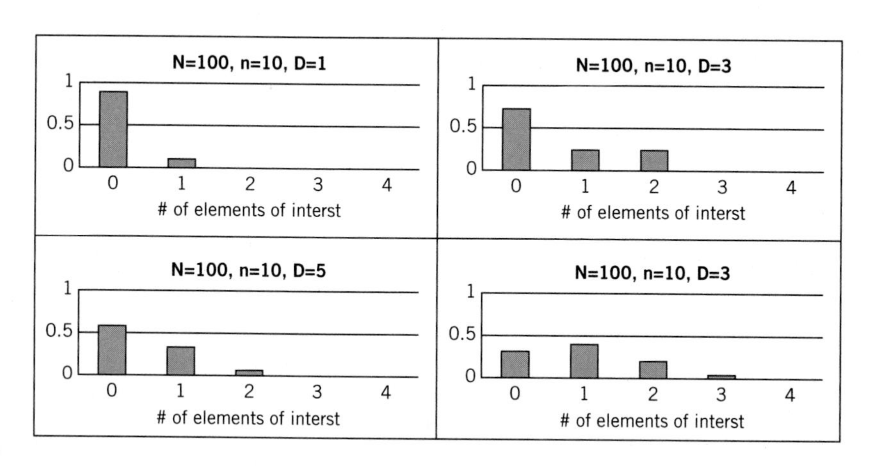

Figure A1-1 Number Nonconforming in a Sample

Consider a process that consists of a sequence of n independent trials. By independent trials, we mean that the outcome of each trial does not depend in any way on the outcome of previous trials. When the outcome of each trial is either a "success" or a "failure," the trials are called **Bernoulli trials.** If the probability of "success" on any trial—say, p—is constant, then the number of "successes" x in n Bernoulli trials has the **binomial distribution** with parameters n and p, defined as follows:

The **binomial distribution** with parameters $n \geq 0$ and $0 < p < 1$ is

$$p(x) = \binom{n}{x} p^x (1 - p)^{n-x} \quad x = 0, 1, \ldots, n$$

The mean and variance of the binomial distribution are

$$\mu = np$$

and

$$\sigma^2 = np(1 - p)$$

example A.2

A fair coin is tossed four times. What is the probability of getting four heads in a row?

$$p(x) = \binom{n}{x} p^4 \times (1 - p)^{n-x}$$

$$p(4) = \binom{4}{4} 0.5^4 (1 - 0.5)^{4-4} = 0.0625$$

Computer packages can be utilized to calculate probabilities from a binomial distribution. Below is an output from SPC XL calculating the probability of obtaining four heads in a row in four tosses of a fair coin.

Binomial Distribution

User defined parameters	
Number of trials (n)	4
Probability of success (p)	0.5
Number of successes in n trials	4
Probabilities	
P(X = 4)	0.0625
P(X < 4)	0.9375
P(X <= 4)	1
P(X > 4)	0
P(X >= 4)	0.0625

The binomial distribution has important applications for quality control applications. To expand on the above example, let's consider a manufacturing process that adds components to a product as it moves down the assembly line. The components operate independently so that the assembly of one nonconforming component will not impact the function of another component. If there are 20 such steps in an assembly process, each with a 5% chance of nonconformance, then the chance of a product have 0 nonconforming components is only 36%!

The binomial distribution can also be used for sampling from an infinitely large population, where p is the fraction of defective or nonconforming items in the population. In these applications, x usually represents the number of nonconforming items found in a random sample of n items.

For example, a steady stream of items are coming from a work center. The probability of nonconformance is 10%, and 15 items have been obtained in a quality control sample. The probability of finding a nonconforming value is shown graphically in Figure A1-3.

Several binomial distributions are shown graphically in Figure A1-4. The shape of those examples is typical of all binomial distributions. For a fixed n, the distribution becomes more symmetric as p increases from 0 to 0.5 or decreases from 1 to 0.5. For a fixed p, the distribution becomes more symmetric as n increases.

A random variable that arises frequently in statistical quality control is

$$\hat{p} = \frac{x}{n}$$

where x has a binomial distribution with parameters n and p. Often $\hat{p}$ is the ratio of the observed number of defective or nonconforming items in a sample (x) to the sample size (n). The mean and standard deviation of this distribution is

$$\mu = \hat{p}$$

$$\sigma_{\hat{p}}^2 = \frac{p(1-p)}{n}$$

A.1.3 THE POISSON DISTRIBUTION

A useful discrete distribution in statistical quality control is the Poisson distribution, defined as follows:

The Poisson distribution is

$$p(x) = \frac{e^{-\lambda}\lambda^x}{x!} \quad x = 0, 1, \ldots$$

where the parameter $\lambda > 0$. The **mean** and **variance** of the Poisson distribution are

$$\mu = \lambda$$

and

$$\sigma^2 = \lambda$$

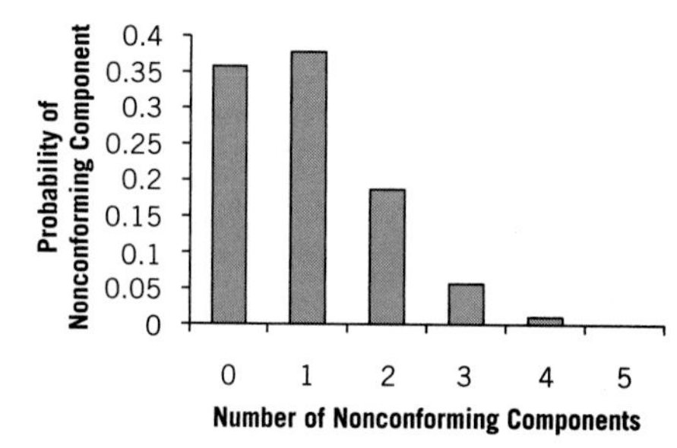

Figure A1-2 Number of Nonconforming Items from a 20-Step Process, Each with a 5% Probability of Nonconformance

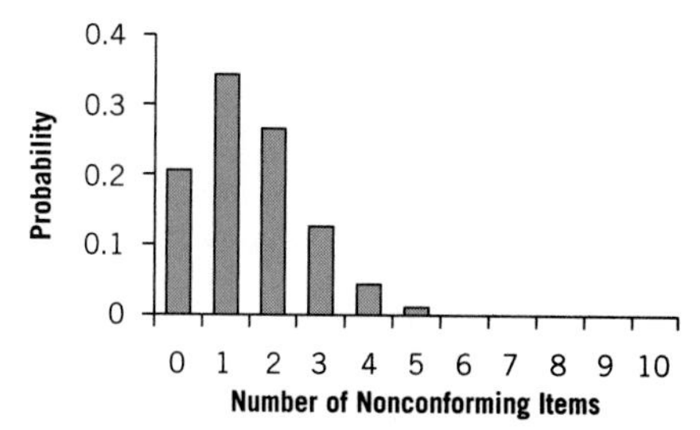

Figure A1-3 Number of Nonconforming Items from a Quality Control Sample of 15 Parts

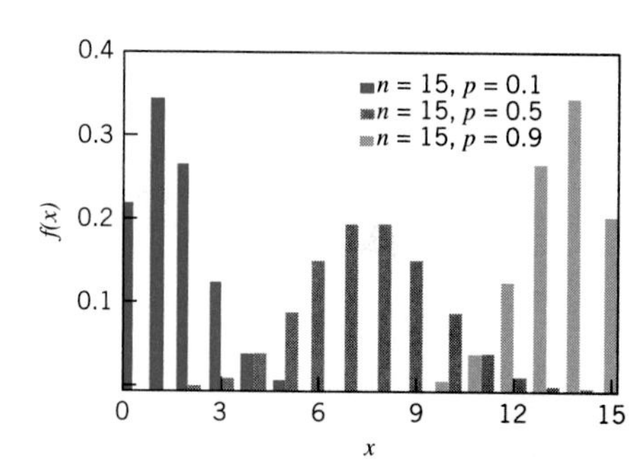

(a) Binomial distributions for different values of p with $n = 15$.

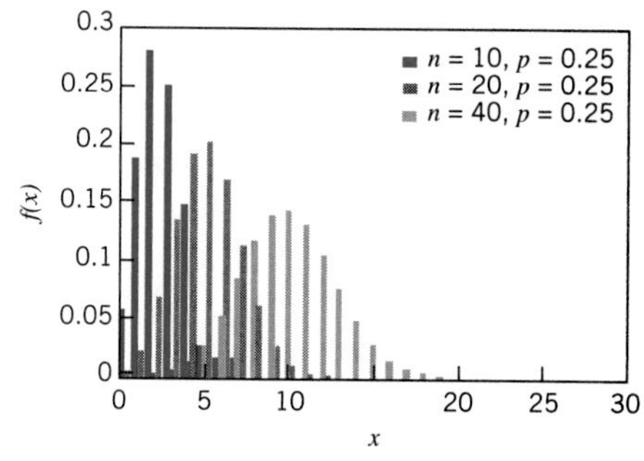

(b) Binomial distributions for different values of n with $p = 0.25$.

Figure A1-4 Shapes of Binomial Distributions

Note that the mean and variance of the Poisson distribution are both equal to the parameter λ. A typical application of the Poisson distribution in quality control is as a model of the number of defects or nonconformities that occur in a unit of product. In fact, any random phenomenon that occurs on a per unit (or per unit area, per unit volume, per unit time, etc.) basis is often well approximated by the Poisson distribution.

example A.3

Suppose that the number of wire–bonding defects per unit that occur in a device is Poisson distributed with parameter $\lambda = 4$ (this means on average there are four defects per unit). The probability that a randomly selected device will contain two or fewer wire–bonding defects is

$$p\{x \leq 2\} = \sum_{x=0}^{2} \frac{e^{-4}4^x}{x!} = 0.018316 + 0.073263 + 0.146525 = 0.238104$$

Computer packages can also perform these calculations. The display below is the output from SPC XL for calculating cumulative Poisson probabilities with $\lambda = 4$.

Poisson Distribution

User-defined parameters	
Mean (lambda)	4
Number of occurrences	2
Probabilities	
P(X = 2)	0.146525111
P(X < 2)	0.091578194
P(X <= 2)	0.238103306
P(X > 2)	0.761896694
P(X >= 2)	0.908421806

Several Poisson distributions are shown in Figure A1-5. Note that the distribution is *skewed;* that is, it has a long tail to the right. As the parameter becomes larger, the Poisson distribution becomes symmetric in appearance.

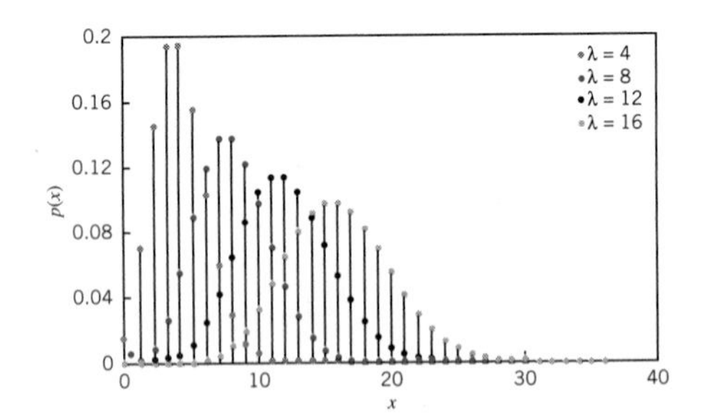

Figure A1-5 Figure Poisson Probability Distributions for Selected Values of l

A.1.4 THE PASCAL AND RELATED DISTRIBUTIONS

The Pascal distribution, like the binomial distribution, has its basis in Bernoulli trials. Consider a sequence of independent trials, each with probability of success p, and let x denote the trial on which the rth success occurs. Then x is a Pascal random variable with probability distribution defined as follows:

The **Pascal distribution** is

$$p(x) = \binom{x-1}{r-1} p^r (1-p)^{x-r} x = r, r+1, r+2, \ldots$$

where $r \geq 1$ is an integer. The ***mean*** and ***variance*** of the Pascal distribution are

$$\mu = \frac{r}{p}$$

and

$$\sigma^2 = \frac{r(1-p)}{p^2}$$

respectively.

Two special cases of the Pascal distribution are of interest.

The first of these occurs when $r > 0$ and is not necessarily an integer. The resulting distribution is called the negative binomial distribution. The negative binomial distribution, like the Poisson distribution, is sometimes useful as the underlying statistical model for various types of "count" data, such as the occurrence of nonconformities in a unit of product. There is an important duality between the binomial and negative binomial distributions. In the binomial distribution, we fix the sample size (number of Bernoulli trials) and observe the number of successes; in the negative binomial distribution, we fix the number of successes and observe the sample size (number of Bernoulli trials) required to achieve them. This concept is particularly important in various kinds of sampling problems.

The other special case of the Pascal distribution is if $r = 1$, in which case we have the geometric distribution. It is the distribution of the number of Bernoulli trials until the first success.

A.2 Important Continuous Distributions

The exponential and Weibull distributions are often used in reliability testing. These distributions are described below:

A.2.1 THE EXPONENTIAL DISTRIBUTION

The exponential distribution is widely used in the field of reliability engineering as a model of the time to failure of a component or system. In these applications, the parameter λ is called the failure rate of the system, and the mean of the distribution $1/\lambda$ is called the mean time to failure. The exponential distribution is described as follows:

The **exponential distribution** is

$$f(x) = \lambda e^{-\lambda x} \quad x \geq 0$$

where $\lambda > 0$ is a constant. The ***mean*** and ***variance*** of the exponential distribution are

$$\mu = \frac{1}{\lambda}$$

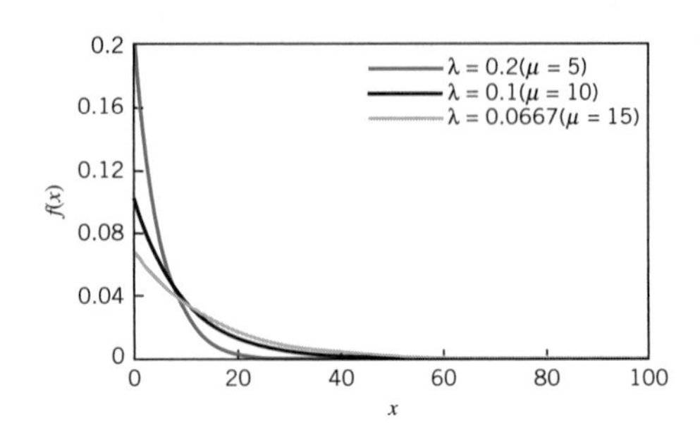

Figure A1-6 Exponential Distributions for Selected Values of λ

and

$$\sigma^2 = \frac{1}{\lambda^2}$$

respectively.

Figure A1-6 illustrates an exponential distribution.

The exponential distribution is useful for modeling the lifetime of electronic components, which have a tendency to fail early in life. For example, suppose that an electronic component in an airborne radar system has a useful life described by an exponential distribution with failure rate 10^{-4}/h; that is, $\lambda = 10^{-4}$. The mean time to failure for this component is $1/\lambda = 10{,}000$ hours.

If we wanted to determine the probability that this component would fail before its expected life, we would evaluate

$$P\left\{x \leq \frac{1}{\lambda}\right\} = \int_0^{1/\lambda} \lambda e^{-\lambda t} dt = 1 - e^{-1} = 0.63212$$

This result holds regardless of the value of λ, that is, the probability that a value of an exponential random variable will be less than its mean is 0.63212. This happens, of course, because the distribution is not symmetric.

Computer packages can do the integration work for you. Below is a screenshot of SPC XL's solution to this example:

You may have noticed that the Poisson and exponential distributions have the same parameter λ. This is because there is a relationship between these two distributions such that if the number of occurrences of an event has a Poisson distribution with parameter λ, then the distribution of the interval between occurrences is exponential with parameter $1/\lambda$.

Exponential Distribution

User defined parameters	
Lambda	0.0001
x_1 (lower limit)	0
x_2 (upper limit)	10000
Probabilities	
P(X <= 0)	0
P(X >= 0)	1
P(X <= 10000)	0.632120559
P(X >= 10000)	0.367879441
P(0 <= X <= 10000)	0.632120559

A.2.2 THE WEIBULL DISTRIBUTION

The Weibull distribution is defined as follows:

The **Weibull distribution** is

$$f(x) = \frac{\beta}{\theta}\left(\frac{x}{\theta}\right)^{\beta-1} exp\left[-\left(\frac{x}{\theta}\right)^{\beta}\right]$$

$$x \geq 0$$

where $\theta > 0$ is the **scale parameter**, and $\beta > 0$ is the **shape parameter**. The **mean** and **variance** of the Weibull distribution are

$$\mu = \theta\Gamma\left(1 + \frac{1}{\beta}\right)$$

and

$$\sigma^2 + \theta^2\left[\Gamma\left(1 + \frac{2}{\beta}\right) - \left\{\Gamma\left(1 + \frac{1}{\beta}\right)\right\}^2\right]$$

respectively.

The Weibull distribution is very flexible and by appropriate selection of the parameters the distribution can assume a wide variety of shapes. Several Weibull distributions are shown in A1-7. Note that when β=1 the Weibull distribution reduces to the exponential distribution with mean $1/\theta$.

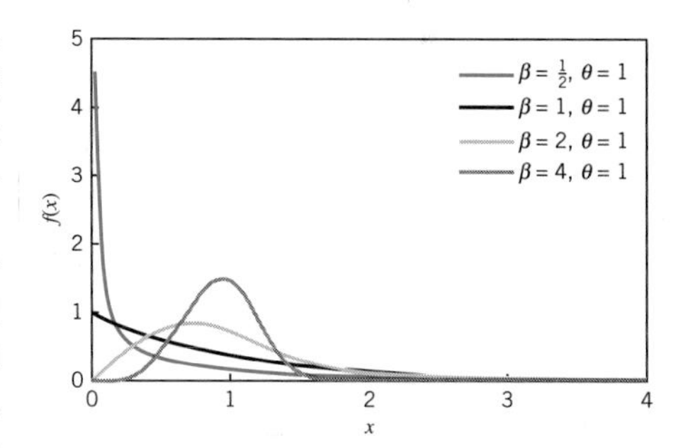

Figure A1-7 Weibull Distributions with Various Parameters of β and $\theta = 1$

The Weibull distribution has been used extensively in reliability engineering as a model of time to failure for electrical and mechanical components and systems. Examples of situations in which the Weibull has been used include electronic devices such as memory elements, mechanical components such as bearings, and structural elements in aircraft and automobiles.

The probabilities from a Weibull distribution can be calculated using statistical software packages. Consider the following example with output from SPC XL.

The time to failure for an electronic component used in a flat panel display unit is satisfactorily modeled by a Weibull distribution with $\beta = \frac{1}{2}$ and $\theta = 5000$. Find the mean time to failure and the fraction of components that are expected to survive beyond 20,000 hours.

Weibull Distribution

User-defined parameters	
Alpha	0.5
Beta	5000
x	20000
Probabilities	
P(X <= 20000)	0.864664717
P(X >= 20000)	0.135335283

Note that the parameters alpha and beta are inverted from the typical definition of the Weibull distribution for this particular software package.

Appendix Two

Cumulative Standard Normal Distribution

$$\phi(z) = \int_{-\infty}^{z} \frac{1}{\sqrt{2\pi}} e^{-x^2/2\,dx}$$

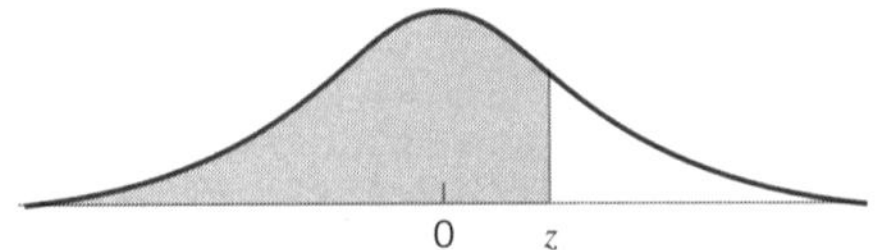

z	Φ(z)	z	Φ(z)	z	Φ(z)	z	Φ(z)	z	Φ(z)
−4.00	0.00003	−2.28	0.01130	−1.71	0.04363	−1.14	0.12714	−0.57	0.28434
−3.90	0.00005	−2.27	0.01160	−1.70	0.04457	−1.13	0.12924	−0.56	0.28774
−3.80	0.00007	−2.26	0.01191	−1.69	0.04551	−1.12	0.13136	−0.55	0.29116
−3.70	0.00011	−2.25	0.01222	−1.68	0.04648	−1.11	0.13350	−0.54	0.29460
−3.60	0.00016	−2.24	0.01255	−1.67	0.04746	−1.10	0.13567	−0.53	0.29806
−3.50	0.00023	−2.23	0.01287	−1.66	0.04846	−1.09	0.13786	−0.52	0.30153
−3.40	0.00034	−2.22	0.01321	−1.65	0.04947	−1.08	0.14007	−0.51	0.30503
−3.30	0.00048	−2.21	0.01355	−1.64	0.05050	−1.07	0.14231	−0.50	0.30854
−3.20	0.00069	−2.20	0.01390	−1.63	0.05155	−1.06	0.14457	−0.49	0.31207
−3.10	0.00097	−2.19	0.01426	−1.62	0.05262	−1.05	0.14686	−0.48	0.31561
−3.00	0.00135	−2.18	0.01463	−1.61	0.05370	−1.04	0.14917	−0.47	0.31918
−2.98	0.00144	−2.17	0.01500	−1.60	0.05480	−1.03	0.15151	−0.46	0.32276
−2.96	0.00154	−2.16	0.01539	−1.59	0.05592	−1.02	0.15386	−0.45	0.32636
−2.94	0.00164	−2.15	0.01578	−1.58	0.05705	−1.01	0.15625	−0.44	0.32997
−2.92	0.00175	−2.14	0.01618	−1.57	0.05821	−1,00	0.15866	−0.43	0.33360
−2.90	0.00187	−2.13	0.01659	−1.56	0.05938	−0.99	0.16109	−0.42	0.33724
−2.88	0.00199	−2.12	0.01700	−1.55	0.06057	−0.98	0.16354	−0.41	0.34090
−2.86	0.00212	−2.11	0.01743	−1.54	0.06178	−0.97	0.16602	−0.40	0.34458
−2.84	0.00226	−2.10	0.01786	−1.53	0.06301	−0.96	0.16853	−0.39	0.34827
−2.82	0.00240	−2.09	0.01831	−1.52	0.06426	−0.95	0.17106	−0.38	0.35197
−2.80	0.00256	−2.08	0.01876	−1.51	0.06552	−0.94	0.17361	−0.37	0.35569
−2.78	0.00272	−2.07	0.01923	−1.50	0.06681	−0.93	0.17619	−0.36	0.35942
−2.76	0.00289	−2.06	0.01970	−1.49	0.06811	−0.92	0.17879	−0.35	0.36317
−2.74	0.00307	−2.05	0.02018	−1.48	0.06944	−0.91	0.18141	−0.34	0.36693
−2.72	0.00326	−2.04	0.02068	−1.47	0.07078	−0.90	0.18406	−0.33	0.37070
−2.70	0.00347	−2.03	0.02118	−1.46	0.07215	−0.89	0.18673	−0.32	0.37448
−2.68	0.00368	−2.02	0.02169	−1.45	0.07353	−0.88	0.18943	−0.31	0.37828
−2.66	0.00391	−2.01	0.02222	−1.44	0.07493	−0.87	0.19215	−0.30	0.38209
−2.64	0.00415	−2.00	0.02275	−1.43	0.07636	−0.86	0.19489	−0.29	0.38591
−2.62	0.00440	−1.99	0.02330	−1.42	0.07780	−0.85	0.19766	−0.28	0.38974
−2.60	0.00466	−1.98	0.02385	−1.41	0.07927	−0.84	0.20045	−0.27	0.39358
−2.58	0.00494	−1.97	0.02442	−1.40	0.08076	−0.83	0.20327	−0.26	0.39743
−2.56	0.00523	−1.96	0.02500	−1.39	0.08226	−0.82	0.20611	−0.25	0.40129
−2.54	0.00554	−1.95	0.02559	−1.38	0.08379	−0.81	0.20897	−0.24	0.40517
−2.52	0.00587	−1.94	0.02619	−1.37	0.08534	−0.80	0.21186	−0.23	0.40905
−2.50	0.00621	−1.93	0.02680	−1.36	0.08691	−0.79	0.21476	−0.22	0.41294
−2.49	0.00639	−1.92	0.02743	−1.35	0.08851	−0.78	0.21770	−0.21	0.41683
−2.48	0.00657	−1.91	0.02807	−1.34	0.09012	−0.77	0.22065	−0.20	0.42074
−2.47	0.00676	−1.90	0.02872	−1.33	0.09176	−0.76	0.22363	−0.19	0.42465
−2.46	0.00695	−1.89	0.02938	−1.32	0.09342	−0.75	0.22663	−0.18	0.42858
−2.45	0.00714	−1.88	0.03005	−1.31	0.09510	−0.74	0.22965	−0.17	0.43251
−2.44	0.00734	−1.87	0.03074	−1.30	0.09680	−0.73	0.23270	−0.16	0.43644
−2.43	0.00755	−1.86	0.03144	−1.29	0.09853	−0.72	0.23576	−0.15	0.44038
−2.42	0.00776	−1.85	0.03216	−1.28	0.10027	−0.71	0.23885	−0.14	0.44433
−2.41	0.00798	−1.84	0.03288	−1.27	0.10204	−0.70	0.24196	−0.13	0.44828
−2.40	0.00820	−1.83	0.03362	−1.26	0.10383	−0.69	0.24510	−0.12	0.45224
−2.39	0.00842	−1.82	0.03438	−1.25	0.10565	−0.68	0.24825	−0.11	0.45620
−2.38	0.00866	−1.81	0.03515	−1.24	0.10749	−0.67	0.25143	−0.1	0.46017
−2.37	0.00889	−1.80	0.03593	−1.23	0.10935	−0.66	0.25463	−0.09	0.46414
−2.36	0.00914	−1.79	0.03673	−1.22	0.11123	−0.65	0.25785	−0.08	0.46812
−2.35	0.00939	−1.78	0.03754	−1.21	0.11314	−0.64	0.26109	−0.07	0.47210

z	$\Phi(z)$	z	$\Phi(z)$	z	$\Phi(z)$	z	$\Phi(z)$	z	$\Phi(z)$
−2.34	0.00964	−1.77	0.03836	−1.20	0.11507	−0.63	0.26435	−0.06	0.47608
−2.33	0.00990	−1.76	0.03920	−1.19	0.11702	−0.62	0.26763	−0.05	0.48006
−2.32	0.01017	−1.75	0.04006	−1.18	0.11900	−0.61	0.27093	−0.04	0.48405
−2.31	0.01044	−1.74	0.04093	−1.17	0.12100	−0.60	0.27425	−0.03	0.48803
−2.30	0.01072	−1.73	0.04182	−1.16	0.12302	−0.59	0.27760	−0.02	0.49202
−2.29	0.01101	−1.72	0.04272	−1.15	0.12507	−0.58	0.28096	−0.01	0.49601
0.00	0.50000	0.57	0.71566	1.14	0.87286	1.71	0.95637	2.28	0.98870
0.01	0.50399	0.58	0.71904	1.15	0.87493	1.72	0.95728	2.29	0.98899
0.02	0.50798	0.59	0.72240	1.16	0.87698	1.73	0.95818	2.30	0.98928
0.03	0.51197	0.6	0.72575	1.17	0.87900	1.74	0.95907	2.31	0.98956
0.04	0.51595	0.61	0.72907	1.18	0.88100	1.75	0.95994	2.32	0.98983
0.05	0.51994	0.62	0.73237	1.19	0.88298	1.76	0.96080	2.33	0.99010
0.06	0.52392	0.63	0.73565	1.20	0.88493	1.77	0.96164	2.34	0.99036
0.07	0.52790	0.64	0.73891	1.21	0.88686	1.78	0.96246	2.35	0.99061
0.08	0.53188	0.65	0.74215	1.22	0.88877	1.79	0.96327	2.36	0.99086
0.09	0.53586	0.66	0.74537	1.23	0.89065	1.80	0.96407	2.37	0.99111
0.10	0.53983	0.67	0.74857	1.24	0.89251	1.81	0.96485	2.38	0.99134
0.11	0.54380	0.68	0.75175	1.25	0.89435	1.82	0.96562	2.39	0.99158
0.12	0.54776	0.69	0.75490	1.26	0.89617	1.83	0.96638	2.40	0.99180
0.13	0.55172	0.70	0.75804	1.27	0.89796	1.84	0.96712	2.41	0.99202
0.14	0.55567	0.71	0.76115	1.28	0.89973	1.85	0.96784	2.42	0.99224
0.15	0.55962	0.72	0.76424	1.29	0.90147	1.86	0.96856	2.43	0.99245
0.16	0.56356	0.73	0.76730	1.30	0.90320	1.87	0.96926	2.44	0.99266
0.17	0.56749	0.74	0.77035	1.31	0.90490	1.88	0.96995	2.45	0.99286
0.18	0.57142	0.75	0.77337	1.32	0.90658	1.89	0.97062	2.46	0.99305
0.19	0.57535	0.76	0.77637	1.33	0.90824	1.90	0.97128	2.47	0.99324
0.20	0.57926	0.77	0.77935	1.34	0.90988	1.91	0.97193	2.48	0.99343
0.21	0.58317	0.78	0.78230	1.35	0.91149	1.92	0.97257	2.49	0.99361
0.22	0.58706	0.79	0.78524	1.36	0.91309	1.93	0.97320	2.50	0.99379
0.23	0.59095	0.80	0.78814	1.37	0.91466	1.94	0.97381	2.52	0.99413
0.24	0.59483	0.81	0.79103	1.38	0.91621	1.95	0.97441	2.54	0.99446
0.25	0.59871	0.82	0.79389	1.39	0.91774	1.96	0.97500	2.56	0.99477
0.26	0.60257	0.83	0.79673	1.40	0.91924	1.97	0.97558	2.58	0.99506
0.27	0.60642	0.84	0.79955	1.41	0.92073	1.98	0.97615	2.60	0.99534
0.28	0.61026	0.85	0.80234	1.42	0.92220	1.99	0.97670	2.62	0.99560
0.29	0.61409	0.86	0.80511	1.43	0.92364	2.00	0.97725	2.64	0.99585
0.30	0.61791	0.87	0.80785	1.44	0.92507	2.01	0.97778	2.66	0.99609
0.31	0.62172	0.88	0.81057	1.45	0.92647	2.02	0.97831	2.68	0.99632
0.32	0.62552	0.89	0.81327	1.46	0.92785	2.03	0.97882	2.70	0.99653
0.33	0.62930	0.90	0.81594	1.47	0.92922	2.04	0.97932	2.72	0.99674
0.34	0.63307	0.91	0.81859	1.48	0.93056	2.05	0.97982	2.74	0.99693
0.35	0.63683	0.92	0.82121	1.49	0.93189	2.06	0.98030	2.76	0.99711
0.36	0.64058	0.93	0.82381	1.50	0.93319	2.07	0.98077	2.78	0.99728
0.37	0.64431	0.94	0.82639	1.51	0.93448	2.08	0.98124	2.80	0.99744
0.38	0.64803	0.95	0.82894	1.52	0.93574	2.09	0.98169	2.82	0.99760
0.39	0.65173	0.96	0.83147	1.53	0.93699	2.10	0.98214	2.84	0.99774
0.40	0.65542	0.97	0.83398	1.54	0.93822	2.11	0.98257	2.86	0.99788
0.41	0.65910	0.98	0.83646	1.55	0.93943	2.12	0.98300	2.88	0.99801
0.42	0.66276	0.99	0.83891	1.56	0.94062	2.13	0.98341	2.90	0.99813
0.43	0.66640	1.00	0.84134	1.57	0.94179	2.14	0.98382	2.92	0.99825
0.44	0.67003	1.01	0.84375	1.58	0.94295	2.15	0.98422	2.94	0.99836
0.45	0.67364	1.02	0.84614	1.59	0.94408	2.16	0.98461	2.96	0.99846
0.46	0.67724	1.03	0.84849	1.60	0.94520	2.17	0.98500	2.98	0.99856
0.47	0.68082	1.04	0.85083	1.61	0.94630	2.18	0.98537	3.00	0.99865
0.48	0.68439	1.05	0.85314	1.62	0.94738	2.19	0.98574	3.10	0.99903
0.49	0.68793	1.06	0.85543	1.63	0.94845	2.20	0.98610	3.20	0.99931
0.50	0.69146	1.07	0.85769	1.64	0.94950	2.21	0.98645	3.30	0.99952
0.51	0.69497	1.08	0.85993	1.65	0.95053	2.22	0.98679	3.40	0.99966
0.52	0.69847	1.09	0.86214	1.66	0.95154	2.23	0.98713	3.50	0.99977
0.53	0.70194	1.10	0.86433	1.67	0.95254	2.24	0.98745	3.60	0.99984
0.54	0.70540	1.11	0.86650	1.68	0.95352	2.25	0.98778	3.70	0.99989
0.55	0.70884	1.12	0.86864	1.69	0.95449	2.26	0.98809	3.80	0.99993
0.56	0.71226	1.13	0.87076	1.70	0.95543	2.27	0.98840	3.90	0.99995
								4.00	0.99997

Appendix Three

Percentage Points of the χ^2 Distributions[a]

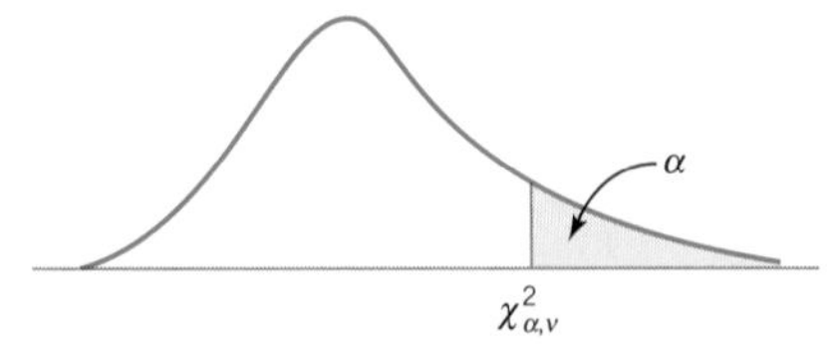

	α								
v	**0.995**	**0.990**	**0.975**	**0.950**	**0.500**	**0.050**	**0.025**	**0.010**	**0.005**
1	0.00+	0.00+	0.00+	0.00+	0.45	3.84	5.02	6.63	7.88
2	0.01	0.02	0.05	0.10	1.39	5.99	7.38	9.21	10.60
3	0.07	0.11	0.22	0.35	2.37	7.81	9.35	11.34	12.84
4	0.21	0.30	0.48	0.71	3.36	9.49	11.14	13.28	14.86
5	0.41	0.55	0.83	1.15	4.35	11.07	12.38	15.09	16.75
6	0.68	0.87	1.24	1.64	5.35	12.59	14.45	16.81	18.55
7	0.99	1.24	1.69	2.17	6.35	14.07	16.01	18.48	20.28
8	1.34	1.65	2.18	2.73	7.34	15.51	17.53	20.09	21.96
9	1.73	2.09	2.70	3.33	8.34	16.92	19.02	21.67	23.59
10	2.16	2.56	3.25	3.94	9.34	18.31	20.48	23.21	25.19
11	2.60	3.05	3.82	4.57	10.34	19.68	21.92	24.72	26.76
12	3.07	3.57	4.40	5.23	11.34	21.03	23.34	26.22	28.30
13	3.57	4.11	5.01	5.89	12.34	22.36	24.74	27.69	29.82
14	4.07	4.66	5.63	6.57	13.34	23.68	26.12	29.14	31.32
15	4.60	5.23	6.27	7.26	14.34	25.00	27.49	30.58	32.80
16	5.14	5.81	6.91	7.96	15.34	26.30	28.85	32.00	34.27
17	5.70	6.41	7.56	8.67	16.34	27.59	30.19	33.41	35.72
18	6.26	7.01	8.23	9.39	17.34	28.87	31.53	34.81	37.16
19	6.88	7.63	8.91	10.12	18.34	30.14	32.85	36.19	38.58
20	7.43	8.26	9.59	10.85	19.34	31.41	34.17	37.57	40.00
25	10.52	11.52	13.12	14.61	24.34	37.65	40.65	44.31	46.93
30	13.79	14.95	16.79	18.49	29.34	43.77	46.98	50.89	53.67
40	20.71	22.16	24.43	26.51	39.34	55.76	59.34	63.69	66.77
50	27.99	29.71	32.36	34.76	49.33	67.50	71.42	76.15	79.49
60	35.53	37.48	40.48	43.19	59.33	79.08	83.30	88.38	91.95
70	43.28	45.44	48.76	51.74	69.33	90.53	95.02	100.42	104.22
80	51.17	53.54	57.15	60.39	79.33	101.88	106.63	112.33	116.32
90	59.20	61.75	65.65	69.13	89.33	113.14	118.14	124.12	128.30
100	67.33	70.06	74.22	77.93	99.33	124.34	129.56	135.81	140.17

v = degrees of freedom.

[a]Adapted with permission from *Biometrika Tables for Statisticians,* Vol. 1, 3rd ed., by E. S. Pearson and H. O. Hartley, Cambridge University Press, Cambridge, UK, 1966.

Appendix Four

Percentage Points of the *t* Distributions[a]

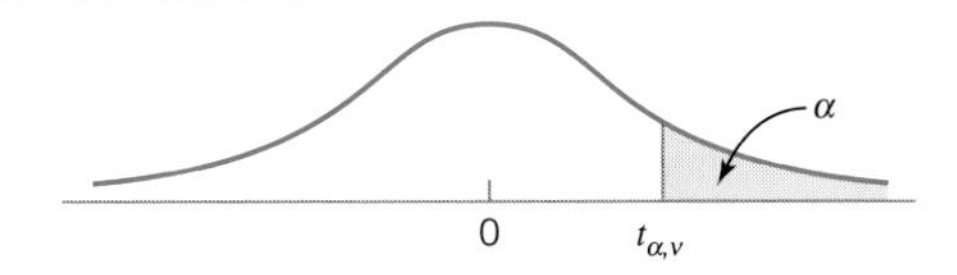

	α									
v	**0.40**	**0.25**	**0.10**	**0.05**	**0.025**	**0.01**	**0.005**	**0.0025**	**0.001**	**0.0005**
1	0.325	1.000	3.078	6.314	12.706	31.821	63.657	127.32	318.31	636.62
2	0.289	0.816	1.886	2.920	4.303	6.965	9.925	14.089	23.326	31.598
3	0.277	0.765	1.638	2.353	3.182	4.541	5.841	7.453	10.213	12.924
4	0.271	0.741	1.533	2.132	2.776	3.747	4.604	5.598	7.173	8.610
5	0.267	0.727	1.476	2.015	2.571	3.365	4.032	4.773	5.893	6.869
6	0.265	0.727	1.440	1.943	2.447	3.143	3.707	4.317	5.208	5.959
7	0.263	0.711	1.415	1.895	2.365	2.998	3.49	4.019	4.785	5.408
8	0.262	0.706	1.397	1.860	2.306	2.896	3.355	3.833	4.501	5.041
9	0.261	0.703	1.383	1.833	2.262	2.821	3.250	3.690	4.297	4.781
10	0.260	0.700	1.372	1.812	2.228	2.764	3.169	3.581	4.144	4.587
11	0.260	0.697	1.363	1.796	2.20	2.718	3.106	3.497	4.025	4.437
12	0.259	0.695	1.356	1.782	2.179	2.681	3.055	3.428	3.930	4.318
13	0.259	0.694	1.350	1.771	2.160	2.650	3.012	3.372	3.852	4.221
14	0.258	0.692	1.345	1.761	2.145	2.624	2.977	3.326	3.787	4.140
15	0.258	0.691	1.341	1.753	2.131	2.602	2.947	3.286	3.733	4.073
16	0.258	0.690	1.337	1.746	2.120	2.583	2.921	3.252	3.686	4.015
17	0.257	0.689	1.333	1.740	2.110	2.567	2.898	3.222	3.646	3.965
18	0.257	0.688	1.330	1.734	2.101	2.552	2.878	3.197	3.610	3.992
19	0.257	0.688	1.328	1.729	2.093	2.539	2.861	3.174	3.579	3.883
20	0.257	0.687	1.325	1.725	2.086	2.528	2.845	3.153	3.552	3.850
21	0.257	0.686	1.323	1.721	2.080	2.518	2.831	3.135	3.527	3.819
22	0.256	0.686	1.321	1.717	2.074	2.508	2.819	3.119	3.505	3.792
23	0.256	0.685	1.319	1.714	2.069	2.500	2.807	3.104	3.485	3.767
24	0.256	0.685	1.318	1.711	2.064	2.492	2.797	3.091	3.467	3.745
25	0.256	0.684	1.316	1.708	2.060	2.485	2.787	3.078	3.450	3.725
26	0.256	0.684	1.315	1.706	2.056	2.479	2.779	3.067	3.435	3.707
27	0.256	0.684	1.314	1.703	2.052	2.473	2.771	3.057	3.421	3.690
28	0.256	0.683	1.313	1.701	2.048	2.467	2.763	3.047	3.408	3.674
29	0.256	0.683	1.311	1.699	2.045	2.462	2.756	3.038	3.396	3.659
30	0.256	0.683	1.310	1.697	2.042	2.457	2.750	3.030	3.385	3.646
40	0.255	0.681	1.303	1.684	2.021	2.423	2.704	2.971	3.307	3.551
60	0.254	0.679	1.296	1.671	2.000	2.390	2.660	2.915	3.232	3.460
120	0.254	0.677	1.289	1.658	1.980	2.358	2.617	2.860	3.160	3.373
∞	0.253	0.674	1.282	1.645	1.960	2.326	2.576	2.807	3.090	3.291

v = degrees of freedom.

[a]Adapted with permission from *Biometrika Tables for Statisticians,* Vol. 1, 3rd ed., by E. S. Pearson and H. O. Hartley, Cambridge University Press, Cambridge, UK, 1966.

Appendix Five

Percentage Points of the *F* Distributions

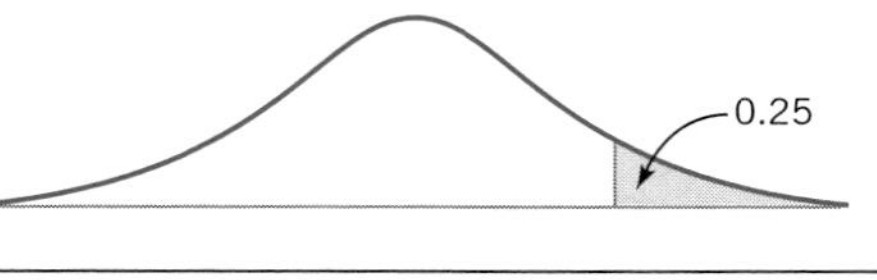

$F_{0.25, v_1, v_2}$

Degrees of freedom for the numerator (v_1); Degrees of freedom for the denominator (v_2)

v_2 \ v_1	1	2	3	4	5	6	7	8	9	10	12	15	20	24	30	40	60	120	∞
1	5.83	7.50	8.20	8.58	8.82	8.98	9.10	9.19	9.26	9.32	9.41	9.49	9.58	9.63	9.67	9.71	9.76	9.80	9.85
2	2.57	3.00	3.15	3.23	3.28	3.31	3.34	3.35	3.37	3.38	3.39	3.41	3.43	3.43	3.44	3.45	3.46	3.47	3.48
3	2.02	2.28	2.36	2.39	2.41	2.42	2.43	2.44	2.44	2.44	2.45	2.46	2.46	2.47	2.47	2.47	2.47	2.47	2.47
4	1.81	2.00	2.05	2.06	2.07	2.08	2.08	2.08	2.08	2.08	2.08	2.08	2.08	2.08	2.08	2.08	2.08	2.08	2.08
5	1.69	1.85	1.88	1.89	1.89	1.89	1.89	1.89	1.89	1.89	1.89	1.89	1.88	1.88	1.88	1.88	1.87	1.87	1.87
6	1.62	1.76	1.78	1.79	1.79	1.78	1.78	1.78	1.77	1.77	1.77	1.76	1.76	1.75	1.75	1.75	1.74	1.74	1.74
7	1.57	1.70	1.72	1.72	1.71	1.71	1.70	1.70	1.70	1.69	1.68	1.68	1.67	1.67	1.66	1.66	1.65	1.65	1.65
8	1.54	1.66	1.67	1.66	1.66	1.65	1.64	1.64	1.63	1.63	1.62	1.62	1.61	1.60	1.60	1.59	1.59	1.58	1.58
9	1.51	1.62	1.63	1.63	1.62	1.61	1.60	1.60	1.59	1.59	1.58	1.57	1.56	1.56	1.55	1.54	1.54	1.53	1.53
10	1.49	1.60	1.60	1.59	1.59	1.58	1.57	1.56	1.56	1.55	1.54	1.53	1.52	1.52	1.51	1.51	1.50	1.49	1.48
11	1.47	1.58	1.58	1.57	1.56	1.55	1.54	1.53	1.53	1.52	1.51	1.50	1.49	1.49	1.48	1.47	1.47	1.46	1.45
12	1.46	1.56	1.56	1.55	1.54	1.53	1.52	1.51	1.51	1.50	1.49	1.48	1.47	1.46	1.45	1.45	1.44	1.43	1.42
13	1.45	1.55	1.55	1.53	1.52	1.51	1.50	1.49	1.49	1.48	1.47	1.46	1.45	1.44	1.43	1.42	1.42	1.41	1.40
14	1.44	1.53	1.53	1.52	1.51	1.50	1.49	1.48	1.47	1.46	1.45	1.44	1.43	1.42	1.41	1.41	1.40	1.39	1.38
15	1.43	1.52	1.52	1.51	1.49	1.48	1.47	1.46	1.46	1.45	1.44	1.43	1.41	1.41	1.40	1.39	1.38	1.37	1.36
16	1.42	1.51	1.51	1.50	1.48	1.47	1.46	1.45	1.44	1.44	1.43	1.41	1.40	1.39	1.38	1.37	1.36	1.35	1.34
17	1.42	1.51	1.50	1.49	1.47	1.46	1.45	1.44	1.43	1.43	1.41	1.40	1.39	1.38	1.37	1.36	1.35	1.34	1.33
18	1.41	1.50	1.49	1.48	1.46	1.45	1.44	1.43	1.42	1.42	1.40	1.39	1.38	1.37	1.36	1.35	1.34	1.33	1.32
19	1.41	1.49	1.49	1.47	1.46	1.44	1.43	1.42	1.41	1.41	1.40	1.38	1.37	1.36	1.35	1.34	1.33	1.32	1.30
20	1.40	1.49	1.48	1.47	1.45	1.44	1.43	1.42	1.41	1.40	1.39	1.37	1.36	1.35	1.34	1.33	1.32	1.31	1.29
21	1.40	1.48	1.48	1.46	1.44	1.43	1.42	1.41	1.40	1.39	1.38	1.37	1.35	1.34	1.33	1.32	1.31	1.30	1.28
22	1.40	1.48	1.47	1.45	1.44	1.42	1.41	1.40	1.39	1.39	1.37	1.36	1.34	1.33	1.32	1.31	1.30	1.29	1.28
23	1.39	1.47	1.47	1.45	1.43	1.42	1.41	1.40	1.39	1.38	1.37	1.35	1.34	1.33	1.32	1.31	1.30	1.28	1.27
24	1.39	1.47	1.46	1.44	1.43	1.41	1.40	1.39	1.38	1.38	1.36	1.35	1.33	1.32	1.31	1.30	1.29	1.28	1.26
25	1.39	1.47	1.46	1.44	1.42	1.41	1.40	1.39	1.38	1.37	1.36	1.34	1.33	1.32	1.31	1.29	1.28	1.27	1.25
26	1.38	1.46	1.45	1.44	1.42	1.41	1.39	1.38	1.37	1.37	1.35	1.34	1.32	1.31	1.30	1.29	1.28	1.26	1.25
27	1.38	1.46	1.45	1.43	1.42	1.40	1.39	1.38	1.37	1.36	1.35	1.33	1.32	1.31	1.30	1.28	1.27	1.26	1.24
28	1.38	1.46	1.45	1.43	1.41	1.40	1.39	1.38	1.37	1.36	1.34	1.33	1.31	1.30	1.29	1.28	1.27	1.25	1.24
29	1.38	1.45	1.45	1.43	1.41	1.40	1.38	1.37	1.36	1.35	1.34	1.32	1.31	1.30	1.29	1.27	1.26	1.25	1.23
30	1.38	1.45	1.44	1.42	1.41	1.39	1.38	1.37	1.36	1.35	1.34	1.32	1.30	1.29	1.28	1.27	1.26	1.24	1.23
40	1.36	1.44	1.42	1.40	1.39	1.37	1.36	1.35	1.34	1.33	1.31	1.30	1.28	1.26	1.25	1.24	1.22	1.21	1.19
60	1.35	1.42	1.41	1.38	1.37	1.35	1.33	1.32	1.31	1.30	1.29	1.27	1.25	1.24	1.22	1.21	1.19	1.17	1.15
120	1.34	1.40	1.39	1.37	1.35	1.33	1.31	1.30	1.29	1.28	1.26	1.24	1.22	1.21	1.19	1.18	1.16	1.13	1.10
∞	1.32	1.39	1.37	1.35	1.33	1.31	1.29	1.28	1.27	1.25	1.24	1.22	1.19	1.18	1.16	1.14	1.12	1.08	1.00

Note: $F_{0.75, v_1, v_2} = 1/F_{0.25, v_2, v_1}$.

Source: Adapted with permission from *Biometrika Tables for Statisticians,* Vol. 1, 3rd ed., by E. S. Pearson and H. O. Hartley, Cambridge University Press, Cambridge, UK, 1966.

Percentage Points of the *F* Distributions (*Continued*)

$F_{0.10, v_1, v_2}$

Degrees of freedom for the numerator (n_1)

v_2 \ v_1	1	2	3	4	5	6	7	8	9	10	12	15	20	24	30	40	60	120	∞
1	39.86	49.50	53.59	55.83	57.24	58.20	58.91	59.44	59.86	60.19	60.71	61.22	61.74	62.00	62.26	62.53	62.79	63.06	63.33
2	8.53	9.00	9.16	9.24	9.29	9.33	9.35	9.37	9.38	9.39	9.41	9.42	9.44	9.45	9.46	9.47	9.47	9.48	9.49
3	5.54	5.46	5.39	5.34	5.31	5.28	5.27	5.25	5.24	5.23	5.22	5.20	5.18	5.18	5.17	5.16	5.15	5.14	5.13
4	4.54	4.32	4.19	4.11	4.05	4.01	3.98	3.95	3.94	3.92	3.90	3.87	3.84	3.83	3.82	3.80	3.79	3.78	3.76
5	4.06	3.78	3.62	3.52	3.45	3.40	3.37	3.34	3.32	3.30	3.27	3.24	3.21	3.19	3.17	3.16	3.14	3.12	3.10
6	3.78	3.46	3.29	3.18	3.11	3.05	3.01	2.98	2.96	2.94	2.90	2.87	2.84	2.82	2.80	2.78	2.76	2.74	2.72
7	3.59	3.26	3.07	2.96	2.88	2.83	2.78	2.75	2.72	2.70	2.67	2.63	2.59	2.58	2.56	2.54	2.51	2.49	2.47
8	3.46	3.11	2.92	2.81	2.73	2.67	2.62	2.59	2.56	2.54	2.50	2.46	2.42	2.40	2.38	2.36	2.34	2.32	2.29
9	3.36	3.01	2.81	2.69	2.61	2.55	2.51	2.47	2.44	2.42	2.38	2.34	2.30	2.28	2.25	2.23	2.21	2.18	2.16
10	3.29	2.92	2.73	2.61	2.52	2.46	2.41	2.38	2.35	2.32	2.28	2.24	2.20	2.18	2.16	2.13	2.11	2.08	2.06
11	3.23	2.86	2.66	2.54	2.45	2.39	2.34	2.30	2.27	2.25	2.21	2.17	2.12	2.10	2.08	2.05	2.03	2.00	1.97
12	3.18	2.81	2.61	2.48	2.39	2.33	2.28	2.24	2.21	2.19	2.15	2.10	2.06	2.04	2.01	1.99	1.96	1.93	1.90
13	3.14	2.76	2.56	2.43	2.35	2.28	2.23	2.20	2.16	2.14	2.10	2.05	2.01	1.98	1.96	1.93	1.90	1.88	1.85
14	3.10	2.73	2.52	2.39	2.31	2.24	2.19	2.15	2.12	2.10	2.05	2.01	1.96	1.94	1.91	1.89	1.86	1.83	1.80
15	3.07	2.70	2.49	2.36	2.27	2.21	2.16	2.12	2.09	2.06	2.02	1.97	1.92	1.90	1.87	1.85	1.82	1.79	1.76
16	3.05	2.67	2.46	2.33	2.24	2.18	2.13	2.09	2.06	2.03	1.99	1.94	1.89	1.86	1.84	1.81	1.78	1.75	1.72
17	3.03	2.64	2.44	2.31	2.22	2.15	2.10	2.06	2.03	2.00	1.96	1.91	1.86	1.84	1.81	1.78	1.75	1.72	1.69
18	3.01	2.62	2.42	2.29	2.20	2.13	2.08	2.04	2.00	1.98	1.93	1.89	1.84	1.81	1.78	1.75	1.72	1.69	1.66
19	2.99	2.61	2.40	2.27	2.18	2.11	2.06	2.02	1.98	1.96	1.91	1.86	1.81	1.79	1.76	1.73	1.70	1.67	1.63
20	2.97	2.59	2.38	2.25	2.16	2.09	2.04	2.00	1.96	1.94	1.89	1.84	1.79	1.77	1.74	1.71	1.68	1.64	1.61
21	2.96	2.57	2.36	2.23	2.14	2.08	2.02	1.98	1.95	1.92	1.87	1.83	1.78	1.75	1.72	1.69	1.66	1.62	1.59
22	2.95	2.56	2.35	2.22	2.13	2.06	2.01	1.97	1.93	1.90	1.86	1.81	1.76	1.73	1.70	1.67	1.64	1.60	1.57
23	2.94	2.55	2.34	2.21	2.11	2.05	1.99	1.95	1.92	1.89	1.84	1.80	1.74	1.72	1.69	1.66	1.62	1.59	1.55
24	2.93	2.54	2.33	2.19	2.10	2.04	1.98	1.94	1.91	1.88	1.83	1.78	1.73	1.70	1.67	1.64	1.61	1.57	1.53
25	2.92	2.53	2.32	2.18	2.09	2.02	1.97	1.93	1.89	1.87	1.82	1.77	1.72	1.69	1.66	1.63	1.59	1.56	1.52
26	2.91	2.52	2.31	2.17	2.08	2.01	1.96	1.92	1.88	1.86	1.81	1.76	1.71	1.68	1.65	1.61	1.58	1.54	1.50
27	2.90	2.51	2.30	2.17	2.07	2.00	1.95	1.91	1.87	1.85	1.80	1.75	1.70	1.67	1.64	1.60	1.57	1.53	1.49
28	2.89	2.50	2.29	2.16	2.06	2.00	1.94	1.90	1.87	1.84	1.79	1.74	1.69	1.66	1.63	1.59	1.56	1.52	1.48
29	2.89	2.50	2.28	2.15	2.06	1.99	1.93	1.89	1.86	1.83	1.78	1.73	1.68	1.65	1.62	1.58	1.55	1.51	1.47
30	2.88	2.49	2.28	2.14	2.03	1.98	1.93	1.88	1.85	1.82	1.77	1.72	1.67	1.64	1.61	1.57	1.54	1.50	1.46
40	2.84	2.44	2.23	2.09	2.00	1.93	1.87	1.83	1.79	1.76	1.71	1.66	1.61	1.57	1.54	1.51	1.47	1.42	1.38
60	2.79	2.39	2.18	2.04	1.95	1.87	1.82	1.77	1.74	1.71	1.66	1.60	1.54	1.51	1.48	1.44	1.40	1.35	1.29
120	2.75	2.35	2.13	1.99	1.90	1.82	1.77	1.72	1.68	1.65	1.60	1.55	1.48	1.45	1.41	1.37	1.32	1.26	1.19
∞	2.71	2.30	2.08	1.94	1.85	1.77	1.72	1.67	1.63	1.60	1.55	1.49	1.42	1.38	1.34	1.30	1.24	1.17	1.00

Degrees of freedom for the denominator (v_2)

Note: $F_{0.90, v_1, v_2} = 1/F_{0.10, v_2, v_1}$.

Percentage Points of the *F* Distributions (*Continued*)

$F_{0.05, v_1, v_2}$

Degrees of freedom for the numerator (v_1); Degrees of freedom for the denominator (v_2)

v_2 \ v_1	1	2	3	4	5	6	7	8	9	10	12	15	20	24	30	40	60	120	∞
1	161.4	199.5	215.7	224.6	230.2	234.0	236.8	238.9	240.5	241.9	243.9	245.9	248.0	249.1	250.1	251.1	252.2	253.3	254.3
2	18.51	19.00	19.16	19.25	19.30	19.33	19.35	19.37	19.38	19.40	19.41	19.43	19.45	19.45	19.46	19.47	19.48	19.49	19.50
3	10.13	9.55	9.28	9.12	9.01	8.94	8.89	8.85	8.81	8.79	8.74	8.70	8.66	8.64	8.62	8.59	8.57	8.55	8.53
4	7.71	6.94	6.59	6.39	6.26	6.16	6.09	6.04	6.00	5.96	5.91	5.86	5.80	5.77	5.75	5.72	5.69	5.66	5.63
5	6.61	5.79	5.41	5.19	5.05	4.95	4.88	4.82	4.77	4.74	4.68	4.62	4.56	4.53	4.50	4.46	4.43	4.40	4.36
6	5.99	5.14	4.76	4.53	4.39	4.28	4.21	4.15	4.10	4.06	4.00	3.94	3.87	3.84	3.81	3.77	3.74	3.70	3.67
7	5.59	4.74	4.35	4.12	3.97	3.87	3.79	3.73	3.68	3.64	3.57	3.51	3.44	3.41	3.38	3.34	3.30	3.27	3.23
8	5.32	4.46	4.07	3.84	3.69	3.58	3.50	3.44	3.39	3.35	3.28	3.22	3.15	3.12	3.08	3.04	3.01	2.97	2.93
9	5.12	4.26	3.86	3.63	3.48	3.37	3.29	3.23	3.18	3.14	3.07	3.01	2.94	2.90	2.86	2.83	2.79	2.75	2.71
10	4.96	4.10	3.71	3.48	3.33	3.22	3.14	3.07	3.02	2.98	2.91	2.85	2.77	2.74	2.70	2.66	2.62	2.58	2.54
11	4.84	3.98	3.59	3.36	3.20	3.09	3.01	2.95	2.90	2.85	2.79	2.72	2.65	2.61	2.57	2.53	2.49	2.45	2.40
12	4.75	3.89	3.49	3.26	3.11	3.00	2.91	2.85	2.80	2.75	2.69	2.62	2.54	2.51	2.47	2.43	2.38	2.34	2.30
13	4.67	3.81	3.41	3.18	3.03	2.92	2.83	2.77	2.71	2.67	2.60	2.53	2.46	2.42	2.38	2.34	2.30	2.25	2.21
14	4.60	3.74	3.34	3.11	2.96	2.85	2.76	2.70	2.65	2.60	2.53	2.46	2.39	2.35	2.31	2.27	2.22	2.18	2.13
15	4.54	3.68	3.29	3.06	2.90	2.79	2.71	2.64	2.59	2.54	2.48	2.40	2.33	2.29	2.25	2.20	2.16	2.11	2.07
16	4.49	3.63	3.24	3.01	2.85	2.74	2.66	2.59	2.54	2.49	2.42	2.35	2.28	2.24	2.19	2.15	2.11	2.06	2.01
17	4.45	3.59	3.20	2.96	2.81	2.70	2.61	2.55	2.49	2.45	2.38	2.31	2.23	2.19	2.15	2.10	2.06	2.01	1.96
18	4.41	3.55	3.16	2.93	2.77	2.66	2.58	2.51	2.46	2.41	2.34	2.27	2.19	2.15	2.11	2.06	2.02	1.97	1.92
19	4.38	3.52	3.13	2.90	2.74	2.63	2.54	2.48	2.42	2.38	2.31	2.23	2.16	2.11	2.07	2.03	1.98	1.93	1.88
20	4.35	3.49	3.10	2.87	2.71	2.60	2.51	2.45	2.39	2.35	2.28	2.20	2.12	2.08	2.04	1.99	1.95	1.90	1.84
21	4.32	3.47	3.07	2.84	2.68	2.57	2.49	2.42	2.37	2.32	2.25	2.18	2.10	2.05	2.01	1.96	1.92	1.87	1.81
22	4.30	3.44	3.05	2.82	2.66	2.55	2.46	2.40	2.34	2.30	2.23	2.15	2.07	2.03	1.98	1.94	1.89	1.84	1.78
23	4.28	3.42	3.03	2.80	2.64	2.53	2.44	2.37	2.32	2.27	2.20	2.13	2.05	2.01	1.96	1.91	1.86	1.81	1.76
24	4.26	3.40	3.01	2.78	2.62	2.51	2.42	2.36	2.30	2.25	2.18	2.11	2.03	1.98	1.94	1.89	1.84	1.79	1.73
25	4.24	3.39	2.99	2.76	2.60	2.49	2.40	2.34	2.28	2.24	2.16	2.09	2.01	1.96	1.92	1.87	1.82	1.77	1.71
26	4.23	3.37	2.98	2.74	2.59	2.47	2.39	2.32	2.27	2.22	2.15	2.07	1.99	1.95	1.90	1.85	1.80	1.75	1.69
27	4.21	3.35	2.96	2.73	2.57	2.46	2.37	2.31	2.25	2.20	2.13	2.06	1.97	1.93	1.88	1.84	1.79	1.73	1.67
28	4.20	3.34	2.95	2.71	2.56	2.45	2.36	2.29	2.24	2.19	2.12	2.04	1.96	1.91	1.87	1.82	1.77	1.71	1.65
29	4.18	3.33	2.93	2.70	2.55	2.43	2.35	2.28	2.22	2.18	2.10	2.03	1.94	1.90	1.85	1.81	1.75	1.70	1.64
30	4.17	3.32	2.92	2.69	2.53	2.42	2.33	2.27	2.21	2.16	2.09	2.01	1.93	1.89	1.84	1.79	1.74	1.68	1.62
40	4.08	3.23	2.84	2.61	2.45	2.34	2.25	2.18	2.12	2.08	2.00	1.92	1.84	1.79	1.74	1.69	1.64	1.58	1.51
60	4.00	3.15	2.76	2.53	2.37	2.25	2.17	2.10	2.04	1.99	1.92	1.84	1.75	1.70	1.65	1.59	1.53	1.47	1.39
120	3.92	3.07	2.68	2.45	2.29	2.17	2.09	2.02	1.96	1.91	1.83	1.75	1.66	1.61	1.55	1.55	1.43	1.35	1.25
∞	3.84	3.00	2.60	2.37	2.21	2.10	2.01	1.94	1.88	1.83	1.75	1.67	1.57	1.52	1.46	1.39	1.32	1.22	1.00

Note: $F_{0.95, v_1, v_2} = 1/F_{0.05, v_2, v_1}$.

0.025

Percentage Points of the F Distributions (*Continued*)

$F_{0.025, v_1, v_2}$

Degrees of freedom for the numerator (v_1); Degrees of freedom for the denominator (v_2)

v_2 \ v_1	1	2	3	4	5	6	7	8	9	10	12	15	20	24	30	40	60	120	∞
1	647.8	799.5	864.2	899.6	921.8	937.1	948.2	956.7	963.3	968.6	976.7	984.9	993.1	997.2	1001.0	1006.0	1010.0	1014.0	1018.0
2	38.51	39.00	39.17	39.25	39.30	39.33	39.36	39.37	39.39	39.40	39.41	39.43	39.45	39.46	39.46	39.47	39.48	39.49	39.50
3	17.44	16.04	15.44	15.10	14.88	14.73	14.62	14.54	14.47	14.42	14.34	14.25	14.17	14.12	14.08	14.04	13.99	13.95	13.90
4	12.22	10.65	9.98	9.60	9.36	9.20	9.07	8.98	8.90	8.84	8.75	8.66	8.56	8.51	8.46	8.41	8.36	8.31	8.26
5	10.01	8.43	7.76	7.39	7.15	6.98	6.85	6.76	6.68	6.62	6.52	6.43	6.33	6.28	6.23	6.18	6.12	6.07	6.02
6	8.81	7.26	6.60	6.23	5.99	5.82	5.70	5.60	5.52	5.46	5.37	5.27	5.17	5.12	5.07	5.01	4.96	4.90	4.85
7	8.07	6.54	5.89	5.52	5.29	5.12	4.99	4.90	4.82	4.76	4.67	4.57	4.47	4.42	4.36	4.31	4.25	4.20	4.14
8	7.57	6.06	5.42	5.05	4.82	4.65	4.53	4.43	4.36	4.30	4.20	4.10	4.00	3.95	3.89	3.84	3.78	3.73	3.67
9	7.21	5.71	5.08	4.72	4.48	4.32	4.20	4.10	4.03	3.96	3.87	3.77	3.67	3.61	3.56	3.51	3.45	3.39	3.33
10	6.94	5.46	4.83	4.47	4.24	4.07	3.95	3.85	3.78	3.72	3.62	3.52	3.42	3.37	3.31	3.26	3.20	3.14	3.08
11	6.72	5.26	4.63	4.28	4.04	3.88	3.76	3.66	3.59	3.53	3.43	3.33	3.23	3.17	3.12	3.06	3.00	2.94	2.88
12	6.55	5.10	4.47	4.12	3.89	3.73	3.61	3.51	3.44	3.37	3.28	3.18	3.07	3.02	2.96	2.91	2.85	2.79	2.72
13	6.41	4.97	4.35	4.00	3.77	3.60	3.48	3.39	3.31	3.25	3.15	3.05	2.95	2.89	2.84	2.78	2.72	2.66	2.60
14	6.30	4.86	4.24	3.89	3.66	3.50	3.38	3.29	3.21	3.15	3.05	2.95	2.84	2.79	2.73	2.67	2.61	2.55	2.49
15	6.20	4.77	4.15	3.80	3.58	3.41	3.29	3.20	3.12	3.06	2.96	2.86	2.76	2.70	2.64	2.59	2.52	2.46	2.40
16	6.12	4.69	4.08	3.73	3.50	3.34	3.22	3.12	3.05	2.99	2.89	2.79	2.68	2.63	2.57	2.51	2.45	2.38	2.32
17	6.04	4.62	4.01	3.66	3.44	3.28	3.16	3.06	2.98	2.92	2.82	2.72	2.62	2.56	2.50	2.44	2.38	2.32	2.25
18	5.98	4.56	3.95	3.61	3.38	3.22	3.10	3.01	2.93	2.87	2.77	2.67	2.56	2.50	2.44	2.38	2.32	2.26	2.19
19	5.92	4.51	3.90	3.56	3.33	3.17	3.05	2.96	2.88	2.82	2.72	2.62	2.51	2.45	2.39	2.33	2.27	2.20	2.13
20	5.87	4.46	3.86	3.51	3.29	3.13	3.01	2.91	2.84	2.77	2.68	2.57	2.46	2.41	2.35	2.29	2.22	2.16	2.09
21	5.83	4.42	3.82	3.48	3.25	3.09	2.97	2.87	2.80	2.73	2.64	2.53	2.42	2.37	2.31	2.25	2.18	2.11	2.04
22	5.79	4.38	3.78	3.44	3.22	3.05	2.93	2.84	2.76	2.70	2.60	2.50	2.39	2.33	2.27	2.21	2.14	2.08	2.00
23	5.75	4.35	3.75	3.41	3.18	3.02	2.90	2.81	2.73	2.67	2.57	2.47	2.36	2.30	2.24	2.18	2.11	2.04	1.97
24	5.72	4.32	3.72	3.38	3.15	2.99	2.87	2.78	2.70	2.64	2.54	2.44	2.33	2.27	2.21	2.15	2.08	2.01	1.94
25	5.69	4.29	3.69	3.35	3.13	2.97	2.85	2.75	2.68	2.61	2.51	2.41	2.30	2.24	2.18	2.12	2.05	1.98	1.91
26	5.66	4.27	3.67	3.33	3.10	2.94	2.82	2.73	2.65	2.59	2.49	2.39	2.28	2.22	2.16	2.09	2.03	1.95	1.88
27	5.63	4.24	3.65	3.31	3.08	2.92	2.80	2.71	2.63	2.57	2.47	2.36	2.25	2.19	2.13	2.07	2.00	1.93	1.85
28	5.61	4.22	3.63	3.29	3.06	2.90	2.78	2.69	2.61	2.55	2.45	2.34	2.23	2.17	2.11	2.05	1.98	1.91	1.83
29	5.59	4.20	3.61	3.27	3.04	2.88	2.76	2.67	2.59	2.53	2.43	2.32	2.21	2.15	2.09	2.03	1.96	1.89	1.81
30	5.57	4.18	3.59	3.25	3.03	2.87	2.75	2.65	2.57	2.51	2.41	2.31	2.20	2.14	2.07	2.01	1.94	1.87	1.79
40	5.42	4.05	3.46	3.13	2.90	2.74	2.62	2.53	2.45	2.39	2.29	2.18	2.07	2.01	1.94	1.88	1.80	1.72	1.64
60	5.29	3.93	3.34	3.01	2.79	2.63	2.51	2.41	2.33	2.27	2.17	2.06	1.94	1.88	1.82	1.74	1.67	1.58	1.48
120	5.15	3.80	3.23	2.89	2.67	2.52	2.39	2.30	2.22	2.16	2.05	1.94	1.82	1.76	1.69	1.61	1.53	1.43	1.31
∞	5.02	3.69	3.12	2.79	2.57	2.41	2.29	2.19	2.11	2.05	1.94	1.83	1.71	1.64	1.57	1.48	1.39	1.27	1.00

Note: $F_{0.975, v_1, v_2} = 1/F_{0.025, v_2, v_1}$.

Percentage Points of the *F* Distributions (*Continued*)

$F_{0.01, v_1, v_2}$

v_2 \ v_1 (Degrees of freedom for the numerator)	1	2	3	4	5	6	7	8	9	10	12	15	20	24	30	40	60	120	∞
1	4052.0	4999.5	5403.0	5625.0	5764.0	5859.0	5928.0	5982.0	6022.0	6056.0	6106.0	6157.0	6209.0	6235.0	6261.0	6287.0	6313.0	6339.0	6366.0
2	98.50	99.00	99.17	99.25	99.30	99.33	99.36	99.37	99.39	99.40	99.42	99.43	99.45	99.46	99.47	99.47	99.48	99.49	99.50
3	34.12	30.82	29.46	28.71	28.24	27.91	27.67	27.49	27.35	27.23	27.05	26.87	26.69	26.00	26.50	26.41	26.32	26.22	26.13
4	21.20	18.00	16.69	15.98	15.52	15.21	14.98	14.80	14.66	14.55	14.37	14.20	14.02	13.93	13.84	13.75	13.65	13.56	13.46
5	16.26	13.27	12.06	11.39	10.97	10.67	10.46	10.29	10.16	10.05	9.89	9.72	9.55	9.47	9.38	9.29	9.20	9.11	9.02
6	13.75	10.92	9.78	9.15	8.75	8.47	8.26	8.10	7.98	7.87	7.72	7.56	7.40	7.31	7.23	7.14	7.06	6.97	6.88
7	12.25	9.55	8.45	7.85	7.46	7.19	6.99	6.84	6.72	6.62	6.47	6.31	6.16	6.07	5.99	5.91	5.82	5.74	5.65
8	11.26	8.65	7.59	7.01	6.63	6.37	6.18	6.03	5.91	5.81	5.67	5.52	5.36	5.28	5.20	5.12	5.03	4.95	4.86
9	10.56	8.02	6.99	6.42	6.06	5.80	5.61	5.47	5.35	5.26	5.11	4.96	4.81	4.73	4.65	4.57	4.48	4.40	4.31
10	10.04	7.56	6.55	5.99	5.64	5.39	5.20	5.06	4.94	4.85	4.71	4.56	4.41	4.33	4.25	4.17	4.08	4.00	3.91
11	9.65	7.21	6.22	5.67	5.32	5.07	4.89	4.74	4.63	4.54	4.40	4.25	4.10	4.02	3.94	3.86	3.78	3.69	3.60
12	9.33	6.93	5.95	5.41	5.06	4.82	4.64	4.50	4.39	4.30	4.16	4.01	3.86	3.78	3.70	3.62	3.54	3.45	3.36
13	9.07	6.70	5.74	5.21	4.86	4.62	4.44	4.30	4.19	4.10	3.96	3.82	3.66	3.59	3.51	3.43	3.34	3.25	3.17
14	8.86	6.51	5.56	5.04	4.69	4.46	4.28	4.14	4.03	3.94	3.80	3.66	3.51	3.43	3.35	3.27	3.18	3.09	3.00
15	8.68	6.36	5.42	4.89	4.36	4.32	4.14	4.00	3.89	3.80	3.67	3.52	3.37	3.29	3.21	3.13	3.05	2.96	2.87
16	8.53	6.23	5.29	4.77	4.44	4.20	4.03	3.89	3.78	3.69	3.55	3.41	3.26	3.18	3.10	3.02	2.93	2.84	2.75
17	8.40	6.11	5.18	4.67	4.34	4.10	3.93	3.79	3.68	3.59	3.46	3.31	3.16	3.08	3.00	2.92	2.83	2.75	2.65
18	8.29	6.01	5.09	4.58	4.25	4.01	3.84	3.71	3.60	3.51	3.37	3.23	3.08	3.00	2.92	2.84	2.75	2.66	2.57
19	8.18	5.93	5.01	4.50	4.17	3.94	3.77	3.63	3.52	3.43	3.30	3.15	3.00	2.92	2.84	2.76	2.67	2.58	2.59
20	8.10	5.85	4.94	4.43	4.10	3.87	3.70	3.56	3.46	3.37	3.23	3.09	2.94	2.86	2.78	2.69	2.61	2.52	2.42
21	8.02	5.78	4.87	4.37	4.04	3.81	3.64	3.51	3.40	3.31	3.17	3.03	2.88	2.80	2.72	2.64	2.55	2.46	2.36
22	7.95	5.72	4.82	4.31	3.99	3.76	3.59	3.45	3.35	3.26	3.12	2.98	2.83	2.75	2.67	2.58	2.50	2.40	2.31
23	7.88	5.66	4.76	4.26	3.94	3.71	3.54	3.41	3.30	3.21	3.07	2.93	2.78	2.70	2.62	2.54	2.45	2.35	2.26
24	7.82	5.61	4.72	4.22	3.90	3.67	3.50	3.36	3.26	3.17	3.03	2.89	2.74	2.66	2.58	2.49	2.40	2.31	2.21
25	7.77	5.57	4.68	4.18	3.85	3.63	3.46	3.32	3.22	3.13	2.99	2.85	2.70	2.62	2.54	2.45	2.36	2.27	2.17
26	7.72	5.53	4.64	4.14	3.82	3.59	3.42	3.29	3.18	3.09	2.96	2.81	2.66	2.58	2.50	2.42	2.33	2.23	2.13
27	7.68	5.49	4.60	4.11	3.78	3.56	3.39	3.26	3.15	3.06	2.93	2.78	2.63	2.55	2.47	2.38	2.29	2.20	2.10
28	7.64	5.45	4.57	4.07	3.75	3.53	3.36	3.23	3.12	3.03	2.90	2.75	2.60	2.52	2.44	2.35	2.26	2.17	2.06
29	7.60	5.42	4.54	4.04	3.73	3.50	3.33	3.20	3.09	3.00	2.87	2.73	2.57	2.49	2.41	2.33	2.23	2.14	2.03
30	7.56	5.39	4.51	4.02	3.70	3.47	3.30	3.17	3.07	2.98	2.84	2.70	2.55	2.47	2.39	2.30	2.21	2.11	2.01
40	7.31	5.18	4.31	3.83	3.51	3.29	3.12	2.99	2.89	2.80	2.66	2.52	2.37	2.29	2.20	2.11	2.02	1.92	1.80
60	7.08	4.98	4.13	3.65	3.34	3.12	2.95	2.82	2.72	2.63	2.50	2.35	2.20	2.12	2.03	1.94	1.84	1.73	1.60
120	6.85	4.79	3.95	3.48	3.17	2.96	2.79	2.66	2.56	2.47	2.34	2.19	2.03	1.95	1.86	1.76	1.66	1.53	1.38
∞	6.63	4.61	3.78	3.32	3.02	2.80	2.64	2.51	2.41	2.32	2.18	2.04	1.88	1.79	1.70	1.59	1.47	1.32	1.00

Rows: Degrees of freedom for the denominator (v_2)

Note: $F_{0.99, v_1, v_2} = 1/F_{0.01, v_2, v_1}$.

Appendix Six

Factors for Constructing Variables Control Charts

Observations in Sample, n	Chart for Averages: Factors for Control Limits			Chart for Averages: Factors for Center Line		Chart for Standard Deviation: Factors for Control Limits				Factors for Center Line		Chart for Ranges: Factors for Control Limits				
	A	A_2	A_3	c_4	$1/c_4$	B_3	B_4	B_5	B_6	d_2	$1/d_2$	d_3	D_1	D_2	D_3	D_4
2	2.121	1.880	2.659	0.7979	1.2533	0	3.267	0	2.606	1.128	0.8865	0.853	0	3.686	0	3.267
3	1.732	1.023	1.954	0.8862	1.1284	0	2.568	0	2.276	1.693	0.5907	0.888	0	4.358	0	2.574
4	1.500	0.729	1.628	0.9213	1.0854	0	2.266	0	2.088	2.059	0.4857	0.880	0	4.698	0	2.282
5	1.342	0.577	1.427	0.9400	1.0638	0	2.089	0	1.964	2.326	0.4299	0.864	0	4.918	0	2.114
6	1.225	0.483	1.287	0.9515	1.0510	0.030	1.970	0.029	1.874	2.534	0.3946	0.848	0	5.078	0	2.004
7	1.134	0.419	1.182	0.9594	1.0423	0.118	1.882	0.113	1.806	2.704	0.3698	0.833	0.204	5.204	0.076	1.924
8	1.061	0.373	1.099	0.9650	1.0363	0.185	1.815	0.179	1.751	2.847	0.3512	0.820	0.388	5.306	0.136	1.864
9	1.000	0.337	1.032	0.9693	1.0317	0.239	1.761	0.232	1.707	2.970	0.3367	0.808	0.547	5.393	0.184	1.816
10	0.949	0.308	0.975	0.9727	1.0281	0.284	1.716	0.276	1.669	3.078	0.3249	0.797	0.687	5.469	0.223	1.777
11	0.905	0.285	0.927	0.9754	1.0252	0.321	1.679	0.313	1.637	3.173	0.3152	0.787	0.811	5.535	0.256	1.744
12	0.866	0.266	0.886	0.9776	1.0229	0.354	1.646	0.346	1.610	3.258	0.3069	0.778	0.922	5.594	0.283	1.717
13	0.832	0.249	0.850	0.9794	1.0210	0.382	1.618	0.374	1.585	3.336	0.2998	0.770	1.025	5.647	0.307	1.693
14	0.802	0.235	0.817	0.9810	1.0194	0.406	1.594	0.399	1.563	3.407	0.2935	0.763	1.118	5.696	0.328	1.672
15	0.775	0.223	0.789	0.9823	1.0180	0.428	1.572	0.421	1.544	3.472	0.2880	0.756	1.203	5.741	0.347	1.653
16	0.750	0.212	0.763	0.9835	1.0168	0.448	1.552	0.440	1.526	3.532	0.2831	0.750	1.282	5.782	0.363	1.637
17	0.728	0.203	0.739	0.9845	1.0157	0.466	1.534	0.458	1.511	3.588	0.2787	0.744	1.356	5.820	0.378	1.622
18	0.707	0.194	0.718	0.9854	1.0148	0.482	1.518	0.475	1.496	3.640	0.2747	0.739	1.424	5.856	0.391	1.608
19	0.688	0.187	0.698	0.9862	1.0140	0.497	1.503	0.490	1.483	3.689	0.2711	0.734	1.487	5.891	0.403	1.597
20	0.671	0.180	0.680	0.9869	1.0133	0.510	1.490	0.504	1.470	3.735	0.2677	0.729	1.549	5.921	0.415	1.585
21	0.655	0.173	0.663	0.9876	1.0126	0.523	1.477	0.516	1.459	3.778	0.2647	0.724	1.605	5.951	0.425	1.575
22	0.640	0.167	0.647	0.9882	1.0119	0.534	1.466	0.528	1.448	3.819	0.2618	0.720	1.659	5.979	0.434	1.566
23	0.626	0.162	0.633	0.9887	1.0114	0.545	1.455	0.539	1.438	3.858	0.2592	0.716	1.710	6.006	0.443	1.557
24	0.612	0.157	0.619	0.9892	1.0109	0.555	1.445	0.549	1.429	3.895	0.2567	0.712	1.759	6.031	0.451	1.548
25	0.600	0.153	0.606	0.9896	1.0105	0.565	1.435	0.559	1.420	3.931	0.2544	0.708	1.806	6.056	0.459	1.541

For $n > 25$.

$$A = \frac{3}{\sqrt{n}} \quad A_3 = \frac{3}{C_4\sqrt{n}} = C_4 = \frac{4(n-1)}{4n-3}$$

$$B_3 = 1 - \frac{3}{C_4\sqrt{2(n-1)}} \quad B_4 = 1 + \frac{3}{C_4\sqrt{2(n-1)}}$$

$$B_5 = C_4 - \frac{3}{\sqrt{2(n-1}} \quad B_6 = C_4 + \frac{3}{\sqrt{2(n-1)}}$$

References

The references include all of the books and articles that are cited in the text. We have also included a number of other papers and books that readers may find useful, particularly if they are involved with applications requiring information about topics beyond the scope of the text.

Adams, B. M., C. Lowry, and W. H. Woodall (1992). "The Use (and Misuse) of False Alarm Probabilities in Control Chart Design," in *Frontiers in Statistical Quality Control 4*, H. J. Lenz, G. B. Wetherill, and P.-Th. Wilrich, (eds.) Physica-Verlag, Heidelberg, pp. 155–158.

Alt, F. B. (1985). "Multivariate Quality Control," in *Encyclopedia of Statistical Sciences*, Vol. 6, N. L. Johnson and S. Kotz, (eds.) Wiley, New York.

Alwan, L. C. (1992). "Effects of Autocorrelation on Control Charts," *Communications in Statistics—Theory and Methods*, Vol. 21, (4), pp. 1025–1049.

Alwan, L. C., and H. V. Roberts (1988). "Time Series Modeling for Statistical Process Control," *Journal of Business and Economic Statistics*, Vol. 6(1), pp. 87–95.

Automotive Industry Action Group (1985). *Measurement Systems Analysis*, 2nd ed., Detroit, MI.

Automotive Industry Action Group (2002). *Measurement Systems Analysis*, 3rd ed., Detroit, MI.

Automotive Industry Action Group (2001). *Potential Failure Modes and Effects Analysis*, AIAG, Detroit MI.

Balakrishnan, N. and M. V. Koutras (2002). *Runs and Scans with Applications*, John Wiley & Sons, New York.

Barnard, G. A. (1959). "Control Charts and Stochastic Processes," *Journal of the Royal Statistical Society*, (B), Vol. 21(2), pp. 239–271.

Bather, J. A. (1963). "Control Charts and the Minimization of Costs," *Journal of the Royal Statistical Society*, (B), Vol. 25(1), pp. 49–80.

Berthouex, P. M., W. G. Hunter, and L. Pallesen (1978). "Monitoring Sewage Treatment Plants: Some Quality Control Aspects," *Journal of Quality Technology*, Vol. 10(4), pp. 139–149.

Bisgaard, S., W. G. Hunter, and L. Pallesen (1984). "Economic Selection of Quality of Manufactured Product," *Technometrics*, Vol. 26(1), pp. 9–18.

Bissell, A. F. (1990). "How Reliable Is Your Capability Index?" *Applied Statistics*, Vol. 39(3), pp. 331–340.

Borror, C. M., and C. M. Champ (2001). "Phase I Control Charts for Independent Bernoulli Data," *Quality and Reliability Engineering International*, Vol. 17(5), pp. 391–396.

Borror, C. M., C. W. Champ, and S. E. Rigdon (1998). "Poisson EWMA Control Charts," *Journal of Quality Technology*, Vol. 30(4), pp. 352–361.

Borror, C. M., J. B. Keats, and D. C. Montgomery (2003). "Robustness of the Time between Events CUSUM," *International Journal of Production Research*, Vol. 41(5), pp. 3435–3444.

Borror, C. M., D. C. Montgomery, and G. C. Runger (1997). "Confidence Intervals for Variance Components from Gauge Capability Studies," *Quality and Reliability Engineering International*, Vol. 13(6), pp. 361–369.

Borror, C. M., D. C. Montgomery, and G. C. Runger (1999). "Robustness of the EWMA Control Chart to Nonnormality," *Journal of Quality Technology*, Vol. 31(3), pp. 309–316.

Bourke, P. O. (1991). "Detecting a Shift in the Fraction Nonconforming Using Run-Length Control Chart with 100% Inspection," *Journal of Quality Technology*, Vol. 23(3), pp. 225–238.

Bowker, A. H., and G. J. Lieberman (1972). *Engineering Statistics*, 2nd ed., Prentice-Hall, Englewood Cliffs, NJ.

Box, G. E. P. (1957). "Evolutionary Operation: A Method for Increasing Industrial Productivity," *Applied Statistics*, Vol. 6(2), pp. 81–101.

Box, G. E. P. (1991). "The Bounded Adjustment Chart," *Quality Engineering*, Vol. 4(2), pp. 331–338.

Box, G. E. P. (1991–1992). "Feedback Control by Manual Adjustment," *Quality Engineering*, Vol. 4(1) pp. 143–151.

Box, G. E. P., S. Bisgaard, and C. Fung (1988). "An Explanation and Critique of Taguchi's Contributions to Quality Engineering." *Quality and Reliability Engineering International*, Vol. 4(2), pp. 123–131.

Box, G. E. P., and N. R. Draper (1969). *Evolutionary Operation*, Wiley, New York.

Box, G. E. P., and N. R. Draper (1986). *Empirical Model Building and Response Surfaces*, Wiley, New York.

Box, G. E. P., G. M. Jenkins, and G. C. Reinsel (1994). *Time Series Analysis, Forecasting, and Control*, 3rd ed. Prentice-Hall, Englewood Cliffs, NJ.

Box, G. E. P., and T. Kramer (1992). "Statistical Process Monitoring and Feedback Adjustment: A Discussion," *Technometrics*, Vol. 34(3), 251–267

Box, G. E. P., and A. LuceÒo (1997). *Statistical Control by Monitoring and Feedback Adjustment*, Wiley, New York.

Box, G. E. P., and J. G. Ramirez (1992). "Cumulative Score Charts," *Quality and Reliability Engineering International*, Vol. 8(1), pp. 17–27.

Boyd, D. F. (1950). "Applying the Group Control Chart for $\overline{x}$ and *R*," *Industrial Quality Control*, Vol. 7(3), pp. 22–25.

Boyles, R. A. (1991). "The Taguchi Capability Index," *Journal of Quality Technology*, Vol. 23(2), pp. 107–126.

Boyles, R. A. (2000). "Phase I Analysis for Autocorrelated Processes," *Journal of Quality Technology*, Vol. 32(4), pp. 395–409.

Boyles, R. A. (2001). "Gauge Capability for Pass-Fail Inspection," *Technometrics*, Vol. 43(2), pp. 223–229.

Brook, D., and D. A. Evans (1972). "An Approach to the Probability Distribution of CUSUM Run Length," *Biometrika,* Vol. 59(3), pp. 539–549.

Bryce, G. R., M. A. Gaudard, and B. L. Joiner (1997–1998). "Estimating the Standard Deviation for Individuals Charts," *Quality Engineering,* Vol. 10(2), pp. 331–341.

Buckeridge, D. L., H. Burkom, M. Campbell, W. R. Hogan, and A. W. Moore (2005). "Algorithms for Rapid Outbreak Detection: A Research Synthesis," *Journal of Biomedical Informatics,* Vol. 38, pp. 99–113.

Burdick, R. K., C. M. Borror, and D. C. Montgomery (2003). "A Review of Methods for Measurement Systems Capability Analysis," *Journal of Quality Technology,* Vol. 35(4), pp. 342–354. Available at www.asq.org/pub/jqt.

Burdick, R. K., and G. A. Larsen (1997). "Confidence Intervals on Measures of Variability in Gauge R & R Studies," *Journal of Quality Technology,* Vol. 29(2), pp. 261–273.

Burdick, R. K., C. M. Borror, and D. C. Montgomery (2005). *Design and Analysis of Gauge R&R Studies: Making Decisions with Confidence Intervals in Random and Mixed ANOVA Models,* ASA-SIAM Series on Statistics and Applied Probability, SIAM, Philadelphia, PA, and ASA, Alexandria, VA.

Burdick, R. K., Y.-J. Park, D. C. Montgomery, and C. M. Borror (2005), "Confidence Intervals for Misclassification Rates in a Gauge R&R Study," *Journal of Quality Technology,* Vol. 37(4), pp. 294–303.

Burr, I. J. (1967). "The Effect of Nonnormality on Constants for $\bar{x}$ and *R* Charts," *Industrial Quality Control,* Vol. 23(11), pp. 563–569.

Byrne, D. M., and S. Taguchi (1987). "The Taguchi Approach to Robust Parameter Design," *40th Annual Quality Progress Transactions* December, pp. 19–26, Milwaukee, Wisconsin.

Champ, C. M., and S.-P. Chou (2003). "Comparison of Standard and Individual Limits Phase I Shewhart, $\bar{X}$, *R*, and *S* Charts," *Quality and Reliability Engineering International,* Vol. 19(1), pp. 161–170.

Champ, C. W., and W. H. Woodall (1987). "Exact Results for Shewhart Control Charts with Supplementary Runs Rules," *Technometrics,* Vol. 29(4), pp. 393–399.

Chan, L. K., S. W. Cheng, and F. A. Spiring (1988). "A New Measure of Process Capability: C_{pm}," *Journal of Quality Technology,* Vol. 20(3), pp. 160–175.

Chan, L. K., K. P. Hapuarachchi, and B. D. Macpherson (1988). "Robustness of $\bar{x}$ and *R* Charts," *IEEE Transactions on Reliability,* Vol. 37(3), pp. 117–123.

Chang, T. C., and F. F. Gan (1995). "A Cumulative Sum Control Chart for Monitoring Process Variance," *Journal of Quality Technology,* Vol. 27(2), pp. 109–119.

Chakraborti, S., P. Van Der Laan, and S. T. Bakir (2001). "Nonparametric Control Charts: An Overview and Some Results," *Journal of Quality Technology,* Vol. 33(3), pp. 304–315.

Chiu, W. K., and G. B. Wetherill (1974). "A Simplified Scheme for the Economic Design of $\bar{x}$ -Charts," *Journal of Quality Technology,* Vol. 6(2), pp. 63–69.

Chiu, W. K., and G. B. Wetherill (1975). "Quality Control Practices," *International Journal of Production Research,* Vol. 13(2), pp. 175–182.

Chrysler, Ford, and GM (1995). *Measurement Systems Analysis Reference Manual,* AIAG, Detroit, MI.

Chua, M., and D. C. Montgomery (1992). "Investigation and Characterization of a Control Scheme for Multivariate Quality Control," *Quality and Reliability Engineering International,* Vol. 8(1), pp. 37–44.

Clements, J. A. (1989). Process Capability Calculations for Non-Normal Distributions," *Quality Progress,* Vol. 22(2), pp. 95–100.

Clifford, P. C. (1959). "Control Charts without Calculations," *Industrial Quality Control,* Vol. 15(11), pp. 40–44.

Coleman, D. E., and D. C. Montgomery (1993). "A Systematic Method for Planning for a Designed Industrial Experiment" (with discussion), *Technometrics,* Vol. 35(1), pp. 1–12.

Cornell, J. A., and A. I. Khuri (1996). *Response Surfaces,* 2nd ed., Dekker, New York.

Cowden, D. J. (1957). *Statistical Methods in Quality Control,* Prentice-Hall, Englewood Cliffs, NJ.

Croarkin, C. and R. Varner (1982). "Measurement Assurance for Dimensional Measurements on Integrated-Circuit Photomasks," NBS Technical Note 1164, U.S. Department of Commerce, Washington, DC.

Crosier, R. B. (1988). "Multivariate Generalizations of Cumulative Sum Quality Control Schemes," *Technometrics,* Vol. 30(3), pp. 291–303.

Crowder, S. V. (1987a). "A Simple Method for Studying Run-Length Distributions of Exponentially Weighted Moving Average Charts," *Technometrics,* Vol. 29(4), pp. 401–407.

Crowder, S. V. (1987b). "Computation of ARL for Combined Individual Measurement and Moving Range Charts," *Journal of Quality Technology,* Vol. 19(1), pp. 98–102.

Crowder, S. V. (1989). "Design of Exponentially Weighted Moving Average Schemes," *Journal of Quality Technology,* Vol. 21(2), pp. 155–162.

Crowder, S. V. (1992). "An SPC Model for Short Production Runs: Minimizing Expected Costs," *Technometrics,* Vol. 34(1), pp. 64–73.

Crowder, S. V., and M. Hamilton (1992). "An EWMA for Monitoring a Process Standard Deviation," *Journal of Quality Technology,* Vol. 24(1), pp. 12–21.

Cruthis, E. N., and S. E. Rigdon (1992–1993). "Comparing Two Estimates of the Variance to Determine the Stability of a Process," *Quality Engineering,* Vol. 5(1), pp. 67–74.

Davis, R. B., and W. H. Woodall (1988), "Performance of the Control Chart Trend Rule under Linear Shift," *Journal of Quality Technology,* Vol. 20(4), pp. 260–262.

Del Castillo, E., and D. C. Montgomery (1994). "Short-Run Statistical Process Control: *Q*-Chart Enhancements and Alternative Methods," *Quality and Reliability Engineering International,* Vol. 10(1), pp. 87–97.

Del Castillo. E. (2002), *Statistical Process Adjustment for Quality Control,* John Wiley & Sons, New York.

De Mast, J. and W. N. Van Wieringen (2004). "Measurement System Analysis for Bounded Ordinal Data," *Quality and Reliability Engineering International,* Vol. 20, pp. 383–395.

De Mast, J. and W. N. Van Wieringen (2007). "Measurement Systems for Categorical Measurements: Agreement and Kappa-Type Indices," *Journal of Quality Technology,* Vol. 39, pp. 191–202.

Dodge, H. F. (1943). "A Sampling Plan for Continuous Production," *Annals of Mathematical Statistics,* Vol. 14(3), pp. 264–279.

Dodge, H. F. (1955). "Chain Sampling Inspection Plans," *Industrial Quality Control,* Vol. 11(4), pp. 10–13.

Dodge, H. F. (1956). "Skip-Lot Sampling Plan," *Industrial Quality Control,* Vol. 11(5), pp. 3–5.

Dodge, H. F., and H. G. Romig (1959). *Sampling Inspection Tables, Single and Double Sampling,* 2nd ed., Wiley, New York.

Dodge, H. F., and M. N. Torrey (1951). "Additional Continuous Sampling Inspection Plans," *Industrial Quality Control,* Vol. 7(1), pp. 5–9.

Duncan, A. J. (1956). "The Economic Design of $\bar{x}$-Charts Used to Maintain Current Control of a Process," *Journal of the American Statistical Association,* Vol. 51(274), pp. 228–242.

Duncan, A. J. (1986). *Quality Control and Industrial Statistics,* 5th ed., Irwin, Homewood, IL.

Duncan, A. J. (1978). "The Economic Design of *p*-Charts to Maintain Current Control of a Process: Some Numerical Results," *Technometrics,* Vol. 20(3), pp. 235–244.

English, J. R., and G. D. Taylor (1993). "Process Capability Analysis: A Robustness Study," *International Journal of Production Research,* Vol. 31(7), pp. 1621–1635.

Ewan, W. D. (1963). "When and How to Use Cu-Sum Charts," *Technometrics,* Vol. 5(1), pp. 1–22.

Farnum, N. R. (1992). "Control Charts for Short Runs: Nonconstant Process and Measurement Error," *Journal of Quality Technology,* Vol. 24(2), pp. 138–144.

Fienberg, S. E., and G. Shmueli (2005). "Statistical Issues and Challenges Associated with the Rapid Detection of Terrorist Outbreaks," *Statistics in Medicine,* Vol. 24, pp. 513–529.

Ferrell, E. B. (1953). "Control Charts Using Midranges and Medians," *Industrial Quality Control,* Vol. 9(5), pp. 30–34.

Fisher, R. A. (1925). "Theory of Statistical Estimation," *Proceedings of the Cambridge Philosophical Society,* Vol. 22, pp. 700–725.

Freund, R. A. (1957). "Acceptance Control Charts," *Industrial Quality Control,* Vol. 14(4), pp. 13–23.

Fricker, R. D., Jr. (2007). "Directionally Sensitive Multivariate Statistical Process Control Procedures with Application to Syndromic Surveillance," *Advances in Disease Surveillance,* Vol. 3, pp. 1–17.

Gan, F. F. (1991). "An Optimal Design of CUSUM Quality Control Charts," *Journal of Quality Technology,* Vol. 23(4), pp. 279–286.

Gan, F. F. (1993). "An Optimal Design of CUSUM Control Charts for Binomial Counts," *Journal of Applied Statistics,* Vol. 20, pp. 445–460.

Gardiner, J. S. (1987). *Detecting Small Shifts in Quality Levels in a Near-Zero Defect Environment for Integrated Circuits,* Ph.D. Dissertation, Department of Mechanical Engineering, University of Washington, Seattle, WA.

Gardiner, J. S., and D. C. Montgomery (1987). "Using Statistical Control Charts for Software Quality Control," *Quality and Reliability Engineering International,* Vol. 3(1), pp. 15–20.

Garvin, D. A. (1987). "Competing in the Eight Dimensions of Quality," *Harvard Business Review,* Sept.–Oct., 87(6), pp. 101–109.

George, M. L. (2002). *Lean Six Sigma,* McGraw-Hill, New York.

Girshick, M. A., and H. Rubin (1952). "A Bayesian Approach to a Quality Control Model," *Annals of Mathematical Statistics,* Vol. 23(1), pp. 114–125.

Glaz, J., J. Naus, and S. Wallenstein (2001). *Scan Statistics,* Springer, New York.

Grant, E. L., and R. S. Leavenworth (1980). *Statistical Quality Control,* 5th ed., McGraw-Hill, New York.

Grigg, O., and V. Farewell (2004). "An Overview of Risk-Adjusted Charts," *Journal of the Royal Statistical Society,* Series A, Vol. 167, pp. 523–539.

Grigg, O., and D. Spiegelhalter (2007). "A Simple Risk-Adjusted Exponentially Weighted Moving Average," *Journal of the American Statistical Association,* Vol. 102, pp. 140–152.

Grubbs, F. E. (1946). "The Difference Control Chart with an Example of Its Use," *Industrial Quality Control,* Vol. 3(1), pp. 22–25.

Guenther, W. C. (1972). "Tolerance Intervals for Univariate Distributions," *Naval Research Logistics Quarterly,* Vol. 19(2), pp. 309–334.

Gupta, S., D. C. Montgomery, and W. H. Woodall (2006), "Performance Evaluation of Two Methods for Online Monitoring of Linear Calibration Profiles," *International Journal of Production Research,* Vol. 44, pp. 1927–1942.

Hahn, G. J., and S. S. Shapiro (1967). *Statistical Models in Engineering,* Wiley, New York.

Hamada. M. (2003). "Tolerance Interval Control Limits for the $\bar{x}$, *R*, and *S* Charts," *Quality Engineering,* Vol. 15(3), pp. 471–487.

Harris, T. J., and W. H. Ross (1991). "Statistical Process Control Procedures for Correlated Observations," *Canadian Journal of Chemical Engineering,* Vol. 69 (Feb), pp. 48–57.

Hawkins, D. M. (1981). "A CUSUM for a Scale Parameter," *Journal of Quality Technology,* Vol. 13(4), pp. 228–235.

Hawkins, D. M. (1991). "Multivariate Quality Control Based on Regression Adjusted Variables," *Technometrics,* Vol. 33(1), pp. 61–75.

Hawkins, D. M. (1992). "A Fast, Accurate Approximation of Average Run Lengths of CUSUM Control Charts," *Journal of Quality Technology,* Vol. 24(1), pp. 37–43.

Hawkins, D. M. (1993a). "Cumulative Sum Control Charting: An Underutilized SPC Tool," *Quality Engineering,* Vol. 5(3), pp. 463–477.

Hawkins, D. M. (1993b). "Regression Adjustment for Variables in Multivariate Quality Control," *Journal of Quality Technology,* Vol. 25(3), pp. 170–182.

Hawkins, D. M. (1987). "Self-starting Cusums for Location and Scale," *The Statistician,* Vol. 36, pp. 299–315.

Hawkins, D. M., S. Choi, and S. Lee (2007). "A General Multivariate Exponentially Weighted Moving-Average Control Chart," *Journal of Quality Technoloyg,* Vol. 39, pp. 118–125.

Hawkins, D. M., and D. H. Olwell (1998). *Cumulative Sum Charts and Charting for Quality Improvement,* Springer Verlag, New York.

Hawkins, D. M., P. Qiu, and C. W. Kang (2003). "The Changepoint Model for Statistical Process Control," *Journal of Quality Technology,* Vol. 35(4), pp. 355–366. Available at www.asq.org/pub/jqt.

Hayter, A. J., and K.-L. Tsui (1994). "Identification and Quantification in Multivariate Quality Control Problems," *Journal of Quality Technology,* Vol. 26(3), pp. 197–208.

Hicks, C. R. (1955). "Some Applications of Hotelling's T^2," *Industrial Quality Control,* Vol. 11(9), pp. 23–29.

Hill, D. (1956). "Modified Control Limits," *Applied Statistics,* Vol. 5(1), pp. 12–19.

Hillier, F. S. (1969). " $\bar{x}$ and R Chart Control Limits Based on a Small Number of Subgroups," *Journal of Quality Technology,* Vol. 1(1), pp. 17–25.

Hines, W. W., D. C. Montgomery, D. M. Goldsman, and C. M. Borror (2004). *Probability and Statistics in Engineering,* 4th ed., Wiley, New York.

Ho, C., and K. E. Case (1994). "Economic Design of Control Charts: A Literature Review for 1981–1991," *Journal of Quality Technology,* Vol. 26(1), pp. 39–53.

Holmes, D. S., and A. E. Mergen (1993). "Improving the Performance of the T^2 Control Chart," *Quality Engineering,* Vol. 5(4), pp. 619–625.

Hotelling, H. (1947). "Multivariate Quality Control," *Techniques of Statistical Analysis,* Eisenhart, Hastay, and Wallis (eds.), McGraw-Hill, New York.

Houf, R. F., and D. B. Berman (1988). "Statistical Analysis of Power Module Thermal Test Equipment Performance," *IEEE Transactions on Components, Hybrids, and Manufacturing Technology,* Vol. 11(4), pp. 516–520.

Howell, J. M. (1949). "Control Charting Largest and Smallest Value," *Annals of Mathematical Statistics,* Vol. 20(2), pp. 305–309.

Hunter, J. S. (1985). "Statistical Design Applied to Product Design," *Journal of Quality Technology,* Vol. 17(4), pp. 210–221.

Hunter, J. S. (1986). "The Exponentially Weighted Moving Average," *Journal of Quality Technology,* Vol. 18(4), pp. 203–210.

Hunter, J. S. (1987). "Signal-to-Noise Ratio Debate," Letter to the Editor, *Quality Progress,* May 1987, pp. 7–9.

Hunter, J. S. (1989). "A One-Point Plot Equivalent to the Shewhart Chart with Western Electric Rules," *Quality Engineering,* Vol. 2(1), pp. 13–19.

Hunter, J. S. (1997). "The Box-Jenkins Manual Bounded Adjustment Chart," *Annual Quality Progress Proceedings,* Orlando, FL, Vol. 51(0), pp. 158–169.

Hunter, W. G., and C. P. Kartha (1977). "Determining the Most Profitable Target Value for a Production Process," *Journal of Quality Technology,* Vol. 9(4), pp. 176–181.

Iglewitz, B., and D. Hoaglin (1987). "Use of Boxplots for Process Evaluation," *Journal of Quality Technology,* Vol. 19(4) pp. 180–190.

Jackson, J. E. (1956). "Quality Control Methods for Two Related Variables," *Industrial Quality Control,* Vol. 12(7), pp. 4–8.

Jackson, J. E. (1959). "Quality Control Methods for Several Related Variables," *Technometrics,* Vol. 1(4), pp. 359–377.

Jackson, J. E. (1972). "All Count Distributions Are Not Alike," *Journal of Quality Technology,* Vol. 4(2), pp. 86–92.

Jackson, J. E. (1980). "Principal Components and Factor Analysis: Part I—Principal Components," *Journal of Quality Technology,* Vol. 12(4), pp. 201–213.

Jackson, J. E. (1985). "Multivariate Quality Control," *Communications in Statistics—Theory and Methods,* Vol. 14(110), pp. 2657–2688.

Jensen, W. A., L. A. Jones-Farmer, C. W. Champ, and W. H. Woodall (2006). "Effects of Parameter Estimation on Control Chart Properties: A Literature Review," *Journal of Quality Technology,* Vol. 38, pp. 95–108.

Jin, J. and J. Shi (2001). "Automatic Feature Extraction of Waveform Signals for In-process Diagnostic Performance Improvement," *Journal of Intelligent Manufacturing,* Vol. 12, pp. 257–268.

Johnson, N. L. (1961). "A Simple Theoretical Approach to Cumulative Sum Control Charts," *Journal of the American Statistical Association,* Vol. 56(296), pp. 835–840.

Johnson, N. L., and S. Kotz (1969). *Discrete Distributions,* Houghton Mifflin, Boston.

Johnson, N. L., and F. C. Leone (1962a). "Cumulative Sum Control Charts—Mathematical Principles Applied to Their Construction and Use," Part I, *Industrial Quality Control,* Vol. 19(2), pp. 15–21.

Johnson, N. L., and F. C. Leone (1962b). "Cumulative Sum Control Charts—Mathematical Principles Applied to Their Construction and Use," Part II, *Industrial Quality Control,* Vol. 19(2), pp. 22–28.

Johnson, N. L., and F. C. Leone (1962c). "Cumulative Sum Control Charts—Mathematical Principles Applied to Their Construction and Use," Part III, *Industrial Quality Control,* Vol. 19(2), pp. 29–36.

Jones, L. A., W. H. Woodall, and M. D. Conerly (1999). "Exact Properties of Demerit Control Charts," *Journal of Quality Technology,* Vol. 31(2), pp. 221–230.

Jones, M. C., and J. A. Rice (1992). "Displaying the Important Features of Large Collections of Similar Curves," *The American Statistician,* Vol. 46, pp. 140–145.

Juran, J. M., and F. M. Gryna, Jr. (1980). *Quality Planning and Analysis,* 2nd ed., McGraw-Hill, New York.

Kaminski, F. C., J. C. Benneyan, R. D. Davis (1992) and R. J. Burke (1992), "Statistical Control Charts Based on a Geometric Distribution," *Journal of Quality Technology,* Vol. 24, pp. 63–69.

Kane, V. E. (1986). "Process Capability Indices," *Journal of Quality Technology,* Vol. 18(1), pp. 41–52.

Kang, L., and S. L. Albin (2000). "On-Line Monitoring When the Process Yields a Linear Profile," *Journal of Quality Technology,* Vol. 32(4), pp. 418–426. Available at www.asq.org/pub/jqt.

Keats, J. B., E. Del Castillo, E. von Collani, and E. M. Saniga (1997). Economic Modeling for Statistical Process Control," *Journal of Quality Technology,* Vol. 29(2), pp. 144–147.

Kim, K., M. A. Mahmoud, and W. H. Woodall (2003). "On the Monitoring of Linear Profiles," *Journal of Quality Technology,* Vol. 35(3), pp. 317–328.

Kittlitz, R. J. (1999). "Transforming the Exponential for SPC Applications," *Journal of Quality Technology,* Vol. 31(3), pp. 301–308.

Kotz, S., and N. L. Johnson (2002). "Process Capability Indices: A Review 1992–2000" (with Discussion), *Journal of Quality Technology,* Vol. 34(1), pp. 2–19. Available at www.asq.org/pub/jqt.

Kotz, S., and C. R. Lovelace (1998). *Process Capability Indices in Theory and Practice,* Arnold, London.

Kovach, J. (2007). "Designing Effective Six Sigma Experiments for Service Process Improvement Projects," *International Journal of Six Sigma and Competitive Advantage,* Vol. 3(1), pp. 72–90.

Kulldorff, M. (1997). "A Spatial Scan Statistic," *Communications in Statistics—Theory and Methods,* Vol. 26, pp. 1481–1496.

Kulldorff, M. (2001). "Prospective Time Periodic Geographical Disease Surveillance Using a Scan Statistic," *Journal of the Royal Statistical Society,* Series A, Vol. 164, pp. 61–72.

Kulldorff, M. (2003). Information Management Services, Inc. *SaTScan*™ Version 3.1: Software for the Spatial and Space-Time Scan Statistics. Available at www.satscan.org.

Kulldorff, M. (2005). "Scan Statistics for Geographical Disease Surveillance: An Overview," Chapter 7 in *Spatial & Syndromic Surveillance,* by A. B. Lawson and K. Kleinman, John Wiley & Sons, New York, pp. 115–132.

Kushler, R. H., and P. Hurley (1992). "Confidence Bounds for Capability Indices," *Journal of Quality Technology,* Vol. 24(4), pp. 188–195.

Langenberg, P., and B. Inglewitz (1986). "Trimmed Mean $\bar{x}$ and R Charts," *Journal of Quality Technology,* Vol. 18(3), pp. 152–161.

Lanning, J. W. (1998). *Methods for Monitoring Fractionally Sampled Multiple Stream Processes,* Ph.D. Dissertation, Department of Industrial Engineering, Arizona State University, Tempe, AZ.

Lanning, J. W., D. C. Montgomery, and G. C. Runger (2002–2003). "Monitoring a Multiple-Stream Filling Operation Using Fractional Samples," *Quality Engineering,* Vol. 15(2), pp. 183–195.

Ledolter, J., and A. Swersey (1997). "An Evaluation of Pre-Control," *Journal of Quality Technology,* Vol. 29(2), pp. 163–171.

Lieberman, G. J., and G. J. Resnikoff (1955). "Sampling Plans for Inspection by Variables," *Journal of the American Statistical Association,* Vol. 50(270), pp. 457–516.

Lieberman, G. J., and H. Solomon (1955). "Multi-Level Continuous Sampling Plans," *Annals of Mathematical Statistics,* Vol. 26(4), pp. 686–704.

Linderman, K., and T. E. Love (2000a). "Economic and Economic Statistical Design for MEWMA Control Charts," *Journal of Quality Technology,* Vol. 32(4), pp. 410–417. Available at www.asq.org/pub/jqt.

Linderman, K., and T. E. Love (2000b). "Implementing Economic and Economic Statistical Design for MEWMA Control Charts," *Journal of Quality Technology,* Vol. 32(4), pp. 457–463. Available at www.asq.org/pub/jqt.

Lorenzen, T. J., and L. C. Vance (1986). "The Economic Design of Control Charts: A Unified Approach," *Technometrics,* Vol. 28(1), pp. 3–10.

Lowry, C. A., C. W. Champ, and W. H. Woodall (1995). "The Performance of Control Charts for Monitoring Process Variation," *Communications in Statistics—Simulation and Computation,* Vol. 24(2), pp. 409–431.

Lowry, C. A., and D. C. Montgomery (1995). "A Review of Multivariate Control Charts," *IIE Transactions,* Vol. 27(6), pp. 800–810.

Lowry, C. A., W. H. Woodall, C. W. Champ, and S. E. Rigdon (1992). "A Multivariate Exponentially Weighted Moving Average Control Chart," *Technometrics,* Vol. 34(1), pp. 46–53.

Lu, C.-W., and M. R. Reynolds, Jr. (1999a). "EWMA Control Charts for Monitoring the Mean of Autocorrelated Processes," *Journal of Quality Technology,* Vol. 31(2), pp. 189–206.

Lu, C.-W., and M. R. Reynolds, Jr. (1999b). "Control Charts for Monitoring the Mean and Variance of Autocorrelated Processes," *Journal of Quality Technology,* Vol. 31(3), pp. 259–274.

Lucas, J. M. (1973). "A Modified V-Mask Control Scheme," *Technometrics,* Vol. 15(4), pp. 833–847.

Lucas, J. M. (1976). "The Design and Use of V-mask Control Schemes," *Journal of Quality Technology,* Vol. 8(1), pp. 1–12.

Lucas, J. M. (1982). "Combined Shewhart-CUSUM Quality Control Schemes," *Journal of Quality Technology,* Vol. 14(2), pp. 51–59.

Lucas, J. M. (1985). "Counted Data CUSUM's," *Technometrics,* Vol. 27(3), pp. 129–144.

Lucas, J. M., and R. B. Crosier (1982). "Fast Initial Response for CUSUM Quality Control Schemes," *Technometrics,* Vol. 24(3), pp. 199–205.

Lucas, J. M., and M. S. Saccucci (1990). "Exponentially Weighted Moving Average Control Schemes: Properties and Enhancements," *Technometrics,* Vol. 32(1), pp. 1–29.

Luceío, A. (1996). "A Process Capability Ratio with Reliable Confidence Intervals," *Communications in Statistics—Simulation and Computation,* Vol. 25(1), pp. 235–246.

Luceío, A. (1999). "Average Run Lengths and Run Length Probability Distributions for Cuscore Charts to Control Normal Mean," *Computational Statistics and Data Analysis,* Vol. 32, pp. 177–195.

MacGregor, J. F. (1987). "Interfaces between Process Control and On-Line Statistical Process Control," *A.I.Ch.E. Cast Newsletter,* 9–19.

MacGregor, J. F., and T. J. Harris (1993). "The Exponentially Weighted Moving Variance," *Journal of Quality Technology,* Vol. 25(2), pp. 106–118.

MacGregor, J. F., and T. J. Harris (1990). "Discussion of: "EWMA Control Schemes: Properties and Enhancement," by Lucas and Sacucci," *Technometrics,* Vol. 32(9), pp. 23–26.

Mahmoud, M. A., P. A. Parker, W. H. Woodall, and D. M. Hawkins (2007). "A Change Point Method for Linear Profile Data," *Quality and Reliability Engineering International,* Vol. 23, pp. 247–268.

Mandel, J. (1969). "The Regression Control Chart," *Journal of Quality Technology,* Vol. 1(1), pp. 1–9.

Manuele, J. (1945). "Control Chart for Determining Tool Wear," *Industrial Quality Control,* Vol. 1(6), pp. 7–10.

Maragah, H. D., and W. H. Woodall (1992). "The Effect of Autocorrelation on the Retrospective $\bar{x}$ -chart," *Journal of Statistical Computation and Simulation,* Vol. 40(1), pp. 29–42.

Mason, R. L., N. D. Tracy, and J. C. Young (1995). "Decomposition of T^2 for Multivariate Control Chart Interpretation," *Journal of Quality Technology,* Vol. 27(2), pp. 109–119.

Mastrangelo, C. M., and D. C. Montgomery (1995). "SPC with Correlated Observations for the Chemical and Process Industries," *Quality and Reliability Engineering International,* Vol. 11(2), pp. 79–89.

Mastrangelo, C. M., G. C. Runger, and D. C. Montgomery (1996). "Statistical Process Monitoring with Principal Components," *Quality and Reliability Engineering International,* Vol. 12(3), pp. 203–210.

Mazza, J. (2000). "Cost Reduction for Document Processing in Major Litigations," presented at the 54th Annual Quality Congress, Charlotte, North Carolina.

Molina, E. C. (1942). *Poisson's Exponential Binomial Limit,* Van Nostrand Reinhold, New York.

Molnau, W. E., D. C. Montgomery, and G. C. Runger (2001). "Statistically Constrained Economic Design of the Multivariate Exponentially Weighted Moving Average Control Chart," *Quality and Reliability Engineering International,* Vol. 17(1), pp. 39–49.

Molnau, W. E., G. C. Runger, D. C. Montgomery, K. R. Skinner, E. N. Loredo, and S. S. Prabhu (2001). "A Program for ARL Calculation for Multivariate EWMA Control Charts," *Journal of Quality Technology,* Vol. 33(4), pp. 515–521.

Montgomery, D. C. (1980). "The Economic Design of Control Charts: A Review and Literature Survey," *Journal of Quality Technology,* Vol. 12(2), pp. 75–87.

Montgomery, D. C. (1982). "Economic Design of an $\bar{x}$ Control Chart," *Journal of Quality Technology,* Vol. 14(1), pp. 40–43.

Montgomery, D. C. (1999). "Experimental Design for Product and Process Design and Development" (with commentary), *Journal of the Royal Statistical Society Series D (The Statistician),* Vol. 48(2), pp. 159–177.

Montgomery, D. C. (2009). *Design and Analysis of Experiments,* 7th ed., Wiley, New York.

Montgomery, D. C. (2009). *Introduction to Statistical Quality Control,* 6th ed., Wiley, New York.

Montgomery, D. C., and J. J. Friedman (1989). "Statistical Process Control in a Computer-Integrated Manufacturing Environment," *Statistical Process Control in Automated Manufacturing,* J. B. Keats and N. F. Hubele (eds.) Dekker, Series in Quality and Reliability, New York.

Montgomery, D. C., C. L. Jennings, and M. Kulahci (2008). *Introduction to Time Series Analysis and Forecasting,* Wiley, New York.

Montgomery, D. C., J. B. Keats, G. C. Runger, and W. S. Messina (1994). "Integrating Statistical Process Control and Engineering Process Control," *Journal of Quality Technology,* Vol. 26(2), pp. 79–87.

Montgomery, D. C., and C. M. Mastrangelo (1991). "Some Statistical Process Control Methods for Aurocorrelated Data" (with discussion), *Journal of Quality Technology,* Vol. 23(3), pp. 179–204.

Montgomery, D. C., E. A. Peck, and G. G. Vining (2006). *Introduction to Linear Regression Analysis,* 4th ed., Wiley, New York.

Montgomery, D. C., and G. C. Runger (1993a). "Gauge Capability and Designed Experiments: Part I: Basic Methods," *Quality Engineering,* Vol. 6(1), pp. 115–135.

Montgomery, D. C., and G. C. Runger (1993b). "Gauge Capability Analysis and Designed Experiments: Part II: Experimental Design Models and Variance Component Estimation," *Quality Engineering,* Vol. 6(2), pp. 289–305.

Montgomery, D. C., and G. C. Runger (2007). *Applied Statistics and Probability for Engineers,* 4th ed., Wiley, New York.

Montgomery, D. C., and H. M. Wadsworth, Jr. (1972). "Some Techniques for Multivariate Quality Control Applications," *ASQC Technical Conference Transactions,* Washington, DC, May, pp. 427–435.

Montgomery, D. C., and W. H. Woodall, eds. (1997). "A Discussion of Statistically-Based Process Monitoring and Control," *Journal of Quality Technology,* Vol. 29(2), pp. 121–162.

Mortell, R. R., and G. C. Runger (1995). "Statistical Process Control of Multiple Stream Processes," *Journal of Quality Technology,* Vol. 27(1), pp. 1–12.

Murphy, B. J. (1987). "Screening Out-of-Control Variables with T^2 Multivariate Quality Control Procedures," *The Statistician,* Vol. 36(5), pp. 571–583.

Myers, R. H., D. C. Montgomery C. M. Anderson–Cook (2009). *Response Surface Methodology: Process and Product Optimization Using Designed Experiments,* 3rd ed., Wiley, New York.

Nair, V. N., editor (1992). "Taguchi's Parameter Design: A Panel Discussion," *Technometrics,* Vol. 34(2) pp. 127–161.

Naus, J., and S. Wallenstein (2006). "Temporal Surveillance Using Scan Statistics," *Statistics in Medicine,* Vol. 25, pp. 311–324.

Nelson, L. S. (1978). "Best Target Value for a Production Process," *Journal of Quality Technology,* Vol. 10(2), pp. 88–89.

Nelson, L. S. (1984). "The Shewhart Control Chart—Tests for Special Causes," *Journal of Quality Technology,* Vol. 16(4), pp. 237–239.

Nelson, L. S. (1986). "Control Chart for Multiple Stream Processes," *Journal of Quality Technology,* Vol. 18(4), pp. 255–256.

Nelson, L. S. (1994). "A Control Chart for Parts-per-Million Nonconforming Items," *Journal of Quality Technology,* Vol. 26(3), pp. 239–240.

Nelson, P. R., and P. L. Stephenson (1996). "Runs Tests for Group Control Charts," *Communications in Statistics—Theory and Methods,* Vol. 25(11), pp. 2739–2765.

NIST/SEMATECH e-Handbook of Statistical Methods. Available at www.itl.nist.gov/div898/handbook/mpc/section3/mpc37.htm.

Ott, E. R. (1975). *Process Quality Control,* McGraw-Hill, New York.

Ott, E. R., and R. D. Snee (1973). "Identifying Useful Differences in a Multiple-Head Machine," *Journal of Quality Technology,* Vol. 5(2), pp. 47–57.

Page, E. S. (1954). "Continuous Inspection Schemes," *Biometrics,* Vol. 41(1), pp. 100–115.

Page, E. S. (1961). "Cumulative Sum Control Charts," *Technometrics*, Vol. 3(1), pp. 1–9.

Page, E. S. (1963). "Controlling the Standard Deviation by Cusums and Warning Lines," *Technometrics*, Vol. 5(3), pp. 307–316.

Pearn, W. L., S. Kotz, and N. L. Johnson (1992). "Distributional and Inferential Properties of Process Capability Indices," *Journal of Quality Technology*, Vol. 24(4), pp. 216–231.

Perry, R. L. (1973). "Skip-Lot Sampling Plans," *Journal of Quality Technology*, Vol. 5(3), pp. 123–130.

Pignatiello, J. J., Jr., and J. S. Ramberg (1991). "Top Ten Triumphs and Tragedies of Genichi Taguchi," *Quality Engineering*, Vol. 4(2), pp. 211–225.

Pignatiello, J. J., Jr., and G. C. Runger (1990). "Comparison of Multivariate CUSUM Charts," *Journal of Quality Technology*, Vol. 22(3), pp. 173–186.

Pignatiello, J. J., Jr., and G. C. Runger (1991). "Adaptive Sampling for Process Control," *Journal of Quality Technology*, Vol. 22(3), pp. 135–155.

Pignatiello, J. J., Jr., and T. R. Samuel (2001). "Estimation of the Change Point of a Normal Process Mean in SPC Applications," *Journal of Quality Technology*, Vol. 33(1), pp. 82–95.

Prabhu, S. S., D. C. Montgomery, and G. C. Runger (1994). "A Combined Adaptive Sample Size and Sampling Interval $\bar{x}$ Control Scheme," *Journal of Quality Technology*, Vol. 26(3), pp. 164–176.

Prabhu, S. S., D. C. Montgomery, and G. C. Runger (1995). "A Design Tool to Evaluate Average Time to Signal Properties of Adaptive $\bar{x}$ Charts," *Journal of Quality Technology*, Vol. 27(1), pp. 74–76.

Prabhu, S. S., and G. C. Runger (1997). "Designing a Multivariate EWMA Control Chart," *Journal of Quality Technology*, Vol. 29(1), pp. 8–15.

Quesenberry, C. P. (1988). "An SPC Approach to Compensating a Tool-Wear Process," *Journal of Quality Technology*, Vol. 20(4), pp. 220–229.

Quesenberry, C. P. (1991a). "SPC *Q* Charts for Start-Up Processes and Short or Long Runs," *Journal of Quality Technology*, Vol. 23(3), pp. 213–224.

Quesenberry, C. P. (1991b). "SPC *Q* Charts for a Binomial Parameter *p*: Short or Long Runs," *Journal of Quality Technology*, Vol. 23(3), pp. 239–246.

Quesenberry, C. P. (1991c). "SPC *Q* Charts for a Poisson Parameter+: Short or Long Runs," *Journal of Quality Technology*, Vol. 23(4), pp. 296–303.

Quesenberry, C. P. (1993). "The Effect of Sample Size on Estimated Limits for $\bar{x}$ and x Control Charts," *Journal of Quality Technology*, Vol. 25(4), pp. 237–247.

Quesenberry, C. P. (1995a). "On Properties of *Q* Charts for Variables," *Journal of Quality Technology*, Vol. 27(3), pp. 184–203.

Quesenberry, C. P. (1995b). "On Properties of Binomial *Q*-charts for Attributes," *Journal of Quality Technology*, Vol. 27(3), pp. 204–213.

Quesenberry, C. P. (1995c). "Geometric *Q* Charts for High Quality Processes," *Journal of Quality Technology*, Vol. 27(4), pp. 304–315.

Quesenberry, C. P. (1995d). "On Properties of Poisson *Q* Charts for Attributes" (with discussion), *Journal of Quality Technology*, Vol. 27(4), pp. 293–303.

Ramìrez, J. G. (1998). "Monitoring Clean Room Air Using Cuscore Charts," *Quality and Reliability Engineering International*, Vol. 14(3), pp. 281–289.

Reynolds, M. R., Jr., R. W. Amin, J. C. Arnold, and J. A. Nachlas (1988). "$\bar{x}$ Charts with Variable Sampling Intervals," *Technometrics*, Vol. 30(2), pp. 181–192.

Reynolds, M. R., Jr., and G.-Y. Cho (2006). "Multivariate Control Charts for Monitoring the Mean Vector and Covariance Matrix," *Journal of Quality Technology*, Vol. 38, pp. 230–253.

Rhoads, T. R., D. C. Montgomery, and C. M. Mastrangelo (1996). "Fast Initial Response Scheme for the EWMA Control Chart," *Quality Engineering*, Vol. 9(2), pp. 317–327.

Roberts, S. W. (1958). "Properties of Control Chart Zone Tests," *Bell System Technical Journal*, Vol. 37, pp. 83–114.

Roberts, S. W. (1959). "Control Chart Tests Based on Geometric Moving Averages," *Technometrics*, Vol. 42(1), pp. 97–102.

Rocke, D. M. (1989). "Robust Control Charts," *Technometrics*, Vol. 31(2), pp. 173–184.

Rodriguez, R. N. (1992). "Recent Developments in Process Capability Analysis," *Journal of Quality Technology*, Vol. 24(4), pp. 176–187.

Rolka, H., H. Burkom, G. F. Cooper, M. Kulldorff, D. Madigan, and W. K. Wong (2007). "Issues in Applied Statistics for Public Health Bioterrorism Surveillance Using Multiple Data Streams: Some Research Needs," *Statistics in Medicine*, Vol. 26, pp. 1834–1856.

Ross, S. M. (1971). "Quality Control Under Markovian Deterioration," *Management Science*, Vol. 17(9), pp. 587–596.

Runger, G. C., and T. R. Willemain (1996). "Batch Means Control Charts for Autocorrelated Data," *IIE Transactions*, Vol. 28(6), pp. 483–487.

Runger, G. C., F. B. Alt, and D. C. Montgomery (1996a). "Controlling Multiple Stream Processes with Principal Components," *International Journal of Production Research*, Vol. 34(11), pp. 2991–2999.

Runger, G. C., F. B. Alt, and D. C. Montgomery (1996b). "Contributors to a Multivariate Statistical Process Control Signal," *Communications in Statistics—Theory and Methods*, Vol. 25(10), pp. 2203–2213.

Runger, G. C., and M. C. Testik (2003). "Control Charts for Monitoring Fault Signatures: Cuscore Versus GLR," *Quality and Reliability Engineering International*, Vol. 19(4), pp. 387–396.

Saniga, E. M. (1989). "Economic Statistical Control Chart Design with an Application to $\bar{x}$ and *R* Charts," *Technometrics*, Vol. 31(3), pp. 313–320.

Saniga, E. M., and L. E. Shirland (1977). "Quality Control in Practice—A Survey," *Quality Progress*, Vol. 10(5), pp. 30–33.

Savage, I. R. (1962). "Surveillance Problems," *Naval Research Logistics Quarterly*, Vol. 9(384), pp. 187–209.

Schilling, E. G., and P. R. Nelson (1976). "The Effect of Nonnormality on the Control Limits of $\bar{x}$ Charts," *Journal of Quality Technology,* Vol. 8(4), pp. 183–188.

Schmidt, S. R., and J. R. Boudot (1989). "A Monte Carlo Simulation Study Comparing Effectiveness of Signal-to-Noise Ratios and Other Methods for Identifying Dispersion Effects," presented at the 1989 Rocky Mountain Quality Conference.

Scranton, R., G. C. Runger, J. B. Keats, and D. C. Montgomery (1996). "Efficient Shift Detection Using Exponentially Weighted Moving Average Control Charts and Principal Components," Quality and Reliability Engineering International, Vol. 12(3), pp. 165–172.

Shapiro, S. S. (1980). *How to Test Normality and Other Distributional Assumptions,* Vol. 3, The ASQC Basic References in Quality Control: Statistical Techniques, ASQC, Milwaukee, WI.

Sheaffer, R. L., and R. S. Leavenworth (1976). "The Negative Binomial Model for Counts in Units of Varying Size," *Journal of Quality Technology,* Vol. 8(3), pp. 158–163.

Siegmund, D. (1985). *Sequential Analysis: Tests and Confidence Intervals,* Springer-Verlag, New York.

Snee, R. D. and R. W. Hoerl (2005). *Six Sigma Beyond the Factory Floor,* Pearson Prentice Hall, Upper Saddle River, NJ.

Sonesson, C. and D. Bock (2003). "A Review and Discussion of Prospective Statistical Surveillance in Public Health," *Journal of the Royal Statistical Society,* Series A, Vol. 166, pp. 5–21.

Somerville, S. E., and D. C. Montgomery (1996). "Process Capability Indices and Nonnormal Distributions," *Quality Engineering,* Vol. 9(2), pp. 305–316.

Spiring, F., B. Leung, S. Cheng, and A. Yeung (2003). "A Bibliography of Process Capability Papers," *Quality and Reliability Engineering International,* Vol. 19(5), pp. 445–460.

Staudhammer, C., T. C. Maness, and R. A. Kozak (2007). "Profile Charts for Monitoring Lumber Manufacturing Using Laser Range Sensor Data," *Journal of Quality Technology,* Vol. 39, pp. 224–240.

Steiner, S. H. (1999). "EWMA Control Charts with Time-Varying Control Limits and Fast Initial Response," *Journal of Quality Technology,* Vol. 31(1), pp. 75–86.

Steiner, S. H., and R. J. MacKay (2000). "Monitoring Processes with Highly Censored Data," *Journal of Quality Technology,* Vol. 32(3), pp. 199–208.

Stephens, K. S. (1979). *How to Perform Continuous Sampling (CSP),* Vol. 2, The ASQC Basic References in Quality Control: Statistical Techniques, ASQC, Milwaukee, WI.

Stover, F. S. and R. V. Brill (1998). "Statistical Quality Control Applied to Ion Chromatography Calibrations," *Journal of Chromatography A,* Vol. 804, pp. 37–43.

Stoumbos, Z. G, and J. H. Sullivan (2002). Robustness to Non-normality of the Multivariate EWMA Control Chart," *Journal of Quality Technology,* Vol. 34(3), pp. 260–276.

Stover, F. S., and R. V. Brill (1998). "Statistical Quality Control Applied to Ion Chromatography Calibrations," *Journal of Chromatography,* A 804, pp. 37–43.

Sullivan, J. H., and W. H. Woodall (1995). "A Comparison of Multivariate Quality Control Charts for Individual Observations," *Journal of Quality Technology,* Vol. 28(4), pp. 398–408.

Svoboda, L. (1991). "Economic Design of Control Charts: A Review and Literature Survey (1979–1989)," in *Statistical Process Control in Manufacturing,* J. B. Keats and D. C. Montgomery, (eds.) Dekker, New York.

Taguchi, G. (1986). *Introduction to Quality Engineering,* Asian Productivity Organization, UNIPUB, White Plains, NY.

Taguchi, G., and Y. Wu (1980). *Introduction to Off-Line Quality Control,* Japan Quality Control Organization, Nagoya, Japan.

Taylor, H. M. (1965). "Markovian Sequential Replacement Processes," *Annals of Mathematical Statistics,* Vol. 36(1), pp. 13–21.

Taylor, H. M. (1967). "Statistical Control of a Gaussian Process," *Technometrics,* Vol. 9(1), pp. 29–41.

Taylor, H. M. (1968). "The Economic Design of Cumulative Sum Control Charts," *Technometrics,* Vol. 10(3), pp. 479–488.

Testik, M. C., G. C. Runger, and C. M. Borror (2003). "Robustness Properties of Multivariate EWMA Control Charts," *Quality and Reliability Engineering International,* Vol. 19(1), pp. 31–38.

Testik, M. C., and C. M. Borror (2004). "Design Strategies for the Multivariate EWMA Control Chart," *Quality and Reliability Engineering International,* Vol. 20, pp. 571–577.

Tobias, P. A., and Trindade, D. C. (1995). *Applied Reliability,* Chapman and Hall/CRC, Boca Raton, FL.

Tracy, N. D., J. C. Young, and R. L. Mason (1992). "Multivariate Control Charts for Individual Observations," *Journal of Quality Technology,* Vol. 24(2), pp. 88–95.

Tseng, S., and B. M. Adams (1994). "Monitoring Autocorrelated Processes with an Exponentially Weighted Moving Average Forecast," *Journal of Statistical Computation and Simulation,* Vol. 50(3–4), pp. 187–195.

Tsui, K. and S. Weerahandi (1989). "Generalized *p*-values in Significance Testing of Hypotheses in the Presence of Nuisance Parameters," *Journal of the American Statistical Association,* Vol. 84, pp. 602–607.

U.S. Department of Defense (1960). *Quality Control and Reliability Handbook,* Government Printing Office, Washington, DC.

U.S. Department of Defense (1977). *Reliability Design Qualification and Production Acceptance Tests: Exponential Distribution, Military Standard 781C,* Government Printing Office, Washington, DC.

U.S. Department of Defense (1980). *Procedures for Performing a Failure Mode, Effects, and Criticality Analysis, Military Standard 1629A,* Government Printing Office, Washington, DC.

United States Department of Defense (1957). *Sampling Procedures and Tables for Inspection by Variables for Percent Defective,* MIL STD 414, U.S. Government Printing Office, Washington, DC.

United States Department of Defense (1989). *Sampling Procedures and Tables for Inspection by Attributes,* MIL STD 105E, U.S. Government Printing Office, Washington, DC.

Vance, L. C. (1986). "Average Run Lengths of Cumulative Sum Control Charts for Controlling Normal Means," *Journal of Quality Technology,* Vol. 18(3), pp. 189–193.

Vander Weil, S., W. T. Tucker, F. W. Faltin, and N. Doganaksoy (1992). "Algorithmic Statistical Process Control: Concepts and an Application," *Technometrics,* Vol. 34(3), pp. 286–288.

Van Wieringen, W. N. (2003). *Statistical Models for the Precision of Categorical Measurement Systems.* Ph.D. thesis, University of Amsterdam.

Wadsworth, H. M., K. S. Stephens, and A. B. Godfrey (2002). *Modern Methods for Quality Control and Improvement,* 2nd ed., Wiley, New York.

Wald, A. (1947). *Sequential Analysis,* Wiley, New York.

Walker, E., J. W. Philpot, and J. Clement (1991). "False Signal Rates for the Shewhart Control Chart with Supplementary Runs Tests," *Journal of Quality Technology,* Vol. 23(3), pp. 247–252.

Walker, E., and S. P. Wright (2002). "Comparing Curves Using Additive Models," *Journal of Quality Technology,* Vol. 34(1), pp. 118–129.

Wang, C.–H., and F. S. Hillier (1970). "Mean and Variance Control Chart Limits Based on a Small Number of Subgroups," *Journal of Quality Technology,* Vol. 2(1), pp. 9–16.

Wang, K. B. and F. Tsung (2005). "Using Profile Monitoring Techniques for a Data-Rich Environment with Huge Sample Size," *Quality and Reliability Engineering International,* Vol. 21, pp. 677–688.

Wardell, D. G., H. Moskowitz, and R. D. Plante (1994). "Run Length Distributions of Special-Cause Control Charts for Correlated Processes," *Technometrics,* Vol. 36(1), pp. 3–18.

Weerahandi, S. (1993). "Generalized Confidence Intervals," *Journal of the American Statistical Association,* Vol. 88, pp. 899–905.

Weiler, H. (1952). "On the Most Economical Sample Size for Controlling the Mean of a Population," *Annals of Mathematical Statistics,* Vol. 23(2), pp. 247–254.

Western Electric (1956). *Statistical Quality Control Handbook,* Western Electric Corporation, Indianapolis, IN.

Wetherill, G. B., and D. W. Brown (1991). *Statistical Process Control: Theory and Practice,* Chapman and Hall, New York.

White, C. C. (1974). "A Markov Quality Control Process Subject to Partial Observation," *Management Science,* Vol. 23(8), pp. 843–852.

White, C. H., J. B. Keats, and J. Stanley (1997). "Poisson Cusum Versus *c* Chart for Defect Data," *Quality Engineering,* Vol. 9(4), pp. 673–679.

White, E. M., and R. Schroeder (1987). "A Simultaneous Control Chart," *Journal of Quality Technology,* Vol. 19(1), pp. 1–10.

Willemain, T. R., and G. C. Runger (1996). "Designing Control Charts Based on an Empirical Reference Distribution," *Journal of Quality Technology,* Vol. 28(1), pp. 31–38.

Williams, J. D., W. H. Woodall, J. B. Birch, and J. H. Sullivan (2006). "Distribution of Hotelling's T^2 Statistic Based on the Successive Difference Estimator," *Journal of Quality Technology,* Vol. 38, pp. 217–229.

Winkel, P., and N. F. Zhang (2007). *Statistical Development of Quality in Medicine,* John Wiley and Sons, New York.

Woodall, W. H. (1986). "Weakness of the Economic Design of Control Charts," Letter to the Editor, *Technometrics,* Vol. 28(4), pp. 408–409.

Woodall, W. H. (1987). "Conflicts between Deming's Philosophy and the Economic Design of Control Charts," in *Frontiers in Statistical Quality Control,* 3rd ed., H. J. Lenz, G. B. Wetherill, and P.–T. Wilrich, Physica-Verlag, Vienna, pp. 242–248.

Woodall, W. H. (1997). "Control Charts Based on Attribute Data: Bibliography and Review," *Journal of Quality Technology,* Vol. 29(2), pp. 172–183.

Woodall, W. H. (2000). "Controversies and Contradictions in Statistical Process Control," *Journal of Quality Technology,* Vol. 20(4), pp. 515–521. Available at www.asq.org/pub/jqt.

Woodall, W. H. (2007). "Profile Monitoring," entry in *Encyclopedia of Statistics in Quality and Reliability,* edited by Fabrizio Ruggeri, Frederick Faltin, and Ron Kenett, John Wiley & Sons, New York.

Woodall, W. H., and B. M. Adams (1993). "The Statistical Design of CUSUM Charts," *Quality Engineering,* Vol. 5(4), pp. 559–570.

Woodall, W. H., and D. C. Montgomery (1999). "Research Issues and Ideas in Statistical Process Control," *Journal of Quality Technology,* Vol. 31(4), pp. 376–386.

Woodall, W. H., and D. C. Montgomery (2000–2001). "Using Ranges to Estimate Variability," *Quality Engineering,* Vol. 13(2), pp. 211–217.

Woodall, W. H., D. J. Spitzner, D. C. Montgomery, and S. Gupta (2004). "Using Control Charts to Monitor Process and Product Quality Profiles," *Journal of Quality Technology,* Vol. 36, pp. 309–320.

Woodall, W. H. (2006). "Use of Control Charts in Health Care Monitoring and Public Health Surveillance" (with discussion), *Journal of Quality Technology,* Vol. 38, pp. 89–104. Available at www.asq.org/pub/jqt.

Woodall, W. H. and M. A. Mahmoud (2005), "The Inertial Properties of Quality Control Charts," *Technometrics,* Vol. 47, pp. 425–436.

Yourstone, S. A., and D. C. Montgomery (1989). "Development of a Real-Time Statistical Process-Control Algorithm," *Quality and Reliability Engineering International,* Vol. 5(4), pp. 309–317.

Yourstone, S., and W. Zimmer (1992). "Non-normality and the Design of Control Charts for Averages," *Decision Sciences,* Vol. 23(5), pp. 1099–1113.

Zhang, N. F., G. A. Stenback, and D. M. Wardrop (1990). "Interval Estimation of Process Capability Index C_{pk}," *Communications in Statistics—Theory and Methods,* Vol. 19(12), pp. 4455–4470.

Zou, C., Y. Zhang, and Z. Wang (2006). "A Control Chart Based on a Change-Point Model for Monitoring Profiles," *IIE Transactions,* Vol. 38, pp. 1093–1103.

Glossary

α-error (or α-risk). In hypothesis testing, an error incurred by rejecting a null hypothesis when it is actually true (also called a type I error).

β-error (or β-risk). In hypothesis testing, an error incurred by failing to reject a null hypothesis when it is actually false (also called a type II error).

Acceptance region. In hypothesis testing, a region in the sample space of the test statistic such that if the test statistic falls within it, the null hypothesis is accepted (better terminology is that the null hypothesis cannot be rejected, since rejection is always a strong conclusion and acceptance is generally a weak conclusion).

Addition rule. A formula used to determine the probability of the union of two (or more) events from the probabilities of the events and their intersection(s).

Additivity property of χ^2. The property of the chi-squared distribution that if two independent random variables X_1 and X_2 are distributed as chi-squared with v_1 and v_2 degrees of freedom respectively, then $Y = X_1 + X_2$ is a chi-squared random variable with $u = v_1 + v_2$ degrees of freedom. This generalizes to any number of independent chi-squared random variables.

Adjusted R^2. A variation of the R^2 statistic that compensates for the number of parameters in a regression model. Essentially, the adjustment is a penalty for increasing the number of parameters in the model.

Alias. In a fractional factorial experiment, when certain factor effects cannot be estimated uniquely, they are said to be aliased.

Alternative hypothesis. In statistical hypothesis testing, this is a hypothesis other than the one that is being tested. The alternative hypothesis contains feasible conditions, whereas the null hypothesis specifies conditions that are under test.

Analysis of variance. A method of decomposing the total variability in a set of observations, as measured by the sum of the squares of these observations from their average, into component sums of squares that are associated with specific defined sources of variation.

Analytic study. A study in which a sample from a population is used to make inference to a future population. Stability needs to be assumed. *See* enumerative study.

Arithmetic mean. The arithmetic mean of a set of numbers $x_1, x_2, ..., x_n$ is their sum divided by the number of observations, or $(1/n)\sum_{i=1}^{n} x_i$. The arithmetic mean is usually denoted by $\bar{x}$ and is often called the **average**.

Assignable cause. The portion of the variability in a set of observations that can be traced to specific causes, such as operators, materials, or equipment. Also called a special cause.

Attribute. A qualitative characteristic of an item or unit, usually arising in quality control. For example, classifying production units as defective or nondefective results in attributes data.

Average run length (ARL). The average number of samples taken in a process monitoring or inspection scheme until the scheme signals that the process is operating at a level different from the level at which it began.

Average. *See* arithmetic mean.

Bernoulli trials. Sequences of independent trials with only two outcomes, generally called "success" and "failure," in which the probability of success remains constant.

Bias. An effect that systematically distorts a statistical result or estimate, preventing it from representing the true quantity of interest.

Biased estimator. *See* unbiased estimator.

Bimodal distribution. A distribution with two modes.

Binomial random variable. A random variable that is the number of successes in a specified number of Bernoulli trials.

Block. In experimental design, a group of experimental units or material that is relatively homogeneous. The purpose of dividing experimental units into blocks is to produce an experimental design wherein variability within blocks is smaller than variability between blocks. This allows the factors of interest to be compared in a environment that has less variability than in an unblocked experiment.

Box plot (or box and whisker plot). A graphical display of data in which the box contains the middle 50 percent of the data (the interquartile range) with the median dividing it, and the whiskers extend to the smallest and largest values (or some defined lower and upper limits).

c chart. *See* control chart for nonconformities.

Categorical data. Data consisting of counts or observations that can be classified into categories. The categories may be descriptive.

Causal variable. When $y = f(x)$ and y is considered to be caused by x, then x is sometimes called a causal variable.

Center line. *See* control chart.

Central composite design (CCD). A second-order response surface design in k variables consisting of a two-level factorial, $2k$ axial runs, and one or more center points. The two-level factorial portion of a CCD can be a fractional factorial design when k is large. The CCD is the most widely used design for fitting a second-order model.

Central limit theorem. The simplest form of the central limit theorem states that the sum of n independently distributed random variables will tend to be normally distributed as n becomes large. It is a necessary and sufficient condition that none of the variances of the individual random variables are large in comparison to their sum. There are more general forms of the central theorem that allow infinite variances and correlated random variables, and there is a multivariate version of the theorem.

Central tendency. The tendency of data to cluster around some value. Central tendency is usually expressed by a measure of location such as the mean, median, or mode.

Chain sampling. A sampling procedure that makes use of information collected over several lots. It is usually a good alternative to a single sampling plan with a zero acceptance number.

Chance cause of variation. The portion of the variability in a set of observations that is due to only random forces and that cannot be traced to specific sources, such as operators, materials, or equipment. Also called a common cause.

Chi-squared (or chi-square) random variable. A continuous random variable that results from the sum squares of independent standard normal random variables. It is a special case of a gamma random variable.

Chi-squared test. Any test of significance based on the chi-square distribution. The most common chi-square tests are (1) testing hypotheses about the variance or standard deviation of a normal distribution and (2) testing goodness of fit of a theoretical distribution to sample data.

Completely randomized design. A type of experimental design in which the treatments or design factors are assigned to the experimental units in a random manner. In designed experiments, a completely randomized design results from running all of the treatment combination in random order.

Components of variance. The individual components of the total variance that are attributable to specific sources. This usually refers to the individual variance components arising from a random or mixed model analysis of variance.

Confidence coefficient. The probability $1 - \alpha$ associated with a confidence interval expressing the probability that the stated interval will contain the true parameter value.

Confidence interval. If it is possible to write a probability statement of the form $P(L \leq \theta \leq U) = 1 - \alpha$ where L and U are functions of only the sample data and θ is a parameter, then the interval between L and U is called a confidence interval (or a $100(1 - \alpha)\%$ confidence interval). The interpretation is that a statement that the parameter θ lies in this interval will be true $100(1 - \alpha)\%$ of the times that such a statement is made.

Confidence level. Another term for the confidence coefficient.

Confounding. When a factorial experiment is run in blocks and the blocks are too small to contain a complete replicate of the experiment, one can run a fraction of the replicate in each block, but this results in losing information on some effects. These effects are linked with or confounded with the blocks. In general, when two factors are varied such that their individual effects cannot be determined separately, their effects are said to be confounded.

Continuity correction. A correction factor used to improve the approximation to binomial probabilities from a normal distribution.

Continuous distribution. A probability distribution for a continuous random variable.

Continuous sampling. An acceptance sampling procedure that can be applied to a continuous production system where there is no natural way to construct lots.

Continuous random variable. A random variable with an interval (either finite or infinite) of real numbers for its range.

Continuous uniform random variable. A continuous random variable with a range of a finite interval and constant probability density function.

Contour plot. A two-dimensional graphic used for a bivariate probability density function that displays curves for which the probability density function is constant.

Control chart. A graphical display used to monitor a process. It usually consists of a horizontal center line corresponding to the in-control value of the parameter that is being monitored and lower and upper control limits. The control limits are determined by statistical criteria and are not arbitrary; nor are they related to specification limits. If sample points fall within the control limits, the process is said to be in control, or free from assignable causes. Points beyond the control limits indicate an out-of-control process; that is, assignable causes are likely present. This signals the need to find and remove the assignable causes.

Control chart for average number of nonconformities per unit. A control chart that plots the average number of nonconformities found in an inspection unit versus the sample number or time.

Control chart for fraction nonconforming. A control chart that plots the fraction of nonconforming items in a sample versus the sample number or time. The fraction of nonconforming items is also called the fraction defective.

Control chart for nonconformities. A control chart that plots the total number of nonconformities on a unit vcrsus the sample number or time. Nonconformities are often referred to as defects.

Control limits. *See* control chart.

Consumer's risk. The probability that a consumer will accept a lot of poor quality.

Correction factor. A term used for the quantity $(1/n)(\sum_{i=1}^{n} x_i)^2$ that is subtracted from $\sum_{i=1}^{n} x_i^2$ to give the corrected sum of squares defined as $(1/n)\sum_{i=1}^{n}(x_i - \bar{x})^2$. The correction factor can also be written as $n\bar{x}^2$.

Correlation coefficient. A dimensionless measure of the interdependence between two variables, usually lying in the interval from -1 to $+1$, with 0 indicating the absence of correlation (but not necessarily the independence of the two variables). The most common form of the correlation coefficient used in practice is

$$r = \sum_{i=1}^{n}[(y_i - \bar{y})(x_i - \bar{x})] / \sqrt{\sum_{i=1}^{n}(y_i - \bar{y})^2 \sum_{i=1}^{n}(x_i - \bar{x})^2}$$

which is also called the product moment correlation coefficient. It is a measure of the linear association between the two variables y and x.

Correlation. In the most general usage, a measure of the interdependence among data. The concept may include more than two variables. The term is most commonly used in a narrow sense to express the relationship between quantitative variables or ranks.

Covariance matrix. A square matrix that contains the variances and covariances among a set of random variables, say, $X_1, X_2, \ldots, X_k$. The main diagonal elements of the matrix are the variances of the random variables and the off-diagonal elements are the covariances between X_i and X_j. Also called the **variance–covariance matrix**. When the random variables are standardized to have unit variances, the covariance matrix becomes the correlation matrix.

Covariance. A measure of association between two random variables obtained as the expected value of the product of the two random variables around their means; that is, $Cov(X,Y) = E[(X - \mu_X)(Y - \mu_Y)]$.

Critical region. In hypothesis testing, the portion of the sample space of a test statistic that will lead to rejection of the null hypothesis.

Critical value(s). The value of a statistic corresponding to a stated significance level as determined from the sampling distribution. For example, if $P(Z \geq z_{0.025}) = P(Z \geq 1.96) = 0.025$, then $z_{0.025} = 1.96$ is the critical value of z at the 0.025 level of significance.

Crossed factors. Another name for factors that are arranged in a factorial experiment.

Cumulative distribution function. For a random variable X, the function of x defined as $P(X \leq x)$ that is used to specify the probability distribution.

Cumulative normal distribution function. The cumulative distribution of the standard normal distribution, often denoted as $\Phi(x)$ and tabulated in Appendix Table II.

Cumulative sum control chart. A control chart that plots the sum of deviations from the in-control process mean or target. Because it accumulates data over several sequential samples, it has greater sensitivity to small shifts than Shewhart type control charts.

Defect. Used in statistical quality control, a defect is a particular type of nonconformance to specifications or requirements. Sometimes defects are classified into types, such as appearance defects and functional defects.

Degrees of freedom. The number of independent comparisons that can be made among the elements of a sample. The term is analogous to the number of degrees of freedom for an object in a dynamic system, which is the number of independent coordinates required to determine the motion of the object.

Density function. Another name for a probability density function.

Dependent variable. The response variable in regression or a designed experiment.

Discrete distribution. A probability distribution for a discrete random variable.

Discrete random variable. **A random variable with a finite (or countably infinite) range.**

Discrete uniform random variable. A discrete random variable with a finite range and constant probability mass function.

Dispersion. The amount of variability exhibited by data.

Distribution function. Another name for a cumulative distribution function.

Double sampling plan. An acceptance sampling procedure where, based on the results of a first sample, a second sample may be taken to determine the disposition of a lot.

Enumerative study. A study in which a sample from a population is used to make inference to the population. *See* analytic study.

Error mean square. The error sum of squares divided by its number of degrees of freedom.

Error of estimation. The difference between an estimated value and the true value.

Error sum of squares. In analysis of variance, this is the portion of total variability that is due to the random component in the data. It is usually based on replication of observations at certain treatment combinations in the experiment. It is sometimes called the residual sum of squares, although this is really a better term to use only when the sum of squares is based on the remnants of a model fitting process and not on replication.

Error variance. The variance of an error term or component in a model.

Estimate (or point estimate). The numerical value of a point estimator.

Estimator (or point estimator). A procedure for producing an estimate of a parameter of interest. An estimator is usually a function of only sample data values and when these data values are available results in an estimate of the parameter of interest.

Event. A subset of a sample space.

Expected value. The expected value of a random variable X is its long-term average or mean value. In the continuous case, the expected value of X is $E(X) = \int_{-\infty}^{\infty} xf(x)dx$ where $f(x)$ is the density function of the random variable X.

Exponential random variable. A continuous random variable that is the time between counts in a Poisson process. It is often used in reliability engineering to model the time to failure of a system or component.

Exponentially weighted moving average (EWMA) control chart. A control chart that plots the exponentially weighted average of the current sample measurement. The EWMA of a measurement *x-t-* is defined as $Z_t = \lambda x_t + (1 - \lambda)Z_{t-1}$ where $\lambda(0 < \lambda < 1)$ is a smoothing or discount factor. Because it accumulates data over several sequential samples, it is more sensitive to small shifts than Shewhart type control charts.

Factorial experiment. A type of experimental design in which every level of one factor is tested in combination with every level of another factor. In general, in a factorial experiment, all possible combinations of factor levels are tested.

Failure rate. The probability that a unit will fail in the next instance of time; also called the **instantaneous failure rate**, or the **hazard function**.

***F*-distribution.** The distribution of the random variables defined as the ratio of two independent chi-square random variables each divided by their number of degrees of freedom.

First-order model. A model that contains only first-order terms. For example, the first-order response surface model in two variables is $y = \beta_0 + \beta_1 x_1 + \beta_2 x_2 + \varepsilon$. A first-order model is also called a main effects model.

Fixed factor (or fixed effect). In analysis of variance, a factor or effect is considered fixed if all the levels of interest for that factor are included in the experiment. Conclusions are then valid about this set of levels only, although when the factor is quantitative, it is customary to fit a model to the data for interpolating between these levels.

Fraction defective. In statistical quality control, that portion of a number of units or the output of a process that is defective.

Fractional factorial. A type of factorial experiment in which not all possible treatment combinations are run. This is usually done to reduce the size of an experiment with several factors.

Frequency distribution. An arrangement of the frequencies of observations in a sample or population according to the values that the observations take on.

***F*-test.** Any test of significance involving the *F*-distribution. The most common *F*-tests tests are (1) testing hypotheses about the variances or standard deviations of two independent normal distributions, (2) testing hypotheses about treatment means or variance components in the analysis of variance, and (3) testing significance of regression or tests on subsets of parameters in a regression model.

Gaussian distribution. Another name for the normal distribution, based on the strong connection of Karl F. Gauss to the normal distribution; this terminology is often used in physics and electrical engineering applications.

Geometric mean. The geometric mean of a set of n positive data values is the nth root of the product of the data values; that is, $\bar{g} = \left(\prod_{i=1}^{n} x_i\right)^{1/n}$.

Geometric random variable. A discrete random variable that is the number of Bernoulli trials until a success occurs.

Goodness of fit. In general, the agreement of a set of observed values and a set of theoretical values that depend on some hypothesis. The term is often used in fitting a theoretical distribution to a set of observations.

Harmonic mean. The harmonic mean of a set of data values is the reciprocal of the arithmetic mean of the reciprocals of the data values; that is, $\bar{h} = \left(\frac{1}{n}\sum_{i=1}^{n}\frac{1}{x_i}\right)^{-1}$.

Hat matrix. In multiple regression, the matrix $\mathbf{H} = \mathbf{X}(\mathbf{X}'\mathbf{X})^{-1}\mathbf{X}'$. This a projection matrix that maps the vector of observed response values into a vector of fitted values by $\hat{\mathbf{y}} = \mathbf{X}(\mathbf{X}'\mathbf{X})^{-1}\mathbf{X}'\mathbf{y} = \mathbf{H}\mathbf{y}$

Hazard function. *See* failure rate.

Histogram. A univariate data display that uses rectangles proportional in area to class frequencies to visually exhibit features of data such as location, variability, and shape.

Hypergeometric random variable. A discrete random variable that is the number of success obtained from a sample drawn without replacement from a finite populations.

Hypothesis (as in statistical hypothesis). A statement about the parameters of a probability distribution or a model, or a statement about the form of a probability distribution.

Hypothesis testing. Any procedure used to test a statistical hypothesis.

Independence. A property of two (or more) events that allows the probability of the intersection to be calculated as the product of the probabilities.

Independent random variables. Random variables for which the following (and equivalent results) holds: $P(X \in A, Y \in B) = P(X \in A)P(Y \in B)$ for any sets A and B in the range of X and Y, respectively.

Independent variable. The predictor or regressor variables in a regression model.

Indicator variable(s). Variables that are assigned numerical values to identify the levels of a qualitative or categorical response. For example, a response with two categorical levels (yes and no) could be represented with an indicator variable taking on the values 0 and 1.

Interaction. In factorial experiments, two factors are said to interact if the effect of one variable is different at different levels of the other variables. In general, when variables operate independently of each other, they do not exhibit interaction.

Intercept. The constant term in an experimental design or regression model.

Interquartile range. The difference between the third and first quartiles of a sample of data. The interquartile range is less sensitive to extreme data values than the usual sample range.

Interval estimation. The estimation of a parameter by a range of values between lower and upper limits, in contrast to point estimation, where the parameter is estimated by a single numerical value. A confidence interval is a typical interval estimation procedure.

Kurtosis. A measure of the degree to which a unimodal distribution is peaked.

Lack of memory property. A property of a Poisson process. The probability of a count in an interval depends only on the length of the interval (and not on the starting point of the interval). A similar property holds for a series of Bernoulli trials. The probability of a success in a specified number of trials depends only on the number of trials (and not on the starting trial).

Least significance difference test (or Fisher's LSD test). An application of the *t*-test to compare pairs of means following rejection of the null hypothesis in an analysis of variance. The error rate is difficult to calculate exactly, because the comparisons are not all independent.

Least squares (method of). A method of parameter estimation in which the parameters of a system are estimated by minimizing the sum of the squares of the differences between the observed values and the fitted or predicted values from the system.

Least squares estimator. Any estimator obtained by the method of least squares.

Level of significance. If Z is the test statistic for a hypothesis, and the distribution of Z when the hypothesis is true is known, then we can find the probabilities $P(Z \leq z_L)$ and $P(Z \geq z_U)$. Rejection of the hypothesis is usually expressed in terms of the observed value of Z falling outside the interval from z_L to z_U. The probabilities $P(Z \leq z_L)$ and $P(Z \geq z_U)$ are usually chosen to have small values, such as 0.01, 0.025, 0.05, or 0.10, and are called levels of significance. The actual levels chosen are somewhat arbitrary, and are often expressed in percentages, such as a 5% level of significance.

Linear combination. A random variable that is defined as a linear function of several random variables.

Linear model. A model in which the observations are expressed as a linear function of the unknown parameters. For example, $y = \beta_0 + \beta_1 x + \varepsilon$ and $y = \beta_0 + \beta_1 x + \beta_2 x^2 + \varepsilon$ are linear models.

Location parameter. A parameter that defines a central value in a sample or a probability distribution. The mean and the median are location parameters.

Lower control limit. *See* control chart.

Main effect. An estimate of the effect of a factor (or variable) that independently expresses the change in response due to a change in that factor, regardless of other factors that may be present in the system.

Mean square. In general, a mean square is determined by dividing a sum of squares by the number of degrees of freedom associated with the sum of squares.

Mean. The mean usually refers either to the expected value of a random variable or to the arithmetic average of a set of data.

Mean time to failure (MTTF). The mean of a time-to-failure distribution. In the case of a repairable unit, this is sometimes called the **mean time between failures (MTBF).**

Median. The median of a set of data is that value that divides the data into two equal halves. When the number of observations is even, say $2n$, it is customary to define the median as the average of the nth and $(n + 1)$st rank-ordered values. The median can also be defined for a random variable. For example, in the case of a continuous random variable X the median M can be defined as $\int_{-\infty}^{M} f(x)dx = \int_{M}^{\infty} f(x)dx = \frac{1}{2}$.

Method of steepest ascent. A technique that allows an experimenter to move efficiently toward a set of optimal operating conditions by following the gradient direction. The method of steepest ascent is usually employed in conjunction with fitting a first-order response surface and deciding that the current region of operation is inappropriate.

Mixed model. In an analysis of variance context, a mixed model contains both random and fixed factors.

Mode. The mode of a sample is that observed value that occurs most frequently. In a probability distribution $f(x)$ with continuous first derivative, the mode is a value of x for which $df(x)/dx = 0$ and $df(x)/dx < 0$. There may be more than one mode of either a sample or a distribution.

Multiplication rule. A probability result used to determine the probability of the intersection of two (or more) events.

Mutually exclusive events. A collection of events whose intersections are empty.

Negative binomial random variable. A discrete random variable that is the number of trials until a specified number of successes occur in Bernoulli trials.

Normal approximation. A method to approximate probabilities for binomial and Poisson random variables.

Normal equations. The set of simultaneous linear equations arrived at in parameter estimation using the method of least squares.

Normal probability plot. A specially constructed plot for a variable x (usually on the abscissa) in which y (usually on the ordinate) is scaled so that the graph of the normal cumulative distribution is a straight line.

Normal random variable. A continuous random variable that is the most important one in statistics because it results from the central limit theorem. *See* central limit theorem.

Nuisance factor. A factor that probably influences the response variable but that is of no interest in the current study. When the levels of the nuisance factor can be controlled, blocking is the design technique that is customarily used to remove its effect.

Null hypothesis. This term generally relates to a particular hypothesis that is under test, as distinct from the alternative hypothesis (which defines other conditions that are feasible but not being tested). The null hypothesis determines the probability of type I error for the test procedure.

One-way model. In an analysis of variance context, this involves a single variable or factor with a different levels.

Operating characteristic curves (OC curves). A plot of the probability of type II error versus some measure of the extent to which the null hypothesis is false. Typically, one OC curve is used to represent each sample size of interest.

Orthogonal design. *See* orthogonal.

Orthogonal. There are several related meanings, including the mathematical sense of perpendicular, or two variables being said to be orthogonal if they are statistically independent, or, in experimental design, when a design admits statistically independent estimates of effects.

Outcome. An element of a sample space.

Outlier(s). One or more observations in a sample that are so far from the main body of data that they give rise to the question that they may be from another population.

Overfitting. Adding more parameters to a model than is necessary.

***p* chart.** *See* control chart for fraction nonconforming.

Parameter estimation. The process of estimating the parameters of a population or probability distribution. Parameter estimation, along with hypothesis testing, is one of the two major techniques of statistical inference.

Parameter. An unknown quantity that may vary over a set of values. Parameters occur in probability distributions and in statistical models, such as regression models.

Percentage point. A particular value of a random variable that defines a probability expressed as a percentage. For example, the upper five percentage points of the standard normal random variable is $z_{0.05} = 1.645$.

Percentile. The set of values that divide the sample into 100 equal parts.

Poisson random variable. A discrete random variable that is the number of counts that occur in a Poisson process.

Pooling. When several sets of data can be thought of as having been generated from the same model, it is possible to combine them, usually for purposes of estimating one or more parameters. Combining the samples for this purpose is usually called pooling.

Population standard deviation. *See* standard deviation.

Population variance. *See* variance.

Population. Any finite or infinite collection of individual units or objects.

Power. The power of a statistical test is the probability that the test rejects the null hypothesis when the null hypothesis is indeed false. Thus the power is equal to 1 minus the probability of type II error.

Prediction interval. The interval between a set of upper and lower limits associated with a predicted value designed to show on a probability basis the range of error associated with the prediction.

Prediction. The process of determining the value of one or more statistical quantities at some future point in time. In a regression model, predicting the response y for some specified set of regressors or predictor variables also leads to a predicted value, although there may be no temporal element to the problem.

Predictor variable(s). The independent or regressor variables in a regression model.

Probability density function. A function used to calculate probabilities and to specify the probability distribution of a continuous random variable.

Probability distribution. For a sample space, a description of the set of possible outcomes along with a method to determine probabilities. For a random variable, a description of the range along with a method to determine probabilities.

Probability mass function. A function that provides probabilities for the values in the range of a discrete random variable.

***P*-value.** The exact significance level of a statistical test; that is, the probability of obtaining a value of the test statistic that is at least as extreme as that observed when the null hypothesis is true.

Qualitative (data). Data derived from nonnumeric attributes, such as sex, ethnic origin or nationality, or other classification variable.

Quality control. Systems and procedures used by an organization to assure that the outputs from processes satisfy customers.

Quantiles. The set of $n - 1$ values of a variable that partition it into a number n of equal proportions. For example, $n - 1 = 3$ values partition data into four quartiles, with the central value usually called the median and the lower and upper values usually called the lower and upper quartiles, respectively.

Quantitative (data). Data in the form of numerical measurements or counts.

Quartile(s). The three values of a variable that partition it into four equal parts. The central value is usually called the median, and the lower and upper values are usually called the lower and upper quartiles, respectively. Also see quantiles.

R^2. A quantity used in regression and experimental design models to measure the proportion of total variability in the response accounted for by the model. Computationally, $R^2 = SS_{Model}/SS_{Total}$, and large values of R^2 (near unity) are considered good. However, it is possible to have large values of R^2 and yet find that the model is unsatisfactory. R^2 is also called the coefficient of determination (or the coefficient of multiple determination in multiple regression).

Random effects model. In an analysis of variance context, this refers to a model that involves only random factors.

Random error. An error (usually a term in a statistical model) that behaves as if it were drawn at random from a particular probability distribution.

Random factor. In analysis of variance, a factor whose levels are chosen at random from some population of factor levels.

Random order. A sequence or order for a set of objects that is carried out in such a way that every possible ordering is equally likely. In experimental design the runs of the experiment are typically arranged and carried out in random order.

Random sample. A sample is said to be random if it is selected in such a way so that every possible sample has the same probability of being selected.

Random variable. A function that assigns a real number to each outcome in the sample space of a random experiment.

Random. Nondeterministic, occurring purely by chance, or independent of the occurrence of other events.

Randomization. The arrangement of a set of objects in random order.

Randomized block design. A type of experimental design in which treatment (or factors levels) are assigned to blocks in a random manner.

Range. The largest member minus the smallest member of a set of data values. The range is a simple measure of variability and is widely used in quality control.

Range or *R* chart. A control chart used to monitor the variability (dispersion) in a process. *See* control chart.

Rank. In the context of data, the rank of a single observation is its ordinal number when all data values are ordered according to some criterion, such as their magnitude.

Reference distribution. The distribution of a test statistic when the null hypothesis is true. Sometimes a reference distribution is called the null distribution of the test statistic.

Rejection region. In hypothesis testing, the region in the sample space of the test statistic that leads to rejection of the null hypothesis when the test statistic falls in this region.

Reliability. The probability that a specified mission is completed. It usually refers to the probability that the lifetime of a continuous random variable exceeds a specified time limit.

Replicates. One of the independent repetitions of one or more treatment combinations treatments in an experiment.

Replication. The independent execution of an experiment more than once.

Reproductive property of the normal distribution. A linear combination of independent, normal random variables is a normal random variable.

Residual analysis. Any technique that uses the residuals, usually to investigate the adequacy of the model that was used to generate the residuals.

Residual sum of squares. *See* error sum of squares.

Residual. Generally this is the difference between the observed and the predicted value of some variable. For example, in regression, a residual is the difference between the observed value of the response and the corresponding predicted value obtained from the regression model.

Response (variable). The dependent variable in a regression model or the observed output variable in a designed experiment.

Response surface designs. Experimental designs that have been developed to work well in fitting response surfaces. These are usually designs for fitting a second-order model. The central composite design is a widely used response surface design.

Response surface. When a response y depends on a function of k quantitative variables $x_1, x_2, \ldots, x_k$ the values of the response may be viewed as a surface in $k + 1$ dimensions. This surface is called a response surface. Response surface methodology is a subset of experimental design concerned with approximating this surface with a model and using the resulting model to optimize the system or process.

Rotatable design. In a rotatable design the variance of the predicted response is the same at all points that are the same distance from the center of the design.

Sample mean. The arithmetic average or mean of the observations in a sample. If the observations are $x_1, x_2, \ldots, x_n$ then the sample mean is $(1/n)\sum_{i=1}^{n} x_i$. The sample mean is usually denoted by $\bar{x}$.

Sample moment. The quantity $(1/n)\sum_{i=1}^{n} x_i^k$ is called the kth sample moment.

Sample range. *See* range.

Sample size. The number of observations in a sample.

Sample standard deviation. The positive square root of the sample variance. The sample standard deviation is the most widely used measure of variability of sample data.

Sample variance. A measure of variability of sample data, defined as $S^2 = [1/(n - 1)]\sum_{i=1}^{n}(x_i - \bar{x})^2$, where $\bar{x}$ is the sample mean.

Sample. Any subset of the elements of a population.

Sampling distribution. The probability distribution of a statistic. For example, the sampling distribution of the sample mean $\bar{x}$ is the normal distribution.

Scatter diagram. A diagram displaying observations on two variables, x and y. Each observation is represented by a point showing its x-coordinate and its y-coordinate. The scatter diagram can be very effective in revealing the joint variability of x and y, or the nature of the relationship between them.

Screening experiment. An experiment designed and conducted for the purpose of screening out or isolating a promising set of factors for future experimentation. Many screening experiments are fractional factorials, such as two-level fractional factorial designs.

Second-order model. A model that contains second-order terms. For example, the second-order response surface model in two variables is $y = \beta_0 + \beta_1 x_1 + \beta_2 x_2 + \beta_{12} x_1 x_2 + \beta_{11} x_1^2 + \beta_{22} x_2^2 + \varepsilon$. The second-order terms in this model are $\beta_{12} x_1 x_2$, $\beta_{11} x_1^2$, and $\beta_2 x_2^2$.

Sequential sampling. An acceptance sampling procedure that operates by drawing single items from the lot one at a time and then, based on the results, deciding either to accept or reject the lot or draw another unit.

Significance level. *See* level of significance.

Significance. In hypothesis testing, an effect is said to be significant if the value of the test statistic lies in the critical region.

Single sampling plan. An acceptance sampling procedure where a decision about the disposition of a lot is made based on drawing a single random sample from the lot.

Skewness. A term for asymmetry usually employed with respect to a histogram of data or a probability distribution.

Standard deviation. The positive square root of the variance. The standard deviation is the most widely used measure of variability.

Standard error. The standard deviation of the estimator of a parameter. The standard error is also the standard deviation of the sampling distribution of the estimator of a parameter.

Standard normal random variable. A normal random variable with mean 0 and variance 1 that has its cumulative distribution function tabulated in Appendix Table II.

Standardize. The transformation of a normal random variable that subtracts its mean and divides by its standard deviation to generate a standard normal random variable.

Statistic. A summary value calculated from a sample of observations. Usually, a statistic is an estimator of some population parameter.

Statistics. The science of collecting, analyzing, interpreting, and drawing conclusions from data.

Stem and leaf display. A method of displaying data in which the stem corresponds to a range of data values and the leaf represents the next digit. It is an alternative to the histogram but displays the individual observations rather than sorting them into bins.

Studentized range. The range of a sample divided by the sample standard deviation.

Sufficient statistic. An estimator is said to be a sufficient statistic for an unknown parameter if the distribution of the sample given the statistic does not depend on the unknown parameter. This means that the distribution of the estimator contains all of the useful information about the unknown parameter.

***t*-distribution.** The distribution of the random variable defined as the ratio of two independent random variables. The numerator is a standard normal random variable and the denominator is the square root of a chi-squared random variable divided by its number of degrees of freedom.

Test statistic. A function of a sample of observations that provides the basis for testing a statistical hypothesis.

Time series. A set of ordered observations taken at different points in time.

Tolerance interval. An interval that contains a specified proportion of a population with a stated level of confidence.

Tolerance limits. A set of limits between which some stated proportion of the values of a population must fall with specified level of confidence.

Total probability rule. Given a collection of mutually exclusive events whose union is the sample space, the probability of an event can be written as the sum of the probabilities of the intersections of the event with the members of this collection.

Treatment sum of squares. In analysis of variance, this is the sum of squares that accounts for the variability in the response variable due to the different treatments that have been applied.

Treatment. In experimental design, a treatment is a specific level of a factor of interest. Thus if the factor is temperature, the treatments are the specific temperature levels used in the experiment.

***t*-test.** Any test of significance based on the t distribution. The most common t tests are (1) testing hypotheses about the mean of a normal distribution with unknown variance, (2) testing hypotheses about the means of two normal distributions, and (3) testing hypotheses about individual regression coefficients.

Type I error. In hypothesis testing, an error incurred by rejecting a null hypothesis when it is actually true (also called an α-error).

Type II error. In hypothesis testing, an error incurred by failing to reject a null hypothesis when it is actually false (also called a β-error).

u chart. *See* control chart for average number of nonconformities per unit.

Unbiased estimator. An estimator that has its expected value equal to the parameter that is being estimated is said to be unbiased.

Uniform random variable. Refers to either a discrete or continuous uniform random variable.

Uniqueness property of moment generating function. Refers to the following property. Random variables with the same moment generating function have the same distribution.

Universe. Another name for population.

Upper control limit. *See* control chart.

Variance component. In analysis of variance models involving random effects, one of the objectives is to determine how much variability can be associated with each of the potential sources of variability defined by the experimenters. It is customary to define a variance associated with each of these sources. These variances in some sense sum to the total variance of the response, and are usually called variance components.

Variance. A measure of variability defined as the expected value of the square of the random variable around its mean.

Weibull random variable. A continuous random variable that is often used to model the time until failure of a physical system or a component. The parameters of the distribution are flexible enough that the probability density function can assume many different shapes.

$\overline{X}$ control chart. A control chart for monitoring the mean or average level of a process.

Answers to Selected Problems

Chapter 3

3.6 No, the last four points are located at distance greater than 1σ from the center line.

3.7 Yes, the pattern is random.

3.8 Yes, the pattern is random.

Chapter 4

4.1 $\bar{x} = 16.0292,\ s = 0.0202$

4.3 $\bar{x} = 952.8889,\ s = 3.7231$

4.5 $\bar{x} = 121.2500,\ s = 22.6258$

4.9 It can be assumed the data follows a Normal distribution.

4.11 Normality assumption is reasonable.

4.13 $\bar{x} = 89.4756,\ s = 4.1578$

4.17
$$p(x) = \begin{matrix} \frac{1}{36};\ x=2 & \frac{2}{36};\ x=3 & \frac{3}{36};\ x=4 & \frac{4}{36};\ x=5 & \frac{5}{36};\ x=6 & \frac{6}{36};\ x=7 \\ \frac{5}{36};\ x=8 & \frac{4}{36};\ x=9 & \frac{3}{36};\ x=10 & \frac{2}{36};\ x=11 & \frac{1}{36};\ x=12 & 0;\ \textit{otherwise} \end{matrix}$$

4.18 $E(X) = 7,\ V(X) = 5.8\overline{3}$

4.19 Less than 35 lb: 7,932.76 $\approx$ 7933 *parts*, Greater than 48 lb: 2,739.97 $\approx$ 2740 *parts*

4.23 $LSL = 4{,}871.2085$

4.25 **(a)** $\mu \geq 125.8369$, **(b)** 95% of the time, the mean tensile strength is expected to be equal to or exceed 125.84 psi.

4.27 **(a)** $25.0581 \leq \mu \leq 26.9419$, **(b)** Normality assumption is reasonable, and the manufacturer can expect the mean battery life to exceed 25 hrs 90% of the time.

4.29 $457.4239 \leq \mu \leq 508.5761$, there is no evidence billboard sales have increased.

4.31 $0.50456 \leq \mu \leq 0.50464$, the 95% CI suggests μ is greater than 0.5025.

4.33 $-0.8514 \leq \mu_{CM} - \mu_{LM} \leq -0.1494$, there is evidence that payments are increasing.

4.35 (a) Assuming equal variances: $-6.71 \leq \mu_{SQ} - \mu_{OQ} \leq 3.11$, (b) $0.21 \leq \sigma_{SQ}^2/\sigma_{OQ}^2 \leq 3.34$, (c) Yes, (d) No significance difference in means or variances.

4.37 $p \leq 0.1514$, 90% of the time, the true defective fraction is less than 15.14% (evidence suggests p is larger than 0.10)

4.41 (a) $0.2278 \leq \sigma_1^2/\sigma_2^2 \leq 4.6113$, conclude variances are equal; (b) $-0.8 \leq \mu_1 - \mu_2 \leq 4.34$, can't conclude the new device reduced the mean percentage of impurities.

4.43 $-0.0020 \leq \mu_{micro} - \mu_{vernier} \leq 0.0011$, no significant difference

4.45 (a) $9.35 \leq \mu_{Northbrook} - \mu_{Southbrook} \leq 42.21$, evidence suggests recovery times at Northbrook are larger; (b) $0.3394 \leq \sigma_1^2/\sigma_2^2 \leq 5.5006$, no evidence to suggest a significant difference in the variances of recovery times; (c) $-0.3208 \leq p_1 - p_2 \leq 0.4708$, no evidence to suggest significant difference in the percentage of motion restored.

4.47 (a) 95% CI: $-258.4 \leq \mu_{10} - \mu_{15} \leq 85$ (No significant difference due to rodding level), (b) There is more variability in rodding level 15; yet, there is no significant difference because the graphs overlap.

Chapter 5

5.14 No. The last four runs appear to plot at a distance of one-sigma or beyond from the center line.

5.15 Yes, the pattern appears to be random.

5.16 Yes, the pattern appears to be random.

5.17 The last four runs appear to plot at a distance of one-sigma or beyond from the center line.

5.19 (a) 2, (b) 4, (c) 5, (d) 1, (e) 3

5.21 (a) $\bar{x}$ chart: $UCL = 16.5420$, $CL = 16.268$, $LCL = 15.9940$
R *chart*: $UCL = 1.004$, $CL = 0.475$, $LCL = 0$.
The process appears to exhibit statistical control.
(b) $\hat{\mu} = \bar{\bar{x}} = 16.268$, $\hat{\sigma} = \bar{R}/d_2 = 0.475/2.326 = 0.2042$
(c) Normality assumption is reasonable.

5.23 (a) $\bar{x}$ chart: $UCL = 16.5484$, $CL = 16.268$, $LCL = 15.9876$
s *chart*: $UCL = 0.4104$, $CL = 0.1965$, $LCL = 0$.

The process appears to exhibit statistical control.
(b) $\hat{\mu} = \bar{\bar{x}} = 16.268$, $\hat{\sigma} = \bar{s}/c_4 = 0.1965/0.94 = 0.2092$

5.25 **(a)** $\bar{x}$ chart: $UCL = 154.45$, $CL = 130.88$, $LCL = 107.31$
R chart: $UCL = 86.40$, $CL = 40.86$, $LCL = 0$. The process appears to exhibit statistical control. **(b)** All new samples exceed the UCL indicating a significant increase in the time to service a customer.

5.29 *MR chart*: $UCL = 14.22$, $CL = 6.73$, $LCL = 0$
Individuals chart: $UCL = 61.82$, $CL = 53.27$, $LCL = 44.72$
Normality assumption is reasonable.

5.30 **(a)** Normality assumption is reasonable.
(b) $\bar{x}$ *chart*: $UCL = 3322.9$, $CL = 2928.9$, $LCL = 2534.9$
R chart: $UCL = 484.1$, $CL = 148.2$, $LCL = 0$. The process appears to exhibit statistical control.
(c) $\hat{\mu} = \bar{\bar{x}} = 2{,}928.9$, $\hat{\sigma} = 131.346 = \overline{MR}_2 = 148.158$

5.33 $ARL_0 = 370.84$, the theoretical performance of these two CUSUM schemes is the same.

5.34 **(a)** $\overline{MR}_2 = 13.72, \hat{\sigma} = \overline{MR}_2/d_2 = 13.72,/1.128 = 12.16$
(b) *CUSUM chart*: $UCL = 60.8$, $CL = 0$, $LCL = -60.8$
The process signals out of control at sample 12. The assignable cause occurred after sample $12 - 10 = 2$.

5.36 EWMA chart: $UCL = 8.0700$, $CL = 8.02$, $LCL = 7.9700$
The process is in control.

5.38 *EWMA chart*: $UCL = 957.53$, $CL = 950$, $LCL = 942.47$
Process is out of control at samples 8, 12, and 13.

5.40 *EWMA chart*: $UCL = 177.30$, $CL = 175$, $LCL = 172.70$.
Process is out of control where the process mean ($\hat{\mu} = 183.594$) is remarkably larger than the process target of $\mu_0 = 175$.

5.41 **(a)** $\ddot{C}_p = 2.98$, **(b)** $\ddot{C}_{pk} = 1.49$, **(c)** $\hat{p}_{potential} = 0.000000$

5.43 **(a)** $\ddot{C}_p = 0.75$, **(b)** $\ddot{C}_{pk} = 0.71$, **(c)** $\hat{p}_{potential} = 0.02382$

5.45 Normality assumption is reasonable. The process capability is 0.1350.

5.47 Normality assumption is reasonable. The process capability is 0.0551.

5.49 **(a)** Ignoring the three outliers in the upper of the probability plot, the normality assumption is reasonable. The process capability is 73.2. **(b)** $C_{pu} = 0.58$

Chapter 6

6.1 *p chart*: $UCL = 0.1289$, $CL = 0.0585$, $LCL = 0$ (Sample 12 exceeds the UCL). *Revised p chart*: $UCL = 0.1213$, $CL = 0.0537$, $LCL = 0$

6.3 *p chart*: $UCL = 0.1331$, $CL = 0.06$, $LCL = 0$. The process appears to be in statistical control.

6.5 **(a)** *p chart*: $UCL = 0.1425$, $CL = 0.1228$, $LCL = 0.1031$ **(b)** Cannot use this data to setup the control chart. The process should be investigated for the causes of wild swings in *p*.

6.8 $n = 81$

6.9 **(a)** *p chart*: $UCL = 0.2751$, $CL = 0.164$, $LCL = 0.0529$, **(b)** If the sample size will remain constant use the limits in (a). Otherwise, consider plotting exact limits whenever the sample size differs from 100. If is expected for the sample size to change frequently, consider using standardized control limits and plotting Z_i.

6.11 **(a)** *p chart*: $UCL = 0.062$, $CL = 0.02$, $LCL = 0$, **(b)** Sample 4 exceeds the upper control limit, signaling a potentially unstable process. For the new data, $\overline{p} = 0.038$, and $\hat{\sigma} = 0.0191$.

6.12 *np chart*: $UCL = 7.213$, $CL = 2.505$, $LCL = 0$, The process is in statistical control.

6.15 *np chart*: $UCL = 0.7007 + 3\sqrt{0.7007/n_i}$, $CL = 0.7007$, $LCL = 0.7007 - 3\sqrt{0.7007/n_i}$. Process appears to be in statistical control. For controlling production, consider using a standardized *u* chart.

6.18 *c chart*: $UCL = 17.38$, $CL = 8.59$, $LCL = 0$. Process is not in statistical control; three subgroups exceed the UCL. For controlling current production; remove samples 11, 15, 22, and recompute the limits. *Revised c chart*: $UCL = 13.6165$, $CL = 6.1667$, $LCL = 0$

6.20 **(a)** *c chart*: $UCL = 27.21$, $CL = 15.43$, $LCL = 3.65$ **(b)** *u chart*: $UCL = 27.20$, $CL = 15.42$, $LCL = 3.64$

6.22 *u chart*: $UCL = 7 + 3\sqrt{7/n_i}$, $CL = 7$, $LCL = 7 - 3\sqrt{7/n_i}$ The process is in statistical control.

6.24 *u chart*: $UCL = 9$(*where* $X \sim Poisson(4)$ and $P(X \leq UCL) = 0.99$), $CL = 4$

6.30 *u chart*: $UCL = 35.95$, $CL = 25.86$, $LCL = 15.77$. The rate of monthly CAT scans is out of control because sample 15 exceeds the UCL.

Chapter 7

7.3 **(a)** Using a hypergeometric distribution: $P_a(d = 35) = 0.9521$, or $\alpha \approx 0.05$ and $P_a(d = 375) = 0.1010$, or $\beta \approx 0.10$, **(b)** Using a binomial distribution: $P_a(p = 0.007) = 0.9521$, or $\alpha \approx 0.05$ and $P_a(p = 0.075) = 0.1013$, or $\beta \approx 0.10$, **(c)** Based on values for α and β, the difference between the two curves is small; either is appropriate.

7.5 Sampling plan: $n = 80$ and $c = 7$

7.7 Different sample sizes offer different levels of protection. $P_a(d = 10;\ N = 5{,}000;\ p = 0.02) = 0.2940$, $P_a(d = 20;\ N = 10{,}000;\ p = 0.02) = 0.1822$. Samples of 5,000 (10,000) units where 0.025% of the sample is defective will be accepted only 29.40% (18.22%) of the time. If the LTPD = 0.05, this sampling scheme will often lead to reject acceptable lots.

7.9 $n = 35;\ c = 1;\ N = 2{,}000$

$$ATI = 2{,}000 - 1{,}965P_a;\ AOQ = \frac{1965}{2000}P_a p;\ AOQL = 0.0234$$

7.11 **(b)** $p = 0.1030$, **(c)** Generally, quality improvement begins with the manufacturing process control, not with a sampling plan that is too harsh on the vendor. **(d)** The OC curve for this single-sampling plan with $c = 0$ is much steeper. **(e)** $ATI_{c=0} = 495$; $ATI_{c=2} = 60$. The $c = 2$ plan is preferred because the $c = 0$ plan will reject good lots 10% of the time as opposed to 0.2% of the time with $c = 2$.

7.13 **(a)** Sampling plan: $n = 49$; $Acc = 1$; $Rej = 4$, **(b)** Three points on the graph are: $p_1 = 0.01$, $P_a = 0.95$; $p = 0.0397$, $P_a = 0.5621$; $p_2 = 0.10$, $P_a = 0.10$

7.15 $AOQ = [P_a \times p \times (N - n)]/[N - P_a \times (np) - (1 - P_a) \times (Np)]$

7.17 Changing level II for level I reduces the sample size by 50%, while also reduces the Accept and Reject numbers. Level I inspection may be used when less discrimination is needed.

7.19 **(a)** Normal: $n = 200$, $Acc = 3$, $Rej = 4$. Tightened: $n = 200$, $Acc = 2$, $Rej = 3$. Reduced: $n = 80$, $Acc = 1$, $Rej = 4$

7.21 **(a)** Sampling plan: $50{,}001 \leq N \leq 100{,}000$; $n = 65$; $c = 3$, **(b)** 82 units, **(c)** LTPD = 10.3%

7.24 The MIL STD 414 sample sizes are considerably smaller than those for MIL STD 105E.

7.26 Normal sampling: $n = 100$, $k = 2.00$. Tightened sampling: $n = 100$, $k = 2.14$. Assuming normal sampling, the lot would be accepted.

7.28 **(a)** Normal: $n = 50$, $M = 2.49$. Tightened: $n = 50$, $M = 1.71$. Reduced: $n = 20$, $M = 4.09$, **(b)** Sampling plan: $n = 60$, $c = 2$ The sample size is slightly larger in attribute sampling. Variables sampling would be more efficient if σ were known. **(c)** Normal: $n = 200$, $Acc = 5$, $Rej = 6$. Tightened: $n = 200$, $Acc = 3$, $Rej = 4$ Reduced: $n = 80$, $Acc = 2$, $Rej = 5$. The sample sizes required are much larger than for the other plans.

7.32 Average process fallout, $p = 0.15\% = 0.0015$ and $q = 1 - p = 0.9985$

1. $f = \frac{1}{2}$ and $i = 140$: $u = 155.915$, $v = 1333.3$, $AFI = 0.5523$, $P_a = 0.8953$
2. $f = 1/10$ and $i = 550$: $u = 855.530$, $v = 6666.7$, $AFI = 0.2024$, $P_a = 0.8863$
3. $f = 1/100$ and $i = 1302$: $u = 4040.000$, $v = 66666.7$, $AFI = 0.0666$, $P_a = 0.9429$

7.34 Prefer Plan B over Plan A since it has a lower P_a at the unacceptable level of p.

Chapter 8

8.1 The effects of glass (p-value = 0.000) and phosphor (p-value = 0.004) are significant, while the effect of the interaction is not. Normality assumption is reasonable, no data to test for the independence assumption, and some issues with the constant variance assumption across all levels of phosphor.

8.3 Constant variance and normality assumptions are reasonable.

8.5 The assumptions of normality and constant variance across all levels of factors A, B, C, and D are reasonable.

8.7 A is significant at $\alpha = 0.10$. The assumptions of normality and constant variance are reasonable for the reduced model using A alone.

8.9 Block 1: (1), *ab, ac, bc, ad, bd, cd, abcd, ae, be, ce, abce, de, abde, acde, bcde*, Block 2: *a, b, c, abc, d, abd, acd, bcd, e, abe, ace, bce, ade, bde, cde, abcde*

8.11 **(a)** $ACE = +1$, $BDE = +1$ for all treatments, **(b)** Defining relation: $I = ACE = BDE = ABCD$. Aliases: $A = CE = ABDE = BCD$, $B = ABCE = DE = ACD$, $C = AE = BCDE = ABD, \ldots,$ $AB = BCE = ADE = CD$. **(c)** $[A] = -1.525$, $[B] = -5.175$, $[C] = 2.275$, $[D] = -0.675$, $[E] = 2.275$

8.13 **(a)** Only factor A and curvature are significant. **(b)** Add axial runs in a sequential experiment to estimate quadratic terms.

8.15 **(a)** A non-rotatable CCD with $k = 2$ and $\alpha = 1.5$.
(b) $y = 160.9 - 58.3x_1 + 2.4x_2 - 10.9x_1^2 + 6x_2^2 - 0.75x_1x_2$, x_2 and x_1x_2 are not significant using $\alpha = 0.10$, **(c)** Response is maximized using: $x_1 = 1.5$ and $x_2 = [-1.5, 1.5]$, **(d)** $Temp = 50x_1 + 750 = 50 \times 1.50 + 750 = 825$. $Time = 15x_2 + 30 = 15 \times \; - 0.1061 + 30 = 28.4085$

8.16 **(a)** A rotatable CCD with $k = 2$ and $\alpha = 1.4$. **(b)** $y_{uncoded} = 13.7273 + 0.2980x_1 - 0.4071x_2 + 0.0550x_1x_2 - 0.1249x_1^2 - 0.0790x_2^2$, x_1x_2 is not significant using $\alpha = 0.10$. Response is maximized using: $x_1 = [0, 1.4]$ and $x_2 = [-1, -1.4]$

Chapter 9

9.1 $E(T) = 25{,}000$; $V(T) = 6.25 \times 10^8$; $R(30{,}000) = 0.3012$

9.3 $E(T) = 800$; $V(T) = 640{,}000$; $R(8{,}000) = 4.54 \times 10^{-5}$

9.5 $1 - R(1{,}500) = 0.3935$; $1 - R(2{,}500) = 0.5654$

9.7 $R(4{,}000) = 0.4088$; $R(7{,}500) = 0.2938$

9.9 $\delta = 75{,}811.95$

9.11 $E(T) = 449.191$; $V(T) = 21{,}288.6$; $R(600) = 0.1559$

9.13 The slope of the graph changes.

9.15 The assumption of an exponential distribution is reasonable. $MTTF = 747.75$ *hrs*

9.17 Assuming a Weibull distribution is reasonable.

9.19 Assuming a Weibull distribution is reasonable.

9.21 $R(S) = 0.9849$

9.23 $R(10{,}000) = 0.1353$

9.24 $\lambda = 3.1266 \times 10^{-8}$, $MTTF = 3.20 \times 10^7$ hours

9.26 $R(S) \approx 1$

9.28 $R(1{,}500) = 0.1421$

9.30 $R(S) = 0.9\overline{9}$

9.32 $h(t) = 0.0028$

9.34 $h(t) = 0.001$

9.36 $h(t) = 8.4746 \times 10^{-5}$

Chapter Opener Photo Credits

Index

A

B

C

D

E

F

G

H

I

J

K

L

M

N

O

P

Q

R

S

T

U

V

W

X

Z